Advances in

Nutritional Research

Volume 10

Immunological Properties of Milk

Advances in

Nutritional Research

A Continuation Order Plan is available for this series. A continuation order will bring delivery of each new volume immediately upon publication. Volumes are billed only upon actual shipment. For further information please contact the publisher.

Advances in

Nutritional Research

Volume 10

Immunological Properties of Milk

Edited by

Bill Woodward

and

Harold H. Draper

University of Guelph
Guelph, Ontario, Canada

Springer Science+Business Media, LLC

The Library of Congress cataloged the first volume of this title as follows:

Advances in nutritional research, v. 1–
New York, Plenum Press, c 1977–
1 v. ill. 24 cm.
Key title: Advances in nutritional research. ISSN 0149-9483
1. Nutrition—Yearbooks

QP141.A1A3 613.2'05 78-640645

ISBN 978-1-4613-5182-5 ISBN 978-1-4615-0661-4 (eBook)
DOI 10.1007/978-1-4615-0661-4

Originally published by Kluwer / Plenum Publishers, New York in 2001
Softcover reprint of the hardcover 1st edition 2001

10 9 8 7 6 5 4 3 2 1
A C.I.P. record for this book is available from the Library of Congress

Contributors

Mohsen Araghi-Niknam, Arizona Prevention Center, P.O. Box 245155, 1501 N. Campbell Avenue, University of Arizona, Tucson, AZ 85724

Patrick F. W. Chien, Department of Obstetrics and Gynaecology, University of Dundee, Ninewells Hospital, Dundee, Scotland DD1 9SY

Roberta Chierici, Department of Clinical and Experimental Medicine, Section of Neonatology, Arcispedale "S. Anna", Ferrara, Italy

Thomas G. Cleary, Department of Pediatrics, Infectious Disease Division, University of Texas-Houston Health Science Center, 6431 Fannin Street, JFB 1.739, Houston, TX 77030

Mark D. Cregan, Department of Biochemistry, The University of Western Australia, Nedlands, Western Australia 6907, Australia

U. I. Dahlgren, Department of Clinical Immunology, Göteborg University, Guldhedsgatan 10, S-413 46 Göteborg, Sweden

Catherine M. Fetherston, Department of Biochemistry, The University of Western Australia, Nedlands, Western Australia 6907, Australia

Suzanne M. Filteau, Centre for International Child Health, Institute of Child Health, 30 Guilford Street, London WC1N 1EH, England

Armond S. Goldman, The Division of Immunology/Allergy/Rheumatology, The Department of Pediatrics, The University of Texas Medical Branch, Galveston, TX 77555-0369

Lars Å. Hanson, Department of Clinical Immunology, Göteborg University, Guldhedsgatan 10, S-413 46 Göteborg, Sweden

Peter E. Hartmann, Department of Biochemistry, The University of Western Australia, Nedlands, Western Australia 6907, Australia

Saeed Hosseini, Arizona Prevention Center, P.O. Box 245155, 1501 N. Campbell Avenue, University of Arizona, Tucson, AZ 85724

Peter W. Howie, Department of Obstetrics and Gynaecology, University of Dundee, Ninewells Hospital, Dundee, Scotland DD1 9SY

Alan J. Husband, Department of Veterinary Anatomy and Pathology, The University of Sydney, NSW 2006, Australia

Paula Inserra, Arizona Prevention Center, P.O. Box 245155, 1501 N. Campbell Avenue, University of Arizona, Tucson, AZ 85724

Charles E. Isaacs, Department of Developmental Biochemistry, Institute for Basic Research, 1050 Forest Hill Road, Staten Island, NY 10314

Shannon L. Kelleher, Department of Nutrition, University of California, Davis, CA 95616

Margaret A. Keller, Harbor-UCLA Medical Center, Torrance, CA 90509-2910

Ann M. Kosloske, Departments of Surgery and Pediatrics, Texas Tech University School of Medicine, Lubbock, TX 79415

Ruth A. Lawrence, Department of Pediatrics, University of Rochester School of Medicine and Dentistry, Rochester, NY 14642

Chee Seong Lee, Department of Veterinary Science, The University of Melbourne, Parkville, Vic 3052, Australia

Bo Lönnerdal, Department of Nutrition, University of California, Davis, CA 95616

Leon R. Mitoulas, Department of Biochemistry, The University of Western Australia, Nedlands, Western Australia 6907, Australia

Marita Noguera-Obenza, Department of Pediatrics, Infectious Disease Division, University of Texas-Houston Health Science Center, 6431 Fannin Street, JFB 1.739, Houston, TX 77030

Gerd Riedel-Caspari, PlantaVet, Postfach 1339, Finkenweg 13, D-88339 Bad Waldsee, Germany

H. Schroten, University Children's Hospital, Moorenstrasse 5, 40225 Düsseldorf, Germany

E. Richard Stiehm, Department of Pediatrics, UCLA Children's Hospital, Los Angleles, CA 90095-1752

E. Telemo, Department of Clinical Immunology, Göteborg University, Guldhedsgatan 10, S-413 46 Göteborg, Sweden

Ian R. Tizard, Department of Veterinary Pathobiology, Texas A&M University, College Station, TX 77843

Ronald R. Watson, Arizona Prevention Center, P.O. Box 245155, 1501 N. Campbell Avenue, University of Arizona, Tucson, AZ 85724

Chun W. Wong, Department of Animal Science, University of New England, Armidale, NSW 2351, Australia

Contents of Earlier Volumes

Volume 8:

Volume 9:

Preface

This volume of *Advances in Nutritional Research* focuses on colostrum and milk as agents of defense against infection both for the suckling offspring and for the lactating mammary gland. The scope of the volume includes positive and negative influences of the consumption of mother's milk on the risk of infection, immunobiological roles of individual milk components, activities of milk and its components in promoting development of neonatal immunocompetence, the potential of milk and its components as therapeutic agents and as functional foods that support immune competence, and external influences that determine the immunological activity of milk. The volume is intended to provide a critical assessment of the limits of available information pertaining to humans and animals, together with authoritative comment regarding newer directions and unproven ideas.

Part I provides a foundation for the volume. Readers unfamiliar with immunology will find, in **Chapter 1**, a selective outline of the anatomy and ontogeny of the mammalian immune system and of the types and regulation of immune defenses in mammals. Some emphasis is given to the place of the mammary gland within the common mucosal defense system, and to important species peculiarities in this regard. **Chapter 2** is an authoritative and forward-looking perspective on the development of knowledge pertaining to the immunobiology of milk as a fluid with both anti-infectious and anti-inflammatory roles. The chapter poses the provocative possibility of a tolerogenic role for milk. Thus, future research into this subject not only will address critical points of detail such as the identification of an increasingly diverse array of immunomodulatory, anti-inflammatory and possibly tolerogenic agents in milk, but also will continue to revise understanding, at the most basic level, of the role of milk as an immunological link between generations. **Chapter 3** then highlights the major immunological components that have been identified in milk and that are the subject of subsequent chapters in the volume.

Part II focuses on the influence of the consumption of mother's milk on the risk of infection without regard for any particular immunological component of milk. **Chapter 4** provides a systematic and critical analysis of epidemiological evidence regarding the impact of breast-feeding on the risk of major opportunistic infections in infancy. The review addresses infections of the gastrointestinal, respiratory and urinary tracts as well as infection with *Haemophilus influenzae*, a leading cause of bacterial meningitis in infants. Compelling evidence exists that breast-feeding protects against gastrointestinal infections in both industrialized and non-industrialized settings, and a reasonable evidence base also points to protective efficacy against respiratory infections in an industrialized setting. Many readers may be surprised by the dearth of quality evidence pertaining to the risk of respiratory, urinary tract and *H. influenzae* infection in non-industrialized settings. Although milk undoubtedly confers anti-infectious benefits, it can also serve as a vehicle for the transmission of viral and bacterial pathogens to the suckling offspring. Evidence is evaluated in **Chapter 5** with respect to the transmission of fourteen different pathogenic viruses through human milk. In addition, practical clinical recommendations specific to each organism are provided. As is pointed out subsequently in this volume (**Chapter 19**), the risk of transmitting infection through milk is an important reason that cross nursing is no longer recommended and that banked human donor milk should be pasteurized. **Chapter 6** comprises a discussion of the epidemiology, clinical presentation and pathogenesis of neonatal necrotizing enterocolitis, a leading cause of infant mortality in some industrialized countries. Although the pathogenesis of this condition is poorly understood, an infectious component is suspected and breast-feeding remains the only proven preventive strategy. **Chapter 7** provides much direct evidence that passive transfer of resistance to infection through colostrum and mature milk is essential to the offspring of domestic species and rodents. Compelling indirect evidence is also discussed in relation both to these species and to an impressive diversity of additional species, and derives from studies of placentation, intestinal absorption of intact immunoglobulins and the timing and mechanisms of gut closure. In addition, this chapter is a rich source of information describing the immunoglobulins and cellular components of milk in a plethora of non-human species. Finally, **Chapter 8** illustrates the importance of milk for the protection of the mammary gland against infection throughout lactation. By demonstrating that knowledge of this role is much more advanced in relation to the bovine species than in relation to humans, the chapter presents a fascinating example of the impact of commercial and economic considerations in determining research priorities.

Part III centers on individual immunological components of milk that appear to function in the passive transfer of defense against infection. The most widely recognized anti-infectious agent in milk is immunoglobulin and, in human milk, this is primarily secretory IgA. The main point of **Chapter 9** is the significance ascribed to the immunoglobulin fraction of milk as reflected in efforts to use bovine colostrum and hyper-immune bovine milk as therapeutic agents in the management of human gastrointestinal infections. The specific immunoglobulin content of some preparations has been clearly implicated in their efficacy, but immunoglobulin has appeared irrelevant to the effectiveness of other preparations. **Chapter 9** also briefly summarizes the more compelling evidence implicating breast milk IgA in the protection of infants against some enteric bacterial and viral pathogens as well as against some respiratory viruses. It is noteworthy that breast-milk IgA does not appear to confer passive protection against all viral agents of respiratory disease. Continuing this theme, **Chapter 10** constitutes an in-depth critique of evidence implicating milk IgA in the protection of infants against bacterial diarrhea. The important point is made that evidence is not abundant to unequivocally relate breast milk antibody to the protection of a suckling infant against infection. **Chapter 11** presents a description of the chemistry, biosynthesis and possible non-specific immunological actions of some high molecular weight milk glycoproteins (mucins). The emphasis of the chapter is on the human species, but laboratory and domestic species are discussed where information exists. Suggestive evidence implicates milk mucins as anti-infectious blocking agents, as prebiotics and as participants in the anti-inflammatory function of milk. However, direct evidence of an immunological role for these macromolecules is not yet available. Stronger and more extensive evidence points to lactoferrin as a non-specific anti-infectious agent, at least in human milk in which this protein is found in high concentration. **Chapter 12** describes the substantial evidence that lactoferrin can exert anti-bacterial, anti-fungal and anti-viral effects *in vitro*. Direct evidence of a role as a milk-borne anti-infectious agent derives from studies demonstrating that oral doses of lactoferrin can protect laboratory rodents against enteric pathogens. Moreover, a prebiotic function favoring colonic *Bifidobacteria* is suggested through studies of rodents and infants. Additional non-specific defense against infection may be conferred by some of the lipids in milk, notably the medium-chain saturated and long-chain unsaturated fatty acids together with their monoglycerides that are released by digestive lipolysis. These components inactivate enveloped viruses and kill bacteria and protozoans *in vitro*, and a discussion of their potential as passive agents of anti-infectious defense for humans is found in **Chapter 13**. Finally, it is often overlooked that milk contains large numbers of lymphocytes, mononuclear phagocytes and neutrophils. **Chapter 14** summarizes information regarding the numbers of these cellular elements in the colostrum and milk of humans and domestic animals. The chapter provides an assessment of the importance of milk leukocytes in protecting both the mammary gland and the suckling offspring. With regard to the latter, evidence remains only suggestive and is based primarily on a body of information describing immunological activities of milk leukocytes *in vitro*. Particularly intriguing possibilities relate to the transfer of cell-mediated immune sensitization from mother to offspring and the capacity of colostral and milk leukocytes to penetrate the intestinal epithelial barrier of the suckling neonate. In this connection, the largely unsubstantiated concern persists that milk leukocytes can serve as vectors for the passive transfer of intracellular parasites.

Part IV focuses on the immunomodulatory activities of milk. **Chapter 15** summarizes evidence that consumption of mother's milk is important to the development of immune competence in infants and neonatal animals. This chapter and **Chapter 16** then examine the evidence implicating particular colostral and milk components in this capacity. The anti-inflammatory action of milk appears critical to its immunomodulatory role, and a disperse literature implicates components such as lactoferrin and a number of cytokines, most notably transforming growth factor beta. Anti-idiotypic antibodies and nucleotides are also strongly implicated as exerting a direct immunomodulatory influence. Species differences are likely, e.g., as indicated by the low levels of lactoferrin and nucleotides in bovine milk relative to that of humans. In addition, an indirect prebiotic effect may contribute to the immunomodulatory action of milk, although the impact of breast-feeding on the gut microflora is probably less than has been widely surmised. Mastitis exerts obvious effects on the composition of milk, but the influence of this condition on the anti-inflammatory and immunomodulatory capacity of milk is unknown. More broadly, there is a need to conduct studies that go beyond the investigation of putative immunomodulators in isolation from one another and from possible gastrointestinal inhibitors. Although long-term influences on the anti-infectious immunocompetence of the neonate are likely, evidence from studies of breast-feeding is equivocal regarding an impact on the risk of developing atopic and autoimmune diseases. In a different vein, **Chapter 17** outlines evidence pointing to bovine milk whey as a functional food that promotes development of adaptive immune competence in the early post-weaning stage of life. Evidence is limited to rodents, and studies with resistance to infectious disease as an outcome measure are almost completely lacking. Recent mechanistic speculation centers on the absorption of digestion-resistant whey components such as lactoferrin. However, older literature demonstrating the importance of the amino acid profile of active whey protein concentrates is clearly relevant.

Part V centers on factors that influence the immunological properties of colostrum and mature milk. **Chapter 18** addresses the influence of maternal dietary and physiological factors on the quantity of milk produced and on the concentrations of milk components considered likely to affect the immunological development of the offspring. In effect, this extends the theme of the preceding chapter, viz. the concept of milk as a functional food. Information is disperse and incomplete. For example, the chapter points out that the composition of human milk beyond six months of lactation is largely undocumented. The importance of early postnatal ontogeny in determining the risk of developing chronic diseases of adulthood places this subject in poignant perspective. **Chapter 19** rounds out the volume with a discussion of the influence of storage conditions, processing procedures and feeding methods on the immunologic components of banked human milk. Information on this subject is based on compositional analyses pertaining to biochemical and cellular constituents considered important to the immunological properties of milk. Ultimately, however, banking procedures must be judged on the basis of their impact on the capacity of colostrum and milk to provide passive protection against infection and to promote the development of immune competence in the infant. This remains a matter for future investigation.

Bill Woodward

Acknowledgment

This volume was produced using funds provided by the Dairy Farmers of Canada.

Contents

Part I

Introduction

Chapter 1

Overview of the Mammalian Immune System

A. J. Husband

1. Introduction

The success of mammalian evolution has depended on the development of an effective defense system against environmental pathogens and foreign antigens. While physical barriers to microbial invasion are a feature mainly of higher life forms, even some primitive invertebrate organisms evolved phagocytic cellular defense networks capable of seeking out and destroying unwanted foreign material, albeit in an unselective and non-adaptive manner. This evolutionary process was first documented in the landmark studies of Metchnikoff published over 150 years ago (Petrov and Ulyankina, 1996). Mammals have evolved humoral and cellular defense networks employing an array of non-adaptive and adaptive antigen-specific defense strategies. Adaptive defense elements have the capacity to activate recognition and response mechanisms that result in immunological memory on re-exposure. The structural and functional features of this defense system have been the subject of intense scrutiny over the past century, but only in the last 25 years has its true nature been revealed by unraveling the mechanisms by which it is controlled at the genetic and molecular level. This introductory chapter will present a brief overview of the essential elements of the immune system that has enabled mammals to evolve in a hostile microbial environment.

A. J. Husband • Department of Veterinary Anatomy and Pathology, The University of Sydney, NSW 2006, Australia

2. Innate Immunity

The importance of physical barriers and innate immune defense mechanisms is often overlooked, but the fact that any impairment of these protective devices rapidly leads to recurrent infections is testimony to their importance. For example, the absence of effective mucus production in cystic fibrosis (Shapiro, 1999), defective saliva production in Sjögrens syndrome (Fox and Kang, 1992) and lack of complement proteins in acquired or inherited complement deficiency diseases (Villiers *et al.*, 1999) all lead to recurrent and often life threatening infections, even when all other components of the adaptive immune system are functionally intact.

The skin, the acid secreted in the stomach secretions, lysozyme in external secretions, the mucus coat secreted at mucosal surfaces and the establishment of a resident population of microorganisms (commensal organisms) on the skin and at mucosal surfaces, all contribute in a non-antigen-specific manner to preventing or limiting pathogenic colonization. Other more complex innate immune defenses involve the phagocytic cells of the granulocyte lineages, in particular the neutrophils, and the monocytes and macrophages, which form a network of non-antigen-specific cellular defenses. Just as the absence of physical barriers to colonization can cause serious disease, defects in these defenses, such as neutropenia lead to recurrent and often overwhelming infections.

The family of proteins known as the complement proteins perhaps contributes the most important element of innate immunity. These proteins form part of a primitive defense system, appearing early in the evolution of the immune system, by which activation of one of these proteins (designated C1) by attachment either to an antibody-antigen complex (classical pathway) or to the surface of certain microorganisms even in the absence of antibody (alternate pathway) initiates a cascade of reactions in which each step is catalyzed by the products of the previous step. This complement cascade leads to a multitude of host defense mechanisms. One of these is complement-mediated opsonization which enhances phagocytosis via attachment of the antigen-complement complex to phagocytes by adhesion of the third component of complement (C3) to C3 receptors on phagocytes. In addition to enabling more rapid removal of the offending material by phagocytosis, this process also initiates the formation of complement breakdown products with proinflammatory properties, establishing a chemotactic gradient that enables more phagocytes to become involved in the reaction zone. Further, direct cytotoxicity may be provoked via the terminal event in the complement cascade, the formation of the membrane attack complex, in which the last component of complement is polymerized to form pores in the offending cell membranes, thus compromising membrane integrity and causing the destruction of the cells. When pathogens become opsonized by C3, the combined involvement of the complement cascade, engagement with phagocytic cells and subsequent cytotoxicity provides a very efficient mechanism for eliminating pathogenic organisms via antibody dependent cell mediated cytotoxicity (ADCC) (Forthal *et al.*, 1999).

3. Specific, Adaptive or Acquired Immunity

The clever adaptation of the mammalian immune system to the demands of higher and more complex organisms for protection from pathogen attack has involved the development of cells capable of antigen-specific recognition and memory. This capability ensures that the response is tailored to the nature of the invading organisms and it elicits a faster, more effective response whenever a repeat exposure occurs. This is accomplished by cells of the lymphoid series that give rise to a network of responses involving both T and B lymphocytes, with the final effector response provided either by antibody produced by B cells (humoral immunity) or by antigen-specific T effector cells capable of direct cellular cytoxicity (cell-mediated immunity). In general terms humoral immunity is most effective against bacterial infections whereas viral infections and some bacterial infections, having an intracellular habitat within the host, are relatively inaccessible to antibody so that cell-mediated immunity is the major defence in this case.

Antigen recognition is effected by surface receptors that are unique to each clone of cells but are highly diverse between clones. In the case of B cells, antigen recognition is effected via surface-bound antibody (immunoglobulin) molecules in which the hypervariable N-terminal end of the molecule is displayed externally to the cell. It triggers cellular differentiation and effector activity (antibody production) whenever receptor (surface antibody) and ligand (antigen) interactions occur. Whenever B cells encounter antigen for which they are specific, receptor binding triggers large scale production and secretion of copies of the receptor molecules, secreted as soluble antibody. These molecules bind to recognition elements (epitopes) on the pathogen surface causing their agglutination, triggering the complement cascade, enhancing phagocytosis and enabling ADCC reactions to proceed.

Similarly, T lymphocytes display surface recognition molecules capable of recognizing antigenic epitopes. These are fixed within a clone, but a large diversity of clones ensures a sufficiently large pool of potentially reactive cells. In the case of B cells, receptor binding triggers differentiation of effector cells in the reactive clone to become plasma cells capable of large scale production and secretion of copies of the receptor as soluble antibody. T cells, after engaging antigen for which they are specific, differentiate, producing a range of regulatory molecules (cytokines) which have a quantitative and qualitative influence on the B cell effector response as well as on the activities of mononuclear phagocytes. Other

T cells may differentiate to become cytotoxic cells capable of lysing invading organisms and cells infected by them.

In quantitative terms, cytokines manage the rates of proliferation of both B and T effector cells by providing growth factors (via T helper cells) during the initiation of a response (via T helper cells) and inhibitory factors (via T suppressor cells) when the response has served its purpose. In qualitative terms, cytokines have a key role in determining whether a response will be primarily cellular or humoral in nature. In humoral responses, the cytokines direct the production of immunoglobulins dictating which of the available variants of the antibody molecule will be produced. This class commitment process ensures that the IgM antibody molecule characteristic of a primary immune response is produced after first encounter with a new antigen but is produced in the monomeric form (IgG) after subsequent exposure antibody. Alternatively, cytokine regulation allows a dimeric variant (IgA) to be produced if the encounter occurs at a mucosal surface such as the gut, lung, urogenital tract or mammary gland. Whenever parasite antigens are encountered another monomeric variant of the antibody molecule (IgE) is produced and is capable of promoting anti-parasitic reactions. Occasionally IgE is also produced in response to some otherwise benign environmental antigens, resulting in immediate-type allergic reactions.

All of the response elements of the adaptive immune system require that antigen is presented to the potentially antigen-reactive B and T cell clones by professional antigen-presenting cells (APC) such as dendritic cells or macrophages, although a range of other immune cells can perform this role to some degree. These cells take up antigen and present it in processed form, together with molecules of the major histocompatibility complex (MHC). The choice of which MHC molecule is used by the APC for this purpose imposes a restriction on which class of T cells is capable of responding. In this context there are two main classes of T cells, T helper cells and T cytotoxic or suppressor cells. All T cells express a common surface marker, designated CD3 (a molecule which forms part of the T cell antigen recognition site), but helper and cytotoxic/suppressor subsets are characterized by additionally expressing surface molecules designated CD4 or CD8, respectively. CD4+ helper T cells require antigen to be presented in association with MHC Class II molecules on the APC, whereas CD8+ suppressor/cytotoxic cells require antigen to be presented in association with Class I MHC. Thus any immune response normally requires the combined involvement of a cellular triumvirate comprising APC's, T cells and B cells, and the interaction and subsequent performance of these cells is determined by the cytokines which each produces.

The final feature which defines specific immunity as adaptive to the needs of the organism is the phenomenon of immunological memory. On first encounter with antigen clones of antigen reactive B and T cells are selected from the vast library of virgin lymphocyte clones. Thus whenever a perfect match is found between antigen and lymphocyte specificity, expansion of the reactive clone ensues giving rise to effector cells (plasma cells making antibody of appropriate specificity derived from reactive B cells, or T effector cells with helper, suppressor or cytotoxic function from reactive T cells). A minority of cells derived from this clonal expansion phase will not, however, develop effector functionality but, instead, will undergo a DNA rearrangement before returning to a resting state, ready to respond more rapidly and with a more appropriately fine-tuned response on next exposure with a higher affinity for the antigen to which the cells are specific. These latter cells are memory cells responsible for the qualitative and quantitative differences between a primary and secondary immune response and form the real foundation of permanent protective immunity in vaccinated or previously exposed individuals.

4. Anatomical Compartmentalization in the Immune System

In considering the anatomical framework within which these activities occur the organs of the immune system can be broadly divided into primary and secondary lymphoid tissue. Primary sites are those in which cellular differentiation occurs in the absence of antigen and secondary sites are those in which antigen-induced clonal selection and proliferation occur. The first site of primary lymphopoiesis in the developing embryo is the fetal liver, but by the end of gestation this function is passed to the bone marrow. Virgin lymphoid cells leave the bone marrow and, in the case of B cells, traffic to the spleen from where they are seeded to peripheral lymphoid tissues such as the regional lymph nodes. Alternatively some cells traffic to the thymus where they acquire the characteristics and functionality of T cells, prior to seeding to the periphery.

Antigen encounter usually occurs when antigen is taken by APC's to regional lymph nodes where appropriate clones of antigen-reactive cells are selected from the vast numbers of cells constantly trafficking through the tissue from the blood supply. In the case of mucosal sites an additional regional specialization occurs in the form of lymphoid nodules such as the Peyer's patches in the gut, or the bronchus-associated lymphoid tissue in the lungs, where antigen sampled from the lumen is presented to cells trafficking through these sites and appropriate effector cells are produced from this encounter. While the process of clonal selection is essentially the same in regional lymph nodes (systemic sites) and mucosal associated lymphoid tissues (mucosal sites), a qualitative difference occurs in respect of the nature of the effector B cells produced. Whereas in systemic sites the response produces cells ultimately capable of making antibodies of the IgG class, B cells encountering antigen in mucosal sites have a propensity for manufacturing IgA antibodies. This differential isotype class commitment is determined by the action of cytokines produced locally by T cells in these microenvironments.

5. Regulation of Immunity: T Cells and Cytokines

The regulation of induction and delivery of the immune response and the specialization that occurs in different anatomical compartments is orchestrated by an array of soluble cytokine molecules, which may act in an autocrine, paracrine or endocrine fashion. At the point of initial antigen presentation by macrophages, dendritic cells or other APC's, IL-1 produced by the APC stimulates the production of IL-2 by responding T cells. In turn, IL-2 acts via autocrine activation of the T cells, thereby leading to growth of the T cell clone and differentiation into either T helper, T suppressor or T cytotoxic effector cells. T helper cells produce a variety of cytokines which either act as B cell growth factors or promote cell-mediated cytotoxic T cell responses. When the cytokine profile consists predominantly of IL-4, IL-5, IL-6 and/or IL-10 (Th2 cytokine profile) a humoral response is encouraged resulting in large scale antibody production. When the T cell cytokine profile swings toward interferon (IFN)-γ production (Th1 cytokine profile) a predominantly cell-mediated response occurs.

In the microenvironment of a mucosal induction site, Th2 cytokines promote IgA class switching and a mucosal variant of the humoral response ensues (Hodgkin *et al.*, 1991). In the induction phase of an IgA response IL-4 and transforming growth factor (TGF)β are important, in addition to signals derived by T cell membrane contact, and promote switching to IgA isotype expression among antigen-reactive B cells (Kunimoto *et al.*, 1988). After the IgA-committed B cells migrate from inductive sites such as the Peyer's patches and enter submucosal effector regions, subsequent terminal differentiation and IgA antibody secretion appear to be entirely cytokine dependent, since IL-5, IL-6 and IL-10 amplify IgA secretion and encourage proliferation among IgA committed B cells in the absence of T cell contact (Beagley *et al.*, 1989; Defrance *et al.*, 1992).

6. Regional Adaptations within the Immune System

The fact that responses to antigens encountered at mucosal surfaces characteristically result in IgA antibody production among responding B cells, reflects an evolutionary adaptation designed to ensure that antibodies can gain access to external secretions, and thereby prevent invading organisms from gaining a foothold through surface colonisation of the epithelium. This volume is focused on the immunological properties of the secretions of the mammary gland, and their role in reducing the risk of infectious diseases affecting mucosal surfaces such as the gut, lungs and urogenital tract which are the most important causes of mortality and morbidity in the young. Therefore, more detailed attention is paid to the peculiarities of the immune system operating at mucosal sites and, in particular, the mammary gland.

Mucosal immune defense appears to have arisen phylogenetically from a primitive immune system that evolved first in fishes to protect the skin and mucosal surfaces from invasion by microbes in the environment (Phillips-Quagliata and Lamm, 1988). In these species a unique antibody molecule is produced in cutaneous mucous secretions, the intestine and bile, that is analogous to the IgA antibodies which characterise mammalian mucosal secretory sites. As mammals moved to a land based habitat they developed a tougher skin coating that provided a more effective barrier against microbial invasion, and evolutionary pressure led to development of increasingly sophisticated systemic and mucosal immune systems. The existence of a linkage between the mammalian mucosal immune system and the skin and gut immune system in primitive fishes is supported by observations that mammalian lymphoid cells arise from stem cells of mesenchymal origin that mature during embryogenesis in the yolk sac and fetal liver, organs derived during embryogenesis from the same endodermal tissues as the gills in primitive chordates.

Mucosal organs in mammals collectively present a vast surface area estimated at $400\,m^2$ in humans (McGhee and Kiyono, 1992). This interface is a common entry point for many infectious agents. IgA antibodies are the predominant isotype produced by plasma cells underlying mucosal epithelia, and dimeric IgA molecules are rapidly and selectively transported to the lumen by a transport protein termed the polymeric immunoglobulin receptor (pIgR) manufactured by the epithelial cells and displayed on their basolateral cell surface. After IgA binds to the pIgR molecule the complex is internalized and transferred intact across the epithelium. Part of the pIgR molecule termed secretory component (SC) remains bound to the IgA dimer, forming a new variant known as secretory IgA (sIgA). These antibodies not only enjoy assisted selective transport across the mucosal epithelium but after secretion they are highly resistant to proteolytic degradation. They are well placed to provide the first line of defence against mucosal pathogens by interfering with their attachment, colonisation and virulence, by binding to specific sites (Williams and Gibbons, 1972) or by encouraging antibody-dependent cell-mediated cytotoxicity at the epithelial surface (Geffner *et al.*, 1990).

The induction of IgA antibody responses is dependent on precursor B cells that engage antigen in specialized mucosal-associated lymphoid tissue (MALT) such as Peyer's patches (PP) in the intestine (Husband, 1982) and bronchus associated lymphoid tissue (BALT) in the respiratory tract (Rudzik *et al.*, 1975). In contrast to the regular cuboidal epithelium overlying the normal surface of the intestine where nutrient absorption occurs, PP are endowed with a flattened epithelial covering containing cells adept at antigen uptake (M cells) (Owen, 1994). After their activation by antigen, the IgA cell precursors leave this inductive site via the lymphatics and enter the blood, where they have the ability to migrate to any of the mucosal subepithelial sites, thus creating an interactive network among mucosal organs. In addition, IgA molecules secreted by submucosal plasma cells may escape secretion at the site of their production and be returned to the circulation via the lymph to be available for secretion at remote mucosal sites (Husband, 1982, 1985).

A common mucosal immune system supported by migration of effector cells and transepithelial secretion of IgA molecules derived from remote mucosal sites ensures that the antigenic experience of one mucosal surface can be reflected in immune effector expression at a distant surface. There has been a great deal of interest in exploiting this phenomenon to achieve remote site mucosal immunisation by oral antigen delivery. With respect to the mammary gland it may not only improve local protection, which is of particular importance in dairy ruminants, but also alter the immunoglobulin profile of milk to provide improved immune protection to suckling neonates. Studies in sheep have confirmed that serum polymeric IgA is selectively transported into the mammary secretion (Sheldrake *et al.*, 1984). Comparison of the rates of selective transport at various stages of lactation with the level of IgA produced locally revealed that when production declines selective transport from serum increases, local production being greatest during involution. In contrast, in the intestine, where the bulk of luminal IgA in the secretions is locally produced, there is no selective transport of serum-derived IgA.

These studies suggest that there is an inverse relationship between the extent to which IgA in mucosal secretions is derived from locally situated plasma cells and the extent to which it is selectively transported into secretions from the plasma (Sheldrake *et al.*, 1984). This relationship probably reflects pIgR availability. When local production of IgA is insufficient to cause saturation of the available pIgR binding sites these are more likely to be available to transport IgA derived from the circulation. Selective removal of IgA from serum is a feature of all sites with pIgR-bearing epithelia. Removal is subject to receptor availability in non-saturating conditions and provides a compensatory mechanism at sites where IgA production is low to ensure that IgA antibody reaches mucosal secretions even in the absence of extensive local production.

7. Ontogeny and Perinatal Immunity

One of the remarkable paradoxes surrounding the development of immune function is the repeated observation that, at birth, animals of most species are incapable of mounting an effective immune response despite the relatively well developed state of the immune apparatus anatomically (Silverstein, 1964). Yet it is at this time that the neonate is expelled from a sterile uterine environment to face a vast range of environmental pathogens. In particular, mucosal responses lag well behind the rest of the immune defence network in achieving adult levels of functional maturity. To bridge the gap between the immune deficit at birth and the onset of immune competence, most species have evolved strategies to ensure that passive immune protection is provided from maternal sources. This protection consists predominantly of transferred antibody, although lymphoid cells may also be transferred to the neonate either by the transplacental route or via colostrum and milk ingested after birth. The route of acquisition of passive immunity is determined largely by the uterine architecture. Whereas in ruminants epitheliochorial placentation prevents effective antibody transport via the placental route, in humans hemochorial placentation enables large scale transport of IgG into the fetal circulation during the last trimester of pregnancy (Brambell, 1966, 1970). Ruminants therefore are dependent on postnatal uptake of antibodies from colostrum. As an adaptation to this demand ruminant colostrum is rich in IgG, which is absorbed intact across the neonatal gut during the first 24 hours after birth (Butler, 1969). Humans do not require IgG from milk, and their colostrum is rich in IgA, which plays a local protective role as a blocking antibody in the gut lumen (Butler, 1979).

In the ruminant mammary gland the IgA system is less well developed than in other species and defense mechanisms in the mammary gland depend less on IgA. Thus there is less compelling evidence for involvement of the ruminant mammary gland in a common mucosal immune system. However, the gut contributes to the IgA response in the mammary gland because IgA molecules produced by plasma cells situated in GALT are transported to the gland via the circulation. The contribution of gut-derived IgA molecules to local immunity at other mucosal surfaces is probably also of more importance in ruminants than in other species because of the relatively small amount of IgA removed from the circulation by the liver (Sheldrake *et al.*, 1985).

Systemic immunization of the maternal ruminant provides the suckling neonate with immunity against some enteric diseases because of the massive selective accumulation of IgG in ruminant colostrum. However, systemic immunization of the adult animal has not proved to be effective in mastitis control. Intestinal immunization by antigens from organisms that infect the mammary gland or neonatal intestine may increase local IgA immunity and other immune defenses (e.g., cell-mediated immunity) in the mammary gland. This may assist in the control of mastitis and provide additional cellular and molecular defense for the suckling neonate, especially during later lactation. At this stage, the transport of IgG from plasma to the glandular secretions is diminished and, therefore, is less likely to provide protection to the gland or, indeed, to the suckling neonate.

It is often overlooked that the mammary gland secretions of most species contain large numbers of cells in addition to soluble factors and antibodies, and that some of these may cross the neonatal gut intact (Sheldrake and Husband, 1985). Whether they contribute to immune function in the neonate is debatable, but it is of interest that in a study of human recipients of kidney grafts from maternal donors, those who had been suckled on maternal breast milk were less likely to reject the grafts than those who had been bottle-fed. This outcome suggests that maternal cells may tolerize the neonate to maternal alloantigens and enhance the survival of subsequently grafted tissue (Campbell *et al.*, 1984). Regardless of whether the cells present in milk cross the neonatal gut intact they may contribute to local protection in the gut lumen of the suckling neonate.

8. Conclusions

The mammalian immune system represents a highly evolved and complex array of cellular and molecular elements that act to protect the internal milieu against a potentially hostile microbial environment. However, immunity can be a two-edged sword and the system does not always work in the best interests of the host. Occasionally an overactive immune response may lead to immunopathology, as in immediate hypersensitivity reactions involving IgE antibodies, autoimmune conditions resulting from persistent immune complex deposition in host tissue, production of antibodies directed against host tissue or delayed hypersensitivity reactions in which T cells cause host damage through persistent activation or cell-mediated autoimmune reactions. In spite of these dangers, the system has enabled mammals to survive massive global changes in the environment and emerge near the top of the food chain. The development of specialized glandular structures and an immune system specifically designed to protect them have been key elements in their success. This volume explores the immunological properties of an organ unique to mammals: the mammary gland.

References

Beagley, K.W., Eldridge, J.H., Lee, F., Kiyono, H., Everson, M.P., Koopman, W.J., Hirano, T., Kishimoto, T., and McGhee, J.R. 1989. Interleukins and IgA synthesis—human and murine interleukin 6 induce high rate IgA secretion in IgA-committed B cells. *J. Exp. Med.* **169**:2133.

Brambell, F.W.R. 1966. The transmission of immunity from mother to young and the catabolism of immunglobulins. *Lancet* **ii**:1087.

Brambell, F.W.R. 1970. *Frontiers of Biology*, Volume 18 (A. Neuberger and E.L. Tatum, eds.). North-Holland Publishing Co., Amsterdam.

Butler, J.E. 1969. Bovine immunoglobulins: A review. *J. Dairy Sci.* **52**:1895.

Butler, J.E. 1979. Immunologic aspects of breast feeding, antiinfectious activity of breast milk. *Sem. Perinatol.* **3**:255.

Campbell, D.A., Lorber, M.I., Sweeton, J.C., Turcotte, J.G., Niederhuber, J.E., and Beer, A.E. 1984. Breast feeding and maternal-donor renal allografts. *Transplantation* **37**:340.

Defrance, T., Vanbervliet, B., Briere, F., Durand, I., Rousset, F., and Banchereau, J. 1992. Interleukin-10 and transforming growth factor-beta cooperate to induce anti-CD40 activated naive human B-cells to secrete immunoglobulin A. *J. Exp. Med.* **175**:671.

Forthal, D.N., Landucci, G., Haubrich, R., Keenan, B., Kuppermann, B.D., Tilles, J.G., and Kaplan, J. 1999. Antibody-dependent cellular cytotoxicity independently predicts survival in severely immunocompromised human immunodeficiency virus-infected patients. *J. Infect. Dis.* **180**:1338.

Fox, R.I. and Kang, H.I. 1992. Pathogenesis of Sjogren's syndrome. *Rheum. Dis. Clin. North Am.* **18**:517.

Geffner, J.R., Minnucci, F., and Isturiz, M.A. 1990. Neutrophil-mediated cytotoxicity induced by secretory IgA. *Immunol. Lett.* **24**:113.

Hodgkin, P.D., Yamashita, L.C., Seymour, B., Coffman, R.L., and Kehry, M.R. 1991. Membranes from both Thl and Th2 T cell clones stimulate B cell proliferation and prepare B cells for lymphokine-induced differentiation to secrete Ig. *J. Immunol.* **147**: 3696.

Husband, A.J. 1982. Kinetics of extravasation and redistribution of IgA-specific antibody-containing cells in the intestine. *J. Immunol.* **128**:1355.

Husband, A.J. 1985. *Progress in Veterinary Microbiology and Immunology. Vol. 1.* (R. Pandey, ed.), p. 25, S. Karger, Basel.

Kunimoto, D.Y., Harriman, G.R., and Strober, W. 1988. Regulation of IgA differentiation in CH12LX B cells by lymphokines. IL-4 induces membrane IgM-positive CH12LX cells to express membrane IgA and IL-5 induces membrane IgA-positive CH12LX cells to secrete IgA. *J. Immunol.* **141**:713.

McGhee, J.R. and Kiyono, H. 1992. Mucosal immunity to vaccines: Current concepts for vaccine development and immune response analysis. *Adv. Exp. Med. Biol.* **327**:3.

Owen, R.L. 1994. M cells—Entryways of opportunity for enteropathogens. *J. Exp. Med.* **180**:7.

Petrov, R.V. and Ulyankina, T.I. 1996. The genius of E. E. Metchnikoff—discoveries over the centuries. *Biosci. Rep.* **16**:189.

Phillips-Quagliata, J.M. and Lamm, M.E. 1988. *Migration and Homing of Lymphoid Cells, Vol. II* (A.J. Husband, ed.), p. 53, CRC Press, Boca Raton.

Rudzik, O., Clancy, R.L., Perey, D.Y.E., Day, R.P., and Bienenstock, J. 1975. Repopulation with IgA-containing cells of bronchial and intestinal lamina propria after transfer of homologous Peyer's patch and bronchial lymphocytes. *J. Immunol.* **114**:1599.

Shapiro, B.J. 1999. Cystic fibrosis. *Curr. Opin. Pulm. Med.* **5**:363.

Sheldrake, R.F. and Husband, A.J. 1985. Intestinal uptake of intact maternal lymphocytes by neonatal rats and lambs. *Res. Vet. Sci.* **39**:10.

Sheldrake, R.F., Husband, A.J., Watson, D.L., and Cripps, A.W. 1984. Selective transport of serum-derived IgA into mucosal secretions. *J. Immunol.* **132**:363.

Sheldrake, R.F., Scicchitano, R., and Husband, A.J. 1985. The effect of lactation on the transport of serum derived IgA into bile of sheep. *Immunology* **54**:471.

Silverstein, A.M. 1964. Ontogeny of the immune response. *Science* **144**:1423.
Villiers, C.L., Villiers, M.B., and Marche, P.N. 1999. Role of the complement C3 protein in the control of the specific immune response. *Ann. Biol. Clin. Paris.* **57**:127.
Williams, R.C. and Gibbons, R.J. 1972. Inhibition of bacterial adherence by secretory immunoglobulin A: A mechanism of antigen disposal. *Science* **177**:697.

Chapter 2

The Immunological System in Human Milk: The Past—A Pathway to the Future

Armond S. Goldman

1. Introduction

Ever since the late 20th century, it has been accepted that the lactating human mammary gland elaborates a complex immunological system uniquely adapted to protect the recipient infant. It is not well appreciated, however, that many basic discoveries concerning these matters were made long before the era of molecular biology. Observations during the eighteenth and nineteenth centuries in Europe suggested that mortality was lower in breastfed than in non-breastfed infants. The basis of this protection was unknown. Although similar but more structured observations were made in the first part of the twentieth century (Westergaard, 1901; Groth and Hahn, 1910; Davis, 1913; Woodbury, 1922, 1925; Grulee *et al.*, 1934, 1935) there was little scientific information concerning the immunology of human milk or of milk obtained from other species. This was partly because most of the immune system was not uncovered until the latter part of the twentieth century. However, discoveries concerning the protective properties of milk from humans and other mammalian species that were made during the nineteenth and early twentieth centuries provided the foundation for the molecular research that came thereafter. Furthermore, it is striking that these discoveries were made by individuals eminent in the history of immunology and medicine. Then in the mid-twentieth century, critical information was discovered

Armond S. Goldman • The Division of Immunology/Allergy/Rheumatology, Department of Pediatrics, The University of Texas Medical Branch, Galveston, Texas 77555-0369

that led to the paradigm of an immune system in human milk and to the realization that the system comprises not only antimicrobial factors but also anti-inflammatory and immunomodulating agents. Parallel investigations indicated that immune systems were in other mammalian milks. Because of many differences in the immune systems in milks from the various mammalian species, the scope of the information in this review will be limited to key discoveries concerning the immunological aspects of human milk.

2. Van Leeuwenhoek, Donné and Leukocytes in Human Milk

Van Leeuwenhoek (1695) first reported the presence of "animalcules" in milk from humans and other species. It was unclear whether those moving objects were cells or milk fat globules undergoing random movement. His observations lay fallow until Donné (1837) reported the presence of globules and granular bodies in milk from humans and other species. Most of these objects were probably milk fat globules, which are numerous in human milk. That is important because of recent discoveries of the immunological role of these macromolecular structures in human milk (see below). When staining techniques became available toward the end of the nineteenth century, some of the "corpuscles de Donné" were found to be cells, but their identity and function remained uncertain until they were accidentally rediscovered by Smith and Goldman (1968).

In 1966, Smith and Goldman began to study whether the few soluble immune factors known to be present in human milk interacted with one another, or perhaps with unrecognized agents in milk, to enhance bacterial killing. They removed most of the lipids in human colostrum by low speed centrifugation, which produced a high density bottom layer and a buoyant top layer containing milk fat globules and free lipids. When the high density material was examined by phase microscopy, moving refractive bodies were observed. Subsequently, the stained material was examined to reveal some cells that appeared to be neutrophils or macrophages and a few that seemed to be lymphocytes. The identity of most of them, however, was uncertain because their morphology was obscured by large, intracytoplasmic vacuoles. Physical examination of the donor indicated that the cells were not due to mastitis. Moreover, fresh colostrum from several other donors revealed similar findings. Many cells were laden with lipids, highly motile, and phagocytic. It was difficult to completely distinguish which of these cells were neutrophils and which were macrophages. Furthermore, there were lymphocytes that underwent blast transformation after stimulation with the mitogen phytohemagglutinin. Thus, human milk was shown to contain living leukocytes, the majority of which were neutrophils and macrophages and some of which were lymphocytes (Smith and Goldman, 1968, 1970; Smith *et al.*, 1971).

The findings that milk displayed some characteristics of a tissue profoundly affected our thinking about this fluid. Since leukocytes are the hallmark of immune processes, their presence in milk as well as soluble antimicrobial agents strongly suggested that the mammary gland produced an assemblage of immunological agents coordinated to complement the defenses of the recipient infant.

A few years later more correct estimates of the frequencies of neutrophils, macrophages and lymphocytes in human milk were made, first by cytochemistry (Crago *et al.*, 1979) and then by flow cytometry (Bertotto *et al.*, 1990; Wirt *et al.*, 1992; Keeney *et al.*, 1993). In the 1990's, the cells in human milk were found to be activated (Bertotto *et al.*, 1990; Wirt *et al.*, 1992; Keeney *et al.*, 1993), and this finding became part of the evidence that led to the discovery of immunomodulating agents in human milk. Although the *in vivo* fate and action of the cells remain unclear, some transfer studies in experimental animals suggest that they may penetrate into the tissues of the infant (Goldman and Goldblum, 1996) and transfer cellular immunity (Oksenberg *et al.*, 1985; Pabst *et al.*, 1989).

3. Antitoxic Effects of Milk

Ehrlich and Metchnikoff are celebrated for their basic studies of systemic immunity, but these scientists also helped to lay the foundation for understanding the immunological functions of the human mammary gland. Ehrlich's work on antitoxins, the side chain theory that presaged the discovery of antigen receptors on lymphocytes, and his development of the first chemotherapeutic agent against a microbial pathogen are reviewed by Silverstein (1996). Ehrlich also demonstrated the transfer of immunological protection to the fetus and newborn infant from the mother via the placenta and the mammary gland, respectively. He became interested in these questions because it had been observed that breastfed infants were less prone to infections than those who were fed artificially. He designed cross-fostering experiments to test whether maternal immunity to the well defined plant toxins abrin and ricin could be transferred to the offspring (Ehrlich, 1892). Intrauterine protection, presumably by way of the placenta, was shown by demonstrating that rodent pups born from immunized dams but suckled by non-immune dams were defended against the toxins. In addition, protection by breastfeeding was demonstrated by showing that dams immunized against the toxins during pregnancy protected not only their offspring but also pups from non-immunized dams. Similar results were obtained with mice immunized against the bacterial toxin tetanus (Ehrlich and Hubener, 1894). These experiments were carried out close to the inception of Ehrlich's studies in immunology. They were the foundation for pediatric immunology as well as the first indication that the mammary gland was an immunological organ.

4. Protective Enteric Bacteria in Breast-Fed Infants

Metchnikoff took a Darwinian (that is phylogenetic) approach to investigations of embryology and the role of phagocytes in immunology (Tauber and Chernyak, 1991). He, Pasteur and Erhlich, believed that immunology was closely linked to nutrition. In that respect, he provided observations that led to the concept of probiotics. Metchnikoff was impressed that many intestinal bacterial pathogens required an alkaline environment for their growth (Metchnikoff, 1906; Metchnikoff and Wollman, 1912). He therefore postulated that if the lumen of the large intestine was acidified, the growth of harmful bacteria could be inhibited. It had been known for some time that milk containing acid-producing bacteria was resistant to putrefaction.

Metchnikoff was strongly influenced by Pasteur's basic discoveries of the fermentive properties of certain bacteria. In addition he was impressed by anecdotal reports of increased longevity among populations in Eastern Europe who routinely drank sour milk. He became so preoccupied with the idea of acid protection of the intestinal tract against "putrefying" bacteria that at age 63 years he introduced sour milk into his diet for the rest of his life (Metchnikoff, 1906, 1921). He tested the idea in an experimental animal model by simultaneously introducing into rats *Lactobacilli* and *Glycobacter peptonicus*, an innocuous bacterium found in the canine intestine that converts starch to glucose. The strategy succeeded, and "putrefying" bacteria in the large intestine were greatly reduced in number (Metchnikoff and Wollman, 1912).

It had been known since Escherich's discovery (1886) that the stools of breastfed infants contained far more Gram positive bacteria than those of adults. It was later found that the Gram positive bacteria were principally bifidobacilli (Moro, 1900; Tissier, 1900; Gyllenberg and Roine, 1957). The organic acids produced by these *Lactobacilli* were found to be responsible for the high hydrogen ion concentration in the stools of breastfed infants (Norton and Schohl, 1926). In turn, as Metchnikoff predicted and as shown *in vitro* by Bienstock (1899), the high hydrogen ion concentration was toxic to enteric bacterial pathogens such as *Escherichia coli.*

The cause of the growth of Lactobacilli in the large intestine of breastfed infants was not understood prior to the work of György in the latter half of this century (György, 1953; György *et al.*, 1954, 1974). He discovered a strain of *Bifidobacter bifidus* that grew exceptionally well after exposure to human but not bovine milk (György *et al.*, 1954). One principal growth factor for those bacteria was later found to be kappa-casein (Bezkorovainy and Tapouzian, 1981). Through these investigations some of the agents in human milk responsible for the growth of protective bacteria in the lower intestinal tract of the infant were identified.

5. Antibodies in Human Milk

After Erhlich's demonstration of the specific protective effects of murine milk, he discovered that antibodies to tetanus toxin formed in milk from immunized goats and could passively immunize non-immune mice (Brieger and Erhlich, 1892). Brieger and Erhlich (1893) then demonstrated the temporal pattern of the formation of antibodies in milk after primary and secondary immunization. Erhlich and Wasserman (1894) showed that antibodies to diphtheria toxin formed in milk after specific immunization. In the first half of the twentieth century several antibody specificities were found in human milk (Debré *et al.*, 1930; Wong and Wong, 1930; Timmerman, 1931; van Genderen *et al.*, 1934; Lemetayer *et al.*, 1950), but not much progress was made along these lines until the biochemical nature of antibodies was unraveled between 1950 and 1970. In the 1950's Hanson applied the technique of immunoelectrophoresis innovated by Grabar and Williams (Grabar and Williams, 1953, 1955) to human milk. The method combined the discriminating powers of immunodiffusion and electrophoresis. Hanson applied the method to investigations of the immune globulins in human milk that he had previously discovered by immunodiffusion (Hanson, 1960). He immunized animals with human serum or with human milk to produce antibodies that would precipitate serum or milk antigens. After electrophoresis, he examined the shared and unshared antigens in serum and milk by using the antisera to human serum or human milk. It became clear that three types of immunoglobulins were present in human milk (Hanson, 1961), and Hanson noted that IgA was more complex in human milk than in human blood.

Contrary to Ehrlich's experimental observations in mice (1892), it was later ascertained that IgA in human milk was assembled from two distinct proteins produced by two types of cells in the mammary gland, and that the action of these antibodies was restricted to the mucosa of the recipient. IgA dimers produced by plasma cells in the mammary gland became bound to part of the polymeric immunoglobulin receptor, termed the secretory component, produced by the mammary epithelium (Brandtzaeg, 1978). This complex molecule, termed secretory IgA, was found to be common at mucosal sites, and because of its resistance to proteolysis (Shim *et al.*, 1969) it persisted in the gastrointestinal tract of the recipient infant (Schanler *et al.*, 1986). Antibodies in human milk were found to be directed principally against enteric and respiratory microbial antigens. For example, Goldblum *et al.* (1975) found that IgA antibodies appeared within a few days in milk against a strain of *E. coli* administered orally. Other investigators found in animal models that this was due to an enteromammary pathway (Roux *et al.*, 1977; Weisz-Carrington *et al.*, 1978). Upon exposure to enteric antigens, B cells in Peyer's patches of the small intestine are induced to switch their isotype from IgM to IgA and to migrate sequentially to intestinal efferent lymphatics, the thoracic duct and the vascular circulation by way of the superior vena cava. Under the influence of hormone stimulation during lactation, they migrate to the mammary gland where they transform into plasma cells that produce dimeric IgA antibodies against the same antigenic determinants that are recognized by the surface IgM antibodies on the B cells in the Peyer's patches. An analogous bronchomammary gland pathway for the formation of IgA antibodies to respiratory syncytial virus was subsequently discovered by Fishaut *et al.* (1981).

6. Non-Antibody Protein Defense Agents in Human Milk

Two antibacterial proteins that are not antibodies were discovered in human milk before the nature of antibodies in milk was understood. The first was lysozyme. This agent is of historical interest in that it was one of the first non-antibody proteins found to have antibacterial properties. The lytic activity of lysozyme was repeatedly demonstrated by Fleming in nasal secretions and in other secretions and tissues (Fleming, 1922; Fleming and Allison, 1922a,b). A more detailed analysis of lysozyme in human milk and its correlation with the detection of lysozyme activity in the stools of the breastfed infant was reported by Rosenthal and Lieberman (1931). They noted the predominance of lactobacilli and the paucity of *E. coli* in stools of the breastfed infants and concluded that this was due to lysozyme. As previously noted, the control of the bacterial flora of the lower intestinal tract proved to involve many other factors. Lysozyme was found to attack peptidoglycans exposed in the cell walls of susceptible bacteria (Salton, 1956).

Two years after Fleming's initial reports, Bordet reported briefly upon the lytic activity of human milk against certain airborne bacteria (Bordet and Bordet, 1924). The activity was found in colostrum and in milk collected at two and three months of lactation. Bordet's research on the role of the non-specific alexine (complement) and specific sensitizing agents (antibodies) in bacterial killing (Bordet, 1895, 1898, 1900) widened the concept of immunology from the more narrow confines of Metchnikoff's phagocytes and helped set the stage for understanding antibodies in milk.

Much later, Garza, Goldblum and others made quantitative, longitudinal measurements of lysozyme in human milk. In contrast to lactoferrin and secretory IgA, the concentrations of lysozyme increased during the first six months of lactation (Goldman *et al.*, 1982). Butte and colleagues reported the first quantitative estimates of the daily production of lysozyme, lactoferrin and secretory IgA in human milk (Butte *et al.*, 1984). The production of lysozyme continued to gradually rise, although the total amount ingested per kilogram of body weight somewhat declined. The total daily secretion of secretory IgA and lactoferrin as well as the amount ingested per kilogram of body weight gradually declined in the first four months of lactation. This finding provided some of the best evidence of a reciprocal relationship between the production of immune factors by the mammary gland and the developmental status of the immune system in early infancy.

Ten years before Hanson investigated the immunoglobulins in human milk, Schäfer (1951) published electrophoretic observations of a hitherto unrecognized protein in human milk. A decade later the protein was found independently by Johansson (1960), Montreuil *et al.* (1960) and Blanc and Isliker (1964) to be a member of the transferrin family of iron binding proteins. The protein, designated lactoferrin by Blanc and Isliker, was found by Masson, Heremans and their colleagues to be common to mucosal secretions (Masson *et al.*, 1966) and to inhibit the *in vitro* growth of siderophilic bacteria and *Candida albicans* (Masson *et al.*, 1966). Later it was discovered that about 80% of the lactoferrin in human milk does not possess bound iron (Fransson and Lonnerdal, 1980). Thus, this human milk protein is poised to inhibit the growth of enteric pathogens. Arnold and colleagues (1977) had previously observed that lactoferrin possessed bactericidal activity that was independent of its ferric iron binding ability. The latter activity, measured by the ability to disrupt surface membranes on susceptible *E. coli* and *C. albicans*, was found to reside in the N-terminal portion of the lactoferrin molecule (Bellamy *et al.*, 1993; Yamauchi *et al.*, 1993).

7. Non-Protein Defense Agents in Human Milk

György discerned that human milk contains probiotic agents and showed that human milk protected against bacterial infections in experimental animal models. He found that young mice injected parenterally with human milk survived a lethal staphylococcal challenge (György *et al.*, 1962). The protective factor was thermostable, resistant to proteases, and too small to be a protein. He and Zilliken reported that the factor was a C18:2 fatty acid located in the phospholipid fraction of human milk that was distinct from linoleic acid (György, 1971).

The presence of oligosaccharides and glycoconjugates in human milk that interfere with the binding of certain enteric bacteria or their toxins to cells was established by Holmgren and others (Holmgren *et al.*, 1981, 1983; Otnaess and Svennerholm, 1982; Kolstö Otnaess *et al.*, 1983). In 1986 Andersson *et al.* (1986) discovered that human milk oligosaccharides inhibited the adherence of pneumococci to epithelial cells. Further discoveries established that human milk contains a complex mixture of these compounds that act as receptor analogues to prevent binding of enteric or respiratory bacteria or bacterial toxins to human epithelial cells.

Milk fat globules aid in protection by three mechanisms. Firstly, mucins in milk fat globule membranes inhibit the adhesion of S-fimbriated *E. coli* to epithelial cells (Schroten *et al.*, 1992). Secondly, enzymatic liberation of lipids from the core triglyceride protects against enveloped viruses (Welsh *et al.*, 1978; Welsh *et al.*, 1979; Kabara, 1980) and *Giardia lamblia* (Reiner *et al.*, 1986). Thirdly, engulfment of milk fat globules by phagocytes was demonstrated by Keeney *et al.* (1993). This natural phagocytosis may enhance the uptake into neutrophils and macrophages of *E. coli* and perhaps other enteric bacteria bound to mucin on the membranes of the milk fat globule. In addition to the direct oxidative intracellular killing pathway activated by phagocytosis, it is possible that microbicidal lipids are released from the core triglyceride by lipases in phagocytes.

Besides specific secretory IgA antibodies, a diverse group of antiviral agents has been reported in human milk. These include the lipids released by the digestion of neutral fat in human milk; mucin, which prevents experimental rotavirus infections (Yolken *et al.*, 1992); lactoferrin, which inhibits certain retroviruses by mechanisms that do not involve its ability to bind ferric iron (Furmanski *et al.*, 1989); and a more broadly active agent that has yet to be characterized (Kumar *et al.*, 1984).

8. Concept of a Host Defense System in Human Milk

In the early 1970's there were four independent reviews of the immunological agents in human milk (György, 1971; Mata and Wyatt, 1971; Hanson and Winberg, 1972; Goldman and Smith, 1973). Each report cited paucity of these agents in bovine milk. In the 1980's, Goldblum and Goldman presented an analysis that serves as a guide to the principal features of the direct acting antimicrobial agents in human milk (Goldman and Goldblum, 1985, 1989). These agents are 1) biochemically heterogeneous, 2) commonly found at mucosal sites, 3) adapted to persist and function at those sites because of their inherent resistance to digestion, 4) inversely related to the production of the agents by the recipient infant, and 5) not well represented in bovine milk or other non-human milks used in milk formulas for infant nutrition.

9. Anti-Inflammatory Agents in Human Milk

It was noted by Mata and his colleagues in studies conducted in Central America in the 1970's that infants who were breastfed by mothers whose nipples and areolas were contaminated with bacterial enteropathogens such as Shigella species failed to display clinical evidence of gastroenteritis (Mata *et al.*, 1967, 1969). Twenty years later Duffy, Ogra and colleagues observed that breastfed infants infected with rotavirus also had meager gastrointestinal signs (Duffy *et al.*, 1986). The scarcity of clinical findings of infection in these and other cases were at first attributed simply to antiinfectious agents in human milk. This raised the question whether antiinfectious agents in human milk are able to ward off large numbers of pathogens and thus prevent most of the clinical signs of the infections. Classically, immunological protection is achieved by the marshaling of inflammation. However, if that occurred as a result of breastfeeding, the consequent injury to the infant's gastrointestinal and respiratory tracts would be devastating. It thus seemed logical that human milk protected the recipient infant by non-inflammatory mechanisms, and that inflammatory mediators would be poorly represented and anti-inflammatory agents would be prominent in human milk. A review of published papers revealed that these predictions were correct. At that time, it appeared that there was a paucity of inflammatory mediators in human milk, but many anti-inflammatory agents. The anti-inflammatory agents were as heterogeneous as the antiinfectious ones which included secretory IgA, lactoferrin, and lysozyme, antioxidants, polyamines and epithelial growth factors, antiproliferative agents, and enzymes that destroy inflammatory mediators. The significance of these points was not widely appreciated because the investigations that led to their discoveries were not designed to elucidate the protective properties of human milk and the discoveries were not presented in the context of the immunobiology of human milk. However, Goldman and collaborators published a report on the anti-inflammatory properties of human milk (Goldman *et al.*, 1986). Based on their thesis, it was predicted that neutrophils in human milk would display fewer inflammatory features than their counterparts in human blood and that anti-inflammatory agents would be found in human milk. Both predictions proved to be correct (Thorpe *et al.*, 1986; Buescher, 1991; Furukawa *et al.*, 1993). Subsequently, anti-inflammatory effects of human milk were demonstrated *in vitro* and in animal models by Buescher and his colleagues (Buescher and McIlheran, 1992; Grazioso *et al.*, 1997).

10. Immunomodulating Agents in Human Milk

Low birth weight infants fed human milk preparations displayed higher levels of secretory IgA, not only in their stools (Schanler *et al.*, 1986), but also in their urine (Goldblum *et al.*, 1989). This did not appear to be due to absorption of the intact immunoglobulin from the alimentary tract. It was more likely that the increased urinary IgA levels were due to an increase in the assembly and transport of secretory IgA by the kidney. At the same time others were reporting that the levels of secretory IgA in saliva were increased in infants fed human milk (Stephens, 1986). Thus, the notion began to emerge that human milk might promote the growth or development or regulate the functions of certain parts of the immune system of the recipient infant. Soon thereafter, it was reported that children who had been breastfed in infancy were less prone to develop lymphomas (Davis *et al.*, 1988) and certain immunologically mediated diseases, including insulin-dependent diabetes mellitus (Mayer *et al.*, 1988) and Crohn's disease (Koletzko *et al.*, 1989).

Based upon these findings, human milk was tested for immunomodulating agents. The first hint of their presence was a report that interferon was produced by human milk leukocytes stimulated *in vitro* (Emodi and Just, 1974). Subsequently, numerous bioactivities were identified including stimulation of the growth and differentiation of and immunoglobulin production by B cells (Pittard and Bill, 1979; Juto, 1985; Julius *et al.*, 1988), transforming growth factor activity (Noda *et al.*, 1984), enhancement of uropod formation by T cells (Dickey *et al.*, 1981), an ability to augment the proliferation of murine thymocytes (Soder, 1987), and the ability to inhibit the production of IL-2 by stimulated human T cells (Hooton *et al.*, 1991). However, the bioactive factors responsible for these events were not identified (Goldman *et al.*, 1996). In an initial study on this question, it was ascertained that human milk had a chemokinetic effect on human blood monocytes (Ozkaragoz *et al.*, 1988). Schmalstieg suggested that the activating agent could be either tumor necrosis-α (TNF-α) or IL-1 because those cytokines stimulated other functions of monocytes and macrophages. Recombinant human TNF-α was found to be chemokinetic for blood monocytes (Mushtaha *et al.*, 1989). Most of the chemokinetic effect caused by human milk was blocked with antibodies against TNF-α but not with antibodies to IL-1 (Mushtaha *et al.*, 1989). This was probably the first demonstration of a cytokine in human milk. Shortly thereafter, a quantitative enzyme assay was used to verify that TNF-α was a normal component of early human milk (Rudloff *et al.*, 1992). At about the same time, Munoz and his colleagues found that early human milk contained high amounts of immunoreactive IL-1β (Munoz *et al.*, 1990). This probably explained the earlier finding that human milk stimulated the proliferation of murine thymocytes *in vitro* (Söder, 1987). These findings and those of other contemporaries in the field set the stage for investigations that led to the discovery of other cytokines including IL-6 (Saito *et al.*, 1991; Rudloff *et al.*, 1993); IL-10 (Garofalo *et al.*, 1995), interferon-γ (Eglinton *et al.*, 1994), transforming growth factor-β2 (Saito *et al.*, 1993), granulocyte colony stimulating factor (Gilmore *et al.*, 1994), macrophage colony stimulating factor (Hara *et al.*, 1995), IL-8 (Palkowetz *et al.*, 1994), and other chemokines (chemotactic cytokines) in human milk (Goldman *et al.*, 1997). Other classes of immunomodulating agents including prolactin (Ho Yuen, 1988), nucleotides (Janas and

Picciano, 1992; Barness, 1994), and anti-idiotypic antibodies (Hahn-Zoric *et al.*, 1993) were also discovered in human milk during that period. It is also noteworthy that the immunomodulatory factors interact with other parts of the immune system in human milk. For example IL-10 inhibits the production of proinflammatory agents and IgA is antimicrobial.

11. Protection of the Immunological System in Human Milk

As mentioned earlier, many components of the immune system in human milk appear to be innately protected against digestive/denaturing conditions found in the alimentary tract. Other adaptations that may contribute toward the protection of host defense agents in human milk include antiproteases (Linberg *et al.*, 1982), developmental delays in the production of gastric HCl (Agunod *et al.*, 1969; Euler *et al.*, 1979) and of pancreatic proteases (Lebenthal and Lee, 1980; Bujanover *et al.*, 1988) and the inhibition of gastric acid production by epithelial growth factor (Guglietta and Lesch, 1993) and transforming growth factor-α (Goldenring *et al.*, 1993).

12. Evolutionary Relationships Between Developmental Delays in the Immune System and the Production of Immune Factors by the Mammary Gland

It is striking that many immune factors in human milk are developmentally delayed in the recipient infant, and the reciprocal relationship between the development of many immune factors and their production by the mammary gland is found in all mammals, regardless whether they are protherians, metatherians or eutherians. Recently, the hypothesis has been posed that this reciprocal relationship arose through natural selection of mutations that improved the immunological function of the mammary gland (Goldman *et al.*, 1998). An examination of the status of the immune system in suckling infants of all the mammalian orders that have been investigated and the immunological composition of the milks from those many species supports that hypothesis. Indeed, it is likely that the mammary gland secretions of the earliest mammals contained certain immunological agents that provided a survival advantage for the newborn infant. A consideration of the immunological systems of the maternal/infant dyad in the context of biological evolution may help to address questions concerning the development of the immune system in human milk and its role in protecting the health of the recipient during and after the period of breastfeeding.

13. Future Research

13.1. Immunological Components

We are far from understanding the totality of the immune system in human milk, in particular the immunomodulators. A large number of cytokines and chemokines have been discovered, but only a small fraction of them have been looked for in human milk. Moreover, additional immunomodulators and anti-inflammatory agents will be discovered as new sensitive immunological detection reagents become available. Also, we should be prepared to recognize that the known immune components in human milk may have unexpected functions.

13.2. Formation and Physical State of Immune Components in Human Milk

Very little is known about the formation and physical state of immune components in human milk, yet that information is critical to understanding their concentration in human milk, how they are protected from digestive agents and denaturing conditions found in the alimentary tract of the infant, and how they operate in the recipient. For example, consider the large number of immunomodulatory agents in human milk and possible redundancies in their functions as predicted from *in vivo* investigations. The function of each of these agents may be dictated not only by its exact molecular structure, but also by whether its original structure is modified subtly by partial digestion in the alimentary tract or whether the agent is bound to other moieties in milk. In that respect, the precise molecular structure of these molecules in milk and the point at which they initially contact the alimentary tract or respiratory tract of the recipient may be crucial determinants to their functions. For example, it is unclear whether each immunomodulating agent in human milk has exactly the same structure as those produced by cells at other sites. In addition, it appears that certain cytokines in human milk such as TNF-α are bound to soluble receptors (Buescher and Malinowska, 1996). Furthermore, one cytokine in human milk, IL-10, is not fully detectable until the milk is treated with natural detergents such as taurocholate (Garofalo *et al.*, 1995). This suggests that part of the total amount of that cytokine is not exposed to the gastrointestinal tract until it is released from its compartmentalized state by taurocholate secreted into the first part of the duodenum. Finally, it is puzzling that human milk is anti-inflammatory, in spite of the fact that some of the cytokines and chemokines in human milk are proinflammatory when they are found at other sites. Again, it is possible that the precise structure or compartmentalization and initial point of contact of the agent with the mucosa of the alimentary tract or respiratory tract may dictate a function that may not be entirely anticipated from artificial *in vitro* studies. For example, the distribution of specific receptors for some of these agents may correlate with the part of the gastrointestinal tract where the agents are released from their bound form in human milk. Clearly, specific research strategies will be required to investigate the mechanisms that control the formation and secretion of each of the components of the immune system, the exact molecular structure and packaging of them, and where and how they are presented to the infant.

13.3. Fate and Function of Human Milk Immune Factors in the Recipient

The fate and function of the immune factors in human milk in the recipient infant are largely unknown. In the few published investigations, it has been demonstrated that a significant proportion of some antimicrobial factors persist throughout the alimentary tract. The ways in which these factors are processed or interact with the mucosa of the alimentary tract or respiratory system are poorly understood. A number of possibilities should be investigated. One is whether partial digestion of some factors alters their function. A second is whether the agents interact with receptors present on the epithelium of the alimentary tract or respiratory system. A third is whether some agents are absorbed into the tissues of the alimentary tract or respiratory system or even enter the systemic circulation of the recipient. One example that pertains to the above questions is the finding that peptide fragments of human milk lactoferrin created by partial digestion in the alimentary tract are absorbed into the systemic circulation and appear in the urine (Goldman *et al.*, 1990; Hutchens *et al.*, 1991). A fourth issue is that, despite the survival in the alimentary tract of the infant of a significant fraction of certain bioactive proteins (Schanler *et al.*, 1986), human milk meets the nutritional requirements for protein synthesis in the recipient infant. This suggests that *in vivo* degradation of the proteins in human milk is closely controlled to permit the bioactivity of the proteins or their fragments to persist long enough to be protective until more complete degradation occurs.

13.4. Immunological Tolerance/Reduced Immunological Responses

An important related question is what factors in human milk are responsible not only for short term protection against common intestinal and respiratory infections, but also for protection against immunological/inflammatory disorders that occur in infancy or years after breastfeeding is over. In that respect, it is unclear whether the immunomodulatory and anti-inflammatory agents in human milk provide the infant with long-term protection. What changes may be set in motion that culminate in these delayed protection?

Indeed, some of the long-term protection provided by breastfeeding against inflammatory or immune mediated diseases may be due to an induction of immunological tolerance by components of human milk. Children who have been breastfed during infancy accept renal allografts more readily than those who have not been breastfed (Campbell *et al.*, 1983, 1984; Kois *et al.*, 1984). Furthermore, blood lymphocytes from children who were breast-fed are less able to proliferate after being incubated with blood lymphocytes from their mothers than those who were never breast-fed (Zhang *et al.*, 1991). Clearly, some form of immunological tolerance or other immunological suppressive responses to maternal antigens is transmitted by breastfeeding. Does this oral tolerance or suppression extend to foreign proteins that are structurally similar to human antigens? Are tolergens formed from human milk proteins by partial digestion in the recipient's alimentary tract? Would such tolerogenic or suppressive processes also operate to prevent food allergic reactions and other immunologically mediated diseases such as autoimmune disorders? It is unlikely that the anti-inflammatory agents in human milk exert effects that extend so far beyond the suckling period. Are tolergens or other immune suppressive agents a fourth segment of the immunological system in human milk?

14. Coda

The following closure contains a few final impressions of the history of the immunobiology of human milk. It is remarkable that many founders of modern immunology made the initial discoveries that form the basis for our current understanding of the antimicrobial, anti-inflammatory, and immunomodulating components of the host defense system in human milk. In that respect, human milk has been in the mainstream of the science of immunology since its inception.

Those early investigators were strongly influenced by the work of their mentors or contemporaries and were struck by the inextricable linkage of the immunobiology of human milk with the science of nutrition. That theme is resounded today by current leaders in the field. Indeed, the formation of the system is dependent upon the nutrient status of the woman: some nutrients such as ascorbic acid and lipids serve as components of the immune system, and most of the components of the immune system in human milk are eventually used as sources of energy for the infant.

Without exception, the major contributors to this field persisted in studying a few research questions for many years. Their tenacity spurred them to carry their investigations to fruition. In addition, it is impressive that these pioneering investigators maintained lofty humanitarian goals and often worked to improve the general well-being of populations in their country or beyond. Scientists with such motivations may be naturally attracted to study the biological interfaces between mothers and their offspring.

This review denotes a great affinity to Ehrlich and Metchnikoff, not only because of Metchnikoff's preoccupation with phagocytes and Erhlich's stunning discovery of the immunological role of the mammary gland, but also because of Metchnikoff's Darwinian approach to his investigations of phagocytes and, as Silverstein (1996) has emphasized, the Darwinian nature of many of Erhlich's ideas. An understanding of the immunobiology of human milk is incomplete unless the evolutionary relationships between the postnatal development of the immune system and the immune functions of the mammary gland are considered. Metchnikoff and Erhlich would have sensed the coupling of those evolutionary events in the context of natural selection if they had lived during the latter half of this century.

Finally, it is fitting that this review was written close to the beginning of the twenty-first century when the techniques of molecular biology can be applied to the tantalizing questions concerning the immune system in human milk. Technologies, however, are only tools. The scientific pioneers in this field displayed a blend of intellectual dedication, imagination and vigor, and an ability to capitalize on serendipitous observations. A similar profile undoubtedly will be required of future investigators who uncover the more mechanistic aspects of the immunobiology of human milk. So much remains to be discovered! Surely the best is yet to come.

ACKNOWLEDGMENTS. I Thank Lynn Burke, MA, AHIP, a Clinical Librarian at the Moody Library and Sarita B. Oertling, a Library Assistant in the Blocker History of Medicine Collections at The University of Texas Medical Branch in Galveston for finding the original manuscripts of Louis Pasteur, Elie Metchnikoff, Jules Bordet and Paul Ehrlich that I used in the preparation of this manuscript. I am also very appreciative of the reviews of this manuscript by my colleagues Samuel Baron and David K. Rassin, The University of Texas Medical Branch; Daniel Goldman, Austin, Texas; Cutberto Garza, Cornell University; and Lars Å Hanson, Göteborg Universitet.

References

Agunod, M., Yamaguchi, N., Lopez, R., Luhby, A.L., and Glass, G.B. 1969. Correlative study of hydrochloric acid, pepsin, and intrinsic factor secretion in newborns and infants. *Am. J. Dig. Dis.* **14**:400.

Andersson, B., Porras, O., Hanson, L.Å., Lagergärd, T., and G., Svanborg-Edén, C. 1986. Inhibition of attachment of *Streptococcus pneumoniae* and *Haemophilus influenzae* by human milk and receptor oligosaccharides attaching to human pharyngeal epithelial cells. *J. Infect. Dis.* **153**:232.

Arnold, R.R., Cole, M.F., and McGhee, J.R. 1977. A bactericidal effect for human lactoferrin. *Science* **197**:263.

Barness, L.A. 1994. Dietary sources of nucleotides—from breast milk to weaning. *J. Nutr.* **124**:128.

Bellamy, W., Wakabayashi, H., Takase, M., Kawase, K., Shimamura, S., and Tomita, M. 1993. Killing of Candida albicans by lactoferricin B, a potent antimicrobial peptide derived from the N-terminal region of bovine lactoferrin. *Med. Microbiol. Immunol. (Berl.)* **182**:97.

Bertotto, A., Gerli, R., Fabietti, G., Crupi, S., Arcangeli, C., Scalise, F., and Vaccaro, R. 1990. Human breast milk T cells display the phenotype and functional characteristics of memory T cells. *Eur. J. Immunol.* **20**:1877.

Bezkorovainy, A. and Tapouzian, N. 1981. *Bifidobacterium bifidus* var. *Pennsylvanicus* growth promoting activity of human milk casein and its derivatives. *Int. J. Biochem.* **13**:585.

Bienstock, E. 1899. Recherches sur la putrefaction. *Ann de l'Pasteur* **13**:121.

Blanc, B. and Isliker, H. 1961. Isolement et caractérisation de la protéine rouge sidérophiline du lait maternel: La lactotransferrine. *Bull. Soc. Chim. Biol. (Paris)* **43**:929.

Bordet, J. 1895. Les leukocytes et les propertetes actives du serum chez les vaccinés. *Ann. Inst. Pasteur* **9**:462.

Bordet, J. 1898. Sur l'agglutination et la dissoltuion des globules rouges par le serum d'animaux injectés de sang defribriné. *Ann. Inst. Pasteur* **12**:688.

Bordet, J. 1900. Les sérums hemolytiques, leurs antitoxines et les théories des sérums cytolytiques. *Ann. Inst. Pasteur* **14**:257.

Bordet, J. and Bordet, M. 1924. Le pouvoir bactériolytique du colostrum et du lait. *Comptes Rendus des Séaces de L'Academie des Sciences* **179**:1109.

Brandtzaeg, P. 1978. Polymeric IgA is complexed with secretory component (SC) on the surface of human intestinal epithelial cells. *Scand. J. Immunol.* **8**:39.

Brieger, L. and Erhlich, P. 1892. Über die Übertragung von Immunität durch Milch, *Deutsch med Wochenschr* **18**:393.

Brieger, L. and Erhlich, P. 1893. Beitr_ge zur Kenntniss der Milch immunisirte Thiere. *Z. Hyg. InfektKr* **13**:336.

Buescher, S.E. 1991. The effects of colostrum on neutrophil function: decreased deformability with increased cytoskeletal-associated actin, in: *Immunology of Milk and the Neonate.* J. Mestecky, C. Blair, and P. Ogra, ed. Plenum Press, NY, p. 131.

Buescher, E.S. and Malinowska, I. 1996. Soluble receptors and cytokine antagonists in human milk. *Pediatr. Res.* **40**:839.

Buescher, S.E. and McIlheran, S.M. 1992. Colostral antioxidants: separation and characterization of two activities in human colostrum. *J. Pediatr. Gastroenterol. Nutr.* **14**:47.

Bujanover, Y., Harel, A., Gete, R, Blau, H., Yahav, J., and Spirer, Z. 1988. The development of the chymotryptic activity during postnatal life using the bentiromide test. *Int. J. Pancreatol.* **3**:53.

Butte, N.F., Goldblum, R.M., Fehl, L.M., Loftin, K., Smith, E.O., Garza, C., and Goldman, A.S. 1984. Daily ingestion of immunologic components in human milk during the first four months of life. *Acta Paediatr. Scand.* **73**:296–301.

Campbell, D.A. Jr., Lorber, M.I., Sweeton, J.C., Turcotte, J.G., and Beer, A.E. 1983. Maternal donor-related transplants: influence of breast feeding on reactivity to the allograft. *Transplant Proc.* **15**:906.

Campbell, Jr., D.A., Lorber, M.I., Sweeton, J.C., Turcotte, J.G., Niederhuber, J.E., and Beer, A.E. 1984. Breast feeding and maternal-donor renal allografts. *Transplantation* **37**:340.

Crago, S.S., Prince, S.J., Pretlow, T.G., McGhee, J.R., and Mestecky, J. 1979. Human colostral cells. I. Separation and characterization. *Clin. Exp. Immunol.* **38**:585.

Davis, M.K., Savitz, D.A., and Grauford, B. 1988. Infant feeding in childhood cancer. *Lancet* **2**:365.

Davis, W.H. 1913. Statistical comparison of the mortality of breast-fed and bottle-fed infants. *Am. J. Dis. Child.* **5**:234.

Debré, R., Ramon, G., Lévy-Solal, E., and Thiroloix, P.L. 1930. Sur le passage des antitoxines par le lait et colostrum, en particulier dans l'espèce humaine. *Nourrisson* **18**:235.

Dickey, W.D., Rudloff, H.B., Goldman, A.S., and Schmalstieg, F.C. 1981. Human uropod bearing lymphocytes: isolation of a factor from human milk that abrogates the uropod inhibitory protein from human milk. *Biochem. Biophys. Res. Com.* **100**:138.

Donné, A. 1837. Du lait et en particulier de celui de nourrices, considéré sous le rapport de ses bonnes et de ses mavaises qualités nutritives et de ses altérations. Paris/L-'auter, Rue de Condé, No. 15; Les Libraires de Médecine, Ch. Chevalier, opticien, Palais Royal, N. 163.

Duffy, L.C., Riepenhoff-Talty, M., Byers, T.E., La Scolea, L.J., Zielezny, M.A., Dryja, D.M., and Ogra, P.L. 1986. Modulation of rotavirus enteritis during breastfeeding. *Am. J. Dis. Child.* **140**:1164.

Eglinton, B.A., Roberton, D.M., and Cummin, A.G. 1994. Phenotype of T cells, their soluble receptor levels, and cytokine profile of human breast milk. *Immunol. Cell. Biol.* **72**:306.

Ehrlich, P. 1892. Über Immunität durch Verebung und Säugung. *Z. Hyg. Infektkr.* **12**:183.

Ehrlich, P. and Hubener, W. 1894. Über die Vererbung der Immunität bei Tetanus. *Z. Hyg. Infektkr.* **18**:51.

Ehrlich, P. and Wassermann, A. 1894. Über die Gewinnung Diphtherie-Antitoxine aus Blutserum und Milch immunsirte Thiere. *Z. Hyg. Infektkr.* **8**:239.

Emodi, G. and Just, M. 1974. Interferon production by lymphocytes in human milk. *Scand. J. Immunol.* **3**:157.

Escherich, T. 1886. Die Darmbakterien des Säuglings und ihre Beziehungen zur Physiologie der Verdauung. Stuttgart, F Enke, p. 1.

Euler, A.R., Byrne, W.J., Meis, P.J., Leake, R.D., and Ament, M.E. 1979. Basal and pentagastrin-stimulated acid secretion in newborn human infants. *Pediatr. Res.* **13**:36.

Fishaut, M., Murphy, D.S., Neifert, M., McIntosh, K., and Ogra, P.L. 1981. Broncho-mammary axis in the immune response to respiratory syncytial virus. *J. Pediatr.* **99**:186.

Fleming, L.A. 1974. Personal recollections of lysozyme and Fleming, in: *Lysozyme*. E.F. Osserman, R.E. Canfield, and S. Beychok, eds. Academic Press, New York and London, pp. xiii.

Fleming, A. 1922. On a remarkable bacteriolytic element found in tissues and secretions. *Proc. Roy. Soc. London* **B93**:306.

Fleming, A. and Allison, V.D. 1922a. Observations on a bacteriolytic element-lysozyme-found in tissues and secretions. *Br. J. Exp. Path.* **3**:252.

Fleming, A. and Allison, V.D. 1922b. Further observations on a bacteriolytic element found in tissues and secretions. *Proc. Roy. Soc. London* **B94**:142.

Fransson, G.B. and Lönnerdal, B. 1980. Iron in human milk. *J. Pediatr.* **96**:380.

Furmanski, P., Li, Z.P., Fortuna, M.B., Swamy, C.V., and Das, M.A. 1989. Multiple molecular forms of human lactoferrin. Identification of a class of lactoferrins that possess ribonuclease activity and lack iron binding capacity. *J. Exp. Med.* **170**:415.

Furukawa, M., Narahara, H., Yasuda, K., and Johnston, J.M. 1993. Presence of platelet-activating factor-acetylhydrolase in milk. *J. Lipid Res.* **34**:1603.

Garofalo, R., Chheda, S., Mei, F., Palkowetz, K.H., Rudloff, H.E., Schmalstieg, F.C., and Goldman, A.S. 1995. Interleukin-10 in human milk. *Pediatr. Res.* **37**:444.

Gilmore, W.S., McKelvey-Martin, V.J., Rutherford, S., Strain, J.J., Kell, M., and Miller, S. 1994. Human milk contains granulocyte-colony stimulating factor (G-CSF). *Europ. J. Clin. Nutr.* **48**:222.

Goldblum, R.M., Ahlstedt, S., Carlsson, B., Hanson, L.Å., Jodal, U., Lidin-Janson, C., and Sohl, A. 1975. Antibody forming cells in human colostrum after oral immunisation. *Nature (London)* **257**:797.

Goldblum, R.M., Schanler, R.J., Garza, C., and Goldman, A.S. 1989. Human milk feeding enhances the urinary excretion of immunologic factors in low birth weight infants. *Pediatr. Res.* **25**:184.

Goldenring, J.R., Tsunoda, Y., Stoch, S.A., Coffey, R.J., and Modlin, I.M. 1993. Transforming growth factor-alpha (TGF alpha) inhibition of parietal cell secretion: structural requirements for activity. *Regulatory Peptides* **43**:37.

Goldman, A.S. and Goldblum, R.M. 1996. Transfer of maternal leukocytes to the infant by human milk, in: *Reproductive Immunology/Current Topics In Microbiology and Immunology.* L. Olding, ed. Springer Verlag EMBH, and Co. KG, Heidelberg, p. 205.

Goldman, A.S., Garza, C., Johnson, C.A., Nichols, B.L., and Goldblum, R.M. 1982. Immunologic factors in human milk during the first year of lactation. *J. Pediatr.* **100**:563.

Goldman, A.S. and Smith, C.W. 1973. Host resistance factors in human milk. *J. Pediatr.* **82**:1082.

Goldman, A.S. and Goldblum, R.M. 1985. Protective properties of human milk, in: *Nutrition in Pediatrics-Basic Sciences and Clinical Applications.* W.A. Walker, and J.B. Watkins, eds. Little, Brown, and Co., Boston and Toronto, p. 819.

Goldman, A.S. and Goldblum, R.M. 1989. Immunologic System in Human Milk: Characteristics and Effects, in: *Textbook of Gastroenterology and Nutrition in Infancy*, Second Edition. E. Lebenthal, ed. Raven Press, New York, p. 135.

Goldman, A.S., Thorpe, L.W., Goldblum, R.M., and Hanson, L.Å. 1986. Antiinflammatory properties of human milk. *Acta Paediatr. Scand.* **75**:689.

Goldman, A.S., Garza, C., Schanler, R.J., and Goldblum, R.M. 1990. Molecular forms of lactoferrin in stool and urine from infants fed human milk. *Pediatr. Res.* **27**:252.

Goldman, A.S., Garofalo, R., and Chheda, S. 1997. Spectrum of immuno- modulating agents in human milk. *Int. J. Pediatr. Hematol/Oncol.* **4**:491.

Goldman, A.S., Chheda, S., and Garofalo, R. 1998. Evolution of immunological functions of the mammary gland and the postnatal development of immunity. *Pediatr. Res.* **42**:155.

Grabar, P. and Williams, C.A. 1953. Méthode permettant l'étude conjugée des propriétés electrophorétiques et immunochimiques d'un mélange de protéines. Application au serum sanguins. *Biochim. Biophys. Acta* **10**:193.

Grabar, P. and Williams, C.A., Jr. 1955. Méthode immuno-électrophorétique d'analyse de mélange de substances antigeniques. *Biochim. Biophys. Acta* **17**:67.

Grazioso, C.F., Werner, A.L., Alling, D.W., Bishop, P.R., and Buescher, E.S. 1997. Antiinflammatory effects of human milk on chemically induced colitis in rats. *Pediatr. Res.* **42**:639.

Groth, A. and Hahn, M. 1910. Die Säuglingverhältnisse in Bayern. *Zeitschrift d. k. Bayerischen Statistiscen Landesamts, 42d year*. **1**:78.

Grulee, C.G., Sanford, H.N., and Herron, P.H. 1934. Breast and artificial feeding. Influence on morbidity and mortality of twenty thousand infants. *JAMA* **103**:735.

Grulee, C.G., Sanford, H.N., and Schwartz, H. 1935. Breast and artificially-fed infants. A study of the age incidence in the morbidity and mortality in twenty thousand cases. *JAMA* **104**:1986.

Guglietta, A. and Lesch, C.A. 1993. Effect of h-EGF and h-EGF 1–48 on histamine-stimulated gastric acid secretion in rats and monkeys. *J. Physiology, Paris* **87**:343.

Gyllenberg, H. and Roine, P. 1957. The value of colony counts in evaluating the abundance of "Lactobacillus" bifidus in infant faeces. *Acta Pathol. Microb. Scand.* **41**:144.

György, P. 1953. A hitherto unrecognized biochemical difference between human milk and cow's milk. *Pediatrics* **11**:98.

György, P. 1971. The uniqueness of human milk. Biochemical aspects. *Am. J. Clin. Nutr.* **24**:970.

György, P., Jeanloz, R.W., von Nicolai, H., and Zilliken, F. 1974. Undialyzable growth factors for *Lactobacillus bifidus var. Pennsylvanicus*. *Eur. J. Biochem.* **43**:29.

György, P., Kuhn, R., Ross, C.S., and Zilliken, F. 1954. Bifidus factor. II. Its occurrence in milk from different species and other natural products. *Arch. Biochem. Biophys.* **48**:202.

György, P., Norris, R.F., and Rose, C.S. 1954. A variant of *Lactobacillus bifidus* requiring a special growth factor. *Arch. Biochem. Biophys.* **48**:193.

György, P., Dhanamitta, S., and Steers, E. 1962. Protective effects of human milk in experimental Staphylococcus infection. *Science* **137**:338.

Hahn-Zoric, M., Carlsson, B., Jeansson, S., Ekre, O., Osterhaus, A.D., Roberton, D.M., and Hanson, L.Å. 1993. Anti-idiotypic antibodies to poliovirus in commercial immunoglobulin, human serum and human milk. *Pediatr. Res.* **33**:475.

Hanson, L.Å. 1960. The serological relationship between human milk and blood plasma. *Int. Arch. Allergy* **17**:45.

Hanson, L.Å. 1961. Comparative studies of the immune globulins of human milk and blood serum. *Int. Arch. Allergy* **18**:241.

Hanson, L.Å. and Winberg, J. 1972. Breastmilk and defence against infection in the newborn. *Arch. Dis. Child.* **47**:845.

Hara, T., Irie, K., Saito, S., Ichijo, M., Yamadam M., Yanai, N., and Miyazaki, S. 1995. Identification of macrophage colony-stimulating factor in human milk and mammary epithelial cells. *Pediat. Res.* **37**:437.

Holmgren, J., Svennerholm, A.-M., and Åhérn, C. 1981. Non-immunoglobulin fraction of human milk inhibits bacterial adhesion (hemagglutination) and enterotoxin binding of *Escherichia coli* and *Vibrio cholerae*. *Infect. Immun.* **33**:136.

Holmgren, J., Svennerholm, A.-M., and Lindblad, M. 1983. Receptor-like glycocompounds in human milk that inhibit classical, and El Tor *Vibrio cholerae* cell adherence (hemagglutination). *Infect. Immun.* **39**:147.

Ho Yuen, B. 1988. Prolactin in human milk: the influence of nursing and duration of postpartum lactation. *Am. J. Obstet. Gynecol.* **158**:583.

Hooton, J.W.L., Pabst, H.F., Spady, D.W., and Paetkau, V. 1991. Human colostrum contains an activity that inhibits the production of IL-2. *Clin. Exp. Immunol.* **86**:520.

Hutchens, T.W., Henry, J.F., Yip, T.T., Hachey, D.L., Schanler, R.J., Motil, K.J., and Garza, C. 1991. Origin of intact lactoferrin and its DNA-binding fragments found in the urine of human milk-fed preterm infants. Evaluation of stable isotopic enrichment. *Pediatr. Res.* **29**:243.

Janas, I.M. and Picciano, M.F. 1992. The nucleotide profile of human milk. *Pediatr. Res.* **16**:659.
Johanson, B. 1960. Isolation of an iron-containing protein from human milk. *Acta Chem. Scand.* **14**:510.
Julius, M.H., Janusz, M., and Lisowski, J. 1988. A colostral protein that induces the growth and differentiation of resting B lymphocytes. *J. Immunol.* **140**:1366.
Juto, P. 1985. Human milk stimulates B cell function. *Arch. Dis. Child.* **60**:610.
Kabara, J.J. 1980. Lipids as host-resistance factors of human milk. *Nut. Rev.* **38**:65.
Keeney, S.E., Schmalstieg, F.C., Palkowetz, K.H., Rudloff, H.E., Schmalstieg, Jr., F.C., and Goldman, A.S. 1993. Activated neutrophils and neutrophil activators in human milk. Increased expression of CD11b and decreased expression of L-selectin. *J. Leuk. Biol.* **54**:97.
Kois, W.E., Campbell, D.A., Lorber, M.I., Sweeton, J.C., and Dafoe, D.C. 1984. Influence of breast feeding on subsequent reactivity to a related renal allograft. *J. Surg. Res.* **37**:89.
Koletzko, S., Sherman, P., Corey, M., Griffiths, A., and Smith, C. 1989. Role of infant feeding practices in development of Crohn's disease in childhood. *BMJ* **298**:1617.
Kumar, S., McKerlie, M.L., Albrecht, T.B., Goldman, A.S., and Baron, S. 1984. A broadly active viral inhibitor in human and animal organ extracts and body fluids. *Proc. Soc. Exp. Biol. Med.* **177**:104.
Lebenthal, E. and Lee, P.C. 1980. Development of functional response in human exocrine pancreas. *Pediatrics* **66**:556.
Lemétayer, E., Nicol, L., Grasset, J., Gauthier, R., and Pialoux, J. 1950. Teneur en antitoxine spécifique du colostrum chez les femmes vaccinées par le vaccin mixte diphtérique antitétanique. *Bull. Acad. Natl. Med. (Paris)* **134**:22.
Linberg, T., Ohlsson, K., and Westrom, B. 1982. Protease inhibitors and their relation to protease activity in human milk. *Pediatr. Res.* **16**:479.
Masson, P.L., Heremans, J.F., and Dive, C.H. 1966. An iron-binding protein common to many external secretions. *Clin. (Chim.) Acta* **14**:735.
Masson, P.L., Heremans, J.F., Prignot, J.J., and Wauters, G. 1966. Immunochemical localization and bacteriostatic properties of an iron-binding protein from bronchial mucus. *Thorax* **21**:538.
Mata, L.J., Urrutia, J.J., and Gordon, J.E. 1967. Diarrhoeal disease in a cohort of Guatemalan village children observed from birth to age two years. *Trop. Geogr. Med.* **19**:247.
Mata, L.J., Urrutia, J.J., García, B., Fernández, R., and Behar, M. 1969. Shigella infection in breast-fed Guatemalan Indian neonates. *Am. J. Dis. Child.* **117**:142.
Mata, L.J. and Wyatt, R.G. 1971. The uniqueness of human milk. Host resistance to infection. *Am. J. Clin. Nutr.* **24**:976.
Mayer, E.J., Hamman, R.F., Gay, E.C., Lezotte, D.C., Savitz, D.A., and Klingensmith, G.J. 1988. Reduced risk of IDDM among breast fed children. *The Colorado IDDM Registry. Diabetes* **37**:1625.
Metchnikoff, E. 1906. *The New Hygiene: Three Lectures on the Prevention of Infectious Diseases.* W.T. Keener, Chicago.
Metchnikoff, E. and Wollman, E. 1912. Sur quelques essais de désintoxication intestinale. *Annales de L'institute Pasteur* **20**:825.
Metchnikoff, O. 1921. *Life of Elie Metchnikoff.* Houghton Mifflin Company, Boston and New York.
Montreuil, J., Tonnelat, J., and Mullet, S. 1960. Préparation et propriétés de la lactosidérophiline (lactotransferrine) du lait de femme. *Biochim. Biophys. Acta* **45**:413.
Moro, E. 1900. Ûber die nach Gram färbbaren Bacillen des Säuglingstuhles, Vorläufige Mitteilung. *Wien. Klin. Wochenschr.* **13**:114.

Munoz, C., Endres, S., van der Meer, J., Schlesinger, L., Arevalo, M., and Dinarello, C. 1990. Interleukin-1β in human colostrum. *Res. Immunol.* **141**:501.

Murphey, D.K. and Buescher, E.S. 1993. Human colostrum has anti-inflammatory activity in a rat subcutaneous air pouch model of inflammation. *Pediatr. Res.* **34**:208.

Mushtaha, A.A., Schmalstieg, F.C., Hughes, T.K., Rudloff, H.E., and Goldman, A.S. 1989. Chemokinetic effects of exogenous and endogenous tumor necrosis factor-α on human blood monocytes. *Int. Arch. Allergy Appl. Immunol.* **90**:11.

Mushtaha, A.A., Schmalstieg, F.C., Hughes, T.K., Rajaraman, S., Rudloff, H.E., and Goldman, A.S. 1989. Chemokinetic agents for monocytes in human milk: Possible role of tumor necrosis factor-α. *Pediatr. Res.* **25**:629.

Noda, K., Umeda, M., and Ono, T. 1984. Transforming growth factor activity in human colostrum. *Gann.* **75**:109.

Norton, R.C. and Shohl, A.T. 1926. The hydrogen ion concentration of the stools of newborn infants. *Am. J. Dis. Child.* **32**:183.

Oksenberg, J.R., Persitz, E., and Brautbar, C. 1985. Cellular immunity in human milk. *Am. J. Reprod. Immunol. Microbiol.* **8**:125.

Otnaess, A.-B. and Svennerholm, A.-M. 1982. Non-immunoglobulin fraction of human milk protects rabbits against enterotoxin-induced intestinal fluid secretion. *Infect. Immun.* **35**:738.

Ozkaragoz, F., Rudloff, H.E., Rajaraman, S., Mushtaha, A.A., Schmalstieg, F.C., and Goldman, A.S. 1988. The motility of human milk macrophages in collagen gels. *Pediatr. Res.* **23**:449.

Kolstö Otnaess, A.-B., Laegreid, A., and Ertresväg, K. 1983. Inhibition of enterotoxin from *Escherichia coli* and *Vibrio cholerae* by gangliosides from human milk. *Infect. Immun.* **40**:563.

Pabst, H.F., Godel, J., Grace, M., Cho, H., and Spady, D.W. 1989. Effect of breast-feeding on immune response to BCG vaccination. *Lancet* **1**:295.

Palkowetz, K.H., Royer, C.L., Garofalo, R., Rudloff, H.E., Schmalstieg, F.C., Jr., and Goldman, A.S. 1994. Production of interleukin-6 and interleukin-8 by human mammary gland epithelial cells. *J. Reprod. Immunol.* **26**:57.

Pasteur, L. 1857. Mémoire sur la fermentation appelée lactique. Paris: Mallet-Gauchelier, 1958.

Pasteur, L. 1860. Mémoire la fermentation accoolique. Paris: Mallet-Gauchelier, 1958.

Pittard, III, W.B. and Bill, K. 1979. Differentiation of cord blood lymphocytes into IgA-producing cells in response to breast milk stimulatory factor. *Clin. Immunol. Immunopathol.* **13**:430.

Reiner, D.S., Wang, C.S., and Gillin, F.D. 1986. Human milk kills *Giardia lamblia* by generating toxic lipolytic products. *J. Infect. Dis.* **154**:825.

Rosenthal, L. and Lieberman, H. 1931. The role of lysozyme in the development of the intestinal flora of the newborn infant. *J. Infect. Dis.* **48**:226.

Roux, M.E., McWilliams, M., Phillips-Quagliata, J.M., Weisz-Carrington, P., and Lamm, M.E. 1977. Origin of IgA secretory plasma cells in the mammary gland. *J. Exp. Med* **146**: 1311.

Rudloff, H.E., Schmalstieg, Jr, F.C., Mushtaha, A.A., Palkowetz, K.H., Liu, S.K., and Goldman, A.S. 1992. Tumor necrosis factor-α in human milk. *Pediatr. Res.* **31**:29.

Rudloff, H.E., Schmalstieg, F.C., Palkowetz, K.H., Paszkiewicz, E.J., and Goldman, A.S. 1993. Interleukin-6 in human milk. *J. Reprod. Immunol.* **23**:13.

Saito, S., Maruyama, M., Kato, Y., Moriyama. I., and Ichijo, M. 1991. Detection of IL-6 in human milk and its involvement in IgA production. *J. Reprod. Immunol.* **20**:267.

Saito, S., Yoshida, M., Ichijo, M., Ishizaka, S., and Tsujii, T. 1993. Transforming growth factor-beta (TGF-β) in human milk. *Clin. Exp. Immunol.* **94**:220.

Salton, M.R.J. 1956. The properties of lysozyme and its action on microorganisms. *Bacteriol. Rev.* **20**:82.

Sanguansermsri, J., György, P., and Zilliken, F. 1974. Polyamines in human and cow's milk. *Am. J. Clin. Nutr.* **27**:859.

Schäfer, K.H. 1951. Elektrophoretische Untersuchungen zum Milcheiweissproblem. *Monatsschr. Kinderheilkd.* **99**:69.

Schanler, R.J., Goldblum, R.M., Garza, C., and Goldman, A.S. 1986. Enhanced fecal excretion of secreted immune factors in very low birth weight infants fed fortified human milk. *Pediatr. Res.* **20**:711.

Schroten, H., Hanisch, F.G., Plogmann, R., Hacker, J., Uhlenbruck, G., Nobis-Bosch, R., and Wahn, V. 1992. Inhibition of adhesion of S-fimbriated *Escherichia coli* in buccal epithelial cells by human milk fat globule membrane components. A novel aspect of the protective function of mucins in the non-immunoglobulin fraction. *Infect. Immunol.* **60**: 2893.

Shim, B.-S., Kang, Y.-S., Kim, W.-J., Cho, S.-H., and Lee, D.B. 1969. Self-protective activities of colostral IgA against tryptic digestion. *Nature* **222**:787.

Silverstein, A.M. 1996. History of Immunology. Paul Ehrlich: The founding of Pediatric Immunology. *Cell. Immunol.* **174**:1.

Smith, C.W. and Goldman, A.S. 1968. The cells of human colostrum. I. *In vitro* studies of morphology and functions. *Pediatr. Res.* **2**:103.

Smith, C.W. and Goldman, A.S. 1970. Interactions of lymphocytes and macrophages from human colostrum: characteristics of the interacting lymphocyte. *Res. J. Reticuloendothelial Soc.* **8**:91.

Smith, C.W., Goldman, A.S., and Yates, R.D. 1971. Interactions of lymphocytes and macrophages from human colostrum: electron microscopic studies of the interacting lymphocyte. *Exp. Cell Res.* **69**:409.

Söder, O. 1987. Isolation of interleukin-1 from human milk. *Int. Arch. Allergy Appl. Immunol.* **83**:19.

Stephens, S. 1986. Development of secretory immunity in breast fed and bottle fed infants. *Arch. Dis. Child.* **61**:263.

Tauber, A.I. and Chernyak, L. 1991. Metchnikoff and the Origins of Immunology. From Metaphor to Theory. Oxford University Press, New York and Oxford.

Thorpe, L.W., Rudloff, H.E., Powell, L.C., and Goldman, A.S. 1986. Decreased response of human milk leukocytes to chemoattractant peptides. *Pediatr. Res.* **20**:373.

Timmerman, W.A. 1931. Zur Frage der Uebertragung des Typhus "H" und "O"- Agglutinin von Mutter auf Kind. *Z. Immunforsch. Exp. Ther.* **70**:388.

Tissier, H. 1900. Recherches sur la flore intestinale normale et pathologique du nourrison. Thése, Paris, 85.

van Genderen, J. 1934. Diphtherie-Antitoxin in Kolostrum und Muttermilch bei Menschen. *Z. Immunitaetsforsch. Allerg. Klin. Immunol.* **83**:54.

Van Leeuwenhoek, A. 1695. Epistola 106. Arcana naturae detecta delphis batavorum. Apud Henricum a Krooneveld.

Weisz-Carrington, P., Roux, M.E., McWilliams, M., Phillips-Quaglita, J.M., and Lamm, M.E. 1978. Hormonal induction of the secretory immune system in the mammary gland. *Proc. Natl. Acad. Sci. USA* **75**:2928.

Welsh, J.K., Skurrie, I.J., and May, J.T. 1978. Use of semiliki forest virus to identify lipid-mediated antiviral activity and anti-alphavirus immunoglobulin A in human milk. *Infect. Immun.* **19**:395.

Welsh, J.K., Arsenakis, M., Coelen, R.J., and May, J.T. 1979. Effect of antiviral lipids, heat and freezing on the activity of viruses in human milk. *J. Infect. Dis.* **140**:322.

Westergaard, H. 1901. Die Lehre von der Mortalität und der Morbilität. *Anthropologisch. Statistiche Untersuchugen*, Revised Edition, p. 361.

Wirt, D., Adkins, L.T., Palkowetz, K.H., Schmalstieg, F.C., and Goldman, A.S. 1992. Activated-memory T cells in human milk. *Cytometry* **13**:282.

Wong, D.H. and Wong, A.I.H. 1930. Significance of dysentery agglutinins in colostrum and milk. *Natl. Med. J. China* **16**:673.

Woodbury, R.M. 1922. The relation between breast and artificial feeding and infant mortality. *Am. J. Hyg.* **2**:668.

Woodbury, R.M. 1925. *Causal Factors In Infant Mortality*. United States Government Printing Office, Washington, DC.

Yamauchi, K., Tomita, M., Giehl, T.J., and Ellison, R.T. 1993. Antibacterial activity of lactoferrin and a pepsin-derived lactoferrin peptide fragment. *Infect. Immun.* **61**:719.

Yolken, R.H., Peterson, J.A., Vonderfecht, S.L., Fouts, E.T., Midthun, K., and Newburg, D.S. 1992. Human milk mucin inhibits rotavirus replication and prevents experimental gastroenteritis. *J. Clin. Invest.* **90**:1984.

Zhang van Bree, S., van Rood, J.J., and Claas, F.H.J. 1991. Influence of breast feeding on the cytotoxic T cell allorepertoire in man. *Transplantation* **52**:914.

Chapter 3

Immunological Activities Associated with Milk

Shannon L. Kelleher and Bo Lönnerdal

1. Introduction

Milk is a complex biological fluid which has been viewed primarily as a food that provides energy and essential nutrients for optimal health, growth and development of the young mammal. However, it is now realized that milk also contains a variety of bioactive components, usually occurring in small quantities, that includes a mix of cells, bioactive proteins, hormones, growth factors, and other immunological factors. Typically, these immunological components are found in greater concentrations in colostrum and early milk and in decreasing concentrations throughout the course of lactation. It is, of course, not coincidental that these longitudinal changes in the levels of immunoactive compounds parallel the development of the immune system of the newborn animal, which is functionally immature at birth and undergoes extensive differentiation and reorganization during the early postnatal period (Xanthou, 1993). For example, levels of complement components, neutrophil activity, macrophage activation by interferon-γ, production of secretory IgA (sIgA), numbers of T cells displaying the CD45RO effector/memory phenotype and production of IgG antibody to T cell-dependent immunogens all are low during early infancy (Chheda *et al.*, 1996; Goldman *et al.*, 1998). In addition, relative to adults, newborn infants produce lower quantities of cytokines, e.g., granulocyte/macrophage colony-stimulating factor, interferon-γ, interleukin(IL)-3,

Shannon L. Kelleher and Bo Lönnerdal • Department of Nutrition, University of California, Davis, California

IL-4, IL-6 and tumor necrosis factor(TNF)-α Consequently, the capacity of infants to respond to an infectious challenge is low, and provision of immune and immunomodulating factors by way of milk may be critical to the health and survival of the newborn offspring.

In humans, there is considerable evidence both in developing countries (Cunningham *et al.*, 1991) and in the well-educated, upper-middle class of affluent countries (Dewey *et al.*, 1995), that the incidence and prevalence of infectious illness are considerably lower in breast-fed infants than in infants fed other diets. This provides functional evidence that infants receiving diets rich in immunomodulating compounds are better suited to protect themselves from infectious agents. An additional consideration is the role that milk-associated immunological factors play in the protection of the mammary gland and ultimately in the mother-infant dyad, itself. Without a protected "food source", i.e., the mother and her milk supply, the young mammal would certainly perish. This multifunctional role of milk-borne immunological factors most certainly played a role in the evolution of mammals. Therefore, not only are immunologically active milk components essential for the protection of the suckled newborn, but they are also critical for the protection of the mammary gland and, therefore, for the survival of the species.

Breast-feeding protects human infants against respiratory tract infections, otitis media, botulism, urinary tract infections (Mårild *et al.*, 1989; 1990; Pisacane *et al.*, 1992) and necrotizing enterocolitis (Saarinen, 1982; Arnon, 1984; Wright *et al.*, 1989; Lucas and Cole, 1990; Duncan *et al.*, 1993). Nevertheless, assessment of the relative importance of the various immunological factors associated with milk with regard to disease prevention in the newborn and mother is very difficult. Moreover, while environmental exposure of newborns to bacteria, viruses and other microorganisms is an important concern, it has recently become evident that there is also risk of transmission of human immunodeficiency virus (HIV), cytomegalovirus (CMV) and other viruses through the milk of infected mothers. Hence, the immunologically active components in milk are involved in protecting the young animal against both external pathogens and pathogens of maternal origin. The dairy industry is also concerned about prevention and treatment of mastitis, a disease that causes large economic losses among producers. Therefore an understanding of the immunobiology of the mammary gland, and of the role that milk-associated factors play in regulating the defense system of the mammary gland, is of prime importance. The defense system in milk is comprised of antimicrobial, anti-inflammatory, and immunomodulating agents, and these components include both cells and soluble substances. This chapter contains a review of the immunological factors found in the milk of various species and of the role that these factors may play in protecting the mammary gland and newborn offspring from infection.

2. Immunobiological Defense of the Mammary Gland

Preventing mammary gland infection is of primary concern, whether discussing dairy cows or lactating women. Mastitis in cows is the main source of economic loss to the dairy industry (Kingwill *et al.*, 1970). In humans, the consequences both of mastitis and of its treatment are also severe. These include a decrease in the quantity and quality of milk (Sordillo *et al.*, 1987, 1989) and the potential for complete lactation failure. The long-term economic effects of early cessation of breast-feeding begin with an increased incidence and prevalence of infection among offspring, and include the burden associated with parental absenteeism from work and increased health care costs (Ball and Wright, 1999). Thus, the maintenance of a healthy mammary gland is of primary concern from an economic standpoint as well as from a physiologic/evolutionary perspective.

2.1. Anatomical Defenses

Mammary gland infection occurs when bacteria or yeast gain entrance to the gland via the mammary ducts or teat canals. In cows, the teat canal is lined with keratin. This barrier not only traps bacteria, preventing them from migrating into the gland (Hibbitt *et al.*, 1969), but also contains esterified and non-esterified fatty acids, e.g., myristic, palmitoleic and linoleic acids, which are reported to have bacteriostatic effects (Treece *et al.*, 1966). In addition, cationic proteins in the teat canal can bind to pathogens and render them susceptible to microbicide (Murphy and Stuart, 1953). The cationic proteins belong to a family of antibacterial molecules that appear to function by eliciting voltage-dependent anion channels, by affecting cellular calcium flux, or by altering bacterial gene transcription (Boman, 1998).

2.2. Cellular Defenses

Once in the duct, pathogenic microorganisms encounter the immunological defenses of the mammary gland and milk, and the activities of the cellular defense system play a pivotal role in preventing establishment of infection. Macrophages are the predominant type of immune cell in healthy lactating mammary tissue (Lee *et al.*, 1980), whereas milk contains a large number of leukocytes of which 15% are macrophages, 80% are polymorphonuclear leukocytes (PMN), also termed neutrophils, and 5–10% are lymphocytes (Goldman and Frawley, 1996). The numbers of leukocytes in the mammary gland and milk increase during infection.

During an infection, PMN are recruited from the blood in response to a variety of inflammatory mediators. These cells provide the first line of defense against pathogens in the mammary gland, and several studies have shown that the ability of the gland to resist infectious challenge relates directly to the speed of neutrophil recruitment in response to bacterial by-products, enterotoxins, or cell wall components (Grommers *et al.*, 1989; Hill, 1981). PMN protect the gland by phagocytosis of organisms opsonized by the complement component, C3b, or by immunoglobulins. In this connection, C3b has been detected in both human and bovine milk during all stages of lactation (Hill and Hibbitt, 1986), and Di Carlo *et al.* (1996) showed that milk PMN exhibit a higher binding capacity for C3b than blood PMN, although the mechanism behind this difference remains unknown. Neutrophils kill ingested microorganisms by means of toxic metabolites of molecular oxygen as well as through the actions of antibacterial peptides including β-defensins (Selsted *et al.*, 1993; Boman, 1998).

Neutrophils recovered from human milk exhibit signs of activation which may be mediated by phagocytosis of membranous structures that abound in milk. For example, these cells exhibit low motility and do not respond to chemoattractants such as N-formyl methionyl peptides (fMLPs) (Thorpe *et al.*, 1986). In addition, high-level expression of CD11b, an adherence glycoprotein of the integrin family, and low expression of L-selectin, an adhesion protein that is shed from activated neutrophils, also suggest that milk neutrophils are activated cells (Keeney *et al.*, 1993). The significance of activated neutrophils in milk is not known. However, in addition to playing a role in the defense of the mammary gland, these cells may compensate for the newborn's poor capacity for inflammatory recruitment of blood neutrophils, i.e., by providing phagocytic defense to the neonatal gastrointestinal tract without the need for an inflammatory reaction to attract and activate the phagocytes (Keeney *et al.*, 1993). The same hypothesis is reasonable in relation to the macrophages and lymphocytes that are found in milk, and which also exhibit evidence of activation.

T cells dominate (>80%) among the lymphocytes in human milk. Using flow cytometry, Wirt *et al.* (1992) found that the proportion of $CD4^+$ T cells among the leukocytes of human and bovine milk is similar to that in peripheral blood, but that the proportion of $CD8^+$ T cells is much higher. Although the functional significance of this is not known, $CD8^+$ T cells may recognize and eliminate altered cells thereby serving as scavengers and functioning as a class of repair cells in the mammary gland. Further, most intraepithelial lymphocytes in the human small intestine are T cells expressing CD8. Thus, the environment in the infant gut may promote the transfer of maternal T cells. Wirt *et al.* (1992) also found that a large proportion of the T cells in human milk show signs of activation as indicated by their expression of the markers, HLA-DR and CD25

(IL-2 receptor). In addition, most milk T cells (99.8% of $CD4^+$ and 92% of $CD8^+$ cells) were $CD45RO^+$, whereas CD45RA expression was low, a finding that suggests predominance of effector and memory T cells in human milk (Wirt *et al.*, 1992). These observations are interesting in view of evidence that milk macrophages possess a potent capacity for antigen processing and presentation (Politis *et al.*, 1990). In any case, Wirt *et al.* (1992) suggested that the low number of memory T cells in the newborn may be compensated, in part, by the transfer of maternal effector and memory T cells via breast milk. Studies on development of $CD45RO^+$ T cells in exclusively breast-fed infants and formula-fed infants are needed.

3. Gastrointestinal Microflora

The microflora in the gut of breast-fed infants also affects the infant's capacity to resist infection. The bacterial populations differ in the gastrointestinal tract of breast-fed and formula-fed infants. Thus, several components in milk, including oligosaccharides, lactoferrin, nucleotides and lysozyme, appear to stimulate the growth of *Lactobacilli* and *Bifidobacteria*, thereby creating a so-called "bifido-flora" (Lönnerdal, 1999). Lactoferrin, in particular, has been shown to bind *Bifidobacterium* spp. *in vitro*, apparently in a receptor-mediated fashion as indicated by its dose-responsiveness and the ability of unlabelled lactoferrin to displace the labelled protein (Petschow *et al.*, 1999). However, binding of lactoferrin to bifidobacteria appears insufficient, of itself, to stimulate bacterial growth, and the growth-promoting effect of lactoferrin was independent of the extent to which the protein was saturated with iron (Petschow *et al.*, 1999). Interestingly, human lactoferrin promoted the growth of *B. bifidum in vitro*, whereas only bovine lactoferrin promoted the growth of *B. infantis* and *B. breve* (Petschow *et al.*, 1999). The authors postulated that these species differences may relate to either the amino acid or the carbohydrate content of lactoferrin. A predominance of bifidobacteria decreases the pH of the gut because of their lactic acid production which, in turn, inhibits the growth of pathogens such as *E. coli*, *Bacteroides*, and *Staphylococci*. In addition, Tejada-Simon *et al.* (1999) showed that oral administration of multiple strains of *Lactobacillus* and *Bifidobacterium* to mice affected IL-6, IL-12, IFN-γ and nitric oxide production by peritoneal cells subsequently cultured *ex vivo*. However, the physiological relevance of these findings remains to be elucidated. For example, dietary administration of *Bifidobacteria* or *Lactobacillus*, either as spray-dried yogurt or as viable cultures failed to stimulate rejection of chemically induced mammary or skin tumors in a murine model (Rice *et al.*, 1995; McCarthy *et al.*, 1997).

4. Antimicrobial Proteins and Peptides

While the presence of antibodies in breast milk was recognized early, it was also apparent that there are many components in milk that may complement immunoglobulin-mediated defense mechanisms. For example, bovine β-casein has been found to inhibit chemotaxis, but to enhance superoxide production, by ovine neutrophils (Wong *et al.*, 1996b). In addition, bovine whey proteins have been reported to play an immunomodulating role in mice by enhancing cell-mediated immune competence as measured by the delayed-type hypersensitivity response and the concanavalin A-induced spleen cell proliferative response (Wong and Watson, 1995). Further, specific whey protein fractions have been shown to elicit antimicrobial effects *in vivo* and *in vitro*. Lactoferrin, an iron-binding protein present in human milk at remarkably high concentrations (1–2 g/L or 10–20% of total protein content) and in bovine milk at lower concentrations, has been reported to exert a bacteriostatic effect against *E. coli* (Bullen *et al.*, 1972), and to be bactericidal against several pathogens to which infants may be exposed, e.g., *Vibrio cholerae*, *Salmonella* and *Staphylococcus* (Arnold *et al.*, 1980). Although this protein is present in breast milk largely in its unsaturated form and the bacteriostatic effect described by Bullen *et al.* (1972) was abolished by addition of iron, the bactericidal activity of lactoferrin may not depend fully on iron-saturation (Arnold *et al.*, 1982). To date, studies of lactoferrin-fortified infant formula have failed to show significant effects on the microbial flora in the infant gut (Lönnerdal, 1999). However, it should be noted that bovine lactoferrin was used in the studies of infant formula, whereas the intestinal receptor for lactoferrin found in human infant brush border membranes recognizes only human lactoferrin and not the bovine protein (Lönnerdal and Iyer, 1995). Recently, lactoferrin also has been shown to have antiviral activity *in vitro* against both HIV and CMV (Harmsen *et al.* 1995). To date, no clinical studies have been performed to investigate this antiviral effect nor is the mechanism behind it known.

Recently, a peptide called lactoferricin has been isolated from human and bovine lactoferrins. These peptides contain 47 and 25 amino acids, respectively, and have strong bactericidal effects *in vitro* and in animal models (Teraguchi *et al.*, 1995). Lactoferricin is located at the N-terminal end of lactoferrin and does not contain an iron-binding region (Yamauchi *et al.*, 1993). It is unknown whether the bactericidal effect of human lactoferrin is due to the formation of this peptide, or if intact lactoferrin and lactoferricin act in concert. However, a recent study shows that both lactoferrin and lactoferrin peptides have giardicidal activity, suggesting that this activity of lactoferrin is due at least in part to formation of active peptides (Turchany *et al.*, 1995).

Lysozyme is present in breast milk in unusually high concentrations (0.1–0.3 g/L) and is found in bovine milk at lower levels. This enzyme can degrade the cell walls of Gram-positive bacteria and has strong antimicrobial activity *in vitro*. There is little *in vivo* evidence, however, for an involvement of milk lysozyme in the defense against infections in the newborn. Many pathogens are Gram-negative and, therefore, resist the bactericidal effect of lysozyme. However, it is possible that lysozyme and lactoferrin act synergistically. Ellison and Giehl (1991) recently have demonstrated that lactoferrin can bind to, and remove, lipopolysaccharide (LPS) from the outer membrane of Gram-negative bacteria, thereby allowing lysozyme access to the underlying proteoglycan matrix to cause lysis. Whether this potent synergism occurs *in vivo* remains to be explored.

Alpha-lactalbumin shows striking homology to lysozyme with respect to amino acid sequence, protein and gene structure. Pellegrini *et al.* (1999) showed that pancreatic enzyme digestion of bovine α-lactalbumin yielded several peptides with antibacterial activity, *in vitro*, against a variety of microorganisms including *E. coli*, *Staphylococcus aureus*, *S. epidermidis*, *Klebsiella pneumoniae*, *Streptococcus spp.* and *Candida albicans*. However, the effect of α-lactalbumin on pathogenic colonization *in vivo* remains to be examined. Recent evidence suggests that multimeric α-lactalbumin isolated from human milk induces apoptosis of tumor cells but not of untransformed cells, and this influence appears to be mediated through a direct effect within the nucleus (Håkansson *et al.*, 1999). These findings could help to explain epidemiological evidence that breast-feeding not only helps to protect the neonate from pathogenic microorganisms but also protects the mother against breast cancer (Davis *et al.*, 1988).

Many other proteins and peptides in milk have been reported to exert antimicrobial and/or immunostimulatory actions (Lönnerdal, 1985), although it should be recognized that these activities have been demonstrated only *in vitro*, or in animal models. Among these components are lactoperoxidase (Gothefors and Marklund, 1975), κ-casein (Stromquist *et al.*, 1995) and digestive fragments of casein subunits including caseinophosphopeptide (CPP) and caseinoglycomacropeptide (CMP) (Parker *et al.*, 1984). Neeser *et al.* (1988) found that caseinoglycopeptides inhibited the adhesion of *Actinomyces* as well as *Streptococci* taken from the oral cavity in an *ex vivo* erythrocyte model. Further studies by these investigators showed that dietary CPP and CMP can inhibit the adherence of *Streptococcus mutans* to the oral cavity of rats fed diets supplemented with either caseinopeptide. This effect is believed to depend on the incorporation of the peptides into the outer biofilm (salivary pellicle) of the dental enamel (Schupbach *et al.*, 1996). Wong *et al.* (1996a) showed that bovine lactoperoxidase administered intravenously to sheep elicited a transient decline in the blood T lymphocyte count, but not in B lymphocyte numbers. These results suggest an effect of lactoperoxidase on the capacity to produce T cell cytokines, although the physiological significance of the results remains unclear. More recently, Gerson *et al.* (2000) have shown that production of thiocyanate by lactoperoxidase promotes clearance of bacteria from the airways of sheep. This enzyme could function in the same manner in humans, thereby helping to protect breast-fed infants from pulmonary infection.

5. Oligosaccharides

Oligosaccharides are present in human milk in substantial concentrations and in a large number of chemical forms (Kunz and Rudloff, 1993; Newburg, 1997). The predominant species are lacto-N-tetraose, lacto-N-fucopentaose, sialyl-lactose and disialyl-lacto-N-tetraose, but over 900 fucosyloligosaccharides containing up to 32 sugars and up to 15 fucose residues have been detected by time-of-flight mass spectrometry (Stahl *et al.*, 1994). This diversity is due, in part, to genetic differences in enzymes of carbohydrate biosynthetic pathways, particularly fucosyltransferases and sialyltransferases, and in part to longitudinal changes during lactation. Most of the oligosaccharides occur in free form, but many are also linked to glycoproteins (e.g., lactoferrin, κ-casein and sIgA), glycolipids and mucins. Many of these carbohydrate structures are similar, or identical, to surface-bound glycans on the small intestinal mucosa which are receptors for adhesive bacteria (Kunz and Rudloff, 1993). Thus, milk oligosaccharides can act as soluble competitors with the host's mucosal surface for attachment of invading pathogens. The carbohydrate structure, itself, is responsible for this activity. Thus, proteolytic digestion of human milk fractions that inhibited adherence of enterotoxigenic *E. coli* had no effect on the activity of these fractions, whereas oxidation of carbohydrates with periodate resulted in loss of activity (Ashkenazi and Mirelman, 1987).

Andersson *et al.* (1986) demonstrated that oligosaccharides from human milk could inhibit adherence of *Haemophilus influenzae* and *Streptococcus pneumoniae* to carbohydrate structures of human buccal epithelial cells. Similarly, neutral oligosaccharides from human milk were shown to protect the infant's intestinal tract from infection by *Vibrio cholerae* (Holmgren *et al.*, 1983). Cravioto *et al.* (1991) described a milk oligosaccharide that inhibits adhesion of enteropathogenic *E. coli* to its receptor. Sialylated oligosaccharides from human milk were shown to inhibit the binding of *E. coli* species which can cause meningitis and neonatal sepsis in infants (Korhonen *et al.*, 1985), and a fucosylated oligosaccharide was demonstrated to inhibit the binding of *Campylobacter jejuni* to intestinal cells (Cervantes *et al.*, 1995). Finally, a study of experimental gastroenteritis has shown an inhibitory effect of human milk mucin on rotavirus replication, a protective function which is most likely mediated by the carbohydrate moiety of the mucin (Yolken *et al.*, 1992).

In all the cited studies it was shown, or strongly implicated, that milk-derived oligosaccharides inhibited the binding of microorganisms to a mucosal surface. Another type of inhibitory mechanism has been demonstrated by Newburg *et al.* (1990) in relation to the heat-stable enterotoxin produced by *E. coli*. These authors showed that binding of this toxin to T84 epithelial cells *in vitro* is inhibited by fucosylated human milk oligosaccharides. The toxin binds to the extracellular domain of guanylyl cyclase, thereby activating the enzyme, causing inhibition of chloride channel function and resulting in a net flux of fluid into the intestinal lumen. This is believed to be the major mechanism underlying secretory diarrhea (Crane *et al.*, 1994). Since washing oligosaccharide-incubated T84 cells to remove free oligosaccharides did not diminish the inhibition of toxin-mediated guanylyl cyclase activity, it was concluded that the oligosaccharides bind to the toxin receptor, rather than to the toxin. This novel protective mechanism may play an important role in the inhibition of enterotoxin-induced secretory diarrhea in breast-fed infants.

6. Nucleotides

Nucleotides are present in human milk in significant concentrations (Janas and Picciano, 1982) but, until recently, these compounds have not been added to infant formulas. In animal models, nucleotides have been reported to prevent malnutrition- and starvation-induced immunosuppression (Kulkarni *et al.*, 1987; Pizzini *et al.*, 1990), to aid resistance to *Staphylococcus aureus* and *Candida albicans* (Kulkarni *et al.*, 1986; Fanslow *et al.*, 1988), to enhance T-cell maturation and function (Van Buren *et al.*, 1985) and to stimulate gastrointestinal growth and maturation (Uauy *et al.*, 1990). Exogenous nucleotides are believed to boost the "salvage" pathway in the enterocytes of human infants, as d*e novo* synthesis may not provide the rapidly growing infant, particularly the pre-term neonate, with an adequate supply of nucleotide to support optimal cellular proliferation (Cosgrove, 1998). Carver *et al.* (1990) found that infants fed nucleotide-fortified formula had a higher level of blood natural killer cell cytotoxic activity and enhanced IL-2 production by mitogen-stimulated blood mononuclear cells than infants fed formula without added nucleotides. Moreover, the nucleotide-supplemented infants were similar in these respects to exclusively breast-fed infants. Importantly, Brunser *et al.* (1994) subsequently found that Peruvian infants who were fed nucleotide-fortified formula had a lower incidence of infections, and a lower total number of days of infectious illness, than those fed unfortified formula. The hypothesis that nucleotides are conditionally essential nutrients merits further study with particular emphasis on infants either born prematurely or exhibiting fetal growth retardation.

Recently, Tanaka *et al.* (1996) have shown that individual nucleotides, particularly AMP, can induce apoptosis in human fetal intestine in culture. These authors suggest that AMP, and possibly other nucleotides, may reinforce intestinal mucosal integrity by increasing cellular proliferation and differentiation via enhanced apoptosis. It should be noted that the concentrations of nucleotides used in most clinical studies have been far lower than the total concentration of nucleotides in breast milk, i.e., free plus DNA and RNA. Human milk contains significant quantities of DNA and RNA, and Thorell *et al.* (1996) showed that infant pancreatic secretions and intestinal tissue contain ample activities of the enzymes needed to degrade nucleic acids to nucleotides. Thus, the total pool of nucleotides in the small intestine of breast-fed infants may be as high as ten times the concentration of free nucleotides in milk. The effect of fortifying infant formula with concentrations of nucleotides similar to those expected to be available from human milk was evaluated in a recent study (Pickering *et al.*, 1998). Infants receiving nucleotide-fortified formula developed higher antibody titers in response to vaccination against *Hemophilus influenzae* type b and diphtheria than infants receiving unsupplemented formula. Thus, it is evident that the nucleotides of milk, although not traditional nutrients, can support the development of immune competence in infants.

7. Cellular Defenses of Maternal Origin

Numerous studies report results showing that milk leukocytes can be taken up by the offspring. Tuboly *et al.* (1995) observed that milk lymphocytes transferred to the duodenum of newborn lambs were taken up more efficiently than blood lymphocytes. Additionally, intraduodenal transfer of milk lymphocytes from immunized sheep conferred immunity to the sensitizing antigen which was assessed in terms of antibody production and the lymphocytic blastogenesis response. Similarly, Kumar *et al.* (1989) found that oral administration of syngeneic T lymphocytes primed to *Trichinella spiralis* conferred specific immunity to recipient rat pups, whereas T cells from dams immunized with an irrelevant antigen failed to transfer resistance to Trichinella.

8. Immunoglobulins, Anti-Idiotypic Antibodies and Hyperimmunization

Milk contains high concentrations of antibodies, IgG being the predominant class in cow's milk and sIgA the predominant class in human milk (Goldman *et al.*, 1982). Secretory IgA was first isolated and characterized from breast milk and this type of IgA is considerably more resistant to proteolytic degradation than the monomeric form (Lindh, 1985). The sIgA molecule is assembled from J-chain-containing IgA dimers that are bound to a protein called secretory component (SC). The latter protein, in turn, is a portion of the polymeric immunoglobulin receptor that is present in the membrane of mammary epithelial cells of lactating women (Brandtzaeg, 1974; Goldman *et al.*, 1994). These receptors appear to be abundant, as measurable concentrations of free SC are found in human milk (Sato and Lönnerdal, 1993), and this observation could be interpreted to suggest selective uptake of IgA by human mammary tissue during lactation. The sIgA antibodies in human milk are known to recognize a wide variety of microorganisms that are found in the respiratory tract and intestine, e.g., bacterial pathogens such as *E. coli*, *V. cholerae*, *H. influenzae*, *S. pneumoniae*, *C. difficile* and *Salmonella*, viruses including rotavirus, CMV, HIV, influenza virus and RSV, and yeasts such as *Candida albicans* (Goldman, 1993). Thus, the antibodies in human milk are believed to derive from antigen-triggered B cells that have migrated from the Peyer's patches of the small intestine, and from lymphoid tissue in the respiratory tract, to the lamina propria of the mammary gland.

The migration of B cells appears to be cytokine-mediated. When the intestinal mucosa is challenged by microbial antigens, cytokines are released from mononuclear cells in the Peyer's patches. These cytokines induce local IgM^+ B lymphocytes to undergo isotype switching to become IgA^+ lymphocytes (Schultz and Coffman, 1991; Whitmore et al., 1991; Goldman, 1993). Isotype-switched cells are then preferentially transported via lymph vessels to the mammary gland, most likely under the influence of hormones such as prolactin that are produced during late pregnancy and early lactation. Within the mammary gland, the lymphocytes differentiate to become IgA-synthesizing plasma cells that secrete dimeric IgA. Once the dimeric IgA binds to the polymeric immunoglobulin receptor on the basolateral membrane of the epithelial cell, the receptor-dimeric IgA complex is transported to the apical side of the cell (Mostov and Blobel, 1982). Following cleavage of the intracytoplasmic part of the receptor, sIgA is released into the milk. This so-called "enteromammary" pathway thereby provides the newborn, which has little sIgA of its own and whose immune system is relatively immature, with protection against the same mucosal pathogens to which the mother is, or has been, exposed (Telemo and Hanson, 1996). The quantitative significance of milk sIgA in the overall defense of the neonate against infection is difficult to assess. However, protection of breast-fed infants against cholera correlates positively with milk sIgA antibody titers against *V. cholerae* enterotoxin and LPS (Glass *et al.*, 1989), and the same is reported with regard to enterotoxigenic *E. coli* (Cruz *et al.*, 1988). These findings strongly suggest that milk antibody is important in the immune defense of infants. Further to this point, Cruz *et al.* (1991) reported that intestinal infections reduced the levels of antibody against *Shigella* and *Giardia* in the milk of Guatemalan women who had these antibodies in their milk prior to their illness. Factors such as the type and load of pathogen and the duration of infection appeared critical to the influence of maternal infection on the titer of milk antibody.

Antibody avidity is probably an important factor in the protection provided by the immunoglobulins of milk. Roberton *et al.* (1988) explored the avidity of milk sIgA antibodies for *E. coli* O antigen and diphtheria toxin and found a higher relative affinity index in milk samples from Swedish mothers than in milk from comparable Pakistani women. Although compromised nutritional status on the part of the Pakistani women could explain this finding, Cruz *et al.* (1991) reported no effect of a three-month food supplement (440 Kcal/d) on either the avidity or the titer of sIgA antibodies in the milk of Guatemalan mothers. Vaccination of lactating Pakistani women with whole cell cholera vaccine increased the titer, but not avidity, of milk sIgA antibodies directed against cholera endotoxin (Dahlgren *et al.*, 1989). The authors proposed that the avidity of the milk antibody was already high, and that it was not possible to increase it further by immunization. Thus, much milk antibody may be a product of memory cells that predominate in the mammary gland. This could also explain how milk can contain antibodies against many more bacteria and serotypes than the mother can have encountered recently (Hanson *et al.*, 1989).

Every antibody is a new "non-self" protein that induces an appropriate humoral response, i.e., the synthesis of an anti-idiotypic antibody. Anti-idiotypic antibodies of milk may stimulate immunity in the suckling newborn by, in effect, providing additional exposure to antigens expressed by infectious organisms. Consequently, the identification, in human colostrum and mature milk, of anti-idiotypic antibodies reactive with poliovirus (Hahn-Zoric *et al.*, 1993) can be interpreted as consistent with studies showing better responses to vaccination by breast-fed than by formula-fed infants (Hahn-Zoric *et al.*, 1990; Pabst and Spady, 1990). The concept of presentation of maternal anti-idiotypic antibodies to the fetus/newborn was first proposed by Hanson and co-workers in relation to the placental transfer of immunoglobulins (Mellander *et al.*, 1986). This group found sIgA and IgM antibodies to *E. coli* O antigens and to poliovirus type 1 antigen, i.e., classes of antibody that could not have been transferred to the fetus across the placenta, in amniotic fluid, saliva and meconium of infants at birth (Mellander *et al.*, 1986). Subsequently, the same group confirmed this finding in infants born to mothers with either hypogammaglobulinemia or IgA deficiency, i.e., mothers who could not produce IgA and/or IgM antibodies (Hahn-Zoric *et al.*, 1993). It was hypothesized that the antibodies arose in response to maternal IgG class anti-idiotypic antibodies that had crossed the placenta. Studies in animal models have extended this concept by showing that anti-idiotypic antibodies transferred from lactating dams to the offspring, via the milk, can elicit an immune response in the newborn to the original antigen (Okamoto *et al.*, 1989; Stein and Söderström, 1984). In these studies, injection of the anti-idiotypic antibodies into the mothers after delivery showed that these molecules could reach the offspring through the milk. Further to these findings, immunization of mice with an anti-idiotypic antibody towards poliovirus induces an immune response that includes antibodies reacting with the virus (Uytedhaag and Osterhaus, 1985). Finally, Roos *et al.* (1995) reported that ^{15}N-labelled bovine immunoglobulins have a lower ileal digestibility (79%) in adult humans than other milk proteins. Likewise, the digestive destruction of immunoglobulins in the gastrointestinal tract of infants, with their relatively high gastric pH and low enzyme activities, needs to be examined. Anti-idiotypic antibodies of milk that remain intact in the neonatal gastrointestinal tract may elicit intestinal mucosal antibodies that react with wild strains of enteric viruses such as the poliovirus.

Hyperimmunization of lactating animals is currently being examined as a method of boosting passive immunization to offspring as well as for producing immunoglobin-enriched milk products for the infant formula industry. For example, hyperimmunization of pregnant cows against *Helicobacter pylori* resulted, after calving, in high antibody titers specific to this organism in both serum and colostrum (Korhonen *et al.*, 1995). In the same study, however, mature milk failed to exhibit a high titer of antibody specific to H. pylori, even after booster immunization during lactation. Further to this point, Mallard *et al.* (1997), using antibody titers and mastitis as outcome measures, showed that not all cows responded to hyperimmunization with *E. coli* J5. Nevertheless, high titers of antibodies specific for several gastrointestinal pathogens have been detected in the milk of hyperimmunized cows (Akita and Li-Chan, 1998; Cordle *et al.*, 1994).

Maternal hyperimmunization may be a valuable strategy for protecting the suckling offspring against some pathogens. Snodgrass *et al.* (1995) presented evidence, using a baboon model, that maternal hyperimmunization can increase milk antibody titer so as to protect the suckling young against rotavirus. The same strategy was effective in protecting suckling neonates against bacteria as well as viruses in studies of pigs (Rejnek *et al.*, 1968) and mice (Cordle *et al.*, 1991; Winner *et al.*, 1991; Bessen and Fischetti, 1998). At the same time, there is some risk that passive immunization of the suckling neonate can interfere with the response to the same antigen following scheduled vaccinations. In this connection, administration of anti-HIV immunoglobulin to lactating mice inhibited the response of their pups to an HIV vaccine (Jelonek *et al.*, 1998).

Preparations from the milk of hyperimmunized cows have conferred protection against gastrointestinal infections when administered parenterally to human infants and children (Cordle *et al.*, 1994). In addition, administration of immunoglobulin concentrates, by the oral route, can prevent infectious diarrhea in humans. Thus, Guarino *et al.* (1994) and Hilpert *et al.* (1987) reported that orally administered immunoglobulins, both human and bovine, protected human infants against rotaviral enteritis. In addition, Rump *et al.* (1992) showed that an immunoglobulin concentrate from bovine colostrum offered protection against virus-associated diarrhea in immunocompromised human adults. However, such a strategy is not always effective. Brunser *et al.* (1994) were unable to demonstrate a protective outcome of feeding an infant formula containing antibodies from the milk of hyperimmunized cows. Finally, Loimaranta *et al.* (1999) found that the colostral proteins from cows hyperimmunized to *Streptococcus mutans* and *S. sobrinus* opsonized these organisms for phagocytosis by human leukocytes *in vitro*. This is of interest because *S. mutans* and *S. sobrinus* are the most common causative agents of dental caries, and immunological protection against these bacteria has, thus far, not proven feasible.

9. Immunomodulating Agents Including Anti-Inflammatory Components: Cytokines and Non-Cytokine Factors

Human milk contains a wide variety of cytokines, some of which are found in concentrations that suggest physiological activity. The discovery of cytokines in human milk opened new possibilities for effects of breast milk components on diverse cellular targets. Cytokines are polypeptides that act in an autocrine, paracrine or endocrine fashion by binding to specific cellular receptors, thereby affecting cell protein synthesis and/or other cellular activities. The first cytokine to be found in human milk was transforming growth factor, TGF (Noda *et al.*, 1984), which can cause isotype switching to IgA^+ B cells. In other work, it was demonstrated that human milk could enhance the chemokinesis of peripheral blood neutrophils *in vitro* and that this action could be inhibited by antibodies against TNF-α (Mushtaha *et al.*, 1989). This cytokine was subsequently reported in breast milk, particularly in colostrum, in which concentrations of ~600 pg/ml were detected (Rudloff *et al.*, 1992). TNF-α also may promote synthesis of the polymeric immunoglobulin receptor and, in this way, may contribute to the selective uptake of dimeric IgA by the lactating mammary gland (Nilsen *et al.*, 1999). In addition, IL-1 has been found in breast milk at concentrations of ~1,100 pg/ml (Munoz *et al.*, 1990), a level that can activate T cells.

IL-6, IL-8 and IL-10 are reported in human milk, and the concentrations of these cytokines are considerably higher in early milk than later in lactation (Saito et al., 1991; Palkowetz *et al.*, 1994; Garofalo *et al.*, 1995). IL-6 is of interest because this cytokine can affect the synthesis of IgA (Saito *et al.*, 1991) and IgM (Hirano *et al.*, 1990), and because it can also activate T cells. It is also possible that IL-6 exerts an indirect influence on the activation of T cells in milk through negative feedback control of TNF-α synthesis in the mammary gland (Rudloff *et al.*, 1993). IL-6 can also stimulate the terminal differentiation of B cells, including IgA^+ B cells, to become plasma cells (Kono *et al.*, 1991). Finally, this cytokine is reported to enhance the production of α_1-antitrypsin by phagocytes (Perlmutter *et al.*, 1989), a phenomenon that could explain the presence of α_1-antitrypsin in breast milk and in the stools of breast-fed infants (Davidson and

Lönnerdal, 1990). IL-8 is reported in human milk as well as in culture fluids of human mammary gland epithelial cells cultured with epidermal growth factor (EGF), insulin and hydrocortisone (Palkowetz *et al.*, 1994). In this experimental system, the culture fluids contained higher concentrations of IL-8 than were found in lysates of the cultured cells. Consequently, it was suggested that mammary epithelial cells produce IL-8 and that the cytokine is released rapidly after synthesis. Similar results were obtained in relation to IL-6, suggesting that the presence of both cytokines in milk is due, at least in part, to mammary gland biosynthesis (Palkowetz *et al.*, 1994). IL-8 may contribute to the recruitment of T cells and neutrophils into breast milk (Palkowetz *et al.*, 1994).

Despite the presence of cytokines in milk at concentrations that suggest physiological activity, human milk was early found to be anti-inflammatory in that it protected against infection in the absence of clinical evidence of inflammation. One possible explanation for this is that human milk contains soluble cytokine/chemokine receptors and receptor antagonists. For example, breast milk at all stages of lactation contains soluble TNF-α receptors I and II, as well as IL-1 receptor antagonist (Buescher and Malinowska, 1996). Goldman (1993) noted that breast milk contains low concentrations of initiators and mediators of inflammation, whereas it has high concentrations of anti-inflammatory agents. One of these anti-inflammatory agents is lactoferrin, which can inhibit the release of IL-6 (stimulated by either LPS or TNF-α from fresh human monocytes or cultured mononuclear phagocytes (Lönnerdal *et al.*, 1987). Both intact lactoferrin and lactoferricin were effective, the latter being the stronger inhibitor on a molar basis. It is possible that breast-fed infants receive anti-inflammatory protection, by this mechanism, against LPS-releasing Gram-negative bacteria. Such a mechanism may contribute to the well-known smaller physiologic weight loss of breast-fed infants compared to formula-fed infants during the first week of life (Mattsby-Baltzer *et al.*, 1996).

In spite of the low production of inflammatory cytokines *in vitro* by the blood leukocytes of human infants, it is apparent that the leukocytes of sick infants respond to infectious challenge by producing inflammatory cytokines such as TNF-α, IL-1, IL-6 and IL-8 (Chheda *et al.*, 1996). Reports of IL-10 in human milk are of interest in this connection because this cytokine inhibits the production of IL-1, IL-8 and TNF-α, limits participation of T helper-1 cells in cell-mediated responses and promotes the expression of T helper-2 cell responses that enhance synthesis of IgA, a blocking type of antibody that does not initiate inflammation. Strong support for an anti-inflammatory role on the part of IL-10 has been obtained from the study of homozygous IL-10 null-gene mice, which develop a lethal enterocolitis similar to necrotizing enterocolitis (Kühn *et al.*, 1993). Chheda *et al.* (1996) showed that stimulated blood T cells and monocytes from human newborns produce less IL-10 than the comparable cells from adults. Although this might be due to decreased production of TNF-α by infant monocytes, production of IL-10 was not affected when the cells were exposed to exogenous TNF-α. Instead, expression of TNF-α receptors was significantly lower in newborns than in adults, suggesting that low IL-10 expression in newborns may be due to immaturity of the regulatory process involving TNF-α and its receptors. Human milk has been found to contain IL-10 in concentrations sufficient to suggest bioactivity, as indicated by inhibition of thymidine uptake by human blood lymphocytes (Garofalo *et al.*, 1995). Thus, IL-10 from milk may offer anti-inflammatory protection to the neonatal gastrointestinal tract at a stage of life when the infant's capacity to produce this cytokine is low (Garofalo *et al.*, 1995). In addition, this may help to explain why human milk protects against an inflammatory condition such as necrotizing enterocolitis.

Proteins and ligands in human milk are usually compartmentalized, i.e., they are distributed among the lipid fraction (as part of the milk fat globule membrane), the aqueous phase (either as soluble components or bound to soluble components) and the casein/cell pellet (Lönnerdal *et al.*, 1987). This compartmentalization may affect the digestive fate and biological function of these compounds. Infants secrete less gastric acid than older children and adults, and also exhibit lower production of digestive proteases. These physiological conditions, coupled with the presence of protease inhibitors in human milk (Davidson and Lönnerdal, 1990), make it possible for peptide/protein components to remain intact within the gut of the suckling infant. The issue of whether IL-10 and other cytokines reach the mucosal surface of the intestine intact was addressed by Garofalo *et al.* (1995). These authors found that IL-10 was distributed between the aqueous phase and the lipid fraction of breast milk. Since bile salts released some of the IL-10 from the lipid fraction, it was suggested that the amount of IL-10 from milk that is physiologically active in the intestine of the infant may be considerably higher than the quantity present in the aqueous phase of milk. An influence of compartmentalization on the outcome of cytokine assays may also explain some of the discrepancies among reports of cytokine levels in milk.

Leukocyte growth-stimulating factors have been found in human milk, among them macrophage (Hara *et al.*, 1995) and granulocyte (Gilmore *et al.*, 1994) colony stimulating factors. The concentrations of these proteins were ~17,000 U/ml and ~300 pg/ml, respectively. Interferon-γ has also been detected in breast milk (Bocci *et al.*, 1993), but it has not yet been determined whether the concentrations present are at a level that could affect immune functions. The significance of these factors to the immune defense of breast-fed infants remains to be elucidated.

Several hormones, not classified as cytokines, are present in milk as potential immunomodulatory agents, e.g., prolactin, EGF and IGF-1 (Grosvenor *et al.*, 1993). These hormones may act locally in the neonatal intestinal mucosa and/or impose systemic effects that modulate immune competence in the newborn.

Through a receptor that is a member of the IL-2 receptor superfamily, prolactin can induce IL-2 synthesis and the expression of IL-2 receptors on lymphocytes as indicated by studies of hyperprolactinemic rats (Visselli *et al.*, 1991). Therefore, at least in part by this mechanism, prolactin promotes T cell activation (Reber, 1993) as well as T cell-dependent macrophage activation and natural killer cell activity (Bernton *et al.*, 1988). EGF and IGF-1 can promote the growth, differentiation and physical integrity of the intestinal mucosa so that, through milk, these hormones may contribute to neonatal mucosal immune competence. It is beyond the scope of this review to discuss these factors in detail.

10. Allergy Considerations

It is thought that overproduction of IgE is avoided by negative feedback mediated by IgE, itself, and that breakdown of this regulatory mechanism results in allergic reactions which can be severe (Sutton and Gould, 1993: Yu *et al.*, 1994). Both genetic and environmental factors strongly influence the occurrence of allergic disorders, although heredity appears to have the larger effect. Thus, the risk of developing atopic disease, including high serum IgE levels, is linked to chromosomal loci encoding the T cell receptor, as well as to IL-4 and other cytokine-related regions. The influence of duration of breast-feeding and/or avoidance of cow's milk-based formulas during early life on the risk of developing atopic disease has been a subject of considerable controversy. This is because many confounders have not been adequately addressed. Nevertheless, early studies of prolonged breast-feeding demonstrated protection against atopic disease up to 3 years of age (Saarinen *et al.*, 1979), and a more recent study showed that the prevalence of atopy up to the age of 17 years was highest among children that had little or no breast-feeding (Saarinen and Kajosaari, 1995).

Exposure to antigens during early infancy is thought to affect sensitization and subsequent development of atopy because of the immaturity of the neonatal immune system (Saarinen and Kajosaari, 1995). Bovine α_{s1}-casein is one of the major allergens responsible for cow's milk allergy, and high titers of IgE specific for this protein are found in infants and children who are allergic to bovine milk. In this connection, low concentrations of cow's milk proteins can be detected in breast milk, and elimination of dairy products from the maternal diet can reduce the levels of these proteins significantly (Lilja *et al.*, 1989). Thus, the use of an elimination diet during lactation has been shown to decrease the incidence of eczema in infants at high risk (Arshad *et al.*, 1992), whereas this strategy was ineffective if mothers used it only during pregnancy.

The age at which an infant begins to consume solid foods also can affect the risk of atopic disease. Ferguson *et al.* (1990) demonstrated an association between the diversity of the infant's diet during the first months of life and the development of eczema. In a long-term, prospective study, Saarinen and Kajosaari (1995) found that breast-feeding for six months or longer is required for prevention of atopic eczema for the first three years of life. However, exclusive breast-feeding for any period of time exceeding one month reduced the incidence of food allergy, which has a peak prevalence around 3 years of age, as well as the incidence of respiratory allergy which has a peak prevalence around 17 years. Importantly, delayed introduction of a food may postpone, but will not necessarily prevent, the respective food allergy (Sarrinen and Kajosaari, 1995).

The fatty acid content of breast milk may be important to the risk of developing neonatal atopic disease. Thus, milk from mothers of children with atopic eczema contained lower levels of α-linoleic acid and long-chain n-3 polyunsaturated fatty acids, but exhibited a higher n-6/n-3 fatty acid ratio, than milk from mothers of non-atopic children (Businco *et al.*, 1993). A cause-effect relationship is not established, but could be mediated by eicosanoids.

Although several studies suggest a beneficial effect of breast-feeding in reducing the risk of atopic disease among offspring, other studies fail to reveal a protective effect. For example, a recent large Finnish study in which premature infants were fed breast milk or formula in a randomized fashion, revealed no effect of diet on the incidence of atopic disease (Saarinen *et al.*, 1999). It was suggested that unidentified confounders may influence the impact of breast-feeding in studies of this nature. Such confounding factors may include differences in genetic background and predisposition, environmental factors (smoking and indoor pets) and feeding practices. Another potential confounder is antigen load which may strongly influence whether atopy or tolerance develops. Many studies consider only qualitative aspects ("all or none") while including little attention to quantitative aspects, e.g., the quantity and diversity of antigens in the diet. Epidemiological studies with improved rigor are needed regarding the potential of breast-feeding to reduce the risk of allergic disease.

11. Summary

Milk contains a multitude of components that can, or may, provide immune protection to the suckling offspring and that also may promote development of neonatal immune competence. In addition, these specialized factors are essential for the protection of the mammary gland, the offspring's food source, from pathogen colonization and lactation failure. Breast milk also facilitates the establishment of a gut flora that inhibits colonization by many pathogens and stimulates the growth of beneficial microorganisms. Maternal immunity can be transferred to the infant via antibodies, primarily of the sIgA type in humans, as well as by leukocytes including effector and memory T lymphocytes. In this way, protection is provided passively against the pathogens to which the mother has been exposed. Currently, there is much interest in determining the protective efficacy of oral supplementation with immunoglobulins from the milk of lactating animals hyperimmunized against specific pathogens. An array of immunostimulatory components in milk, notably cytokines, may be protected against intestinal proteolysis, thereby providing the offspring with a prepackaged immune response system. These components may help to boost the infant's immature immune system. At the same time, anti-inflammatory factors in breast milk help to modulate cytokine responses to infection, thereby facilitating defense while minimizing tissue damage such as that which occurs in infants with necrotizing enterocolitis. Undoubtedly, the many components constituting the repertoire of immune and immunomodulating agents in milk interact synergistically to protect both the mammary gland and the offspring from invading pathogenic microorganisms.

References

Akita, E.M. and Li-Chan, E.C. 1998. Isolation of bovine immunoglobulin G subclasses from milk, colostrum and whey using egg yolk antibodies. *J. Dairy Sci.* **81**:54.

Andersson, B., Porras, O., Hanson, L.Å., Lagergård, T., and Svanborg-Eden, C. 1986. Inhibition of attachment of *Streptococcus pneumoniae* and *Haemophilus influenzae* by human milk and receptor oligosaccharides. *J. Infect. Dis.* **153**:232.

Arnold, R.R., Brewer, M., and Gauthier, J.J. 1980. Bactericidal activity of human lactoferrin: sensitivity of a variety of microorganisms. *Infect. Immun.* **28**:893.

Arnold, R.R., Russell, J.E., Champion, W.J., Brewer, M., and Gauthier, J.J. 1982. Bactericidal activity of human lactoferrin: differentiation from the stasis of iron deprivation. *Infect. Immun.* **35**:792.

Arnon, S.S. 1984. Breast-feeding and toxigenic intestinal infections: missing links in crib death? *Rev. Infect. Dis.* **6**:193.

Arshad, S.H., Matthews, S., Gant, C., and Hide, D.W. 1992. Effect of allergen avoidance on development of allergic disorders in infancy. *Lancet* **339**:493.

Ashkenazi, S., and Mirelman, D. 1987. Nonimmunoglobulin fraction of human milk inhibits the adherence of certain enterotoxigenic Escherichia coli strains to guinea pig intestinal tract. *Pediatr. Res.* **22**:130.

Ball, T.M. and Wright, A.L. 1999. Health care costs of formula-feeding in the first year of life. *Pediatrics* **103**:870.

Bernton, E., Meltzer, M., and Holaday, J. 1988. Suppression of macrophage activation and T-lymphocyte function in hypoprolactinemic mice. *Science* **239**:401.

Bessen, D. and Fischetti, V.A. 1998. Passive acquired mucosal immunity to group A streptococci by secretory immunoglobulin A. *Am. J. Exp. Med.* **167**:1945.

Bocci, V., von Bremen, K., Corradeschi, F., Franchi, F., Luzzi, E., and Paulesu, L. 1993. Presence of interferon- and interleukin-6 in colostrum of normal women. *Lymphokine Cytokine Res.* **12**:21.

Boman, H.G. 1998. Gene-encoded peptide antibiotics and the concept of innate immunity: an update review. *Scand. J. Immunol.* **48**:15.
Brandtzaeg, P. 1974. Mucosal and glandular distribution of immunoglobulin components: differential localization of free and bound SC in secretory epithelial cells. *J. Immunol.* **112**:1553.
Brunser, O., Espinoza, J., Araya, M., Cruchet, S., and Gil, A. 1994. Effect of dietary nucleotide supplementation on diarrhoeal disease in infants. *Acta. Paediatr.* **83**:188.
Buescher, E.S. and Malinowska, I. 1996. Soluble receptors and cytokine antagonists in human milk. *Pediatr. Res.* **40**:839.
Bullen, J.J., Rogers, H.J., and Leigh, L. 1972. Iron-binding proteins in milk and resistance to *Escherichia coli* infection in infants. *Br. Med. J.* **1**:69.
Businco, L., Ioppi, M., Morse, N.L., Nisini, R., and Wright, S. 1993. Breast milk from mothers of children with newly developed atopic eczema has low levels of long chain polyunsaturated fatty acids. *J. Allergy Clin. Immunol.* **91**:1134.
Carver, J.D., Cox, W.I., and Barness, L.A. 1990. Dietary nucleotide effects upon murine natural killer cell activity and macrophage activation. *J. Parenter Enteral Nutr.* **14**:18.
Cervantes, L.E., Newburg, D.S., and Ruiz-Palacios, G.M. 1995. α-1-2 Fucosylated chains (H-2 and Lewis^b) are the main human milk receptor analogs for *Campylobacter*. *Pediatr. Res.* **37**:171A.
Chheda, S., Palkowetz, K.H., Garofalo, R., Rassin, D.K., and Goldman, A.S. 1996. Decreased interleukin-10 production by neonatal monocytes and T-cells: relationship to decreased production and expression of tumor necrosis factor-alpha and its receptors. *Pediatr. Res.* **40**:475.
Cordle, C.T., Duska-McEwen, G., Janas, L.M., Malone, W.T., and Hirsch, M.A. 1994. Evaluation of immunogenicity of protein hydrolysate formulas using laboratory hyperimmunization. *Ped. Allergy Immunol.* **5**:14.
Cordle, C.T., Schaller, J.P., Winship, T.R., Candler, E.L., Hilty, M.D., Saif, L.J., Kohler, E.M., and Krakowka, S. 1991. Passive immune protection from diarrhea caused by rotavirus or E. coli in an animal model to demonstrate quantitative efficacy. *Adv. Exp. Med. Biol.* **310**:317.
Cosgrove, M. 1998. Nucleotides. *Nutrition* **14**:748.
Crane, J.K., Azar, S.S., Stam, A., and Newburg, D.S. 1994. Oligosaccharides from human milk block binding and activity of the *Escherichia coli* heat-stable enterotoxin (Sta) in T84 intestinal cells. *J. Nutr.* **124**:2358.
Cravioto, A., Tello, O., Villafari, H., Ruiz, J., del Vedovo, S., and Neeser, J.-R. 1991. Inhibition of localized adhesion of enteropathogenic *Escherichia coli* to Hep-2 cells by immunoglobulin and oligosaccharide fractions of human colostrum and breast milk. *J. Infect. Dis.* **163**:1247.
Cruz, J.R., Cano, F., and Cáceres, P. 1991. Association of human milk SIgA antibodies with maternal intestinal exposure to microbial antigens, in: *Immunology of Milk and the Neonate* (J. Mestecky, C. Blair, P.L. Ogra, eds) Plenum Press, New York, NY, pp. 193–199.
Cruz, J.R., Gil, L., Cano, F., Cáceres, P., and Pareja, G. 1988. Breast milk anti-*Escherichia coli* heat-labile toxin IgA antibodies protect against toxin-induced diarrhea. *Acta. Paediatr. Scand.* **77**:658.
Cunningham, A.S., Jelliffe, D.B., and Jelliffe, E.F.P. 1991. Breast-feeding and health in the 1980s: A global epidemiologic review. *J. Pediatr.* **118**:659.
Dahlgren, U., Carlsson, B., Jalil, F., MacDonald, R., Mascart-Lemone, F., Nilsson, K., Roberton, D., Sennhauser, F., Wold, A., and Hanson, L.A. 1989. Induction of mucosal immune response. *Curr. Top. Microbiol. Immunol.* **146**:155.

Davidson, L.A. and Lönnerdal, B. 1990. Fecal α_1-antitrypsin in breast-fed infants is derived from human milk and is not indicative of enteric protein loss. *Acta. Paediatr. Scand.* **79**:137.

Davis, M.K., Savitz, D.A., and Graubard, B.I. 1988. Infant feeding and childhood cancer. *Lancet* **2**:365.

Dewey, K.G., Heinig, M.J., and Nommsen-Rivers, L.A. 1995. Differences in morbidity between breast-fed and formula-fed infants. *J. Pediatr.* **126**:696.

Di Carlo, A.L., Paape, M.J., and Miller, R.H. 1996. Reactivity of purified complement component 3b with bovine neutrophils and modulation of complement receptor 1. *Am. J. Vet. Res.* **57(2)**:151.

Duncan, B., Ey, J., and Holberg, C.J., Wright, A.L., Martinez, F.D., and Taussig, L.M. 1993. Exclusive breast-feeding for at least 4 months protects against otitis media. *Pediatrics* **91**:867.

Ellison, R.T. III and Giehl, T.J. 1991. Killing of Gram-negative bacteria by lactoferrin and lysozyme, *J. Clin. Invest.* **88**:1080.

Fanslow, W.C., Kulkarni, A.D., Van Buren, C.T., and Rudolph, F.B. 1988. Effect of nucelotide restriciton and supplementaion on resistance to experiemental candidiasis. *JPEN* **12**:49.

Ferguson, D.M., Horwood, L.J., and Shannon, F.T. 1990. Early solid feeding and recurrent childhood eczema: a 10-year longitudinal study. *Pediatrics* **86**:541.

Garofalo, R., Chheda, S., Mei, F., Palkowetz, K.H., Rudloff, H.E., Schmalstieg, F.C., Rassin, D.K., and Goldman, A.S. 1995. Interleukin-10 in human milk. *Pediatr. Res.* **37**:444.

Gerson, C., Sabater, J., Scuri, M., Torbati, A., Coffey, R., Abraham, J.W., Lauredo, T., Forteza, R., Wanner, A., Salathe, M., Abraham, W.M., and Conner G.E. 2000. The lactoperoxidase system functions in bacterial clearance of airways. *Am. J. Respit. Cell Molec. Biol.* **22**:665.

Gilmore, W.S., McKelvey-Martin, V.J., Rutherford, S., Strain, J.J., Loane, P., Kell, M., and Millar, S. 1994. Human milk contains granulocyte-colony stimulating factor (G-CSF). *Eur. J. Clin. Nutr.* **48**:222.

Glass, R.I., Svennerholm, A.M., Stoll, B.J., Khan, M.R., Huda, S., Huq, M.I., and Holmgren, J. 1989. Effects of undernutrition on infection with Vibrio cholerae O1 and on response to oral cholera vaccine. *Ped. Infect. Dis. J.* **8**:105.

Goldman, A.S. 1993. Immune systme of human milk: antimicrobial, antiinflammatory and immunomodulating properties. *Ped. Infec. Dis.* **12**:664.

Goldman, A.S., Chheda, S., and Garofalo, R. 1998. Evolution of immunologic functions in the mammary gland and the postnatal development of immunity. *Ped. Res.* **43**:155.

Goldman, A.S., Chheda, S., Keeney, S.E., Schmalstieg, F.C., Schanler, R.J. 1994. Immunologic protection of the premature newborn by human milk. *Sem. Perinatol.* **18**:495.

Goldman, A.S. and Frawley, S. 1996. Bioactive components of milk. *J. Mam. Gland Biol. Neoplas.* **1**:241.

Goldman, A.S., Garza, C., Johnson, C.A., Nichols, B.L., and Goldblum, R.M. 1982. Immunologic factors in human milk during the first year of lactation. *J. Pediatr.* **100**: 563.

Gothefors, L. and Marklund, S. 1975. Lactoperoxidase activity in human milk and in saliva of newborn infants. *Infect. Immun.* **11**:1210.

Grommers, F.J., van de Geer, D., van de Vliet, H., Henricks, P.A., and Nijkamp, F.P. 1989. Polymorphonuclear leukocyte functions: relationship between induced migration into the bovine mammary gland and in vitro cell activity. *Vet. Immunol.* **23**:75.

Grosvenor, C.E., Picciano, M.F., and Baumrucker, C.R. 1993. Hormones and growth factors in milk. *Endocrine Rev.* **14**:710.

Guarino, A., Canani, R.B., Russo, S., Albano, F., Canani, M.B., Ruggeri, F.M., and Rubino, A. 1994. Oral immunoglobulins for treatment of acute rotaviral gastroenteritis. *Pediatrics* **93**:12.

Hahn-Zoric, M., Carlsson, B., Jeansson, S., Ekre, H.P., Osterhaus, A.D., Roberton, D., and Hanson, L.A. 1993. Anti-idiotypic antibodies to polio virus in commercial immunoglobulin preparations, human serum, and milk. *Pediatr. Res.* **33**:475.

Hahn-Zoric, M., Fulconis, F., Minoli, J., Moro, G., Carlson, B., Böttiger, M., Räihä, N., Hanson, L.Å. 1990. Antibody responses to parenteral and oral vaccines are impaired by conventional and low protein formulas as compared to breast-feeding. *Acta. Paediatr. Scand.* **79**:1137.

Håkansson, A., Andréasson, J., Zhivotovsky, B., Karpman, D., Orrenius, S., and Svanborg, C. 1999. Multimeric α-lactalbumin from human milk induces apoptosis through a direct effect on cell nuclei. *Exp. Cell Res.* **246**:451.

Hanson, L.Å., Carlsson, B., Ekre, H.P., Hahn-Zoric, M., Osterhaus, A.D., and Roberton, D. 1989. Immunoregulation mother-fetus/newborn, a role for anti-idiotypic antibodies. *Acta. Paediatr. Scand. Suppl.* **351**:38.

Hara, T., Irie, K., Saito, S., Ichijo, M., Yamada, M., Yanai, N., and Miyazaki, S. 1995. Identification of macrophage colony-stimulating factor in human milk and mammary epithelial cells. *Pediatr. Res.* **37**:437.

Harmsen, M.C., Swart, P.J., de Bethune, M.P., Pauwels, D., Clercq, E., The, T.H., and Meijer, D.K. 1995. Antiviral effects of plasma and milk proteins: lactoferrin shows potent activity against both human immunodeficiency virus and human cytomegalovirus in vitro. *J. Infect. Dis.* **172**:380.

Hibbitt, K.G., Cole, C.B., and Reiter, B. 1969. Antimicrobial proteins isolated from the teat canal of the cow. *J. Gen. Microbiol.* **56**:365.

Hill, A.W. 1981. Factors influencing the outcome of Escherichia coli mastitis in cows. *Res. Vet. Sci.* **31**:107.

Hill, A.W., and Hibbit, K.G. 1986. Antimicrobial systems within the bovine mammary gland. *Proc. FEMS Symp. Fed. Eur. Microbiol.* Soc **92**:72.

Hilpert, H., Brussow, H., Mietens, C., Sidoti, J., Lerner, L., and Werchau, H. 1987. Use of bovine milk concentrate containing antibody to rotavirus to treat rotavirus gastroenteritis in infants. *J. Infec. Dis.* **156**:158.

Hirano, T., Akira, S., Taga, T., and Kishimoto, T. 1990. Biological and clinical aspects of interleukin 6. *Immunol. Today* **11**:443.

Holmgren. J., Svennerholm, A.M., and Lindblad, M. 1983. Receptor-like glycocompounds in human milk that inhibit classical and El Tor Vibrio cholerae cell adherence (hemagglutination). *Infect. Immun.* **39**:147.

Janas, L.M. and Picciano, M.F. 1982. The nucleotide profile of human milk. *Pediatr. Res.* **16**:659.

Jelonek, M.T., Maskrey, J.L., Cummins, L.M., Arora, Y., Steimer, K.S., White-Scharf, M., and Keller, M.A. 1998. HIV hyperimmune globulin or intravenous immune globulin inhibits response to HIV vaccine. *Biotech. App. Biochem.* **27**:89.

Keeney, S.E., Schmalstieg, F.C., Palkowetz, K.H., Rudloff, H.E., Le, B.M., and Goldman, A.S. 1993. Activated neutrophils and neutrophil activators in human milk. Increased expression of CD11b and decreased expression of L-selectin. *J. Leukocyte Biol.* **54**:97.

Kingwill, R.G., Neave, F.K., Dodd, F.H., Griffin, T.K., Westgarth, D.R., and Wilson, C.D. 1970. The effect of a mastitis control system on levels of subclinical and clinical mastitis in two years. *Vet. Record* **87**:94.

Kono, Y., Beagley, K.W., Fujihasi, K., McGhee, J.R., Taga, T., Hirano, T., Kishimoto, T., and Kiyono, H. 1991. Cytokine regulation of localized inflammation. Induction of activated B cells and IL-6 mediated polyclonal IgG and IgA synthesis in inflamed human gingiva. *J. Immunol.* **146**:1812.

Korhonen, T.K., Valtonen, M.V., and Parkkinen, J., Vaisanen-Rhen, V., Finne, J., Orskor, F., Orskor, I., Svenson, S.B., and Makela, P.H. 1985. Serotypes, hemolysin production and receptor recognition of Escherichia coli strains associated with neonatal sepsis and meningitis. *Infect. Immun.* **48**:486.

Korhonen, H., Syväoja, E.L., Ahola-Luttila, H., Sivelä, S., Kopola, S., and Kosunen, T.U. 1995. Bactericidal effect of bovine normal and immune serum, colostrum and milk against Helicobacter pylori. *J. App. Bacteriol.* **78**:655.

Kühn, R., Löher, J., Rennick, D., Rajewski, K., and Müller, W. 1993. Interleukin-10-deficient mice develop chronic enterocolitis. *Cell* **75**:263.

Kulkarni, A.D., Fanslow, W.C., Rudolph, F.B., and Van Buren, C.T. 1987. Modulation of delayed hypersenitivity in mice by dietary nucleotide restriction. *Transplantation* **44**:847.

Kulkarni, A.D., Fanslow, W.C., Rudolph, F.B., and Van Buren, C.T. 1986. Effect of dietray nucleotides on response to bacterial infections. *JPEN* **10**:169.

Kumar, S.N., Stewart, G.L., Steven, W.M., and Seelig, L.L. 1989. Maternal to neonatal transmission of T-cell mediated immunity to Trichenella spiralis during lactation. *Immunology* **68**:87.

Kunz, C. and Rudloff, S. 1993. Biological functions of oligosaccharides in human milk. *Acta. Paediatr.* **82**:903.

Lee, C.S., Wooding, F.B.P., and Kemp, P. 1980. Identification properties and differential counts of cell populations using electron microscopy of dry cow secretions, colostrum, and milk form normal cows. *J. Dairy Res.* **47**:39.

Lilja, G., Dannaeus, A., Foucard, T., Graff-Lönnevig, V., Johansson, S.G.O., and Öhman, H. 1989. Effects of maternal diet during late pregnancy and lactation on the development of at opic diseases in infants up to eighteen months of age—in vivo results. *Clin. Exp. Allergy* **19**:473.

Lindh, E. 1985. Increased resistance of immunoglobulin dimers to proteolytic degradation after binding of secretory component. *J. Immunol.* **113**:284.

Loimaranta, V., Nuutila, J., Marnila, P., Tenovuo, J., Korhonen, H., and Lilius, E.-M. 1999. Colostral proteins from cows immunised with Streptococcus mutans/S. sobrinus support the phagocytosis and killing of mutans streptococci by human leukocytes. *J. Med. Mcrobiol.* **48**:917.

Lönnerdal, B. 1985. Biochemistry and physiological function of human milk proteins. *Am. J. Clin. Nutr.* **42**:1299.

Lönnerdal, B. 1999. Effect of nutrition on microbial flora in infants—the role of lactoferrin, iron and nucleotides, in: *Probiotics, Other Nutritional Factors and Intestinal Microflora*, 42nd Nestle Nutrition Workshop (L.-A. Hanson and R.H. Yolken, eds.), pp. 189–201, Lippincott-Raven Publishers, Philadelphia, PA.

Lönnerdal, B. and Iyer, S. 1995. Lactoferrin: molecular structure and biological function. *Annu. Rev. Nutr.* **15**:93.

Lönnerdal, B., Woodhouse, L.R., and Glazier, C. 1987. Compartmentalization and quantitation of proteins in human milk. *J. Nutr.* **117**:1385.

Lucas, A. and Cole, T.J. 1990. Breast milk and neonatal necrotising entercolitits. *Lancet* **336**:1519.

Mallard, B.A., Wagter, L.C., Ireland, M.J., and Dekkers, J.C. 1997. Effects of growth hormone, insulin-like growth factor-I, and cortisol on periparturient antibody response profiles of dairy cattle. *Vet. Immunol. Immunopath.* **60**:61.

Marild, S., Jodal, U., and Hanson, L.A. 1990. Breastfeeding and urinary tract infection. *Lancet* **336**:942.

Marild, S., Jodal, U., and Mangelus, L. 1989. Medical histories of children with acut pyleonephritis compared with controls. *Pediatr. Infect. Dis. J.* **8**:511.

Mattsby-Baltzer, I., Roseanu, A., Motas, C., Elverfors, J., Engberg, I., and Hanson, L.Å. 1996. Lactoferrin or a fragment thereof inhibits the endotoxin-induced interleukin-6 response in human monocytic cells. *Pediatr. Res.* **40**:257.

McCarthy, A.C., La, E., Conti, C.J., and Locniskar, M.F. 1997. Effect of spray-dried yogurt and lactic acid bacteria on the initiation and promotion of chemically induced skin carcinogenesis in mice. *Nut. and Cancer* **27**:231.

Mellander, L., Carlsson, B., and Hanson, L.Å. 1986. Secretory IgA and IgM antibodies to E. coli O and poliovirus type 1 antigens occur in amniotic fluid, meconium and saliva from newborns. A neonatal immune response without antigenic exposure: a result of anti-idiotypic induction? *Clin. Exp. Immunol.* **63**:555.

Mostov, K.E. and Blobel, G.A. 1982. A transmembrane precursor of secretory component: the receptor for transcellular transport of polymeric immunoglobulins. *J. Biol. Chem.* **257**:11816.

Munoz, C., Endres, S., van der Meer, J., Schlesinger, L., Arevalo, M., and Dinarello, C. 1990. Interleukin-1 in human colostrum. *Res. Immunol.* **141**:501.

Murphy, J.M. and Stuart, O.M. 1953. The effect of introducing small numbers of Streptococcus agalactiae (Cornell Strain 48) directly into the bovine teat cavity. *Cornell Vet.* **43**:290.

Mushtaha, A.A., Schmalstieg, F.C., Hughes, Jr. T.K., Rajaraman, S., Rudloff, H.E., and Goldman, A.S. 1989. Chemokinetic agents for monocytes in human milk: possible role of tumor necrosis factor-alpha. *Pediatr. Res.* **25**:629.

Neeser, J.R., Chambaz, A., Del Vedovo, S., Prognet, M.-J., and Guggenheim, B. 1988. Specific and non-specific inhibition of adhesion of oral actinomyces and Streptococci to erythrocytes and polystyrene by caseinoglycopeptide derivatives. *Infect. Immun.* **56**:3201.

Newburg, D.S., Pickering, L.K., McCluer, R.H., and Cleary, T.G. 1990. Fucosylated oligosaccharides of human milk protect suckling mice from heat-stable enterotoxin from *Escherichia coli*. *J. Infect. Dis.* **162**:1075.

Newburg, D.S. 1997. Do the binding properties of oligosaccharides in milk protect human infants from gastrointestinal bacteria? *J. Nutr.* **127**:980S.

Nilsen, E.M., Johansen, F.E., Kvale, D., Krajci, P., and Brandtzaeg, P. 1999. Different regulatory pathways employed in cytokine-enhanced expression of secretory component and epithelial HLA class I genes. *Euro. J. Immunol.* **29**:168.

Noda, K., Umeda, M., and Ono, T. 1984. Transforming growth factor activity in human colostrum. *Gann* **75**:109.

Okamoto, J., Tsutsumi, H., Kumar, N.S., and Ogra, P.L. 1989. Effect of breast feeding on the development of anti-idiotype antibody response to F glycoprotein of respiratory synctial virus in infant mice after post-partum maternal immunization. *J. Immunol.* **142**:2507.

Pabst, H.F. and Spady, D.W. 1990. Effect of breast-feeding on antibody response to conjugate vaccine. *Lancet* **336**:269.

Palkowetz, K.H., Royer, C.L., Garofalo, R., Rudloff, H.E., Schmalstieg, F.C., and Goldman, A.S. 1994. Production of interleukin-6 and interleukin-8 by human mammary gland epithelial cells. *J. Reprod. Immunol.* **26**:57.

Parker, F., Migliore-Samour, D., Floch, F., Zerial, A., Werner, G.H., Jolles, J., Cassaretto, M., Zahn, H., and Jolles, P. 1984. Immunostimulating hexapeptide from human casein: Amino acid sequence, synthesis and biological properties. *Eur. J. Biochem.* **145**:677.

Pellegrini, A., Thomas, U., Bramaz, N., Hunziker, P., and von Fellenberg, R. 1999. Isolation and identification of three bactericidal domains in the bovine α-lactalbumin molecule. *Biochim. Biophys. Acta.* **1426**:439.

Perlmutter, D.H., May, L.T., and Sehgal, P.B. 1989. Interferon-γ and interleukin-6 modulates synthesis of α_{-1}-antitrypsin in human mononuclear phagocytes and in human hepatoma cells. *J. Clin. Invest.* **84**:138.

Petschow, B.W., Talbott, R.D., and Batema, R.P. 1999. Ability of lactoferrin to promote the growth of Bifidobacterium spp. In vitro is independent of receptor binding capacity and iron saturation level. *J. Med. Microbiol.* **48**:451.

Pickering, L.K., Granoff, D.M., Erickson, J.R., Masor, M.L., Cordle, C.T., Schaller, J.P., Winship, T.R., Paule, C.L., and Hilty, M.D. 1998. Modulation of the immune system by human milk and infant formula containing nucleotides. *Pediatrics* **101**:242.

Pisacane, A., Graziano, L., Mazzarells, G., Scarpellino, B., and Zona, G. 1992. Breastfeeding and urinary tract infection. *J. Pediatr.* **120**:87.

Pizzini, R.P., Kumar, S., Kulkarni, A.D., Rudolph, F.B., and Van Buren, C.T. 1990. Dietary nucleotides reverse malnutrition and starvation-induced immunosuppression. *Arch. Surg.* **125**:86.

Politis, I., Zhao, X., McBride, B.W., and Burton, J.H. 1990. Function of bovine mammary macrophages as antigen-presenting cells. *Vet. Immunol. Immunopathol.* **30**:399.

Reber, P.M. 1993. Prolactin and immunomodulation. *Am. J. Med.* **95**:637.

Rejnek, J., Travieck, J., Kostka, J., Sterzl, J., and Lane, A. 1968. Study of the effect of antibodies in the intestinal tract of germ-free baby pigs, *Microbiol.* **13**:36.

Rice, L.J., Chai, Y.J., Conti, C.J., Willis, R.A., and Locniskar, M.F. 1995. The effect of dietary fermented milk products and lactic acid bacteria on the initiation and promotion stages of mammary carcinogenesis. *Nut. and Cancer* **24**:99.

Roberton, D.M., Carlsson, B., Coffman, K., Hahn-Zoric, M., Jalil, F., Jones, C., and Hanson, L.Å. 1988. Avidity of IgA antibody to *Escherichia coli* polysaccharide and diphtheria toxin in breast-milk from Swedish and Pakistani mothers. *Scand. J. Immunol.* **28**:783.

Roos, N., Mahé, S., Benamouzig, R., Sick, H., Rautureau, J., and Tomé, D. 1995. ^{15}N-labeled immunoglobulins from bovine colostrum are partially resistant to digestion in human intestine. *J. Nutr.* **125**:1238.

Rudloff, H.E., Schmalstieg, F.C., Mushtaha, A.A., Palkowetz, K.H., Liu, S.K., and Goldman, A.S. 1992. Tumor necrosis factor-alpha in human milk. *Pediatr. Res.* **31**:29.

Rudloff, H.E., Schmalstieg, F.C., Palkowetz, K.H., Paskiewicz, E.J., and Goldman, A.S. 1993. Interleukin-6 in human milk. *J. Reprod. Immunol.* **23**:13.

Rump, J.A., Arndt, R., Arnold, A., Bendick, C., Dichtelmüller, H., Franke, M., Helm, E.B., Jäger, H., Kampmann, B., and Kolb, P. 1992. Treatment of diarrhoea in human immunodeficiency virus-infected patients with immunoglobulins from bovine colostrum. *Clin. Invest.* **70**:588.

Saarinen, U.M., Kajosaari, M., Blackman, A., and Siimes, M.A. 1979. Prolonged breast-feeding as prophylaxis for atopic disease. *Lancet* **ii**:163.

Saarinen, U.M. and Kajosaari, M. 1995. Breastfeeding as prophylaxis against atopic disease: Prospective follow-up study until 17 years old. *Lancet* **346**:1065.

Saarinen, K.M., Juntunen-Backman, K., Jarvenpaa, A.L., Kuitunen, P., Lope, L., Renlund, M., Siivola, M., and Savilahti, E. 1999. Supplementary feeding in maternity hospitals and the risk of cow's milk allergy: A prospective study of 6209 infants. *J. Allergy Clin. Immunol.* **104**:457.

Saarinen, U.M. 1982. Prolonged breast-feeding as a prophylaxis for recurrent otitis media. *Acta. Ped. Scand.* **71**:567.

Saito, S., Maruyama, M., Kato, Y., Moriyama, I., and Ichijo, M. 1991. Detection of IL-6 in human milk and its involvement in IgA production. *J. Reprod. Immunol.* **20**:267.

Sato, K. and Lönnerdal, B. 1993. Uptake and transcytosis of free secretory component from human and bovine milk by Caco-2 cells. *FASEB* **7**:A823.

Schultz, C.L. and Coffman, R.L. 1991. Control of isotype switching by T cells and cytokines. *Curr. Opin. Immunol.* **3**:350.

Schupbach, P., Neeser, J.R., Golliard, M., Rouvet, M., and Guggenheim, B. 1996. Incorporation of caseinoglycomacropeptide and caseinophosphopeptide into the salivary pellicle inhibits adherence of mutans Streptococci. *J. Dent. Res.* **75**:1779.

Selsted, M.E., Tang, Y.Q., Morris, W.L., McGuire, P.A., Nonotny, M.J., Smith, W., Henschen, A.H., and Cullor, H.S. 1993. Purification, primary structures, and antibacterial activities of the beta-defensins, a new family of antimicrobial peptides from bovine neutrophils. *J. Biol. Chem.* **268**:6641.

Snodgrass, D.R., Campbell, I., Mwenda, J.M., Chege, G., Suleman, M.A., Morein, B., and Hart, C.A. 1995. Stimulation of rotavirus IgA, IgG and neutralising antibodies in baboon milk by parenteral vaccination. *Vaccine* **13**:408.

Sordillo, L.M., Nickerson, S.C., and Akers, R.M. 1989. Pathology of Staphylococcus aureus mastitis during lactogenesis: relationship with bovine mammary structur and function. *J. Dairy Sci.* **72**:228.

Sordillo, L.M., Nickerson, S.C., Akers, R.M., and Oliver, S.P. 1987. Secretion composition during bovine mammary involution and the relationship with mastitis. *Int. J. Biochem.* **19**:1165.

Stahl, B., Thurl, S., Zeng, J., Karas, M., Hillenkamp, F., Steup, M., and Sawatzki, G. 1994. Oligosaccharides from human milk as revealed by matrix-associated laser desorption/ionization mass spectrometry. *Anal. Biochem.* **223**:218.

Stein, K.E. and Söderström, T. 1984. Neonatal administration of idiotype or anti-idiotype primes for protection against E. coli K 13 infection in mice. *J. Exp. Med.* **160**:1001.

Strömqvist, M., Falk, P., Bergström, S., Hansson, L., Lönnerdal, B., Normark, S., and Hernell, O. 1995. Human milk κ-casein and inhibition of *Helicobacter pylori* adhesion to human gastric mucosa. *J. Pediatr. Gastroenterol Nutrition* **21**:288.

Sutton, B.J. and Gould, H.J. 1993. The human IgE network. *Nature* **366**:421.

Tanaka, M., Lee, K., Martinez-Augustin, O., He, Y., Sanderson, I.R., and Walker, W.A. 1996. Exogenous nucleotides alter the proliferation, differentiation and apoptosis of human small intestinal epithelium. *J. Nutr.* **126**:424.

Tejada-Simon, M.V., Ustunol, Z., and Pestka, J.J. 1999. *Ex vivo* effects of lactobacilli, streptococci and bifidobacteria ingestion on cytokine and nitric oxide production in a murine model. *J. Food Protect.* **62**:162.

Telemo, E. and Hanson, L.Å. 1996. Antibodies in milk. *J. Mammary Gland Biol. Neopl.* **1**:243.

Teraguchi, S., Shin, K., Ozawa, K., Nakamura, S., Fukuwatari, Y., Tsuyuki, S., Namihira, H., and Shimamura, S. 1995. Bacteriostatic effect of orally administered bovine lactoferrin on proliferation of *Clostridium* species in the gut of mice fed bovine milk. *Appl. Environ. Microbiol.* **61**:501.

Thorell, L., Sjöberg, L.B., and Hernell, O. 1996. Nucleotides in human milk: Sources and metabolism by the newborn infant. *Pediatr. Res.* **40**:845.

Thorpe, L.W., Rudloff, H.E., Powell, L.C., and Goldman, A.S. 1986. Decreased response of human milk leukocytes to chemoattractant peptides. *Pediatr. Res.* **20**:373.

Treece, J.M., Morse, G.E., and Levy, C. 1966. Lipid analysis of bovine teat canal keratin. *J. Dairy Sci.* **49**:1240.

Tuboly, S., Bernath, S., Glavitis, R., Kovacs, A., and Megyeri, Z. 1995. Intestinal absorption of colostral lymphocytes in newborn lambs and their role in development of immune status. *Acta Vet. Hungar.* **43**:105.

Turchany, J.M., Aley, S.B., and Gillin, F.D. 1995. Giardicidal activity of lactoferrin and N-terminal peptides. *Infect. Immun.* **63**:4550.

Uauy, R., Stringel, G., Thomas, R., and Quan, R. 1990. Effect of dietary neucleoside on growth and maturation of the developing gut in the rat. *J. Pediatri. Gastroenterol. Nutr.* **10**:497.

Uytedehaag, F.G.C.M. and Osterhaus, A.D.M.E. 1985. Induction of neutralizing antibody in mice against poliovirus type II with monoclonal anti-idiotypic antibody. *J. Immunol.* **134**:1225.

Van Buren, C.T., Kulkarni, A.D., Fanslow, W.C., and Rudolph, F.B. 1985. Dietary nuecleotides, a requirement helper/inducer T-lymphocytes. *Transplantation* **40**:694.

Viselli, S.M., Stanek, E.M., Mukerjee, P., Hymer, W.C., and Mastro, A.M. 1991. Prolactin-induced mitogenesis of lymphocytes for ovariectomized rats. *Endocrinology* **129**:983.

Whitmore, A.C., Prowse, D.M., Haughton,G., and Arnold, L.W. 1991. Ig isotype switching in B lymphocytes. The effect of T-cell derived interleukines, cytokines, cholera toxin, and antigen on isotype switch frequency of a cloned B cell lymphoma. *Int. Immunol.* **3**:95.

Winner, L.S., Weltzin, R.A., Mekalanos, J.J., Krachenbuhl, J.P., and Neutra, M.R. 1991. New model for analysis of mucosal immunity: Intestinal secretin of specific monoclonal immunoglobulin A from hybridoma tumor protects against Vibrio cholerae. *Infect. Immun.* **59**:977.

Wirt, D.P., Adkins, L.T., Palkowetz, K.H., Schmalstieg, F.C., and Goldman, A.S. 1992. Activated and memory lymphocytes-T in human milk. *Cytometry* **13**:282.

Wong, C.W. and Watson, D.L. 1995. Immunomodulatory effects of dietary whey proteins. *J. Dairy Res.* **62**:359.

Wong, C.W., Regester, G.O., Francis, G.L., and Watson, D.L. 1996a. Immunomodulatory activities of whey fractions in efferent prefemoral lymph of sheep. *J. Dairy Res.* **63**:257.

Wong, C.W., Seow, H.F., Liu, A.H. Husband, A.J., Smithers, G.W., and Watson, D.L. 1996b. Modulation of immune responses by bovine beta-casein. *Immuno. Cell Biol.* **74**:323.

Wright, A.L., Holberg, C.J., Martinez, F.D., Morgan, W.J., and Taussig, L.M. 1989. Breast-feeding and lower respiratory tract illness in the first year of life. *Brit. Med. J.* **299**:946.

Xanthou, M. 1993. The development of the immune system, in: *Neonatal Haemotolgy and Immunology II* (M. Xanthou, R. Bracciand G. Prindull, eds.), Elsevier, Amsterdam, The Netherlands, pp. 113–122.

Yamauchi, K., Tomita, M., Giehl, T.J., and Ellison, R.T. III. 1993. Antibacterial activity of lactoferrin and a pepsin-derived lactoferrin peptide fragment. *Infect. Immun.* **61**:719.

Yolken, R.H., Peterson, J.A., Vonderfecht, S.L., Fouts, E.T., Midthun, K., and Newburg, D.S. 1992. Human milk mucin inhibits rotavirus replication and prevents experimental gastroenteritis. *J. Clin. Invest.* **90**:1984.

Yu, P., Kosco-Vilbios, M., Richards, M., Kohler G., and Lamers, M.C. 1994. Negative feedback regulation of IgE synthesis by murine CD23. *Nature* **369**:753.

Part II

Milk and Risk of Infection During the Suckling Period

Chapter 4

Breast Milk and the Risk of Opportunistic Infection in Infancy in Industrialized and Non-Industrialized Settings

Patrick F. W. Chien and Peter W. Howie

1. Introduction

The purpose of this systematic review is to examine the relationship between infant feeding mode (i.e., breast-feeding *versus* others) and infective illnesses in industrialized and non-industrialized countries. This report focuses on four clinically important research questions:

1. Does breast-feeding confer any protection against gastrointestinal infections?
2. Does breast-feeding provide any protection against respiratory tract infections?
3. Is protection against primary invasive *Haemophilus influenzae* accorded by breast-feeding?
4. Is the incidence of urinary tract infection lower in infants who are breast-fed?

Patrick F. W. Chien and Peter W. Howie • Department of Obstetrics & Gynaecology, University of Dundee, Ninewells Hospital, Dundee, Scotland DD1 9SY

2. Methodological Issues

2.1. Randomized Trials

Randomized controlled trials are the most rigorous means of determining whether a cause-effect relationship exists between an intervention such as infant feeding mode and an outcome such as infantile infections (Sibbald and Roland, 1998). To date, there have been only four reported randomized controlled trials in which expressed maternal or donor milk has been compared with formula feeding with regard to the risk of subsequent neonatal infections (Narayanan *et al.*, 1980, 1981, 1982, 1984). All four trials indicated that the incidence of infections was significantly lower in the groups that received breast milk. However, inadequate descriptions of study design (such as the methods of randomization and concealment) render the estimates of treatment effects in these trials questionable (Schultz *et al.*, 1995). Furthermore, application of these results to healthy infants with normal birth weight is invalid because the studies were conducted on infants with low birth weight who were admitted to a special care baby unit.

It is unlikely that randomized controlled trials will be feasible in the future because random assignment of feeding modes to infants would be considered unethical (Sauls, 1979; Winikoff, 1981; Bauchner *et al.*, 1986). Thus, quality problems pertaining to existing randomized studies, coupled with the impossibility of conducting further experimental investigations, means that we have to rely on observational studies to address the above research questions. This review therefore has been conducted on observational cohort and case-control studies which are ethically more acceptable than randomized trials, but which are more susceptible to bias.

2.2. Time Frame of Evaluated Studies

This evaluation is restricted to the first year of life. Consequently, several types of studies are not included in the review. For example, studies of the association between breast-feeding and childhood infection (from age 1 year to preschool age) were excluded because there is still uncertainty about the duration of protection against infections associated with breast-feeding (Jason *et al.*, 1984; Kovar *et al.*, 1984; Bauchner *et al.*, 1986). It is unusual for children to be breastfed beyond the first year of life in both industrialized and non-industrialized countries (Martinez and Dodd, 1983; Yoon *et al.*, 1996).

2.3. Infective Morbidity as an Outcome Measure

It is clear from breast-feeding and other epidemiological studies that the magnitudes of effects on mortality and morbidity from infections are not necessarily the same (Jason *et al.*, 1984). Because death is a relatively rare outcome event, small sample sizes are unable to reveal statistically significant associations between breast-feeding and protection from this outcome (Riegelman and Hirsch, 1996). On the other hand, data on infectious morbidity are more difficult to collect accurately. They are more susceptible to errors in recall, to subjective judgements in the grading of severity and to bias introduced in the recording of illness or because of differences in access to medical care (Winikoff, 1981). In an attempt to minimize inconsistencies in the results, we have included only those studies in which infective morbidity, such as the number of cases diagnosed or hospitalized with infantile infections, were reported. Studies in which mortality was employed as a combined outcome event were excluded.

2.4. Types of Infection

Not only are there differences in the magnitude of effects observed with mortality and morbidity from infection as outcome measures, but also there are differences in the relationship between breast-feeding and various types of infection. Several studies have indicated that breast-feeding conferred a stronger protection against gastrointestinal than respiratory infections (Fergusson *et al.*, 1978; Holmes *et al.*, 1983; Howie *et al.*, 1990). For this reason, we have organized this review to address separately the four different types of infections indicated by the above research questions. Studies in which all the various types of infections were combined as an outcome measure were excluded.

2.5. Industrialized and Non-Industrialized Settings

Most current evidence suggests that breast milk has significant advantages for infants, particularly in non-industrialized countries where poverty, lack of parental education, overcrowding at home, and lack of safe drinking water and sanitation are common (Sauls, 1979). Studies on industrialized societies, however, have produced conflicting results regarding the protective effect of breast-feeding (Bauchner *et al.*, 1986). To allow for valid assessment of the effects of breast-feeding in different study populations, we have divided the studies according to whether they were conducted in industrialized or non-industrialized countries.

2.6. Study Selection and Method of Review

Before selecting the studies, we performed a computerized literature search using the MEDLINE database for 1966 through January, 1998. The limitations of electronic literature searches to identify relevant articles are well known

(Dickersin *et al.*, 1994). We have therefore supplemented our electronic search with an examination of the reference lists of all relevant primary research and review publications. The Science Citation Index was used to identify frequently cited articles and additional studies that had cited these articles.

The electronic searches yielded 727 citations. After scrutinizing these citations and their abstracts (where available), 83 articles were considered relevant. A further 48 articles were identified from the reference lists of primary research and review articles. Following evaluation of these 131 articles, 39 were judged to be eligible for inclusion in this overview.

We reviewed the results in two ways. Firstly, a narrative summary of the evidence for an association between infant feeding mode and infective illnesses was developed. Clear differences were observed between mothers who chose to breast-feed and those who chose to bottle-feed. Women who breast-feed tended to have a higher socioeconomic status and a higher level of education. Mothers who chose to bottle-feed had a higher incidence of socioeconomic deprivation, a lower educational level and a higher incidence of smoking. The use of non-experimental studies required an evaluation of potential confounding factors. We also evaluated the results of studies in which multivariate analysis was used to detect potential confounding, including information on the method used to assess confounding, the outcome variable and any co-variates in the study. From this information we developed an overall conclusion regarding the results of each study.

Secondly, we pooled the results (weighted according to sample size) in order to assess quantitatively the quality and strength of the evidence. Because it was not possible to pool the adjusted relative measures of association (such as the relative risk and odds ratios) we based our analysis on the infection outcomes without any analysis for potential confounding. However, if associations in the subgroups were sufficiently large and consistent, we assumed that a statistically significant effect was present.

2.7. Infant Feeding Classification

Infant feeding practices are usually categorized according to one of the following three groups: (a) exclusively breast-fed, (b) partially breast-fed (i.e., breast-fed with supplemental artificial feeding) (c) exclusively artificially fed. To generate 2×2 tables of data on infant feeding and infective outcomes, we had to dichotomize infant feeding into 2 groups despite a resultant loss of information. To allow for the most conservative estimate of the effect of breast-feeding on the risk of infection, we grouped infant feeding practices into either (a) exclusive breast-feeding and partial/mixed feeding or (b) exclusive artificial feeding. Inclusion of partial bottle feeding in the breast-feeding group would be expected to dilute any protective effect of breast milk, but it overcomes reverse causality bias arising in the case of babies who begin breast-feeding and are subsequently changed to bottle feeding. It is rare for bottle-fed babies to revert to breast-feeding.

2.8. Statistical Methods

Odds ratios (OR) and their confidence intervals (CI) were calculated for each of the country subgroups and methods of design. Meta-analysis was limited to these subgroups and outcome measures to minimize heterogeneity. Odds ratios were pooled and combined using the fixed effects model (Mantel and Haenszel, 1959). Heterogeneity of treatment effects in the meta-analyses was formally assessed by comparing the OR for each study using the test of Breslow-Day (1980). Results in some studies were reported using relative risk (RR), rather than OR, as the measure of effect. In the narrative section of this review, we have described effects as reported in the primary article. Risk and odds are not synonymous, but these two measures are similar in cases wherein the rate of the outcome event is less than 10% (Chatellier *et al.*, 1996).

3. Industrialized Countries

3.1. Breast-Feeding and Gastrointestinal Infection

Gastrointestinal infection was taken to be any illness associated with vomiting, change in consistency or frequency of stools, or isolation of a known enteropathogenic bacterial or viral agent. Sixteen articles were eligible for inclusion in this section. The results are shown in Table 1.

3.1.1. Cohort Studies. The Research Committee of the Royal College of General Practitioners (RCGP) in the UK conducted a prospective study on 334 infants followed regularly from birth till the end of their first year of life (RCGP Report, 1972). Infant feeding practices were divided into two groups: (a) breast-fed either totally or partially and (b) bottle-fed exclusively. The outcome event, namely, gastrointestinal infection, was not clearly defined. There were no statistically significant differences in the incidence of gastrointestinal infection between breast-fed infants (51/223, 23%) and bottle-fed infants (28/111, 25%) ($\chi^2 = 0.37$, $p > 0.05$). No analysis was made for possible confounding.

A prospective study by Totterdell *et al.* (1976) was performed during an outbreak of rotavirus infection among babies aged 4–9 days in a London

Table 1. Studies of Breast-Feeding Practices and Gastrointestinal Infection During Infancy in Industrialized Countries

Reference	Country of origin	Recruitment source	Adjustment for confounding	Breast-feeding clearly defined	Outcome event clearly defined	Infective pathogen defined
Cohort Studies						
RCGP Report, 1972	England	Community	No	Yes	No	No
Totterdell *et al.*, 1976	England	Community	No	No	Yes	Rotavirus
Banatvala *et al.*, 1978	England	Hospital	No	No	Yes	Rotavirus
Fergusson *et al.*, 1978	New Zealand	Community	Yes	Yes	Yes	No
Cunningham, 1979	United States	Community	No*	Yes	No	No
Crewe and Murphy, 1980	Australia	Hospital	No	No	Yes	Rotavirus
France *et al.*, 1980	United States	Community	No	No	Yes	*Salmonella*
Cushing and Anderson, 1982	United States	Community	No	Yes	Yes	LT/CFA-*E. Coli*, Rotavirus
Holmes *et al.*, 1983	United States	Community	No*	Yes	No	No
Jenista *et al.*, 1984	United States	Community	No	No	Yes	Enterovirus
Agre, 1985	United States	Community	No	Yes	No	No
Duffy *et al.*, 1986	United States	Community	Yes	Yes	Yes	Rotavirus
Eaton-Evans and Dugdale, 1987	Australia	Community	Yes	Yes	Yes	No
Howie *et al.*, 1990	Scotland	Community	Yes	Yes	Yes	No
Case-control Studies						
Weinberg *et al.*, 1984	United States	Hospital	No	Yes	Yes	Rotavirus
Leventhal *et al.*, 1986	United States	Hospital	No*	Yes	No	No

LT = heat-labile enterotoxin; CFA = colonization factor antigen. * Although multivariate analysis was performed, the dependent variable was a combination of respiratory and gastrointestinal infections.

maternity unit. There was no description of the modes of feeding or of possible confounding factors. However, rotavirus infection was less frequent in breast-fed babies (31/70, 44%) than in bottle-fed babies (42/53, 79%) ($\chi^2 = 13.9$, $p < 0.001$).

In another prospective study performed in the UK, feces from 1056 5-day-old newborns were examined for rotaviral particles by electron microscopy (Banatvala *et al.*, 1978). Rotaviral infection was present in 166/751 (22%) of the babies who were breast-fed compared to 177/305 (58%) of those who were bottle-fed ($\chi^2 = 127$; $p < 0.001$). No clear definition of breast-feeding was reported. Moreover, no account was given of possible differences between the breast-fed and bottle-fed infants, e.g., season of the year in which the infant was born, or socioeconomic status. Although rotavirus was prevalent throughout the study period, the infection exhibited seasonal variation, reaching its peak during the winter months and its lowest prevalence during the summer months.

Fergusson *et al.* (1978) prospectively studied 1210 infants in the Christchurch Child Development Study. Each child's mother was interviewed shortly following birth and when the infant was 4 months old. Infant feeding was categorized as (a) totally bottle-fed from birth, (b) supplemented (i.e., the infants were breast-fed with regular cow's milk supplements or were weaned from the breast before 16 weeks), (c) near total breast-fed (i.e., given some cow's milk on an irregular basis or introduced to solid food by 16 weeks), or (d) totally breast-fed (i.e., exclusively breast-fed for the first 16 weeks of life). The main difference between categories (b) and (c) was that cow's milk supplements were given on a regular and irregular basis, respectively, until the infants were 16 weeks old. The diagnosis of gastrointestinal illness included vomiting and diarrhea but excluded crying ascribed to colic. The incidence of such symptoms was 96/282 (34%) in totally bottle-fed infants, 179/518 (35%) in supplement-fed infants, 34/278 (12%) in those near breast-fed and 9/132 (7%) in totally breast-fed infants ($\chi^2 = 81.2$; $p < 0.01$). These differences remained statistically significant even after analysis of the data for the confounding effects of maternal education, single parenting and living standards using multiple linear regression analysis ($p < 0.01$).

Cunningham (1979) reported the incidence of diarrhea and vomiting in a cohort of 503 New York infants who were regularly seen at a pediatric clinic during the first year of life. The relationship between these symptoms and infant feeding mode was assessed. Breast-feeding was defined as exclusive breast-feeding for $\geq 4\frac{1}{2}$ months. Since all the infants were followed for a year from birth, there were 6,202 breast-fed infant-weeks (135 infants × 52 weeks) out of a total of 26,156 infant-weeks (503 infants × 52 weeks). Compared to the artificially fed group, there were significantly fewer episodes of diarrhea and vomiting in the breast-fed group (22 vs. 137 episodes of diarrhea or vomiting per 1,000 infant-weeks of feeding, $p < 0.01$ using the χ^2 test). In addition to symptoms of diarrhea and vomiting, outcome variables in this study included otitis media, lower respiratory tract infections and any illness requiring hospital admission other than trauma or surgery for congenital anomalies. Although stratified analysis was used to examine the confounding effects of factors such as gender, paternal education, maternal age, presence or absence of older siblings and birthweight percentile, the dependent variable employed was all illnesses (i.e., including non-gastrointestinal illnesses) during the first year of life. Hence, the assessment of confounding cannot be applied to gastrointestinal illness, alone.

Crewe *et al.* (1980) conducted a prospective study on 427 babies in Sydney, Australia. Between the day following birth and the end of the first month of life, fecal specimens were collected for detection of rotaviral particles by electron microscopy. During the one month follow-up period, 128/276 (46%) of the neonates who were breast-fed and 76/151 (50%) of those who were bottle-fed developed rotavirus infection. There was no significant difference between the two groups, but the definitions of breast- and bottle-feeding were not reported clearly and no assessment was made of possible confounding.

A large retrospective study on the relationship between infant feeding practices and gastrointestinal *Salmonella* infection was carried out in Arkansas, USA, by France *et al.* (1980). From Jan., 1977 to Dec., 1978, 72,000 babies were born in the state of Arkansas. During the same 2-year period, there were 253 reported cases of infantile *Salmonella* infections. Data on infant feeding practices (either breast- or bottle-feeding) were obtained from medical records and maternal interviews at clinic or by telephone. Breast-feeding was associated with a lower incidence of *Salmonella* infection in infancy (0.05 cases per 1000 population in breast-fed infants vs. 4.8 cases per 1000 population in bottle-fed infants ($p < 0.01$).

In a study conducted in New Mexico, USA, 40 babies were prospectively observed for diarrhea and fecal carriage of heat-labile enterotoxin (LT) producing *Escherichia coli* with or without the presence of colonisation factor antigen (CFA) and rotavirus during the first year of life (Cushing and Anderson, 1982). Strains of LT-positive *E. coli* have been shown to be more likely to cause diarrhea if the surface antigen, CFA, is present than if it is absent (Satterwhite *et al.*, 1978). Each infant was visited weekly at home to ascertain the presence of diarrhea, and a fecal sample was collected at each visit. Mothers were asked to report by telephone if an infant had diarrhea between routine visits. If so, a nurse visited the home to verify the diagnosis and to collect an additional fecal sample. In non-breast-fed babies, diarrhea was defined as passage of ≥3 unformed stools per day, and a diarrhea episode was defined as ≥2 days of diarrhea separated by ≤2 diarrhea-free days. Breast-fed babies often pass feces with a high water content, and were considered to have diarrhea only if they produced ≥2 stools per day more than usual and if all stools were unformed. A baby was considered to be breast-fed if fed in this way more than twice per day, regardless of other food intake. The total number of observation weeks reported referred to the number of weeks associated with a particular mode of infant feeding during the entire study. The breast-fed cohort exhibited 3.8 diarrhea episodes per 100 weeks of observation, and non-breast-feeding was associated with a similar frequency of diarrhea (3.4 episodes per 100 bottle-feeding weeks). There was no significant difference in the frequency of diarrhea between these two modes of infant feeding. The number of weeks of carriage of LT-positive *E coli* ± the presence of CFA was similar for breast-fed and bottle-fed infants (49 *vs* 54 weeks for LT-positive *E. coli* in the absence of CFA in breast-fed and bottle-fed infants, respectively, and 13 *vs* 9 weeks for LT-positive *E. coli* and the presence of CFA, respectively). The proportion of LT-positive *E. coli* ± the presence of CFA-carriage weeks associated with diarrhea was similar in the breast-fed and bottle-fed infants.

Holmes *et al.* (1983) studied the association between first-year postnatal gastrointestinal infection among 251 infants in Kansas City, USA. Data were collected from hospital records and from interviews at regular follow-up at pediatric clinics. The outcome variables included respiratory, gastrointestinal, urinary tract, skin and other infections. Infant feeding was classified as (a) no breast-feeding (n = 124), (b) breast-feeding ≤3 months (n = 70), or (c) breast-feeding >3 months (n = 57). In this study, 26 breast-fed babies (20% of infant feeding groups (b) and (c) combined) had at least one episode of gastroenteritis, whereas among the 124 formula-fed infants, 42 (34%) had at least one episode of the same type of infection. Stepwise linear regression analysis was used to adjust for confounding, but the dependent variable was infantile infections (i.e., all types of infection) rather than gastrointestinal infections, alone. Consequently, no statistical test was applied to compare the rates of gastroenteritis between the two groups.

In another prospective study in New York, USA, 586 neonates were followed up for non-polio enterovirus infection (Jenista *et al.*, 1984). Following birth, the babies were visited regularly at home during the first one month of life to monitor symptoms of infection and to obtain rectal swabs for viral culture. Infant feeding was categorized into breast- and bottle-feeding, although the criteria used to define these two methods of feeding were not reported. Neonates who were bottle-fed were more likely to be infected with enterovirus than those who were breast-fed (53/290 bottle-fed vs. 22/296 breast-fed; χ^2 $p < 0.001$). However, possible confounding was not assessed.

In a study in New York, USA, 200 healthy, full-term infants born during a one-year study period were followed prospectively for frequency of gastroenteritis during the first year of life (Agre, 1985). Infants were categorised by mode of feeding: (a) exclusively breast-fed; (b) breast- or bottle-fed at least 3 times per day (but no more than 60% of feedings from the breast) with supplemental proprietary formula, or (c) exclusively bottle-fed. The frequency of gastroenteritis was 0.16, 0.17 and 0.17 episodes per patient per year for breast-fed, breast- or bottle-fed and bottle-fed infants, respectively, a frequency of infection that was too low to permit meaningful statistical analysis.

Duffy *et al.* (1986) conducted a prospective study on a cohort of 197 infants from birth until 4 months of age in New York, USA. All members of the cohort were followed through a rotavirus winter season using bimonthly maternal reports via telephone and interviews at home or at a clinic. Stool samples were obtained during each clinic visit. In addition, mothers were asked to report all diarrheal episodes in the interim between clinic visits and to submit stool samples during these episodes. Infants were classified at four months of age as being exclusively breast-fed (19), breast- and bottle-fed (31), initially breast-fed but converted to complete bottle-feeding prior to the age of 4 months (28), or exclusively bottle-fed (119). Rotavirus enteritis was identified using direct search electron microscopy with negative staining for identification and quantification of viral particles. The rate of rotavirus-related gastroenteritis in the exclusively breast-fed group was 1/19 (5%). Infants who were fed combined breast and formula milk and those who were initially breast-fed but subsequently converted to full bottle-feeding had the same rates of rotavirus infection (9/31 and 8/28, i.e., 29%) Seventeen infants (14%) of the exclusively bottle-fed group had rotavirus gastroenteritis. Exclusive breast-feeding did not affect the rate of rotavirus infection RR = 0.27, 95% CI 0.28–1.90), and the risk estimate was not substantially altered by analysis for sex of infant, race, or the occupation, education and marital status of the mother using multiple linear regression.

In a study conducted in Brisbane and Ipswich, Australia, 123 infants were followed up from one month to one year of age (Eaton-Evans and Dugdale, 1987). The mothers were interviewed at one-month intervals at maternal and child health clinics. They were asked if their infants were fed breast, bottle, or mixed (breast and bottle) milk and if they were given food other than milk. Infants were considered to have experienced diarrhea and/or vomiting if their mothers reported that their infants required treatment for these problems during the preceding month. They were not considered to have this outcome event if they merely regurgitated following feeds or if they had diarrhoea and/or vomiting but their mothers took no specific action. The data from the monthly interviews were combined into three age groups (0–3, 3–6, 6–9, 9–12 months). Between birth and 3 months of age, the incidence of diarrhea and/or vomiting did not differ significantly between infants who were breast-fed and those who were mixed fed (10% vs. 8%). However, bottle-fed infants had significantly more diarrhea and/or vomiting than infants who were mixed fed (20% vs. 8%, $p = 0.03$). The same outcome was observed between 3 and 6 months of age. However, from 6 to 9 months of age, no differences were observed among the three groups in the incidence of diarrhea and/or vomiting. After 9 months of age, none of the infants were exclusively breast-fed and there was no significant difference in the proportion of infants with diarrhea and/or vomiting. Social class and parity were found not to exert any confounding effect using multi-factorial analyses.

A study was conducted in Dundee, Scotland, on a cohort of 674 infants (Howie *et al.*, 1990). The infants were followed up regularly with planned home visits by health visitors for 24 months after birth, and the data were presented as adjusted rates of infection during 0–13, 14–26, 27–39, and 40–52 postnatal weeks. Mothers who exclusively breast-fed for ≥13 weeks were classified as full breast feeders. Those who breast-fed for ≥13 weeks, but who introduced supplements before that time, were considered to be partial breast feeders and those who started breast-feeding but discontinued it before 13 weeks were categorized as early weaners. Bottle feeders were defined as mothers who exclusively bottle-fed from birth. Multivariate logistic modelling was used to analyze for the possible effects of paternal social class, maternal age and parental smoking. The diagnosis of gastrointestinal infection was based on the presence of vomiting or diarrhea, or both, lasting as discrete illness for ≥48 h. Episodes of vomiting were distinguished from those of regurgitation following feeding, and diarrhea was diagnosed on the basis of frequent unformed stools. Bottle feeders were found to have significantly higher rates of gastrointestinal infection during the first 13 weeks of life than partial and full breast feeders combined (15.7% vs. 4.0% respectively, $p < 0.01$). This trend toward a higher rate of infection among bottle feeders than breast feeders (partial and full) persisted throughout the first year of life.

3.1.2. Case-Control Studies. Weinberg *et al.* (1984) conducted a study in New York, USA, to compare 50 infants who had developed rotavirus gastroenteritis and a group of randomly selected 50 infants without infection. Diarrhea was defined as the presence, during the 24 h prior to a clinic visit, of four or more loose or watery bowel movements. The incidence of breast-feeding in the two groups was obtained by follow-up contact with the parents. Infants were considered to be breast-fed if they received breast milk for at least two feedings during a 24 h period. Infant feeding practices were categorized into three groups: (a) breast-feeding at the time of infection; (b) breast-feeding stopped prior to infection and (c) exclusive formula feeding from birth. Of 50 cases with rotavirus gastrointestinal infection, 32 (64%) were being either breast-fed at the time of infection ($n = 7$) or had been breast-fed in the past ($n = 25$). The remaining 18 infants (36%) had been exclusively bottle-fed from birth. In the control population, 35 infants (70%) had been breast-fed from birth. At 7 months of age 14% of the infants infected with rotavirus and 18% of control infants were still breast-fed.

Leventhal *et al.* (1986) performed a study on 281 matched case-control pairs of infants who were hospitalized for infections under the age of 3 months. The infections included fever/viral syndrome, aseptic meningitis, gastroenteritis, bronchiolitis, pneumonia, otitis media, bacteremia, bacterial meningitis and others. There were 50 pairs matched for gastroenteritis. Six (12%) of the 50 cases were being breast-fed at the time of onset of the infection that led to admission to hospital, whereas 11/50 (22%) of the controls were breast-fed. The association between breast-feeding and the incidence of hospital admissions for gastroenteritis was not statistically significant (OR = 0.29, 95% CI 0.06–1.37, $p = 0.10$). Multivariate logistic regression analysis was used to correct for possible confounding factors, but the dependent variable used was hospital admissions for all the infections combined. The absence of regression modelling based on hospital admissions for gastroenteritis, alone, as the dependent variable made it impossible to analyze the data for confounding effects in relation to gastrointestinal illness.

3.2. Breast-feeding and Respiratory Tract Infection

Both upper and lower respiratory tract infections are included in this review. We have also included acute otitis media in this group of infections as this condition frequently occurs together with upper respiratory tract infections (Paradise, 1981). Sixteen articles were selected (Table 2), and one article provided data on two separate studies (Pisacane *et al.*, 1994).

3.2.1. Cohort Studies. Six studies on gastrointestinal infections also provided relevant data for respiratory infections (RCGP Report, 1972; Fergusson *et al.*, 1978; Cunningham, 1979; Holmes *et al.*, 1983; Agre, 1985; Howie *et al.*, 1990).

In the RCGP study (1972), the incidence of respiratory infections amongst breast-fed infants (152/223) was not significantly different from that in bottle-fed infants (82/111, $\chi^2 = 6.78$, $p > 0.05$). The definition of respiratory infections in this report was unclear and the data were not analyzed for potential confounding.

In the Christchurch Child Development Study (Fergusson *et al.*, 1978), respiratory illness was defined as cough, coryza, bronchitis, bronchiolitis, pneumonia or otitis media. Symptoms of these illnesses were present in 165/282 (58%) of exclusively bottle-fed infants, 303/518 (58%) of supplement-fed, 133/278 (48%) of near breast-fed infants and 68/132 (52%) of exclusively breast-fed infants ($\chi^2 = 10.3$, $p < 0.05$). Multiple linear regression analysis indicated that the association between risk of respiratory symptoms and mode of infant feeding was significantly affected by the confounding effects of differences in maternal education, single parenting and living standards.

In the cohort study of Cunningham (1979) in New York, USA, otitis media and lower respiratory tract infections were used as the outcome events. These infections were not clearly defined. Each episode was considered in the context of the total number of infant-weeks in a given feeding mode (breast-fed or artificially-fed). The method of calculating the total number of infant-weeks for each feeding mode has been described in an earlier section of this chapter. There

Table 2. Studies of Breast-Feeding Practices and Respiratory Infection* During Infancy in Industrialized Countries

Reference	Country of origin	Recruitment source	Adjustment for confounding	Breast-feeding clearly defined	Outcome event clearly defined	Infective pathogen defined
Cohort Studies						
RCGP Report, 1972	England	Community	No	Yes	No	No
Fergusson *et al.*, 1978	New Zealand	Community	Yes	Yes	Yes	No
Cunningham, 1979	United States	Community	No***	Yes	No	No
Watkins, 1979	England	Community	No	Yes	No	No
Frank, 1982	United States	Community	No	No	Yes	RSV**
Saarinen, 1982	Finland	Community	No	Yes	Yes	No
Holmes *et al.*, 1983	United States	Community	No***	Yes	No	No
Forman, 1984	United States	Community	Yes	Yes	No	No
Agre, 1985	United States	Community	No	Yes	No	No
Howie, 1990	Scotland	Community	Yes	Yes	Yes	No
Duncan, 1993	United States	Community	Yes	Yes	No	No
Nafstad, 1996	Norway	Community	Yes	Yes	No	No
Case-control Studies						
Downham, 1976	England	Hospital	No	No	Yes	RSV**
Pullan, 1980	England	Hospital	Yes	No	Yes	RSV**
Leventhal *et al.*, 1986	United States	Hospital	No***	Yes	No	No
Pisacane, 1994	Italy	Hospital	Yes	Yes	Yes	No

* Acute otitis media, upper and lower respiratory tract infections. ** RSV: respiratory syncytial virus. *** Although multivariate analysis was performed, the dependent variable was a combination of respiratory and gastrointestinal infections.

were significantly fewer episodes of otitis media in the breast-fed group (23 vs. 182 episodes per 1,000 infant-weeks, $p < 0.01$). Breast-feeding was also associated with a lower number of episodes of lower respiratory infections (7 vs. 112 episodes per 1,000 infant-weeks; $p < 0.01$). Although stratified analysis was used to examine the effect of other factors such as gender, paternal education, maternal age, presence or absence of older siblings and birth weight percentile, the dependent variable was the frequency of all illnesses (including the non-respiratory illnesses) during the first year of life. Therefore, the multivariate analysis used to assess confounding did not apply to respiratory infection, alone.

Holmes *et al.* (1983) studied the relationship between feeding mode, acute otitis media and lower respiratory tract infections. A cohort of 251 infants in the USA was followed throughout the first year of life. Among 124 formula-fed infants, 15 (12%) had one of these infections compared to 5/127 (4%) breast-fed infants ($p = 0.03$). This study also used other infections (including gastrointestinal, urinary tract, and skin) as outcome events. Although stepwise linear regression analysis was used to analyze for confounding, the dependent variable used in the model was all infantile infections. Hence, the analysis of confounding cannot be applied to the two respiratory infections, alone.

In a study conducted in New York, USA (Agre, 1985) the reported frequency of lower respiratory tract infections by mode of feeding was 0.10, 0.02 and 0.12 episodes per patient per year for breast-fed, breast- and bottle-fed and bottle-fed infants, respectively. The diagnostic criteria for lower respiratory tract infection were not cited, and the frequency of infections was too low to permit a meaningful statistical analysis.

In a cohort study carried out in Dundee, UK, respiratory infection was defined on the basis of coryza accompanied by cough, wheeze, or both, lasting ≥48 h (Howie *et al.*, 1990). After controlling for the confounding effects of paternal social class, maternal age and parental smoking, the adjusted rate of respiratory infections during the first 13 weeks of life was significantly higher in the bottle-fed group (37.0%) than in the partially (24.2%) and fully (25.6%) breast-fed infants ($p < 0.05$). Between 14–26 and 27–39 weeks of age, the difference among the three groups of infants in the adjusted rates of these infections was not statistically significant. However, the adjusted rate of respiratory infection between 40–52 weeks of age was found to be significantly higher in the bottle-feeders (53.3%) than in the partial (38.6%) and full breast-feeders (44.2%), $p < 0.05$.

Six cohort studies have been reported on the association between breast-feeding and respiratory tract infections, alone. These include the reports of Watkins *et al.* (1979), Frank *et al.* (1982), Saarinen (1982), Forman *et al.* (1984), Duncan *et al.* (1993) and Nafstad *et al.* (1996).

Watkins *et al.* (1979) carried out a prospective study in a residential suburb of London, UK. Two thousand and seventy four infants were followed up by their health visitors at home within 14 days of birth and at the end of their first year. The incidence of bronchitis and pneumonia was lower in infants who were breast-fed than in those who were exclusively bottle-fed (8.1 cases of infection annually per 100 infants vs. 14.8; $\chi^2 = 20.35$; $p < 0.01$). The modes of infant feeding and diagnosis of bronchitis and pneumonia were not clearly described, and mothers who breast-fed their infants were younger and less likely to be smokers. In addition, assessment of possible confounding was not reported in this study.

The Houston Family Study in the USA (Frank *et al.*, 1982) enrolled 42 infants who were bottle-fed from birth and 39 infants who had been breast-fed for at least 3 months. These infants were followed up prospectively by weekly contact with their mothers, and home visits were also made at times of respiratory symptoms. In addition, the health records of the infants were examined to ascertain clinic visits for illnesses. The rates of upper respiratory tract infections (afebrile and febrile, otitis media) and lower tract infections (laryngotracheobronchitis, bronchiolitis, pneumonia) did not differ in the two groups either during the first 6 months of life (310 vs. 390 infections per 100 infants for breast- and bottle-fed infants respectively) or during the second 6 months of life (509 vs. 492 infections). The lack of difference between groups was apparent even in the case of infections in which a viral agent was confirmed by serological testing and isolation from nasal wash specimens.

In a prospective study conducted in Helsinki, Finland, 222 infants were followed up until the age of one year. The incidence of otitis media among infants who were breast-fed exclusively for at least 6 months was lower than that among those who were fed exclusively with cow's milk-based formulas from birth or who were weaned before the age of 2 months (6% vs. 19%, respectively) (Saarinen, 1982). In this study, the diagnosis of otitis media was made by otoscopic examination for redness and inflammation of the tympanic membrane, loss of translucency and landmarks of the ear drum or perforation of the ear membrane with pus draining from it.

Forman *et al.* (1984) conducted a retrospective study on 571 Pima and Papago Indian infants on the Gila River Reservation in Arizona, USA. Mothers were interviewed and hospital records for the children were examined for information pertaining to infectious disease episodes. Infant feeding was categorized as follows: (a) breast-fed exclusively until 5 months, at which time solid diet was introduced, (b) mixed breast- and bottle-fed before or by 3 months of age, at which time solids were introduced, or (c) bottle-fed exclusively for 4 months, at which time solids were introduced. After multivariate logistic regression analysis to adjust for birth weight, birth year, sex, birth order, seasonality, maternal age at birth and socioeconomic status, the adjusted odds ratio for an upper respiratory infection among exclusively breast-fed infants was less than unity, albeit not significantly so (OR = 0.63, 95% CI 0.38–1.08, $p = 0.06$). Likewise, the adjusted odds ratio for pneumonia among exclusively breast-fed infants was less than unity but was not statistically significant (OR = 0.67, 95% CI 0.37–1.19, p = 0.18).

In a community based study conducted in Tucson, Arizona, USA, 1013 newborns were followed up for acute otitis media during their entire first year of life (Duncan *et al.*, 1993). Data were incomplete for 76 cases (8%). No specific criteria were established for the diagnosis of acute otitis media. Infant feeding was categorized as (a) artificial feeding, (b) breast-feeding <4 months, (c) breast-feeding ≥4 months with supplemental formula or foods prior to 4 months, (d) breast-feeding ≥4 months with supplementation beginning between 4 and 6 months and (e) exclusive breast-feeding ≥6 months. The cumulative incidence of acute otitis media by 12 months of age in the infants who were exclusively breast-fed for 4 months was 56.0% compared to 68.3% for those who were not ($p < 0.01$). Logistic regression modelling was used to assess the confounding effect of gender, day care, presence of one or more siblings at home, exposure to maternal smoking of ≥11 cigarettes per day and physician-diagnosed history of parental hay fever. The risk of acute otitis media were significantly reduced in those infants who were exclusively breast-fed ≥6 months compared to those who were artificially fed (OR = 0.61; 95% CI 0.40–0.92).

Nafstad *et al.* (1996) followed up 3754 healthy Norwegian infants with birth weights greater than 2 kg. Postal questionnaires were sent to each child's family when the child was 6 and 12 months of age. The primary outcome measure was an episode of lower respiratory tract infection such as pneumonia, bronchitis or bronchiolitis. Information on respiratory health was obtained on 3238 infants (86%), and information on the duration of breast-feeding was obtained for 3189 infants (85%). Only 30 infants (1%) were not breast-fed at all, and 837 (26%) were still breast-fed at age 12 months. Logistic regression analysis to assess the effects of gender, number of siblings, sharing of bedrooms, day-care attendance, maternal education, parental asthma, maternal and paternal smoking habits, increased the adjusted OR of lower respiratory tract infections by 5% (95% CI 2–8%) for each month of decrease in the duration of breast-feeding. Artificial feeding was associated with a significantly higher OR for infection than 12 months of breast-feeding (adjusted OR: 1.7; 95% CI 1.2–2.5).

3.2.2. Case-Control Studies. In a study carried out in Newcastle, UK, 8 out of 115 infants (7%) hospitalized with respiratory syncytial virus infection were found to have been breast-fed up to the age of 4 months whereas 46 out of 167 controls (28%) had a similar history of breast-feeding ($\chi^2 = 18.65$; $P < 0.01$) (Downham *et al.*, 1976). No analysis was made for confounding factors.

A similar study was conducted on infants under the age of 6 months in Newcastle, UK (Pullan *et al.*, 1980). Breast-feeding was less common in cases admitted with respiratory syncytial virus infection (38/127) than in age-matched controls (239/503) ($\chi^2 = 13.93$; $p < 0.01$). In this study, the quality of maternal care, maternal smoking and the presence of another child in the same room at night were significant confounding factors.

The study by Leventhal *et al.* (1986) is described in this chapter under gastrointestinal infections. Out of the 48 pairs matched for bronchiolitis and pneumonia, 13 cases (27%) were being breast-fed at the time of onset of the infection that led to admission to hospital, whereas 21 (44%) of the controls were being breast-fed. No difference was apparent between breast-fed infants and controls in the incidence of hospital admissions for bronchiolitis and pneumonia (OR = 0.48, 95% CI 0.18–1.22, p = 0.14). Although multivariate logistic regression analysis was used to assess possible confounding, the dependent variable was hospital admissions for all infections. Hence, it was not possible to assess confounding for respiratory infections, alone.

Pisacane *et al.* (1994) conducted two studies in Naples, Italy. The first involved a group of 74 infants, aged 0–6 months, admitted with a clinical and radiological diagnosis of pneumonia or bronchiolitis. The control group consisted of 73 infants matched for age and date of admission to the hospital for treatment of other conditions. Thirty-two cases (44%) and 18 controls (25%) had never been breast-fed (OR = 0.42, 95% CI 0.19–0.90, p < 0.05), and there was no evidence of confounding in the stratified analysis. The second study involved a group of 88 infants 0–12 months of age who were admitted with a clinical diagnosis of pertussis. Eighty-eight infants matched for age, gender and month of admission served as a control group. The proportions of infants who have never been breast-fed were similar in the two groups (26% vs. 31% for cases and controls, respectively), and stratified analysis did not reveal any confounding.

3.3. Breast-Feeding and Primary Invasive *Haemophilus influenzae* Infection

Haemophilus influenzae type b (Hib) is a leading cause of bacterial meningitis in some countries, including the United States and Sweden (Cochi *et al.*, 1985b; Silfverdal *et al.*, 1997). It is also the aetiological agent for other non-meningitic invasive diseases such as epiglottitis, arthritis, pneumonia and cellulitis (Granoff and Basden, 1980). Two case-control studies were found relevant to this review (Table 3).

3.3.1. Case-Control Studies. Cochi *et al.* (1986) conducted a study of 89 children less than 5 years of age residing in Atlanta, USA from whom Hib was cultured from a normally sterile body site and who had clinical signs of infection. The results were compared with those of 503 control children of the same age residing in the area. Lack of a clear definition of breast-feeding in this study

Table 3. Studies of Breast-Feeding and the Incidence of Primary Invasive *Haemophilus influenzae* and Urinary Tract Infections During Infancy in Industralized Countries

Reference	Country of origin	Recruitment source	Adjustment for confounding	Breast-feeding clearly defined	Outcome event clearly defined	Infective pathogen defined
Primary Invasive *Haemophilus influenzae* Infection*						
Case-Control Studies						
Lum *et al.*, 1982	United States	Community	No	No	No	*H. influenzae*
Cochi *et al.*, 1986	United States	Community	Yes	No	Yes	*H. influenzae*
Urinary Tract Infection						
Case-Control Study						
Pisacane *et al.*, 1992	Italy	Hospital	No	Yes	Yes	No

*Diagnosed in infants with clinical signs of infection from whom *H. influenzae* was cultured from a normally sterile body site.

has raised concern about the validity of the results (Rubin *et al.*, 1987). Multivariate logistic regression modelling was restricted to children in their first year of life. After controlling for differences in race, daycare or nursery attendance, household size and family income, breast-feeding among infants aged 2–5 months of age was associated with a lower risk of Hib infection (OR = 0.08; 95% CI 0.01–0.59; $p < 0.007$). In infants aged 6–11 months, no association was apparent between breast-feeding and the risk of Hib infection (OR = 0.68; 95% CI 0.15–3.20; $p > 0.05$).

In a study of Alaskan Eskimos, Lum *et al.* (1982) compared 19 infants with invasive Hib infection and 38 controls matched for age and village of residence. Eskimos are known to have a high incidence of invasive Hib infection; 3% develop bacteraemia or meningitis in the first year of life. Breast-feeding was considered to be the mode of feeding if the subject was predominantly breast-fed until an age equal to that of the case at the time of hospitalization. Using this definition, breast-feeding was significantly less prevalent among cases (1/19) than among controls (18/38) ($p < 0.01$). The authors concluded that breast-feeding reduces the risk of developing invasive Hib infection.

3.4. Breast-feeding and Urinary Tract Infection

3.4.1. Case-Control Study. Only one case-control study was identified in which the relationship between infant feeding mode and the incidence of urinary tract infection was investigated (Table 3). Pisacane *et al.* (1992) performed a study in Naples, Italy, involving128 infants less than 6 months of age who were hospitalized with a urinary tract infection. Infection was ascertained on the basis of two urine cultures each of which yielded $>10^5$ colonies ml^{-1} of a single bacterial species. One-hundred and twenty-eight control infants were also examined having been hospitalized with an acute illness but a negative urine culture. The control subjects were matched to cases by gender, age and month and year of hospital admission. Infants with diagnoses of acute diarrhea or respiratory infection were not included in the control group. Breast-fed infants had a relative risk of urinary tract infection of 0.38 (95% CI 0.22–0.65) when compared to exclusively bottle-fed infants.

3.5. Summary of Studies in Industrialized Countries

Studies conducted in industrialized countries have produced conflicting results on the effect of breast-feeding on gastrointestinal and respiratory infections. Only 9/16 studies (56%) yielded a statistically significant protective effect of breast-feeding on gastrointestinal infections. Among 16 studies on respiratory infection, only 8 (50%) signified a beneficial effect of breast-feeding. Previous studies have also yielded inconsistent results (Bauchner *et al.*, 1986).

It is noteworthy that a majority of studies suffer from methodological deficiencies such as detection bias, lack of analysis for confounding factors and lack of a clear definition of infant feeding practices and infectious outcomes. Confounding was evaluated in only 4 out of 16 studies on gastrointestinal infections, and Table 4 shows that 3 out of these 4 studies reported a beneficial effect of breast-feeding. The proportion of studies on respiratory infections involving assessment of confounding was higher (7/16, 44%), and breast-feeding was found to be protective against respiratory infections in only 4 out of these 7 studies (Table 5). A high methodological standard was present in 1 of the 2 studies on primary invasive Hib. In this study the risk of infection was found to be significantly lower in breast-fed infants aged 2–5 months but not in those 6–11 months old (Table 6). No analysis for confounding was employed in the single study on urinary tract infection.

There were 4 studies on gastrointestinal infection which met all the above methodological criteria (Fergusson *et al.*, 1978; Duffy *et al.*, 1986; Eaton-Davies and Dugdale, 1987; Howie *et al.*, 1990). In three studies, breast-feeding was reported to protect against gastrointestinal infection (Fergusson *et al.*, 1978; Eaton-Davies and Dugdale, 1987; Howie *et al.*, 1990). There were three studies on respiratory infection (Fergusson *et al.*, 1978; Howie *et al.*, 1990; Pisacane *et al.*, 1994). Two yielded evidence for a beneficial effect of breast-feeding.

A heterogeneous estimate of the pooled odds ratio was obtained for the association between breast-feeding and risk of gastrointestinal infection in the cohort studies. The typical odds ratio was 0.36 (95% CI 0.32–0.41, $p < 0.01$, heterogeneity $p < 0.01$) (Figure 1). The risk of gastrointestinal infection associated with breast-feeding as compared to exclusive artificial feeding was examined in two case-control studies. The homogeneous odds ratio was 0.54 (95% CI 0.36–0.80, heterogeniety $p = 0.35$) (Figure 1). The pooled estimates of the odds ratio for the effect of breast-feeding on the risk of respiratory tract infections were heterogeneous. The typical odds ratio was 0.56 (95% CI 0.49–0.64, $p < 0.01$; heterogeneity $p < 0.01$) for cohort and 0.46 (95% CI 0.36–0.57, $p < 0.01$; heterogeneity $p < 0.01$) for case-control studies (Figure 2). Given that there are major differences in design among the various studies, it is not surprising that there are significant inconsistencies in the meta-analyses. Nevertheless, presentation of the data in this way shows a distinct trend towards protection from gastrointestinal and respiratory infections by breast milk.

4. Non-Industrialized Countries

Ten studies were identified relating to the association between breast-feeding and gastrointestinal infections in non-industrialized countries (Table 7). Three studies also covered respiratory tract infection (Table 7). No studies on

Table 4. Results of Multivariate Analysis in Studies of Breast-Feeding and Gastrointestinal Infection During Infancy in Industrialized Countries

Reference	Adjustment for confounding	Dependent variable	Co-variates	Conclusions and results of multivariate analysis
Fergusson *et al.*, 1978	Multiple linear regression	Vomiting & diarrhea	Maternal education, single or dual parenting, family living	Breast-feeding protective Total/near breast = 1, $r = 0$, RC = -0.25, $f = 77.2$, standard $p < 0.01$
Duffy *et al.*, 1986	Multiple linear regression	Rotavirus gastroenteritis	Infant sex, race, occupation, education, mother marital status	Breast-feeding not protective
Eaton-Evans *et al.*, 1987	Multi-factorial analysis	Vomiting & diarrhea	Paternal social class, maternal parity	Breast-feeding protective up to 6 months of age
Howie *et al.*, 1990	Multiple logistic regression	Gastrointestinal infection	Paternal social class, maternal age, parental smoking	Breast-feeding protective Bottle vs. breast-feeders χ^2 for adj. rates = 11.89, 95% CI = 6.6–16.8, $p < 0.01$

RC = regression coefficient.

Table 5. Results of Multivariate Analysis in Studies of Breast-Feeding and Respiratory Infection During Infancy in Industrialized Countries

Reference	Adjustment for Confounding	Dependent Variable	Co-variates	Conclusions and Results of Multivariate Analysis
Fergusson *et al.*, 1978	Multiple linear regression	Vomiting & diarrhea	Maternal education, single or dual parenting, family living standard	Breast-feeding not protective Total/near breast = 1, r = 0, RC = –0.05, f = 3.34, p > 0.05
Pullan *et al.*, 1980	Multiple logistic regression	RSV infection	Maternal care score, maternal smoking, presence or absence of another child in the same bedroom	Breast-feeding not protective No data for regression analysis
Forman *et al.*, 1984	Multiple logistic regression	URTI (including otitis media)	Gender, socioeconomic status, birth order, seasonality, birth weight, maternal age	Breast-feeding not protective Breast- and bottle-fed—adj. OR = 1.13, 95% CI = 0.59–1.09 Bottle-fed—adj. OR = 0.63, 95% CI = 0.38–1.08
Howie *et al.*, 1990	Multiple logistic regression	Respiratory infection	Paternal social class, maternal age, parental smoking	Breast-feeding protective Bottle vs. breast-feeders χ^2 for adj. rates = 6.35, 95% CI = 3.9–20.3, p < 0.05
Duncan *et al.*, 1993	Multiple logistic regression	Acute otitis media	Gender, use of day care, ≥1 sibling, maternal smoking, parental allergy	Breast-feeding protective Bottle/breast-feeding <4 mth—adj. OR = 1.0 Breast-feeding ≥6 mth—adj. OR = 0.61, 95% CI = 0.40–0.92
Pisacane *et al.*, 1994	Stratified analysis	Pneumonia/ bronchiolitis	No confounding present	Breast-feeding protective
		Pertussis-like illness	No confounding present	Breast-feeding not protective
Nafstad *et al.*, 1996	Multiple logistic regression	Pneumonia, bronchitis,	Gender, number of siblings, sharing of bedroom, day-care attendance, maternal education, parental asthma and smoking	Breast-feeding protective No breast-feeding vs. 12 months breast-feeding—adj. OR = 1.7, 95%, CI = 1.2–2.5

RC = regression coefficient; OR = odds ratio; URTI = upper respiratory tract infection.

Table 6. Results of Multivariate Analysis in Studies of Breast-Feeding and Primary *Haemophilus influenzae* Infection During Infancy in Industrialized Countries

Reference	Adjustment for confounding	Dependent variable	Co-variates	Conclusions and results of multivariate analysis
Cochi *et al.*, 1986	Multiple logistic regression	Primary *H. influenzae* Infection	Day-care or nursery attendance, race, household size, family income	Breast-feeding protective in infants aged 2–5 months Infant aged 2–5 months—adj. OR = 0.08, 95% CI = 0.01–0.59, p < 0.007 Infant aged 6–11 months—adj. OR = 0.68, 95% CI = 0.15–3.2, p > 0.05

OR = odds ratio.

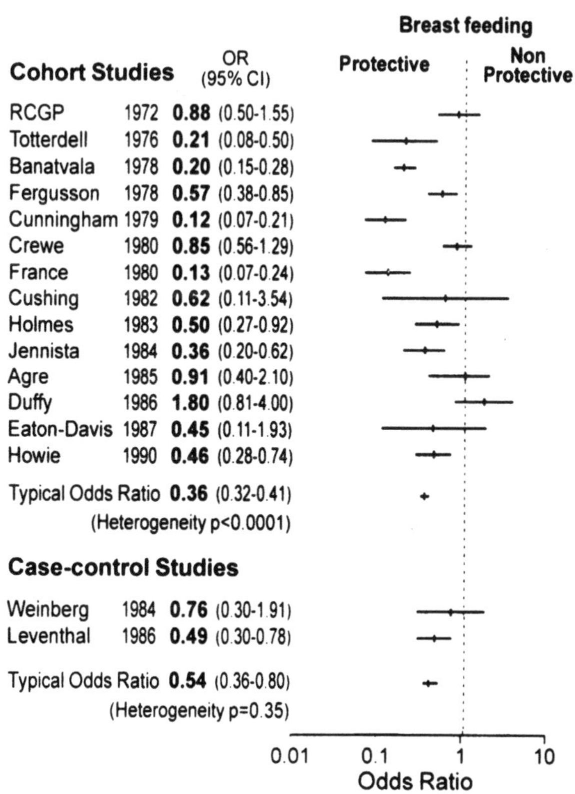

Figure 1. Meta-analysis of the effect of breast-feeding on gastrointestinal infection during infancy in industrialized countries.

Table 7. Studies of Breast-Feeding and Gastrointestinal* and Respiratory** Infections During Infancy in Non-Industrialized Countries

Reference	Country of origin	Recruitment source	Adjustment for confounding	Breast-feeding clearly defined	Outcome event clearly defined	Infective pathogen defined
GASTROINTESTINAL INFECTION*						
Cohort Studies						
Jayasuriya and Soysa, 1974	Sri Lanka	Hospital	No	Yes	No	No
Patel *et al.*, 1981	India	Hospital	No	Yes	No	No
Clavano, 1982	Phillipines	Hospital	No	Yes	Yes	No
Brown *et al.*, 1989	Peru	Community	No	Yes	Yes	No
VanDerslice *et al.*, 1994	Phillipines	Community	Yes	Yes	Yes	No
Long *et al.*, 1994	Mexico	Community	Yes	Yes	Yes	LT—*E. Coli*
Lopez-Alarcon *et al.*, 1997	Mexico	Community	Yes	Yes	Yes	No
Case-Control Studies						
Gunn *et al.*, 1979	Bahrain	Community	No	Yes	Yes	*V. cholerae*
Mittal, 1983	India	Hospital	No	No	Yes	LT/ST—*E. coli, Salmonella, Klebsiella, Shigella*
Richardson *et al.*, 1983	South Africa	Hospital	No	Yes	Yes	*C. jejuni, E. coli*
RESPIRATORY INFECTION**						
Cohort Studies						
Patel *et al.*, 1981	India	Hospital	No	Yes	No	No
Brown *et al.*, 1989	Peru	Community	No	Yes	Yes	No
Lopez-Alarcon *et al.*, 1997	Mexico	Community	Yes	Yes	Yes	No

LT = heat-labile enterotoxin; ST = heat-stable enterotoxin. * Diarrhea ± vomiting. ** Acute otitis media, upper and lower respiratory tract infections.

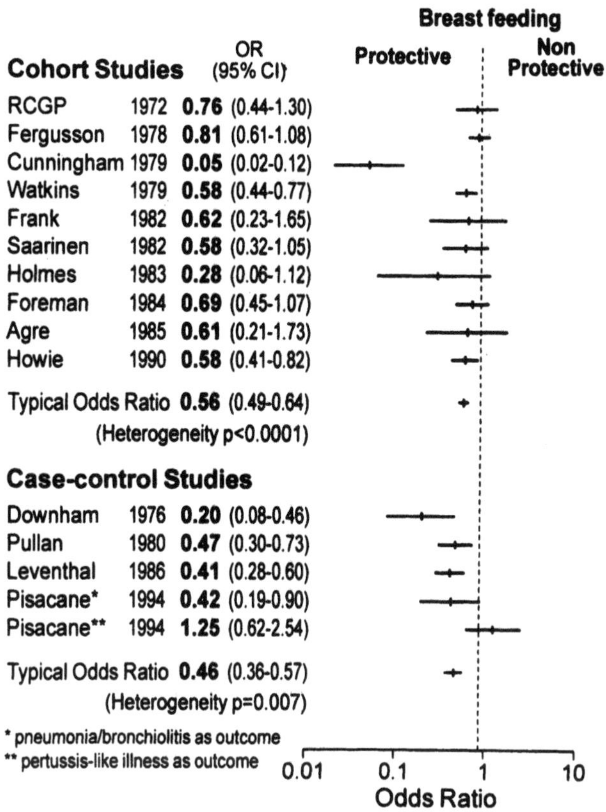

Figure 2. Meta-analysis of the effect of breast-feeding on respiratory infection during infancy in industrialized countries. Studies by Duncan *et al.* (1993) and Nafstad *et al.* (1996) were not included because the data required to calculate the odds ratio could not be extracted.

primary invasive *Haemophilus influenzae* or on urinary tract infections met the criteria for inclusion in this review.

4.1. Breast-Feeding and Gastrointestinal Infections

Initially, eleven articles were selected for review. Two studies (Popkin *et al.*, 1990; VanDerslice *et al.*, 1994) were of the same study population. Consequently, the article by Popkins *et al.* (1990) is not included in this review.

4.1.1. Cohort Studies. Jayasuriya and Soysa (1974) conducted a prospective study on 112 babies admitted to a nursery with birth weights between 1250 and 2057 g. Nineteen (17%) received breast milk and the remaining 93

(83%) were formula-fed. Diarrhea was absent in the breast-fed babies but present in 23/93 (25%) of the artificially-fed infants. The definition of diarrhea was not clearly reported, and there was no analysis for confounding factors.

Patel *et al.* (1981) followed a cohort of 200 Indian low birthweight babies (1–1.8 kg) admitted to a special care nursery. One hundred babies were exclusively breast-fed and the other 100 were exclusively formula-fed. The incidence of gastroenteritis, for which no clear definition was given, was 12/100 (12%) in the breast-fed and 15/100 (15%) in the artificially-fed groups ($p > 0.05$), and no analysis for confounding factors was performed.

Clavano *et al.* (1982) followed a cohort of 9886 infants for the occurrence of diarrhea during the first month of life in Baguio City, Philippines. Diarrhea was defined on the basis of frequent passage of watery stools or recurrent and protracted watery stools, but no information was given regarding the cut-off levels for frequency or recurrence. Information on infant feeding patterns in 9622 cases (97%) was obtained from the case notes prior to discharge from the hospital. Infant feeding was categorized as either exclusively breast-fed, mixed-fed or exclusively formula-fed. Of the 138 infants with diarrhea, 124 (90%) were formula-fed, 8 (6%) were mixed-fed and 6 (4%) were breast-fed. In 9484 infants without diarrhea, 2479 (26%) were formula-fed, 603 (6%) were mixed- fed and 6402 (68%) were breast-fed.

Brown *et al.* (1989) conducted a prospective study in Lima, Peru, on 153 singleton newborns from birth until the age of 12 months. The infants were were subjected to thrice weekly community-based surveillance for diarrheal and other diseases. Although upper and lower respiratory tract infections were also used as outcome events, only the data for diarrhea (number of new episodes per 100 days of observation) were presented. Diarrhea was defined according to the following age-specific criteria: infants <1 month of age, more than five stools per day; infants 1 month of age, more than four stools per day; and infants ≥2 months of age more than three stools per day. An episode of diarrhea was considered to comprise all "continuous" days with diarrhea that were not separated by at least 2 diarrhea-free days. Infant feeding was categorized as (a) exclusive breast-feeding, (b) breast-feeding plus other liquids but no other milk or solids, (c) breast-feeding and artificial milk with or without other liquids, but not solids, (d) breast-feeding and solids with or without other types of milk and liquids and (e) exclusively artificial feeding with or without other liquids and solids. The incidence of diarrhea was determined for subgroups of infants aged 0–2, 3–5, 6–8, and 9–11 months. Infants who were exclusively artificially fed had a higher incidence of diarrhea than their exclusively breast-fed counterparts for the age subgroups 0–2 months (5.30 vs. 1.89 episodes per 100 days of observation, $RR = 2.80$, $p < 0.01$) and 3–5 months (4.46 vs. 1.47 episodes, $RR = 3.05$, $p < 0.01$). For the subgroups 6–8 and 9–11 months, the comparison was possible only between infants who were exclusively artificially fed and those who received breast milk and solids. This was because 83% of the study population was consuming solid food by the age of 6 months. In these two subgroups, the incidence of diarrhea was higher in infants aged 6–8 months who were exclusively artificially-fed than in those of the same age group who received breast milk and solids (3.83 vs. 2.59 episodes per 100 days of observation, RR = 1.48, $p < 0.01$). In contrast, no difference associated with feeding practice was apparent in the group aged 9–11 months (3.30 vs. 2.88 episodes, RR = 1.15, $p > 0.05$). Although microbiologic evaluation for known enteric pathogens was performed in all cases of diarrhea, there was no clear relationship between feeding practice and the incidence of specific infectious causes of diarrhea.

In a study in the Philippines, a cohort of 3080 children living in Cebu City was studied from birth until the age of 2 years (VanDerslice *et al.*, 1994). The mothers of these children were interviewed every 2 months. Only data for diarrheal disease during the first 6 months of life were used. Attrition due to migration, death, or refusal to participate reduced the sample size to 1963 (64%) at the end of 6 months. Diarrhea was defined as frequent and watery stools. Feeding mode was classified as exclusively breast-fed, breast-fed and given only non-nutritive liquids (NNL) such as teas, brews or plain water, mixed- fed (nutritive foods and/or liquids in addition to breast milk), or exclusively artificially-fed. Mode of feeding was assessed by 24 h recall during the interviews. After 6 months of age, the prevalence of diarrhoea was 13% in infants who were exclusively breast-fed, 10% in those who were breast-fed + NNL, 20% in mixed-fed infants and 24% in those not breast-fed at all. After analysis for influences by factors such as quality of drinking water, community sanitation, presence or absence of in-house water supply, private excreta disposal and excreta in yard, the protective effect of full breast-feeding relative to artificial feeding was found to be large and statistically significant.

A cohort study was carried out in poor neighbourhoods of Guadalajara, Mexico (Long *et al.*, 1994). Ninety-eight mother-infant pairs were enrolled upon the birth of a child and were visited twice weekly by the project nurses until the infants were 12 months of age. Diarrhea was defined as ≥3 bowel movements with a liquid or semiliquid consistency in 24 h and/or the presence of blood or mucus in the stools as reported by the mother and confirmed by the project personnel. An episode of diarrhea was considered to have ended when an infant showed no symptoms for 72 h and the stools were normal in consistency. Weekly stool samples were taken from each infant for isolation of heat-labile enterotoxigenic *Escherichia coli* (LT-ETEC), and samples were taken every 2–3 days if a diarrheal episode was reported. The infants were classified as exclusively breast-fed, breast and formula-fed (receiving supplemental foods in addition to breast milk), or exclusively formula-fed. The exclusively formula-fed infants had a diarrheal incidence (number of episodes per child-year) three times higher than those exclusively breast-fed and twice as high as breast- and formula-fed infants. None of the stools from infants exclusively breast-fed were positive for LT-ETEC, as opposed to approximately 10% of the stools of breast- and formula-fed infants and 22% of those from infants exclusively formula-fed. Because of a steady loss of mother-infant pairs during the study period, multivariate modelling using proportional hazards (Cox regression) was used to assess confounding. Individuals who completed the entire follow-up and those who had limited follow-up due to migration out of the study area were included. Analysis was made for socioeconomic covariates such as parental age and education, household income, source of drinking water, type of bathroom, and number and types of animals in the household. The risk of LT-ETEC colonization was increased among infants who consumed *avena*, a barley drink (adjusted OR = 4.49, 95% CI 3.04–5.93, $p = 0.04$) and during the rainy season (adjusted OR = 5.41, 95% CI 4.81–6.00, $p < 0.01$).

López-Alarcón *et al.* (1997) recruited 216 infants into a prospective study in the community of Iztapalapa, a slum neighbourhood located in east Mexico City. Of these infants, 170 (79%) were followed from birth till 6 months of age. Infants were selected that were singleton, full term and weighed >2500 g at birth. Subjects were excluded if they had congenital malformations, or if they moved out of the neighbourhood during their first 6 months of life. Diarrhea was defined as ≥3 liquid or semi-liquid stools per day regardless of accompanying blood, mucus, or fever. For inclusion, the total number of stools had to exceed the usual number of daily bowel movements. An episode was defined as an event separated from another event by at least 2 days. Infants were classified as (a) fully breast-fed, (b) fully formula-fed, or (c) partially breast-fed. Infants were examined and mothers were interviewed every 2 weeks, alternately at the clinic and at home. The confounding effects of years of maternal education, maternal marital status, presence or absence of siblings, number of individuals sleeping in the same room and socioeconomic level (based on the availability of tap water, sewage, refrigerator and paved floor) were assessed by multiple linear regression. After accounting for confounding, the number of episodes of diarrhea was negatively correlated with the duration of breast-feeding ($r = -0.17$, $p < 0.02$).

4.1.2. Case-Control Studies. In a study conducted in Bahrain, Gunn *et al.* (1979) selected 61 culture-confirmed cases (<1 year of age) of cholera caused by *Vibrio choleræ*, biotype El Tor, serotype Ogawa and a similar number of controls matched for age (0–2, 3–6, and 7–11 months) and place of residence (urban or rural). Information on mode of feeding was obtained by interviews with the mothers. Infant feeding was classified as exclusively breast-fed, exclusively bottle-fed or a combination of both. When infants took both types of feed, mothers were asked which was the principle source (>50% intake). Bottle-feeding was significantly more common in cases than in controls (31/40, 78% vs. 15/40, 38%; relative risk = 9.0, $p < 0.01$).

In another study in India, Mittal *et al.* (1983) enrolled 148 infants hospitalized with acute diarrheal illness and 35 age- and socioeconomically-matched controls. The feeding pattern was determined by direct interview with the mothers, and was categorized as either breast- or bottle-feeding. In the diarrheal group, the incidence of artificial feeding was 89/148 (60%) compared to 13/35 (37%) in the control group, and this difference was statistically significant ($p < 0.01$). There was no definition of diarrhea or analysis for confounding factors.

Richardson *et al.* (1983) compared 111 children ≤2 years of age hospitalized with diarrhea in Johannesburg, South Africa, with a control group of 111 healthy children matched for age and sex. Stool specimens were obtained from all cases and controls for isolation of *Campylobacter jejuni*, enterotoxic *Escherichia coli*, *Salmonella* and *Shigella* species. Among the infants ≤8 months of age, partial or exclusive breast-feeding was significantly less frequent among the babies in the diarrheal group than in the controls (31/60, 52% vs. 46/60, 77%; $p < 0.01$). *C. jejuni* was isolated as the sole intestinal pathogen in 10/31 (32%) of the diarrheal cases, but was present in only 1/41 (2%) of the control group. Breast-feeding not only protected against colonization with *C. jejuni*, but was associated with a significant reduction in the incidence of diarrhea caused by other infective agents. No assessment of possible confounding factors was carried out.

4.2. Breast-Feeding and Respiratory Infections

Three cohort studies are available relating breast-feeding and respiratory infections (Table 7).

In a study conducted by Patel *et al.* (1981) in India, the incidence of lower respiratory tract infection was 15/100 (15%) in breast-fed and 19/100 (19%) in artificially-fed groups ($p > 0.05$). A clear definition of the outcome event (i.e., lower respiratory tract infection) was not included in the report.

In the study by Brown *et al.* (1989) conducted in Lima, Peru, upper respiratory tract infection was defined as the presence of coryza or purulent rhinitis in addition to cough, fever and/or pharyngitis. Lower respiratory tract infections included bronchitis (respiratory illness with cough, fever and rhonchi detected by auscultation), bronchiolitis (respiratory illness with evidence of bronchospasm, i.e., wheezing and/or prolonged expiration, assessed by auscultation) and pneumonia (respiratory illness with fever and either dyspnea and crepitant rales on auscultation or radiographic evidence of pulmonary infiltrate). Infant feeding was categorized as described for gastrointestinal infection. The incidence of upper and lower respiratory tract infections was reported for subgroups aged 0–5 and 6–11 months. In the 0–5 months subgroups the incidence of upper respiratory tract infection was significantly higher in artificially-fed infants than in infants who were exclusively breast-fed (12.0% vs. 2.9% of days of observation, RR = 4.14, $p < 0.01$). Infants who were exclusively artificially-fed also had a significantly higher incidence of lower respiratory tract infections than those who were exclusively breast-fed (0.88% vs. 0.16% of days of observation; RR = 5.50, $p < 0.05$). By 7 months of age, none of the infants was exclusively breast-fed. In the infants aged 6–11 months, the incidence of upper respiratory tract infections was statistically higher in those who were fed with formula milk and solids compared to those who were fed breast milk and solids (14.9% vs. 12.5% of days of observation; RR = 1.19; $p < 0.01$). Within this age group, infants fed formula milk and solids also had a significantly higher incidence of lower respiratory tract infections than those who were fed breast milk and solids (1.77% vs. 0.92% of days of observation; RR = 1.92; $p < 0.01$).

In a study by López-Alarcón *et al.* (1997) in Mexico City, acute respiratory infection was defined as the presence of runny nose or cough for at least two consecutive days plus one or more of the following signs regardless of their duration: erythematous mucosa, hoarse cry, respiratory distress or fever. An individual episode of infection was defined as an event separated from another event by at least 2 days. The classification of infant feeding and the co-variates used in the multiple linear regression model have been described in the section on gastrointestinal infection. The number of episodes of acute respiratory infection was found to be negatively correlated with the duration of breast-feeding ($r = -0.17$, $p = 0.05$) and positively correlated with the presence of siblings at home ($r = 0.16$, $p < 0.04$).

4.3. Summary of Studies in Non-Industrialized Countries

In contrast to the inconsistency of the evidence for a relationship between breast-feeding and gastrointestinal infections in industrialized countries, 8 out of the 10 studies conducted in non-industrialized countries indicated that breast-feeding had a protective effect against this type of infection. However, analysis for potential confounding factors was carried out in only three of the studies.

Figure 3 shows the pooled results on the relationship between breast-feeding and gastrointestinal infection. In contrast to the results for industrialized countries (Figure 1), all the odds ratios in the studies conducted in non-industrialized countries were in the direction of a beneficial effect of breast feeding.

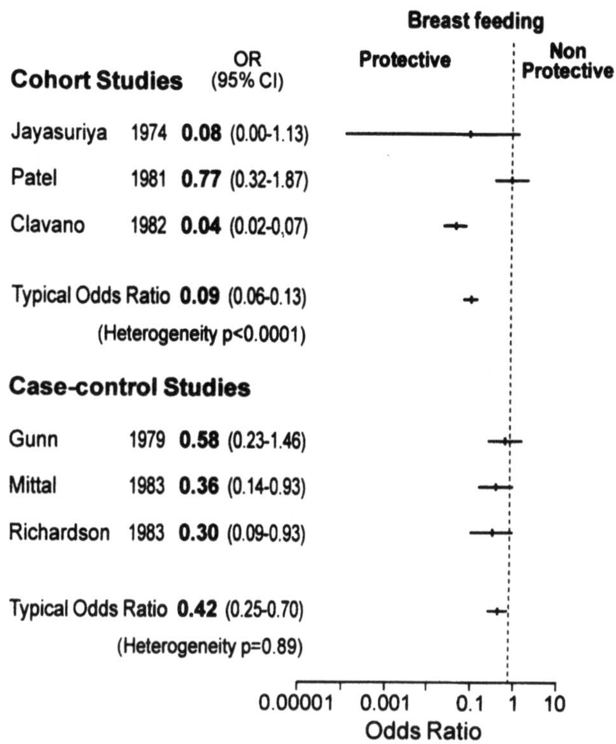

Figure 3. Meta-analysis of the effect of breast-feeding on gastrointestinal infection during infancy in non-industrialized countries. Studies by Brown *et al.* (1989), VanDerslice *et al.* (1994), Long *et al.* (1994) and López-Alarcón *et al.* (1997) were not included because the data necessary to calculate odds ratios could not be extracted from these reports.

5. Overall Perspectives

Although no firm conclusion can be drawn about the magnitude of a protective effect of breast-feeding against infections in industrialized countries, there is no doubt that it is protective in the case of many infants. This trend is most obvious in the studies of the highest quality. The beneficial effect of breast-feeding is more apparent for gastrointestinal infections than for respiratory infections. There may be other beneficial effects of breast milk that are not addressed in this review. Pollitt and Kariger (1996) cited evidence that breast-feeding in some infants has a beneficial effect on child development. There is still debate about the possibility of a relationship between infant feeding method, allergic disease and sudden infant death syndrome, and there is some evidence for other long term benefits of breast-feeding including a reduction in juvenile onset diabetes mellitus and maternal breast cancer (Statement of the Standing Committee on Nutrition of the British Paediatric Association, 1994).

There is no doubt that breast-feeding is a major factor in the prevention of gastrointestinal infections and mortality from diarrhea in non-industrialized societies (Victora *et al.*, 1989, 1992). Most infant mortality from infection occurs in non-industrialized countries (UNICEF, 1993). The World Health Organisation (WHO) estimates that 1.5 million deaths per annum could be prevented if women breast-fed their infants exclusively for 6 months and at least partially until 2 years of age (WHO, 1993). For this reason, the WHO and the United Nations Children's Fund (UNICEF) jointly launched the baby-friendly hospital initiative in 1991 to promote breast-feeding.

An international code for the marketing of breast milk substitutes was adopted by the World Health Assembly in 1981 and affirmed in 1996. This code binds member states to encourage breast-feeding and protect mothers from commercial pressure to use substitutes for breast milk (WHO, 1981). Nevertheless, a rigorously conducted survey of health facilities in 4 countries (Bangladesh, Poland, Thailand and South Africa) recently found clear violations of the code (Taylor, 1998). Among the health facilities surveyed, 8–50% had received free samples of milk, bottles or teats which were not used for research or for professional evaluation. Some of these samples were subsequently passed on to nursing mothers with infants less than 6 months old, whether or not that was the intention of the company donating them. Product information that violated the code was received by 30% of health facilities (range 15–56%), and 11% of health care workers (2–18%) received gifts from companies involved in the manufacturing or distribution of formula milk. There is clear evidence that women who receive free commercial discharge packs containing formula milk, bottles or teats are less likely to be still breast-feeding at one month postpartum (Inch *et al.*, 1988; Pérez-Escamilla *et al.*, 1994). Such violations raise concern that efforts to promote breast-feeding in these countries could be undermined by commercial marketing pressures promoting breast milk substitutes. Much remains to be done to promote breast-feeding and to monitor violations of the code pertaining to breast milk substitutes (Costello and Sachdev, 1998), especially in developing countries where there is evidence that the incidence of breast-feeding has been in decline in recent years (Jolly, 1990).

References

Agre, F. 1985. The relationship of mode of infant feeding and location of care to frequency of infection. *Am. J. Dis. Child.* **139**:809.

Banatvala, J.E., Chrystie, I.L., and Totterdell, B.M. 1978. Rotaviral infections in human neonates. *J. Am. Vet. Med. Assoc.* **173**:527.

Bauchner, H., Leventhal, J.M., and Shapiro, E.D. 1986. Studies of breast-feeding and infections. How good is the evidence? *JAMA* **256**:887.

Breslow, N.E. and Day, N.E. 1980. *Statistical Methods in Cancer Research, Vol. 1 The Analysis of Case Control Studies*. Lyon: France: LARC Scientific Publications.

Brown, K.H., Black, R.E., de Romaña, G.L., and de Kanashiro, H.C. 1989. Infant-feeding practices and their relationship with diarrheal and other diseases in Huascar (Lima), Peru. *Pediatrics* **83**:31.

Chatellier, G., Zapletal, E., Lemaitre, D., Menard, J., and Degoulet, P. 1996. The number needed to treat: A clinically useful nomogram in its proper context. *Br. Med. J.* **312**:426.

Clavano, N.R. 1982. Mode of feeding and its effect on infant mortality and morbidity. *J. Trop. Pediatr.* **28**:287.

Cochi, S.L., Fleming, D.W., Hightower, A.W., Limpakarnjanarat, K., Facklam, R.R., Smith, J.D., Sikes, R.K., and Broome, C.V. 1986. Primary invasive *Haemophilus influenzae* type b disease: A population-based assessment of risk factors. *J. Pediatr.* **108**:887.

Costello, A. and Sachdev, H.S. 1998. Protecting breastfeeding from breast milk substitutes. *Br. Med. J.* **316**:1103.

Crewe, E. and Murphy, A.M. 1980. Further studies on neonatal rotavirus infections. *Med. J. Aust.* **1**:61.

Cunningham, A.S. 1979. Morbidity in breast-fed and artificially fed infants. *J. Pediatr.* **95**:685.

Cushing, A.H. and Anderson, L. 1982. Diarrhoea in breast-fed and non-breast-fed infants. *Pediatrics* **70**:921.

Dickersin, K., Scherer, R., and Lefebvre, C. 1994. Identification of relevant studies for systematic reviews. *Br. Med. J.* **309**:1286.

Downham, M.A.P.S., Scott, R., Sims, D.G., Webb, J.K.G., and Gardner, P.S. 1976. Breast-feeding protects against respiratory syncytial virus infections. *Br. Med. J.* **2**:274.

Duffy, L.C., Byers, T.E., Riepenhoff-Talty, M., La Scolea, L.J., Zielezny, M., and Ogra, P.L. 1986. The effects of infant feeding on rotavirus-induced gastroenteritis: a prospective study. *Am. J. Public Health.* **76**:259.

Duncan, B., Ey, J., Holberg, C.J., Wright, A.L., Martinez, F.D., and Taussig, L.M. 1993. Exclusive breast-feeding for at least 4 months protects against otitis media. *Pediatrics* **91**:867.

Eaton-Evans, J. and Dugdale, A.E. 1987. Effects of feeding and social factors on diarrhoea and vomitting in infants. *Arch. Dis. Child.* **62**:445.

Fergusson, D.M., Horwood, L.J., Shannon, F.T., and Taylor, B. 1978. Infant health and breast-feeding during the first 16 weeks of life. *Aust. Paediatr. J.* **14**:254.

Forman, M.R., Graubard, B.I., Hoffman, H.J., Beren, R., Harley, E.E., and Bennett, P. 1984. The Pima infant feeding study: Breast-feeding and respiratory infections during the first year of life. *Int. J. Epidemiol.* **13**:447.

France, G.L., Marmer, D.J., and Steele, R.W. 1980. Breast-feeding and *Salmonella* infection. *Am. J. Dis. Child.* **134**:147.

Frank, A.L., Taber, L.H., Glezen, W.P., Kasel, G.L., Wells, R.W., and Paredes, A. 1982. Breast-feeding and respiratory virus infection. *Pediatrics* **70**:239.

Granoff, D.M. and Basden, M. 1980. *Haemophilus influenzae* infections in Fresno County, California: A prospective study of the effects of age, race and contact with a case on incidence of disease. *J. Infect Dis.* **141**:40.

Gunn, R.A., Kimball, A.M., Pollard, R.A., Feeley, J.C., Feldman, R.A., Dutta, S.R., Matthew, P.P., and Levine, M.M. 1979. Bottle feeding as a risk factor for cholera in infants. *Lancet* **2**:730.

Holmes, G.E., Hassanein, K.M., and Miller, H.C. 1983. Factors associated with infections among breast-fed babies and babies fed proprietary milks. *Pediatrics* **72**:300.

Howie, P.W., Forsyth, J.S., Ogston, S.A., Clark, A., and du V Florey, C. 1990. Protective effect of breastfeeding against infection. *Br. Med. J.* **300**:11.

Inch, S. and Garforth, S. 1988. Establishing and maintaining breastfeeding, in: *Effective Care in Pregnancy and Childbirth*, (I. Chalmers, M. Enkin, M.J.N.C. Keirse, eds.), Oxford University Press, Oxford.

Jason, J.M., Nieburg, P., and Marks, J.S. 1984. Mortality and infectious disease associated with infant-feeding practices in developing countries. *Pediatrics* **74**:702.

Jayasuriya, D. and Soysa, P. 1974. Feeding studies in Ceylonese babies. *J. Trop. Pediatr. Enviromen. Child. Health* **20**:275.

Jenista, J.A., Powell, K.R., and Menegus, M.A. 1984. Epidemiology of neonatal enterovirus infection. *J. Pediatr.* **104**:685.

Jolly, R. 1990. Breastfeeding and health care serviccs. *Int. J. Gynecol. Obstet.* **31(Suppl. 1)**:7.

Kovar, M.G., Serdula, M.K., Marks, J.S., and Fraser, D.W. 1984. Review of the epidemiologic evidence for an association between infant feeding and infant health. *Pediatrics* **74**:615.

Leventhal, J.M., Shapiro, E.D., Aten, C.B., Berg, A.T., and Egerter, S.A. 1986. Does breast-feeding protect against infections in infants less than 3 months of age? *Pediatrics* **78**:896.

Long, K.Z., Wood, J.W., Gariby, E.V., Weiss, K.M., Mathewson, J.J., de la Cabada, F.J., DuPont, H.L., and Wilson, R.A. 1994. Proportional hazards analysis of diarrhoea due to enterotoxigenic *Escherichia coli* and breastfeeding in a cohort of urban Mexican children. *Am. J. Epidemiol.* **139**:193.

López-Alarcón, M., Villalpando, S., and Fajardo, A. 1997. Breast-feeding lowers the frequency and duration of acute respiratory infection and diarrhoea in infants under six months of age. *J. Nutr.* **127**:436.

Lum, M.K., Ward, J.I., and Bender, T.R. 1982. Protective influence of breastfeeding (BF) on the risk of developing invasive *H. influenzae* type b (Hib) disease [Abstract]. *Pediatr. Res.* **16**:151A.

Mantel, N. and Haenszel, W. 1959. Statistical aspects of the analysis of data from retrospective studies of disease. *J. Natl. Cancer Inst.* **22**:719.

Martinez, G.A. and Dodd, D.A. 1983. 1981 milk feeding patterns in the United States during the first 12 months of life. *Pediatrics* **71**:166.

McKean, K.S., Baum, J.D., and Sloper, K. 1975. Factors influencing breastfeeding. *Arch. Dis. Child.* **50**:165.

Mittal, S.K., Kanwar, A., Varghese, A., and Ramachandran, V.G. 1983. Gut flora in breast and bottle fed infants with and without diarrhoea. *Indian J. Pediatr.* **20**:21.

Nafstad, P., Jaakkola, J.J.K., Hagen, J.A., Botten, G., and Kongerud, J. 1996. Breastfeeding, maternal smoking and lower respiratory tract infections. *Eur. Respir. J.* **9**:2623.

Narayanan, I., Prakash, K., Bala, S., Verma, R.K., and Gujral, V.V. 1980. Partial supplementation with expressed breast-milk for prevention of infection in low-birth-weight infants. *Lancet* **2**:516.

Narayanan, I., Prakash, K, and Gujral, V.V. 1981. The value of human milk in prevention of infection in the high-risk, low-birth-weight infant. *J. Pediatr.* **99**:496.

Narayanan, I., Prakash, K., Prabhakar, A.K., and Gujral, V.V. 1982. A planned prospective evaluation of the anti-infective property of varying quantities of expressed human milk. *Acta Pædiatr. Scand.* **71**:441.

Narayanan, I., Prakash, K., Murthy, N.S., and Gujral, V.V. 1984. Randomised controlled trial of effect of raw and holder pasteurised human milk and of formula supplements on incidence of neonatal infection. *Lancet* **2**:1111.

Patel, R.B., Khanna, S.A., Lahiri, K., and Kulkarni, G.V. 1981. Breast milk in low birth weight babies. *Indian J. Pediatr.* **48**:195.

Paradise, J.L. 1981. Otitis media during early life: How hazardous to development? A critical review of the evidence. *Pediatr.* **68**:869.

Pérez-Escamilla, R., Pollitt, E., Lönnerdal, B., and Dewey, K. 1994. Infant feeding policies in maternity wards and their effect on breast-feeding success: an analytical overview. *Am. J. Pub. Health.* **84**:89.

Pisacane, A., Graziano, L., Mazzarella, G, Scarpellino, B, and Zona, G. 1992. Breast-feeding and urinary infection. *J. Pediatr.* **120**:87.

Pisacane, A., Graziano, L., Zona, G., Granata, G., Dolezalova, H., Cafiero, M., Coppola, A., Scarpellino, B., Ummarino, M., and Mazzarella, G. 1994. Breastfeeding and acute lower respiratory infection. *Acta Pædiatr.* **83**:714.

Pollitt, E. and Kariger, P. 1996. Breastfeeding and child development. *Food and Nutrition Bulletin* **17**:401.

Popkin, B.M., Adair, L., Akin, J.S., Black, R., Briscoe, J., and Flieger, W. 1990. Breast-feeding and diarrhoeal morbidity. *Pediatrics* **86**:874.

Pullan, C.R., Toms, G.L., Martin, A.J., Gardner, P.S., Webb, J.K.G., and Appleton, D.R. 1980. Breast-feeding and respiratory syncytial virus infection. *Br. Med. J.* **281**:1034.

Research Sub-Committee of the South-East England Faculty of RCGP. 1972. The influence of breastfeeding on the incidence of infectious illness during the first year of life. *Practitioner* **209**:356.

Richardson, N.J., Koornhof, H.J., Bokkenheuser, V.D., Mayet, Z., and Rosen, E.U. 1983. Age related susceptibility to *Camphylobacter jejuni* infection in a high prevalence population. *Arch. Dis. Child.* **58**:616.

Riegelman, R.K. and Hirsh, R.P. 1996. *Studying a Study and Testing a Test.* Little Brown & Company, Boston.

Rubin, DH., Palmer, L.S., Menasse, L., and Erenberg, F.G.. 1987. Does breast-feeding protect infants from *Haemophilus influenzae* infection? *J. Pediatr.* **110**:162.

Saarinen, U.M. 1982. Prolonged breastfeeding as prophylaxis for recurrent otitis media. *Acta Pædiatr. Scand.* **71**:567.

Satterwhite, T.K., Evans, D.G., Dupont, HL., and Evans, D.J., Jr. 1978. Role of *Escherichia coli* colonization factor antigen in acute diarrhoea. *Lancet* **2**:181.

Sauls, H.S. 1979. Potential effect of demographic and other variables in studies comparing morbidity of breast-fed and bottle-fed infants. *Pediatrics* **64**:523.

Schultz, K.F., Chalmers, I., Hayes, R.J., and Altman, D.G. 1995. Empirical evidence of bias. Dimensions of methodological quality associated with estimates of treatment effects in controlled trials. *JAMA* **273**:408.

Sibbald, B. and Roland, M. 1998. Understanding controlled trials. Why are randomised controlled trials important? *Br. Med. J.* **316**:201.

Silfverdal, S.A., Bodin, L., Hugosson, S., Garpenholt, Ö., Werner, B., Esbjörner, E., Lindquist, B., and Olcén, P. 1997. Protective effect of breastfeeding on invasive *Haemophilus influenzae* infection: A case-control study in Swedish preschool children. *Int. J. Epidemiol.* **26**:443.

Statement of the Standing Committee on Nutrition of the British Paediatric Association. 1994. Is breastfeeding beneficial in the UK? *Arch. Dis. Child.* **71**:376.

Taylor, A. 1998. Violations of the international code of marketing of breast milk substitutes: prevalence in four countries. *Br. Med. J.* **316**:1117.

Totterdell, B.M., Chrystie, I.L., and Banatvala, J.E. 1976. Rotavirus infections in a maternity unit. *Arch. Dis. Child.* **51**:924.

United Nations Children's Fund (UNICEF). 1993. *The State of The World's Children*. Oxford University Press, New York.

VanDerslice, J., Popkin, B., and Briscoe. J. 1994. Drinking-water quality, sanitation, and breast-feeding: their interactive effects on infant health. *Bull. WHO* **72**:589.

Victora, C.G., Smith, P.G., Vaughan, J.P., Nobre, L.C., Lombardi, C., Teixeira, A.M.B., Fuchs, S.C., Moreira, L.B., Gigante, L.P., and Barros, F.C. 1989. Infant feeding and deaths due to diarrhoea. *Am. J. Epidemiol.* **129**:1032.

Victora, C.G., Huttly, S.R., Fuchs, S.C., Nobre, L., and Barros, F.C. 1992. Death due to dysentery, acute and persistent diarrhoea among Brazilian infants. *Acta Pædiatr. Suppl.* **381**:7.

Watkins, C.J., Leeder, S.R., and Corkhill, R.T. 1979. The relationship between breast and bottle feeding and respiratory illness in the first year of life. *J. Epidemiol. Community Health.* **33**:180.

Weinberg, R.J., Tipton, G., Klish, W.J., and Brown, M.R. 1984. Effect of breast-feeding on morbidity in rotavirus gastroenteritis. *Pediatrics* **74**:250.

Winikoff, B. 1981. Issues in the design of breastfeeding research. *Stud Fam. Plann.* **12**:177.

World Health Organisation. 1981. *International Code of Marketing of Breast Milk Substitutes*. WHO, Geneva.

World Health Organisation. 1993. *Infant and Young Child Nutrition*. WHO, Geneva.

Yoon, P.W., Black, R.E, Moulton, L.H., and Becker, S. 1996. Effect of not breastfeeding on the risk of diarrhoeal and respiratory mortality in children under 2 years of age in Metro Cebu, The Philippines. *Am. J. Epidemiol.* **143**:1142.

Chapter 5

Breast Milk Transmission of Viral Disease

E. Richard Stiehm and Margaret A. Keller

1. Introduction

Breast-feeding is, in general, nutritionally ideal, economically sound, immunologically advantageous, hygienically sterile and emotionally satisfying. Nevertheless, in certain maternal viral illnesses there is a substantial risk of maternal-infant transmission by breast milk. This is particularly true for HIV-1, HTLV-1, HTLV-2 and, to a lesser extent, CMV infection. Further, in the presence of acute infection with many viral agents, breast-feeding may be contraindicated.

We have limited this chapter to viral disease. Occasionally babies can also be infected by bacteria from breast-feeding, particularly if there is infection within the breast tissue, e.g., abscess. Several excellent reviews of breast milk transmission of infectious agents are available (Ruff, 1994; Ogra and Rassin, 1995; American Academy of Pediatrics, 1997).

2. Herpes Viruses

The Herpesviridae family of viruses includes many viruses that cause significant human disease. This family is divided into three groups: α-herpesviruses [herpes simplex virus-1, herpes simplex virus-2, varicella-zoster virus], β-

E. Richard Stiehm • UCLA Department of Pediatrics, UCLA Children's Hospital (ERS), Los Angeles, California Margaret A. Keller • Harbor-UCLA Medical Center (MAK), Torrance, California

herpesviruses [cytomegalovirus, human herpesvirus-6, human herpesvirus-7)] and γ-herpesviruses [Epstein-Barr virus].

All of the herpes viruses are enveloped DNA viruses and induce lifelong latent infection. Latent infection can be reactivated if the host becomes immunodeficient or experiences other stresses. Surprisingly, despite the latency of all herpesviruses, cytomegalovirus is the only herpesvirus known to be efficiently transferred to the infant via human milk. There are a few case reports of possible breast milk transmission of herpes simplex virus and varicella-zoster virus, and there is strong evidence for lack of transmission of Epstein-Barr virus even though the viral genome can be detected in the milk.

2.1. Cytomegalovirus (CMV)

2.1.1. Detection in Human Milk. Cytomegalovirus was first isolated from human milk by Diosi *et al.* in 1967. Subsequently, many investigators have cultured CMV from colostrum, early milk and mature milk. Hayes *et al.* (1972) isolated CMV from the milk of 27% (17/63) of seropositive Australian women. Positive specimens were obtained from 2 days to 10 weeks postpartum. For milk specimens obtained less than 6 days postpartum, the isolation rate was only 11% (4/37), but the rate increased to 50% (13/26) one to 13 weeks postpartum. In a subsequent study by Ahlfors and Ivarsson (1985) in Swedish women, 70% (16/23) of seropositive women excreted CMV in their milk from 8 days to 3 months postpartum. Cheeseman and McGraw (1983) isolated CMV from the milk of 44% (7/16) of seropositive lactators.

Hotsubo *et al.* (1994) studied the breast milk of CMV-seropositive Japanese women. Only 1/177 milk specimens obtained less than 6 days postpartum, 2/115 obtained from 7 days to 1 month postpartum, and 10/33 specimens obtained later than one month postpartum were CMV culture positive. These investigators examined the milk of 13 seropositive mothers by CMV DNA polymerase chain reaction (PCR) at one month postpartum. Twelve had a positive result in the whey (liquid) portion of the milk. However, the significance of these results is not clear since the positive CMV DNA PCR may have resulted from lysis of latently infected cells and does not prove the presence of infectious virus.

Asanuma *et al.* (1996) determined which fraction of milk (whey vs cells) was the source of CMV using PCR to detect a major immediate early gene region of CMV DNA. CMV DNA was not detected in colostrum, but was detected in milk obtained after one month postpartum. Although 39% (25/64) of whey samples were positive by DNA PCR, only 17% (11/64) of milk cell samples were positive, providing evidence that cell-free CMV in the whey may have an important role in breast milk transmission.

In summary, CMV is much less common in early milk and colostrum than in mature milk (obtained after one month postpartum). The reason is not clear, but DNA PCR techniques (Hotsubo *et al.*, 1994) failed to demonstrate CMV in milk whey in the first 3 days after delivery (0/20). Although anti-CMV IgA antibody, which is present in high concentration in colostrum, could explain negative cultures in the early postpartum period, anti-CMV IgA does not explain the negative results with DNA PCR. The factors that control CMV excretion in milk or reactivation of CMV in the mammary gland are not understood.

2.1.2. Maternal-Infant Transmission. In a seminal study by Stagno *et al.* (1980), the role of breast milk vs urine, saliva, and cervical secretions was examined as a source for nontransplacental vertical CMV transmission. When the mother's only site of CMV excretion was breast milk, 58% (11/19) of breast-fed infants became infected within the first year of life as assessed by infant CMV viruria. If a mother's breast milk was culture positive for CMV but the infant was bottle-fed, none of the infants (0/9) became infected within the first year. If the cervix was the only positive site of maternal CMV excretion and was positive only in the third trimester and postpartum, 57% (8/14) of the infants became infected. If cervical secretions were positive only in the third trimester, 26% (18/68) of infants became infected, and if cervical secretions were positive only in the first and second trimesters, only 12% (1/8) of infants became infected. When the mother's urine or saliva was the only site of CMV excretion, none of the infants (0/11 and 0/15 respectively) became infected. In women without any evidence of CMV excretion at any site, none (0/125) of their bottle-fed and 9% (1/11) of their breast-fed infants developed CMV infection within the first year of life.

Since Stagno *et al.* (1980) cultured CMV only from early milk or colostrum, the number of women excreting CMV in mature milk may have been underestimated. Nonetheless, the role of infected breast milk was at least as strong as infected cervical secretions in transmission. CMV viruria in infants usually occurred between 30 and 120 days following exposure to CMV in breast milk, similar to that of infants exposed to infected cervical fluid. This study establishes a role for breast milk in CMV vertical transmission and illustrates the minimal role of maternal salivary CMV excretion. It also suggests that transplacental maternal neutralizing antibody may explain the minimal morbidity that was evident in infants exhibiting CMV viruria.

In an extension of these studies, Dworsky *et al.* (1983) followed 41 seropositive breast-feeding women and their infants, 13 of whom were premature. Thirty-two percent (13/41) of the mothers excreted CMV in their milk, 2% (1/41) excreted CMV in their saliva, 7% (3/41) in their urine, and 10% (4/41) in their vaginal secretions. All patients were cultured multiple times and shed CMV intermittently from all sites. Forty-two percent (13/31) of milk samples obtained between two and 12 weeks postpartum were CMV culture positive. No infant that was breast-fed for less than one month (0/10) became infected in the first year of life. Thirty-nine percent (12/31) of infants breast-fed for more than one month became infected. The most striking finding was that 69% (9/13) of infants breast-fed by a mother who excreted CMV in her milk developed CMV viruria in the first year of life. If the mother was seropositive but not excreting CMV in her milk, only 11% (3/28) of the infants became infected. This confirms the view that the breast is the most likely site for reactivation and secretion of CMV in post-partum women and that consumption of infected milk often leads to infection of the infant.

Larger studies from England (Peckham *et al.*, 1987) and Japan (Minamishima *et al.*, 1994) also support the role of breast-feeding in vertical CMV transmission. In a study of 253 mother-infant pairs followed from birth to one year of age with CMV urine cultures and CMV antibody, CMV seropositivity of the mother was the strongest risk factor for perinatal transmission (Peckham *et al.*, 1987). Breast-feeding increased the risk of infection 2.3 fold (95% confidence limits 1.1–4.9). In total, the peak period of acquisition was 9 to 15 weeks of age. Overall, 33% of infants of seropositive mothers and 4% of infants of seronegative mothers acquired CMV in the first year of life. Thirty-nine percent of breast-fed infants and 24% of bottle-fed infants from seropositive mothers acquired CMV; by contrast, 11% of breast-fed infants and 2% of bottle-fed infants from seronegative mothers acquired CMV.

In the study of Peckham *et al.* (1987), CMV was isolated from the breast milk of 20% of breast-feeding mothers (17/87), and 13 of the 17 infants of CMV-excreting mothers acquired CMV. Surprisingly, four infants did not acquire CMV despite suckling CMV-infected milk. Further analysis demonstrated that 19% (7/36) of bottle-fed infants, 30% (7/23) of infants breast-fed for less than 12 weeks, and 42% (27/64) of infants breast-fed for more than 12 weeks acquired CMV in the first year of life. These results are similar to those of Stagno *et al.* (1980) and Dworsky *et al.* (1983) in which 58% and 69%, respectively, of infants seroconverted if suckled from a mother excreting CMV in her milk.

Minamishima *et al.* (1994) examined the prevalence of CMV antibody in breast- and bottle-fed Japanese infants at varying ages. By age 2, 64% (32/50) of breast-fed infants were seropositive compared to 29% (4/14) of bottle-fed infants, suggesting that 35% of infant CMV infections were transmitted by breast milk.

All of these studies indicate that breast milk CMV transmission is common. Of particular concern are premature infants and infants of seronegative mothers who receive milk from a milk bank or from a CMV-excreting mother. Welsh *et al.* (1979) demonstrated that pasteurization, but not freezing, can eliminate infectious CMV from milk. Friis and Andersen (1982) found that storage of milk at –20°C for 3 days decreases the amount of infectious CMV by 99% but pasteurization at 63°C killed all infectious virus. Dworksy *et al.* (1982) demonstrated that freezing for 7–30 days at –20°C eliminated naturally infectious CMV, but milk inoculated artificially with large amounts of CMV was still infectious after seven days at –20°C. Pasteurization at 63°C but not heat treatment at 56°C eliminated all infectious virus. Thus, to reliably prevent CMV infection in breast milk samples, pasteurization at 63°C is required. Some milk banks use only CMV-antibody negative donors, but this is of unproven reliability.

Recommendation: CMV seropositive mothers may breast-feed since their infants, even if infected, will be protected from developing serious disease by placentally transferred maternal antibody. The risk of a seropositive mother providing breast milk for her premature infant, who may have less transplacental antibody has not been established; such premature infants may be at risk for symptomatic disease. In acute maternal CMV infection, breast-feeding should be avoided. Banked donor milk should be pasteurized to eliminate the risk of CMV transmission.

2.2. Herpes Simplex Virus (HSV)

Dunkle *et al.* (1979) reported that a breast-fed neonate developed HSV-1 infection of the palate and pharynx at age 6 days. The mother was clinically well but her breast milk was HSV culture positive, suggesting vertical HSV transmission by breast-feeding. However, Kibrick (1979) suggested that this infant may have been infected during birth canal passage, then developed oral lesions and then infected the breast, despite the absence of breast lesions.

Herpes simplex virus has been isolated from the breast milk of HSV infected women. Tantivanich *et al.* (1983) examined breast milk after delivery in mothers with asymptomatic or symptomatic genital herpes infection during pregnancy and found that 32% (8/25) of the asymptomatic women and 36% (4/11) of the symptomatic women had a positive breast milk HSV culture and 36 to 44% of their infants had HSV in their throat. They concluded that breast-feeding is a possible route of vertical transmission, but they could not exclude the possibility that the infant acquired HSV during passage through the birth canal and then contaminated the mother's breast.

Thus there is no conclusive evidence to support HSV transmission by breast-feeding, but it is also difficult to exclude the possibility.

Recommendation: Mothers with herpetic oral or genital lesions may breast-feed with proper handwashing and a mask if oral lesions are present. Women with herpetic breast lesions should not breast-feed.

2.3. Epstein-Barr Virus (EBV) and Other Human Herpes Viruses (HHV-6, HHV-7, HHV-8)

To evaluate EBV or HHV-6 transmission via breast milk, Kusuhara *et al.* (1997) examined 331 serum samples from bottle- and breast-fed infants after age 12 months. Over 95% of the mothers were seropositive for both EBV and HHV-6. There was no difference in seropositivity for either virus between breast and bottle-fed infants, suggesting that breast milk transmission is not a significant source of EBV or HHV-6 infection in infancy. Junker *et al.* (1991) demonstrated EBV genome was in milk cells obtained from the mature milk of 46 of 100 seropositive women studied using a DNA hybridization dot-blot assay. However, this assay does not distinguish between latently infected cells and actively replicating virus. By contrast, Dunne and Jevon (1993) found no evidence of HHV-6 DNA in 120 milk cell samples.

As yet there are no studies that have evaluated breast transmission of other herpesviruses, such as HHV-7 or HHV-8.

Recommendation: Mothers who are seropositive for EBV or HHV-6 may breast-feed their infants. However, since the EBV genome has been demonstrated in milk cells, acute EBV infection in the mother may be a contraindication to breast-feeding. There are no reports of such transmission.

2.4. Varicella-Zoster Virus (VZV)

Varicella-zoster virus could not be cultured in the milk from two breast-feeding mothers who developed acute infection (Frederick *et al.*, 1986). Yoshida *et al.* (1992) reported a case of chicken pox in a breast-feeding mother whose infant developed the illness 16 days after the onset of the mother's illness. VZV DNA was detected in the mother's milk by PCR technique. Since other ways of transmission (hands, droplet) were not excluded, this does not prove breast milk transmission.

Recommendation: Mothers with acute varicella may breast-feed their infants since there is no convincing evidence of transmission through the breast milk. Mothers with herpes zoster may also breast-feed if the lesions are not on the breast and can be covered.

3. Rubella

Rubella virus is an enveloped RNA virus of the Togaviridae family. It has been identified in the breast milk of one lactating woman, and attenuated rubella vaccine virus has been found in the milk of recently immunized postpartum women.

Buimovici-Klein *et al.* (1977) studied a mother-infant pair in which the breast-feeding mother received rubella vaccine (HPV-77 DE5 strain) postpartum and developed rubella twelve days later. The mother's throat culture was negative, whereas her breast milk samples were culture-positive twelve days after vaccine but negative by 35 and 70 days. The infant, whose throat culture was positive for rubella on the same day as the milk sample was positive, developed rubella-specific cell-mediated immunity (CMI) but not rubella antibody. When the child was given rubella vaccine at age one year there was an increase in the CMI response and development of IgG rubella antibody, suggesting that the infant was not harmed by neonatal exposure to vaccine virus and responded to subsequent immunization.

Klein *et al.* (1980) reported the isolation of rubella virus from the milk of a woman who developed a rubella rash eight days postpartum. Virus was present in the milk for the first three days of the rash, but not thereafter. The infant remained clinically well but a pharyngeal culture at 15 days of age was positive. The infant developed antibody and a rubella-specific cell-mediated immune response (CMI). This case suggests, but does not prove, breast milk transmission since other routes were not excluded.

Losonsky *et al.* (1982) studied 16 breast-feeding and 10 non-breast-feeding mothers given two different attenuated vaccines 2–4 days postpartum. Sixty-nine percent (11/16) of the breast-feeding immunized mothers excreted rubella virus in their milk for up to seventeen days after immunization and 9 of 16 of their infants had rubella virus or antigen in their nasopharyngeal secretions. Four of these infants transiently seroconverted and/or developed a CMI response. High titers of rubella virus in the milk were correlated with high titers in the infant's nasopharyngeal secretions. No infants of bottle-fed immunized mothers had evidence of infection. This study provides definitive evidence that rubella vaccine virus can be transmitted from the mother to her infant, that the transmission correlates with the amount of virus in the milk, and that definitive evidence of infection (positive cultures, antibody, and CMI) develops in clinically well infants. However, there is no evidence for persistent infection or harmful effects from this exposure.

Recommendation: Neither postpartum immunization with rubella vaccine norural rubella infection should prevent a mother from breast-feeding her infant.

4. Parvovirus B19

Parvovirus B19, a nonenveloped, single-stranded DNA virus, belongs to the Parvoviridae family and is the causative agent in erythema infectiosum (Fifth

Disease). Although transplacental transmission of this virus has occurred, there is no evidence of breast milk transmission (Keyserling, 1997). Transmission usually results from direct contact with respiratory secretions, usually before the appearance of the rash (American Academy of Pediatrics, 1997). If a breast-feeding mother develops Fifth Disease with rash, transmission would have occurred prior to diagnosis of erythema infectiosum.

Recommendation: Breast-feeding mothers with either acute Fifth Disease with rash or previous infection with parvovirus B19 may breast-feed their infants.

5. HIV-1

Transmission of human immunodeficiency virus-1 (HIV-1) via breast-feeding was first described by Ziegler *et al.* (1985). Following that report several other individual cases were identified over the next 6 years (Stiehm and Vink, 1991). However, clinical proof requires that the infected infant be proven uninfected at birth, and subsequently breast-fed by a mother who acquires HIV infection in the post-partum period.

Van de Perre *et al.* (1991) reported that 8 of 16 Rwandan newborns became HIV-infected from breast-feeding by mothers who first seroconverted shortly after giving birth. All infants were uninfected at birth as determined by a negative PCR of their peripheral leukocytes. This experience emphasizes the high risk of breast milk transmission by a mother who acquires HIV infection in the postpartum period.

Dunn *et al.* (1992) conducted a meta-analysis of 9 epidemiological studies to determine the additional risk of vertical HIV transmission by breast-feeding. In four studies in which the mothers acquired HIV-1 postnatally, the transmission rate was 29% (12 of 42) (95% confidence limits, 16–42%). In five studies in which the mothers were infected prenatally, the mean transmission rate was 14% higher (confidence limits 7–22%) in the breast-fed infants than in the bottle-fed infants. Despite these findings, the World Health Organization continues to recommend breast-feeding for all mothers in developing countries because of the great risk of non-HIV infection associated with formula feeding. In the US and Europe, breast-feeding is not recommended for infants of HIV-infected mothers (American Academy of Pediatrics, 1997).

In recent years several groups have studied, in more detail, the role of breast-feeding in HIV transmission. The transmission rate to Kenyan infants breast-fed from birth by HIV-1 infected mothers was 42.8% at 12 months of age (Datta *et al.*, 1994). A comparison of HIV-infected and HIV-uninfected infants indicated that infants who breast-fed for greater than 15 months were at higher risk of transmission. Maternal marital status was also associated with the likelihood of breast milk transmission, possibly because married mothers are more likely to exclusively breast-feed their infants.

Bertolli *et al.* (1996) estimated the contribution of breast-feeding to an overall transmission rate of 26% among breast-fed infants of 323 HIV-infected mothers in Zaire over an 18-month period. Using serial serum HIV-PCR to indicate the time of acquisition of HIV infection, these investigators determined that 16 (23%) of the 69 HIV-infected infants were infected *in utero*, 45 (65%) intrapartum or early postpartum, and 8 (12%) in the late postpartum (breast-feeding) period. The overall risk of acquiring HIV from breast-feeding among infants uninfected prenatally or in the peripartum period was 4% (8 of 189, confidence limits 2–8%).

Using RNA PCR, Karlsson *et al.* (1997) studied 139 breast-fed children who were uninfected at 6 months of age despite being born to HIV-infected mothers. Eight infants became infected six to twenty-three months later, representing 11% of the total number of infants who became infected. The rate of transmission increased from 25% at 6 months of age to 30.2% at 1 year of age. Seroconversion of breast-fed infants did not occur before 11 months of age. These investigators and others (de Martino *et al.*, 1992; Nagelkerke *et al.*, 1995; Kreiss, 1997) suggest that prolonged breast-feeding, a decrease in transplacental antibody, or the presence of teeth in the suckling infant may contribute to late breast milk transmission. Thus breast-feeding is a substantial route of maternal-fetal HIV transmission, particularly when the mother seroconverts during breast-feeding or suckles her infant for a prolonged period.

The benefit of human milk to the infants of HIV-infected mothers has also been questioned. Bobat *et al.* (1997) could find no advantage to infants of HIV-infected breast-fed infants in South Africa in terms of protection against common non-HIV infections, prevention of failure to thrive or delay of their progression of AIDS.

Mechanisms: Numerous studies indicate that breast milk of HIV-infected mothers contains cell-free and cell-associated virus, making the milk potentially contagious to a suckling infant (Oxtoby, 1988; Kreiss, 1997). This is modified by many factors, probably chief of which is the presence of maternal antibody or cytotoxic cells in the breast milk which may neutralize the virus in situ and prevent its infectivity (Table 1). Without either of these protective factors, the rate of transmission approaches 50% as judged by Van de Perre's data on transmission from mothers seroconverting during breast-feeding (Van de Perre *et al.*, 1991).

In attempts to prevent transmission without stopping breast-feeding, several alternatives have been suggested including the following: 1) withholding colostrum since it is rich in cells and possibly more infectious; 2) pasteurization of expressed milk or colostrum; 3) weaning at a relatively early age (5–7 months) while the infant has maternal antibody but no teeth; 4) antiretroviral medication

Table 1. Factors Influencing HIV-1 Transmission by Breast Milk

Maternal Factors Promoting HIV Transmission	
1.	Cell-free or cell-associated virus in breast milk
2.	Specific viral strain, phenotype or genotype of HIV-1
3.	Advanced maternal disease
4.	Poor maternal nutrition
5.	Lack of specific HIV antibodies in breast milk
6.	Lack of specific cytotoxic T cells in breast milk
7.	Nipple or breast abnormalities-cracking, bleeding, mastitis, abscesses, etc.
8.	Vitamin A deficiency
Infant Factors Promoting HIV Transmission	
1.	Prematurity
2.	Poor nutritional status
3.	Mucosal lesions, stomatitis, oral ulcers, thrush, esophagitis, pharyngitis
4.	HLA type
5.	Presence of teeth
6.	Prolonged breast feeding (>6 months)
Maternal Factors Inhibiting HIV Transmission	
1.	HIV specific antibodies or cytotoxic T cells in breast milk
2.	Low viral burden
3.	Antiretroviral therapy
4.	Breast milk lipids, lactoferrin, lysozyme
Infant Factors Inhibiting HIV Transmission	
1.	Salivary inhibitors
2.	Antiretroviral therapy
3.	Low stomach pH

Modified from Kreiss (1997).

to infants and mothers; 5) administering vitamin A to infants and mothers (Morrison and Alp, 1997; Mofenson, 1997).

In summary, breast milk transmission of HIV-1 from mother to infant is not uncommon, particularly when the mother first becomes infected during breast-feeding. Another high risk practice is prolonged breast-feeding beyond 7–9 months of age, which is possibly associated with decreased blood levels of maternal (transplacental) anti-HIV antibody and the presence of teeth in the infants. The additional risk of HIV infection due to breast-feeding approaches 15–20%, above the rate of transmission *in utero* and peripartum.

Recommendation: In developed countries, breast-feeding by HIV-infected mothers should be avoided. In developing countries where the risk of severe non-HIV infection is great, breast-feeding should be limited to 5–6 months. In addition, priority should be given to testing mothers prenatally and providing sterile formula for infants of HIV-infected mothers.

6. HIV-2

HIV-2 is endemic in areas of West Africa and usually produces a slower, more indolent form of acquired immunodeficiency syndrome. Transmission may occur by sexual contact, blood transmission and needle sharing, and rarely by vertical transmission from mother to infant. Andreasson *et al.* (1993) found that none of 51 infants of HIV-2-infected mothers from Guinea-Bissau became infected. However, Morgan *et al.* (1990) identified 8 HIV-2 infected infants in Gambia who probably were infected by their HIV-2-infected mother.

No reported cases of breast milk HIV-2 transmission have been reported. This may be due to the low viral load and lack of immunologic progression among HIV-2 infected mothers.

Recommendation: There are insufficient data to make a firm recommendation; however, breast milk transmission is possible.

7. HTLV-1

Human T cell lymphotropic virus type 1 (HTLV-1) causes Japanese Adult T cell leukemia/lymphoma (ATL), tropical spastic paraparesis and several other diseases including uveitis and arthritis (Centers for Disease Control, 1993). It is endemic in southern Japan, the Caribbean basin, Melanasia and in parts of Africa, but is rare in the United States. Its seroprevalence in US blood donors was .016% in 1989 (Quinn *et al.*, 1989) and is probably similar nowadays.

There is a particularly high seroprevalence in southwestern Japan near Nagasaki (15%), such that 1000 HTLV-1^+ women deliver an infant annually (Hino, 1989; Takezaki *et al.*, 1995). There are 50–100 cases of ATL per year and 40,000 carriers over age 40 of which 5% may develop ATL. Patients acquiring HTLV-1 as adults via blood transfusion or through male-female sexual contact are less likely to develop ATL than are those who have carried the virus for a lifetime.

Twenty to twenty-five percent of infants of HTLV-1 carrier mothers who are breast-fed will become infected, compared to 0–4% of non-breast-fed infants (Hino, 1988; Hino *et al.*, 1995). Hino *et al.* (1987) reported that 17 of 78 breast-fed children of carrier mothers (22%) were HTLV-1 positive whereas only 36 of 3799 control children of non-carrier mothers (1%) were HTLV-1 positive. None of 39 formula-fed infants of HTLV-1^+ carrier mothers became infected.

The factors associated with breast milk transmission of HTLV-1 include advanced maternal age, high titers of anti-HTLV-1 antibody in maternal sera and long duration of breast-feeding (Hino *et al.*, 1995). Thus breast-feeding is the main route of vertical transmission, although intrauterine and peripartum transmission may occur occasionally. HTLV-1 infected lymphocytes have been identified in the breast milk of HTLV-1$^+$ mothers. Concentrated breast milk and infected breast milk cells from HTLV-infected mothers can infect marmoset monkeys via the oral route (Hino, 1989).

Recommendation: HTLV-1-infected mothers should not breast-feed their infants.

8. HTLV-2

HTLV-2 is closely related to HTLV-1 but is serologically distinct. It is present in injecting drug users in the USA and Europe and in some Indian tribes in Panama, Florida and New Mexico (Cortes *et al.*, 1989; Feigal *et al.*, 1991). Earlier serologic testing could not differentiate between HTLV-1 and HTLV-2; it is now evident that a majority of these positive tests in the USA were due to HTLV-2. There are no firm disease associations. However, HTLV-2 has been identified in two (but not subsequent) patients with hairy cell leukemia and in a few patients with unusual lymphomas (Centers for Disease Control, 1993).

HTLV-2 can occasionally be transmitted vertically before or at the time of birth and by breast-feeding. Kaplan *et al.* (1992) studied 20 non-breast-fed infants of 19 HTLV-2$^+$ mothers and none were infected. However, HTLV-2 was identified in 10 of 12 breast milk samples from HTLV-2 infected mothers (Heneine *et al.*, 1992). Van Dyke *et al.* (1995) reported that none of 28 bottle-fed newborn children born to HTLV-2 infected mothers were infected with the virus, whereas 2 of 16 siblings were infected, including 1 of 7 (14%) who was breast-fed.

Recommendation: HTLV-2-infected mothers should not breast-feed their infants.

9. Hepatitis A

Vertical transmission of hepatitis A virus (HAV) is extremely rare even when the mother is icteric (Tong *et al.*, 1981; Zhang *et al.*, 1990). One acutely infected mother spread infection to her breast-fed newborn but breast milk transmission was not proven (Watson *et al.*, 1993).

Recommendation: Breast-feeding can be started or continued by an acutely infected HAV-positive mother. The infant should be given human immunoglobulin to prevent transmission by the usual enteric routes.

10. Hepatitis B

Vertical transmission of Hepatitis B virus (HBV) is not uncommon in mother-infant pairs in which the mother is a chronic carrier or is acutely infected. The vertical transmission rate varies widely (from 1 to 50%) in different areas of the world and is dependent on the titer of HBsAg (surface antigen denoting presence of circulating virus, both replicating and non-replicating) and particularly the titer of HBeAg (a peptide of the viral core antigen, reflecting active viral replication) in the maternal serum. Most vertical transmission occurs peripartum (95%) but occasionally may occur in utero. The infected infant, usually asymptomatic, has a high likelihood of becoming a chronic carrier (American Academy of Pediatrics, 1997).

Transmission by breast-feeding is possible because HBsAg is present in the breast milk of infected mothers (Lin *et al.*, 1993) and serum containing HBsAg is infectious by the oral route (Giles *et al.*, 1969). Indeed, 71% of samples from the breast milk of carrier mothers were positive for HBsAg (Lee *et al.*, 1978). The colostrum may be particularly rich in HBsAg, especially in mothers who were HBeAg positive (Lin *et al.*, 1993).

Despite these laboratory studies, clinical studies do not indicate a high rate of breast milk transmission. Beasley *et al.* (1975) found no difference in the rate of HBV transmission by Taiwanese women to their breast-fed (45/92; 49%) or bottle-fed (29/55; 53%) infants. De Martino *et al.* (1985) found that 21 of 22 breast-fed infants and 61 of 63 bottle-fed infants of $HbsAg^{+}$ mothers responded to HBV vaccine; no infants became carriers. Thus vaccine should be given to all infants.

Recommendation: Breast milk transmission is theoretically possible, so if the carrier woman refuses passive-active immunization for her infant, breast-feeding should not be recommended. If hepatitis B immunoglobulin and HBV vaccines are given, breast-feeding poses no additional risk to the infant.

11. Hepatitis C

Hepatitis C virus (HCV) is the main cause of non-A, non-B hepatitis; about 2% of the US population are carriers, many of whom will develop clinical hepatitis, cirrhosis or hepatoma after several years (Palomba *et al.*, 1996). Transmission occurs by sexual contact, shared needles, blood transfusion, organ donation and occasionally by the vertical route (Nagata *et al.*, 1992). Vertical transmission is variable. Ohto (1994) described transmission to 5 of 54 infants (10%) born to HCV-infected carrier mothers; transmission was correlated with high titers of HCV RNA, as determined by PCR. Vertical transmission is also correlated with the presence of maternal HIV coninfection (Zuccotti *et al.*, 1992), vaginal delivery (as opposed to Cesarean section) and certain HCV genotypes (Paccagnini *et al.*, 1995).

Breast milk transmission may also occur, since HCV has occasionally been identified in colostrum (Lin *et al.*, 1993) and breast milk (Gürakan *et al.*, 1994, Croxson *et al.*, 1997) by RNA-PCR. Ohto *et al.* (1994) provided preliminary evidence that transmission was correlated with duration of breast-feeding.

Recommendation: Although the US Public Health Service and an NIH consensus statement (NIH, 1997) suggest that HCV is not a contraindication to breast-feeding, the occasional presence of virus in colostrum and milk of carrier mothers makes breast milk transmission a possibility. Because of the consequences of lifelong hepatitis C infection, breast-feeding is not recommended.

12. Summary

Breast milk transmission of maternal viral infection is well established for CMV and HIV-1. In the case of CMV, this usually does not pose a risk to the infant since serious disease is prevented by placentally transferred maternal antibody. However, in HIV infection, breast-feeding increases the risk of maternal-fetal transmission by about 25% with late breast-feeding (after six months of age) constituting a particular risk.

In other maternal viral diseases, e.g., other herpes viruses, parvovirus, hepatitis A, B and C, and rubella, the virus is often demonstrated in the breast milk, but transmission is very rare. The highest risk is during an acute viral infection at the time of birth, since the breast milk has a high titer of virus, and a lack of antibody to neutralize the organism.

References

Ahlfors, K. and Ivarsson, S.-A. 1985. Cytomegalovirus in breast milk of Swedish milk donors. *Scand J. Infect Dis.* **17**:11.

American Academy of Pediatrics. 1997. Parvovirus B19, in: *Red Book: Report of the Committee on Infectious Diseases*, 24th Edition. G. Peter, ed. Elk Grove Village, IL, p. 383.

American Academy of Pediatrics. 1997. Recommendation for Care of Children in Special Circumstances, in: *Red Book: Report of the Committee on Infectious Diseases*, 24th Edition. G. Peter, ed. Elk Grove Village, IL, p. 74.

American Academy of Pediatrics. 1997. Hepatitis B, in: *Red Book: Report of the Committee on Infectious Diseases*, 24th Edition. G. Peter, ed. Elk Grove Village, IL, p. 247.

American Academy of Pediatrics. 1997. HIV Infection, in: *Red Book: Report of the Committee on Infectious Diseases*, 24th Edition. G. Peter, ed. Elk Grove Village, IL, p. 279.

Andreasson, P.A., Dias, F., Naucler, A., Andersson, S., and Biberfeld, G. 1993. A prospective study of vertical transmission of HIV-2 in Bissau, Guinea-Bissau. *AIDS* **7**:989.

Asanuma, H., Numazaki, K., Nagata, N., Hotsubo, T., Horino, K., and Chiba, S. 1996. Role of milk whey in the transmission of human cytomegalovirus infection by breast milk. *Microbiol. Immunol.* **40**:201.

Beasley, R.P., Stevens, C.E., Shiao, I.S., and Meng, H.C. 1975. Evidence against breast-feeding as a mechanism for vertical transmission of hepatitis B. *Lancet* **2**:740.

Bertolli, J., St Louis, M.E., Simonds, R.J., Nieburg, P., Kamenga, M., Brown, C., Tarande, M., Quinn, T., and Ou, C.Y. 1996. Estimating the timing of mother-to-child transmission of human immunodeficiency virus in a breast-feeding population in Kinshasa, Zaire. *J. Infect. Dis.* **174**:722.

Bobat, R., Moodley, D., Coutsoudis, A., and Coovadia, H. 1997. Breastfeeding by HIV-1-infected women and outcome in their infants: a cohort study from Durban, South Africa. *AIDS* **11**:1627.

Buimovici-Klein, E., Hite, R.L., Byrne, T., and Cooper, L.Z. 1977. Isolation of rubella virus in milk after postpartum immunization. *J. Pediatr.* **91**:939.

Centers for Disease Control and Prevention and the U.S.P.H.S. Working Group. 1993. Guidelines for counseling persons infected with human T-lymphotropic virus type I (HTLV-I) and type II (HTLV-II). *Ann. Intern. Med.* **118**:448.

Cheeseman, S.H. and McGraw, B.R. 1983. Studies on cytomegalovirus in human milk. *J. Infect. Dis.* **148**:615.

Cortes, E., Detels, R., Aboulafia, D., Li, X.L., Moudgil, T., Alam, M., Bonecker, C., Gonzaga, A., Oyafuso, L., Tondo, M., Boite, C., Hammershlak, N., Capitani, C., Slamon D.J., and Ho, D.D. 1989. HIV-1, HIV-2, and HTLV-I infection in high-risk groups in Brazil. *N. Engl. J. Med.* **320**:953.

Croxson, M., Couper, A., Voss, L., Groves, D., and Gunn, T. 1997. Vertical transmission of hepatitis C virus in New Zealand. *N.Z. Med. J.* **110**:165.

Datta, P., Embree, J.E., Kreiss, J.K., Ndinya-Achola, J.O., Braddick, M., Temmerman, M., Nagelkerke, N.J.D., Maitha, G., Holmes, K.K., Piot, P., Pamba, H.O., and Plummer, F.A. 1994. Mother-to-child transmission of human immunodeficiency virus type 1: report from the Nairobi Study. *J. Infect. Dis.* **170**:1134.

de Martino, M., Tovo, P.A., Tozzi, A.E., Pezzotti, P., Galli, L., Livadiotti, S., Caselli, D., Massironi, E., Ruga, E., Fioredda, F., Plebani, A., Gabiano, C., and Zuccotti, G.V. 1992. HIV-1 transmission through breast-milk: appraisal of risk according to duration of feeding. *AIDS* **6**:991.

de Martino, M., Appendino, C., Resti, M., Rossi, M.E., Muccioli, A.T., and Vierucci, A. 1985. Should hepatitis B surface antigen positive mothers breast feed? *Arch. Dis. Child.* **60**:972.

Diosi, P., Babusceac, L., Nevinglovschi, O., and Kun-Stoiciu, G. 1967. Cytomegalovirus infection associated with pregnancy. *Lancet* **2**:1063.

Dunkle, L.M., Schmidt, R.R., and O'Connor, D.M. 1979. Neonatal herpes simplex infection possibly acquired via maternal breast milk. *Pediatrics* **63**:250.

Dunn, D.T., Newell, M.L., Ades, A.E., and Peckham, C.S. 1992. Risk of human immunodeficiency virus type 1 transmission through breastfeeding. *Lancet* **340**:585.

Dunne, W.M., Jr. and Jevon, M. 1993. Examination of human breast milk for evidence of human herpesvirus 6 by polymerase chain reaction. *J. Infect. Dis.* **168**:250.

Dworsky, M., Stagno, S., Pass, R.F., Cassady, G., and Alford, C. 1982. Persistence of cytomegalovirus in human milk after storage. *J. Pediatr.* **101**:440.

Dworsky, M., Yow, M., Stagno, S., Pass, R.F., and Alford, C. 1983. Cytomegalovirus infection of breast milk and transmission in infancy. *Pediatrics* **72**:295.

Feigal, E., Murphy, E., Vranizan, K., Bacchetti, P., Chaisson, R., Drummond, J.E., Blattner, W., McGrath, M., Greenspan, J., and Moss, A. 1991. Human T cell lymphotropic virus types I and II in intravenous drug users in San Francisco: risk factors associated with seropositivity. *J. Infect. Dis.* **164**:36.

Frederick, I.B., White, R.J., and Braddock, S.W. 1986. Excretion of varicella-herpes zoster virus in breast milk. *Am. J. Obstet. Gynecol.* **154**:1116.

Friis, H. and Anderson, H.K. 1982. Rate of inactivation of cytomegalovirus in raw banked milk during storage at −20°C and pasteurization. *Br. Med. J.* **285**:1604.

Giles, J.P., McCollum, R.W., Berndtson, L.W. Jr., and Krugman, S. 1969. Relation of Australia-SH antigen to the willowbrook MS-2 strain. *N. Engl. J. Med.* **281**:119.

Gürakan, B., Oran, O., and Yigit, S. 1994. Vertical transmission of hepatitis C virus [Letter]. *N. Engl. J. Med.* **331**:399.

Hayes, K., Danks, D.M., Gibas, H., and Jack, I. 1972. Cytomegalovirus in human milk. *N. Engl. J. Med.* **287**:177.

Heneine, W., Khabbaz, R.F., Lal, R.B., and Kaplan, J.E. 1992. Sensitive and specific polymerase chain reaction assays for diagnosis of human T-cell lymphotropic virus type I HTLV-I) and HTLV-II infections in HTLV-I/II-seropositive individuals. *J. Clin. Microbiol.* **30**:1605.

Hino, S., Sugiyama, H., Doi, H., Ishimaru, T., Yamade, T., Tsuji, Y., and Miyamoto, T. 1997. Breaking the cycle of HTLV-1 transmission via carrier mothers' milk. *Lancet* **2**:158.

Hino, S. 1989. Milk-borne transmission of HTLV-1 as a major route in the endemic cycle. *Acta Paediatr. Jpn.* **31**:428.

Hino, S., Katamine, S., Miyamoto, T., Doi, H., Tsuji, Y., Yamabe, T., Kaplan, J.E., Rudolph, D.L., and Lal, R.B. 1995. Association between maternal antibodies to the external envelope glycoprotein and vertical transmission of human T-lymphotropic virus type I. Maternal anti-env antibodies correlate with protection in non-breast-fed children. *J. Clin. Invest.* **95**:2920.

Hotsubo, T., Nagata, N., Shimada, M., Yoshida, K., Fujinaga, K., and Chiba, S. 1994. Detection of human cytomegalovirus DNA in breast milk by means of polymerase chain reaction. *Microbiol. Immunol.* **38**:809.

Junker, A.K., Thomas, E.E., Radcliffe, A., Forsyth, R.B., George, A., Davidson, F., and Rymo, L. 1991. Epstein-Barr virus shedding in breast milk. *Am. J. Med. Sci.* **302**:220.

Kaplan, J.E., Abrams, E., Shaffer, N., Cannon, R.O., Kaul, A., Krasinski, K., Bamji, M., Hartley, T.M., Roberts, B., Kilbourne, B., Thomas, P., Rogers, M., Heneine, W., and the NYC Perinatal HIV Transmission Collaborative Study. 1992. Low risk of mother-to-child transmission of human T lymphotropic virus type II in non-breast-fed infants. The NYC Perinatal HIV Transmission Collaborative Study. *J. Infect. Dis.* **166**:892.

Karlsson, K., Massawe, A., Urassa, E., Kawo, G., Msemo, G., Kazimoto, T., Lyamuyam E., Mbena, E., Urassa, W., Bredberg-Raden, U., Mhalu, F., and Biberfeld, G. 1997. Late postnatal transmission of human immunodeficiency virus type 1 infection from mothers to infants in Dar es Salaam, Tanzania. *Pediatr. Infect. Dis. J.* **16**:963.

Keyserling, H.L. 1997. Other viral agents of perinatal importance. Varicella, parvovirus, respiratory syncytial virus, and enterovirus. *Clin. Perinatol.* **24**:193.

Kibrick, S. 1979. Herpes simplex virus in breast milk. *Pediatrics* **64**:390.

Klein, B., Byrne, T., and Cooper, L.Z. 1980. Neonatal rubella in a breast-fed infant after postpartum maternal infection. *J. Pediatr.* **97**:774.

Kreiss, J. 1997. Breastfeeding and vertical transmission of HIV-1. *Acta Paediatr. Suppl.* **421**:113.

Kusuhara, K., Takabayashi, A., Ueda, K., Hidaka, Y., Minamishima, I., Take, H., Fujioka, K., Imai, S., and Osato, T. 1997. Breast milk is not a significant source for early Esptein-Barr virus or human herpesvirus 6 infection in infants: a seroepidermiologic study. *Microbiol. Immunol.* **41**:309.

Lee, A.K.Y., Ip, H.M., and Wong, V.C.W. 1978. Mechanisms of maternal-fetal transmission of hepatitis B virus. *J. Infect. Dis.* **138**:668.

Lin, H.H., Hsu, H.Y., Chang, M.H., Chen, P.J., and Chen, D.S. 1993. Hepatitis B virus in the colostra of HBeAg-positive carrier mothers. *J. Pediatr. Gastroenterol. Nutr.* **17**:207.

Losonsky, G.A., Fishaut, J.M., Strussenberg, J., and Ogra, P.L. 1982. Effect of immunization against rubella on lactation products. II. Maternal-neonatal interactions. *J. Infect. Dis.* **145**:661.

Minamishima, I., Ueda, K., Minematsu, T., Minamishima, Y., Umemoto, M., Take, H., and Kuraya, K. 1994. Role of breast milk in acquisition of cytomegalovirus infection. *Microbiol. Immunol.* **38**:549.

Mofenson, L.M. 1997. Interaction between timing of perinatal human immunodeficiency virus infection and the design of preventive and therapeutic interventions. *Acta Paediatr. Suppl.* **421**:1.

Morgan, G., Wilkins, H.A., Pepin, J., Jobe, O., Brewster, D., and Whittle, H. 1990. AIDS following mother-to-child transmission of HIV-2. *AIDS* **4**:879.

Morrison, J. and Alp, N.J. 1997. Vertical transmission of human immunodeficiency virus. *QJ Med.* **90**:5.

Nagata, I., Shiraki, K., Tanimoto, K., Harada, Y., Tanaka, Y., and Okada, T. 1992. Mother-to-infant transmission of hepatitis C virus. *J. Pediatr.* **120**:432.

Nagelkerke, N.J.D., Moses, S., Embree, J.E., Jenniskens, F., and Plummer, F.A. 1995. The duration of breastfeeding by HIV-1-infected mothers in developing countries: balancing benefits and risks. *J. Acquir. Immune Defic. Syndr. Hum. Retrovirol.* **8**:176.

National Institutes of Health. 1997. Management of Hepatitis C. National Institutes of Health Consensus Statement 1997, March **15**:3, 1.

Ohto, H., Okamoto, H., and Mishiro, S. 1994. Vertical transmission of hepatitis C virus. *N. Engl. J. Med.* **331**:400.

Ogra, P.L. and Rassin, D.K. 1995. Human breast milk, in: *Infectious Diseases of the Fetus and Newborn.* J.S. Remington and J.O. Klein, eds., W.B. Saunders, Philadelphia, p. 108.

Oxtoby, M.J. 1988. Human immunodeficiency virus and other viruses in human milk: placing the issues in broader perspective. *Pediatr. Infect. Dis. J.* **7**:825.

Paccagnini, S., Principi, N., Massironi, E., Tanzi, E., Romano, L., Muggiasca, M.L., Ragni, M.C., and Salvaggio, L. 1995. Perinatal transmission and manifestation of hepatitis C virus infection in a high risk population. *Pediatr. Infect. Dis. J.* **14**:195.

Palomba, E., Manzini, P., Fiammengo, P., Maderni, P., Saracco, G., and Tovo, P.A. 1996. Natural history of perinatal hepatitis C virus infection. *Clin. Infect. Dis.* **23**:47.

Peckham, C.S., Johnson, C., Ades, A., Pearl, K., and Chin, K.S. 1987. Early acquisition of cytomegalovirus infection. *Arch. Dis. Child.* **62**:780.

Quinn, T.C., Zacarias, F.R., and St. John, R.K. 1989. AIDS in the Americas: an emerging public health crises. *N. Engl. J. Med.* **320**:1005.

Ruff, A.J. 1994. Breastmilk, breastfeeding, and transmission of viruses to the neonate. *Semin. Perinatol.* **18**:510.

Stagno, S., Reynolds, D.W., Pass, R.F., and Alford, C.A. 1980. Breastmilk and the risk of cytomegalovirus infection. *N. Engl. J. Med.* **302**:1073.

Stiehm, E.R. and Vink, P. 1991. Transmission of human immunodeficiency virus infection by breast-feeding. *J. Pediatr.* **118**:410.

Takezaki, T., Tajima, K., Komoda, H., and Imai, J. 1995. Incidence of human T lymphotropic virus type I seroconversion after age 40 among Japanese residents in an area where the virus is endemic. *J. Infect. Dis.* **171**:559.

Tantivanich, S., Supavej, S., Chityothin, O., Katepunya, A., and Tharavanij, S. 1983. Herpes simplex virus and genital infections in Thai pregnant women. *Southeast Asian J. Trop. Med. Pub. Hlth.* **14**:195.

Tong, M.J., Thursby, M., Rakela, J., McPeak, C., Edwards, V.M., and Mosley, J.W. 1981. Studies on the maternal-infant transmission of the viruses which cause acute hepatitis. *Gastroenterology* **80**:999.

Van de Perre, P., Simonon, A., Msellati, P., Hitimana, D.G., Vaira, D., Bazubagira, A., Van Goethem, C., Stevens, A.M., Karita, E., Sondag-Thull, D., Dabis, F., and Lepage P. 1991. Postnatal transmission of human immunodeficiency virus type 1 from mother to infant. A prospective cohort study in Kigali, Rwanda. *N. Engl. J. Med.* **325**:593.

Van Dyke, R.B., Heneine, W., Perrin, M.E., Rudolph, D., Starszak, E., Woods, T., Switzer, W.M., and Kaplan, J.E. 1995. Mother-to-child transmission of human T-lymphotropic virus type II. *J. Pediatr.* **127**:924.

Watson, J.C., Fleming, D.W., Borella, A.J., Olcott, E.S., Conrad, R.E., and Baron, R.C. 1993. Vertical transmission of hepatitis A resulting in an outbreak in a neonatal intensive care unit. *J. Infect. Dis.* **167**:567.

Welsh, J.K., Arsenakis, M., Coelen, R.J., and May, J.T. 1979. Effect of antiviral lipids, heat, and freezing on the activity of viruses in human milk. *J. Infect. Dis.* **140**:322.

Yoshida, N., Yamagami, N., Tezuka, T., and Hondo, R. 1992. Case report: detection of varicella-zoster virus DNA in maternal breast milk. *J. Med. Virol.* **38**:108.

Zhang, R.L., Zeng, J.S., and Zhang, H.Z. 1990. Survey of 34 pregnant women with hepatitis A and their neonates. *Chin. Med. J.* **103**:552.

Ziegler, J.B., Cooper, D.A., Johnson, R.O., and Gold, J. 1985. Postnatal transmission of AIDS-associated retrovirus from mother to infant. *Lancet* **1**:896.

Zimmermann, R., Perucchini, D., Fauchère, J.-C., Joller-Jemelka, H., Geyer, M., Huch, R., and Huch, A. 1995. Hepatitis C virus in breast milk. *Lancet* **345**:928.

Zuccotti, G.V., Ribero, M.L., Giovannini, M., Fasola, M., Riva, E., Portera, G., Biasucci, G., Decarlis, S., Profeta, M.L., and Tagger, A. 1995. Effect of hepatitis C genotype on mother-to-infant transmission of virus. *J. Pediatr.* **127**:278.

Chapter 6

Breast Milk Decreases the Risk of Neonatal Necrotizing Enterocolitis

Ann M. Kosloske

1. Introduction

Necrotizing enterocolitis (NEC), a syndrome that strikes premature infants, has become a leading cause of infant mortality in the United States, Canada, and many other countries. The pathogenesis of NEC is poorly understood, although leading theories have implicated bacteria, intestinal ischemia, formula feedings, immaturity of the neonatal gut and combinations thereof (Santulli *et al.*, 1975, Lawrence *et al.*, 1982, Kosloske, 1990). This article summarizes our current understanding of NEC, including its history, clinical management, epidemiology and outcomes. Current theories of pathogenesis and prevention are outlined, with emphasis on the single agent with proven effectiveness for prevention: breast milk.

2. History of NEC

Case reports of "idiopathic" gastrointestinal perforation in newborn infants as early as 1825 (Siebold) probably included instances of NEC. Rossier *et al.* (1959) described a syndrome which they characterized as "l'enterocolite ulceronecrotique du premature." Fourteen of 15 premature infants in their series died with intestinal gangrene and perforation. In the 1960s and 1970s, NEC

Ann M. Kosloske • Departments of Surgery and Pediatrics, Texas Tech University School of Medicine, Lubbock, TX 79415

emerged as a major cause of infant mortality and morbidity, parallel to the establishment of neonatal intensive care units (NICU). Infants who formerly would have died of hyaline membrane disease, survived because of great improvements in neonatal care, only to develop potentially-fatal NEC. The mortality from NEC in the United States was calculated in 1988 at 1.3 cases per 1,000 live births, or 15–40% of cases (Holman *et al.*, 1989). It remains the most common gastrointestinal emergency of premature infants (Kliegman and Fanaroff, 1984) and one of the most serious of pediatric nosocomial (hospital-acquired) infections.

3. Clinical NEC

The clinical presentation consists of abdominal distention, gastrointestinal bleeding and pneumatosis intestinalis (air in the bowel wall) on abdominal X-ray. In addition, infants with severe NEC may have gas within the portal vein or pneumoperitoneum following intestinal perforation (Figure 1). The afflicted infant develops feeding intolerance, vomiting, and signs of sepsis. Typically, onset occurs near the end of the first week of life or during the second, although the syndrome may appear later, even in the second month. Ninety percent of infants are premature. When NEC is suspected, feedings are stopped and intensive medical therapy, including gastric suction, intravenous fluids and antibiotics, is begun. More than half of infants with NEC recover under medical management; the others, with more severe gut pathology, require emergency operation to prevent the development of intestinal gangrene or perforation (Kliegman and Fanaroff, 1984; Kosloske, 1997a,b).

The operative findings are those of ischemia of the bowel, varying from focal dusky areas to extensive gangrene (Figure 2). Affected loops may be pale gray from infarction or purple from hemorrhage into the gut wall. "Skip areas" of viable intestine may alternate with ischemic or frankly gangrenous segments. The ileocecal area is usually the most severely affected and the site of perforations. The process may involve ileum, colon, or jejunum, in that order, although rarely even stomach, duodenum, or rectum becomes necrotic (Kosloske, 1985). The surgical goal in NEC is resection of all the gangrenous bowel, with preservation of as much length as possible (Kosloske, 1997b; Albanese and Rowe, 1998). The majority of pediatric surgeons exteriorize the marginally viable ends of bowel as an enterostomy; primary anastomosis carries greater risk of complications (e.g., leak, stricture). The enterostomy is closed weeks or months later, after the infant has recovered from acute NEC.

The histologic findings of the resected bowel typically are those of ischemic necrosis, ranging from sloughing of the mucosa to full-thickness infarction (Figure 3). Acute inflammation may be minimal and confined to the margins of

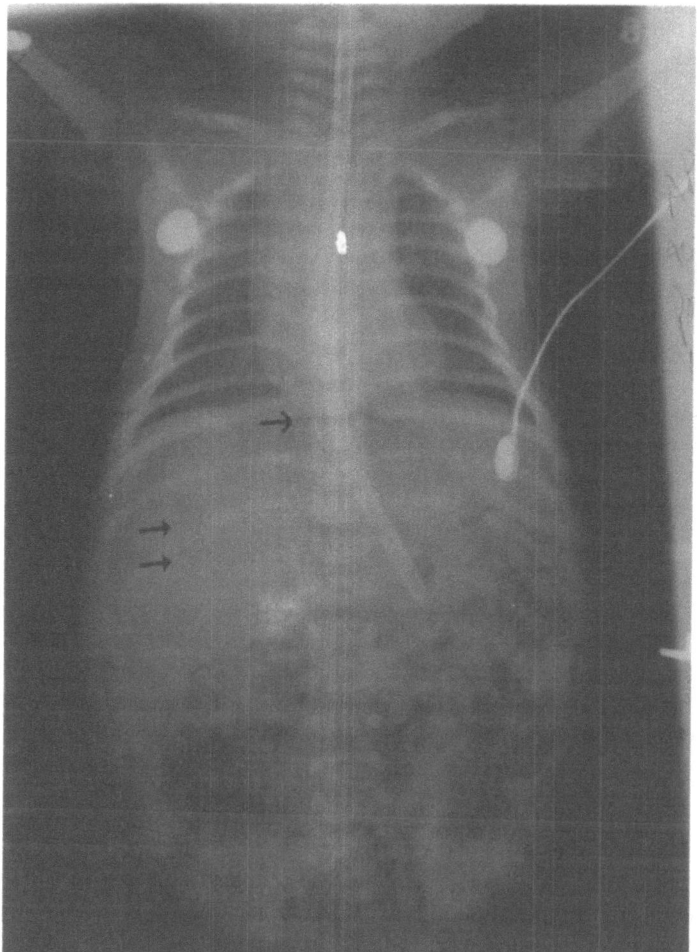

Figure 1. "Babygram" of an infant with NEC. Extensive pneumatosis intestinalis indicated by bubbly appearance of bowel throughout abdomen. Pneumoperitoneum (single arrow) and portal venous gas (double arrow) also are present.

the necrotic segment (Ballance *et al.*, 1990). Microthrombi are frequent, although major vessels are usually patent.

Improved survival from NEC over the past 3 decades has been attributed to earlier diagnosis and more effective supportive treatment of the premature infant. Currently, about two-thirds of surgical infants survive operation (Kosloske, 1997b; Albanese and Rowe, 1998), although some die later from complications

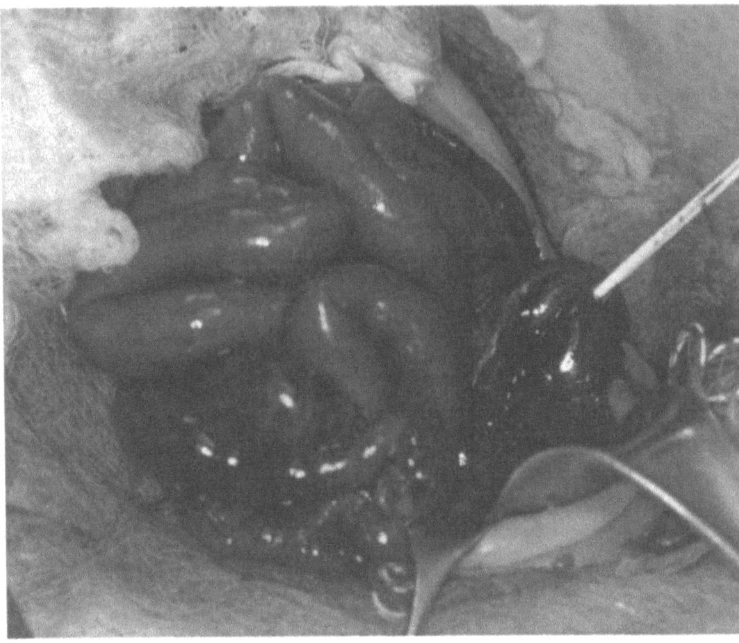

Figure 2. Photograph showing gangrenous loops of bowel in foreground and viable bowel in background. (Reprinted with permission, from: Kosloske, A.M. 1996. Necrotizing enterocolitis, in: *Newborn Surgery*, P. Puri, ed. Butterworth-Heinemann, Oxford, p.358.)

of NEC (e.g., short bowel syndrome, sepsis) or complications of prematurity (e.g., respiratory failure, sepsis). Since virtually all medically-treated infants survive the acute episode of NEC, the overall survival rate approximates 75–80%. The most common late complication of NEC is intestinal stricture, which develops in 15–25% of survivors (Kosloske *et al.*, 1980) from cicatricial healing of a site that sustained ischemic injury. Strictures usually require another operation for resection, although minor strictures distal to an enterostomy may be safely managed by balloon dilatation (Ball *et al.*, 1985).

4. Pathogenesis

Although the pathogenesis of NEC is poorly understood, a multifactorial process is invoked. The theories of Santulli *et al.* (1975) and of Lawrence *et al.* (1982) cite factors that may initiate NEC. Neither theory fully explains the clinical experience, although each has supportive clinical or experimental evidence.

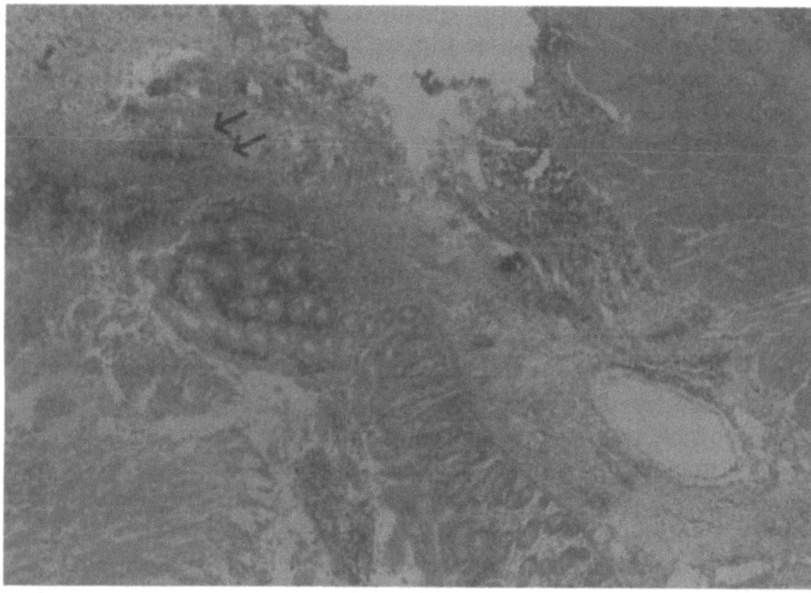

Figure 3. Photomicrograph of intestine resected for NEC, showing severe ischemic change of intestine, and focus of frank gangrene (arrows). (Hematoxylin & eosin; original magnification ×100.) (Reprinted with permission, from: Kosloske, A.M. 1997. Necrotizing enterocolitis, in: *Surgery of Infants and Children*, K.T. Oldham *et al.*, eds. Lippincott-Raven, Philadelphia, p.1202.)

Santulli *et al.* (1975) attributed NEC to the interaction of "three essential components": injury to the intestinal mucosa, the presence of bacteria, and the availability of a metabolic substrate, usually formula feedings, in the lumen of the gut. The hypothesis of Lawrence *et al.* (1982) emphasized the aberrant and delayed gut colonization of babies in the NICU, and the immunologic immaturity of the neonatal gut, which is vulnerable to direct damage from bacterial toxins in the gut lumen.

4.1. Ischemia

The intestinal injury which may precede NEC was linked by Lloyd (1969) with "the diving reflex," a physiologic defense mechanism that occurs in many species of birds and mammals in response to asphyxia (Scholander, 1963; Nowicki and Miller, 1992; Krissinger, 1994). Premature infants are prone to episodes of hypoxia associated with the immaturity of their lungs. During hypoxia, blood is shunted selectively to the heart and brain, away from the gut and other organs. After oxygenation is restored, reactive hyperemia occurs in these organs. The resulting ischemia-reperfusion insult is believed to initiate NEC. Reperfusion of ischemic tissue triggers a chain of biochemical events leading to necrosis; a key step is the release of free radicals of oxygen, whose toxic chemical reactivity is a function of a single unpaired electron in the outer shell (Amoury, 1993). Thus, the enzyme xanthine oxidase, as well as hydrogen peroxide (H_2O_2) and cytotoxic free radicals, including superoxide (O_2^-) and the hydroxyl radical (HO^-), are released by polymorphonuclear neutrophils, and these inflammatory mediators generated during the process of injury may propagate the necrosis (Caplan and MacKendrick, 1994). Pharmacologic agents which are free radical anion scavengers (e.g., superoxide dismutase) or inhibitors of xanthine oxidase (e.g., allopurinol) might limit the necrosis induced by ischemia-reperfusion. Although several such agents have been studied experimentally, none has yet been evaluated in a clinical trial.

Other possible pathways for injury to the neonatal intestine include: (1) micro-emboli from umbilical arterial catheters inserted into the aorta for monitoring of arterial blood gases and blood pressure in the premature infant (Tyson *et al.*, 1976); (2) microthrombi from umbilical venous catheters which may communicate with the portal system; (3) decreased mesenteric flow associated with congenital heart defects (Leung *et al.*, 1988); (4) low-flow states from sepsis or other types of shock; (5) hyperviscosity; (6) a hyperosmolar flush of the mesenteric circulation from contrast media injected during cardiac catheterization (Cooke *et al.*, 1980); (7) hyperosmolar liquid in the bowel lumen (e.g., certain formulas or medications); (8) maternal use of cocaine, a potent vasoconstrictor. (Czyrko *et al.*, 1991); or (9) hypothermia, which experimentally produced intestinal ischemia more prolonged than that following hypoxia (Powell *et al.*, 1999). Episodes of hypoxic or other injury may be documented or inferred in the majority of preterm infants in NICU. It is not clear, however, why NEC develops in 5–8% of such infants, yet the remaining 92–95% of infants never develop the syndrome.

4.2. Bacteria

Although an infectious agent for NEC has been diligently sought, no single microbial pathogen has been found. The bacteria isolated from the blood or peritoneal fluid in NEC are, with few exceptions, members of the normal flora of the neonatal gut, most commonly *Klebsiella, E. coli*, or species of *Clostridia* (Guinan *et al.*, 1979; Kliegman *et al.*, 1979; Kosloske *et al.*, 1985; Speer *et al.*, 1976). *Clostridium perfringens*, the ultimate exotoxin-forming species, is associated with a fulminant, highly-lethal form of NEC (Kosloske *et al.*, 1978; Blakey *et al.*, 1985). The occurrence of NEC is usually sporadic, although occasionally it appears in epidemic-like clusters (Book *et al.*, 1977) during which a predominant organism may be isolated. Salmonella (Stein *et al.*, 1972), rotavirus (Rotbart *et al.*, 1988), and coronavirus (Chany *et al.*, 1982) are among the agents implicated in clusters of cases of NEC. The documentation of asymptomatic carriers in such epidemics of NEC suggests that impaired host resistance may be an important factor in the pathogenesis.

4.3. Feedings: Formula

Most infants with NEC had been fed a formula based on cow's milk. Although hyperosmolar formulae have been implicated in a few instances (de Lemos *et al.*, 1974, Book *et al.*, 1975), the mechanism by which isosmolar milk formulas may lead to NEC has not been established. A study by Engel and associates (1973) implicated feedings in the etiology of the pneumatosis intestinalis, i.e., the gas blebs which occur in the intestinal wall. Analysis of the gas, which was sampled from subserosal blebs at operation, showed that hydrogen predominated, a finding which supported the hypothesis that enteric bacteria, acting on formula as a substrate in the intestinal lumen, produce pneumatosis intestinalis. Conversely, pneumatosis intestinalis was found in only 57% of unfed infants who developed NEC (Marchildon *et al.*, 1982). Although some investigators have attributed NEC to overfeeding of premature infants, prospective clinical studies have failed to demonstrate that postponing enteral feeding prevents NEC. Rather, it may simply delay the onset of disease (Book *et al.*, 1976; Ostertag *et al.*, 1986). The option of total parenteral nutrition (TPN) may spare the gut but carries its own hazards (e.g., sepsis, cholestasis). The current trend in feeding the premature infant is by a combination of parenteral nutrition and low-volume enteral feeding (gut "priming"), which may induce adaptation of the immature gut. The role of feeding in the pathogenesis of NEC has been reviewed by Williams (1997).

4.4. Feedings: Breast Milk

NEC is rare in infants fed breast milk alone. In the animal model of Barlow *et al.* (1974), maternal milk protected neonatal rats from an intestinal necrosis that resembled NEC. Extrapolation of murine models to human infants is hazardous, however, because of immunological differences. In particular, rat pups receive their systemic immunoglobulin antibody (IgG) via the enteral route, whereas human infants receive IgG transplacentally. Human breast milk furnishes mucosal antibody (secretory IgA) to the neonatal intestine, as well as macrophages, lactoferrin, epidermal growth factor and other substances. The prospective multicenter study of Lucas and Cole (1990) documented a low incidence of NEC among premature babies fed breast milk alone. NEC was 6–10 times more common in those who received only formula feedings and three times more common in those who received formula plus breast milk. Pasteurized breast milk was as effective as raw milk, and the protective effect appeared to be strongest among infants 34–36 weeks gestation.

The precise mechanism by which human milk confers protection against NEC has yet to be elucidated. A variety of factors in milk have antimicrobial, anti-inflammatory or immunomodulating properties (Goldman, 1993; Buescher, 1994). In an experimental study (Go *et al.*, 1994), breast milk was compared to other feedings using bacterial translocation as a marker for the permeability of the mucosal barrier. Newborn rabbits fed breast milk had no bacterial translocation to the mesenteric lymph nodes or liver although 9% of the animals exhibited translocation to the spleen. In contrast, animals fed formula alone had incidences of 88%, 60% and 32% bacteria in the mesenteric nodes, liver and spleen, respectively. Stored breast milk was as effective as fresh breast milk in preventing bacterial translocation, suggesting that the cellular component of breast milk is not the protective element. Investigators from Michigan, using the same model, demonstrated a decrease in formula-associated bacterial translocation by the subcutaneous administration of epidermal growth factor (EGF) (Okuyama *et al.*, 1998), a component of breast milk, suggesting that EGF may be one of the protective elements. Improved barrier function was associated with an increase in the number of goblet cells in the mucosa of the intestinal villi.

4.5. Gut Colonization

Bacterial colonization of the neonatal gut begins by contact with the vaginal flora and is propagated by oral feedings and exposure to the environment. By 10 days of life normal infants are colonized by a range of aerobic and anaerobic species (Long and Swenson, 1977). In the aseptic environment of the NICU, however, colonization is delayed and limited to a small number of bacterial species. The use of antibiotics and sterile nursing practices in the NICU may exert the harmful effect of selection of an aberrant bacterial flora, characterized by antibiotic resistance and toxin production (Lawrence *et al.*, 1982). Damage to the mucosal enterocyte from such toxins may initiate NEC.

Bacterial overgrowth in the premature gut has been attributed to three different mechanisms: (1) formula feedings, (2) intestinal stasis, and (3) hypochlorhydria. Both clinical and experimental studies have demonstrated predominance of *Klebsiella* and other members of the family *Enterobacteriaceae* in the gut flora of formula-fed infants, in contrast to the *Lactobacilli* and *Bifidobacteria*, which predominate in the flora of breast-fed infants (Cooperstock and Zedd, 1983; Millar *et al.*, 1992).

Stasis in the neonatal gut may play a role in the pathogenesis. A surprisingly high incidence of postoperative NEC (18.5%) was reported in a series from

Michigan (Oldham *et al.*, 1988) following repair of gastroschisis, an abdominal wall defect often associated with delayed and sluggish gastrointestinal function. Jayanthi and associates (1998) from Great Ormond Street confirmed a protective role for breast milk in a series of 60 infants following repair of gastroschisis. Postoperative NEC developed in 30% of infants who were fed solely on formula, and in 5% of those who received formula plus breast milk. None of the 12 babies fed exclusively with breast milk developed postoperative NEC.

Premature infants normally have little gastric hydrochloric acid, and this may encourage bacterial overgrowth in the stomach. A clinical study by Carrion and Egan (1990) evaluated the effect of acidified feedings on the frequency of NEC. Infants who received acidified formula not only had decreased bacterial counts in the gastric juice, but also exhibited decreased occurrence of NEC. Only 1 of 34 premature infants who received acidified feedings developed NEC, whereas 8 of 34 control infants developed NEC. Experimental studies in germ-free models further support the role of bacteria in the pathogenesis of NEC. For example, Lawrence *et al.* (1982) produced necrotic enteritis by feeding any of six toxin-forming bacteria to germ-free rat pups. Moreover, Musemeche *et al.* (1986), using germ-free and gnotobiotic rats (i.e. animals colonized by a known flora), evaluated the relative importance of ischemia, bacteria and substrate in the pathogenesis of intestinal necrosis. In this important study, an isolated segment of ileum was subjected to ischemia by transient application of a microaneurysm clip. Formula, breast milk, or saline was injected into the lumen. A second, control segment without ischemia was created in each animal. Animals were sacrificed at 48 hrs and the severity of necrosis was graded by a pathologist. The most severe necrosis occurred in the gnotobiotic rats regardless of the duration of ischemia or the nature of the luminal contents. The germ-free rats showed little or no necrosis. Thus the most important of the three factors was the presence of bacteria.

4.6. The Immature Gut

The mucosal barrier to bacteria is incompletely developed in the newborn human infant, particularly in the premature infant. Udall (1990) cited evidence that immaturity of gastrointestinal host defense mechanisms is probably the single most important factor in predisposing infants to NEC. Intestinal B and T lymphocytes are decreased in number in newborn infants, compared with older infants (Udall, 1990). The terminal ileum at birth is permeable to the passage of intact macromolecules. Translocation of whole bacteria may occur with little impediment. Closure of the mucosal barrier coincides with the synthesis at the mucosal border of adequate levels of secretory IgA, and may not occur until 2 to 3 weeks of age. During this early vulnerable period, bacterial toxins generated in the bowel lumen may damage the enterocyte, thus initiating NEC (Lawrence *et al.*, 1982). A study by Bell *et al.* (1985) revealed higher levels of serum IgA in infants with NEC than in a control group of infants. This finding infers that there is a decrease in transport of IgA to the mucosal border in NEC, a conclusion consistent with the hypothesis of Lawrence *et al.* (1982).

The development of normal gastrointestinal motility depends upon the appearance of the interstitial cells of Cajal, a specialized type of cell in the gut wall distinct from the ganglion cell. The tyrosine-kinase receptor *c-kit* is a specific marker for the interstitial cells of Cajal. C-KIT$^+$ cells, such as mast cells and intraepithelial T-lymphocytes, are also important for immune system homeostasis. Investigators from Tokyo (Yamataka *et al.*, 1998) evaluated the role of *c-kit* in both experimental and clinical models. Strains of mice depleted of *c-kit* developed spontaneous intestinal mucosal erosions simulating NEC at 14 days of age. Ten infants with NEC showed a significant decrease in C-KIT$^+$ cells, compared to 10 controls. Six other infants with enteritis from conditions other than NEC had a significant increase of C-KIT$^+$ cells compared to controls.

Prospective clinical studies suggest that enhancement of gastrointestinal host defense may decrease the incidence of NEC. The British study which demonstrated the benefits of breast milk (Lucas and Cole, 1990) revealed a lower incidence of NEC in breastfed versus formula-fed prematures. Eibl *et al.* (1988) fed an immunoglobulin mixture (IgA and IgG) to infants at risk for NEC. They recorded a decreased incidence of NEC compared to control infants, although other investigators failed to confirm their results. Halac *et al.* (1990), in a prospective study in Argentina, evaluated the effect of corticosteroid therapy, which hastens maturation of developing tissues. They found a NEC incidence of 3.4% with prenatal steroid treatment and 6.9% with postnatal steroid treatment, compared to 14.4% in control infants. Substances that are trophic to the immature gut (e.g. epidermal growth factor, interleukin-11) or that function as nitric oxide donors (e.g. nitroglycerin), have shown promise in experimental models of bowel necrosis (Lawrence *et al.*, 1997; Du *et al.*, 1997; Graf *et al.*, 1997), but have yet to be tried clinically.

5. Epidemiology

Epidemiologic studies of risk factors for NEC have failed to identify any consistent cause, except prematurity (Stoll *et al.*, 1980; Ryder *et al.*, 1980). An interaction of risk factors is probable and quantitative aspects may be crucial. For example, severe intestinal ischemia might initiate injury in combination with highly pathogenic bacteria, an excess of potentially-damaging substrate or a very immature mucosal barrier. This unifying hypothesis (Kosloske, 1990) implies a threshhold of injury, which, if exceeded, results in NEC.

Although NEC occurs in developed countries throughout the world, its distribution is uneven (Kosloske, 1997a). For example, NEC is a problem in the United States, Canada, the United Kingdom, and Australia, but is rarely encountered in Switzerland, the Scandinavian countries and especially Japan, where a survey of NICUs showed an incidence that was 4 to 28 times lower than that reported for the United States (Shimura, 1990). In general, countries that enjoy low premature birth rates have few cases of NEC. Moreover, in developing countries in which NICU technology is not available, NEC is not a problem; however, older infants in developing countries may develop a type of necrotic enteritis resembling NEC following an episode of diarrhea and dehydration (Arseculeratne *et al.*, 1980). In some cases, the necrotic enteritis has been linked with overgrowth of a strain of *Clostridium perfringens* in the lumen of the bowel (Lawrence and Walker, 1976; Butler *et al.*, 1987).

6. Prevention

The prevention of NEC awaits a clearer understanding of its pathogenesis. There are few prospective, randomized studies of interventions which might prevent NEC. Randomization and concurrent controls are essential. Because NEC may occur in epidemics, historical controls may yield deceptive data, i.e., an intervention which was introduced as an epidemic subsided could be falsely credited with preventive efficacy (Kosloske, 1994). Prevention of the primary risk factor, prematurity, goes beyond medical issues, and would require societal and behavioral changes, e.g. universal prenatal care and cessation of maternal cigarette smoking and drug abuse. Clinical trials of interventions which might prevent NEC are hampered by its relatively low incidence (3–8% of NICU admissions), thus requiring a large sample size in order to achieve adequate statistical power. As many as 800 infants (400 treated, 400 controls), for example, may be required to achieve a statistical power of 0.8 (Hennekens and Buring, 1987). Thus far, only the British multicenter study of breast milk (Lucas and Cole, 1990), which entered 936 infants, meets such criteria. Their data support the recommendation of the American Academy of Pediatrics, which, for more than 50 years, has remained a staunch advocate of breast feeding as the optimal form of nutrition for infants (Gartner, 1997). Well-designed clinical trials are needed for evaluation of additional measures which might be beneficial to infants at risk for NEC.

References

Albanese, C.T. and Rowe, M.I. 1998. Necrotizing enterocolitis, in: *Pediatric Surgery*, J.A. O'Neill, Jr., M.I. Rowe, J.L. Grosfeld, E.W. Fonkalsrud, and A.G. Coran, eds. Mosby, St. Louis, pp. 1297–1320.

Amoury, R.A. 1993. Necrotizing Enterocolitis, in: *Pediatric Surgery*, K.W. Ashcraft and T.M. Holder, eds. Saunders, Philadelphia, pp. 341–357.

Arseculeratne, S.N., Panabokke, R.G., and Navaratnam, C. 1980. Pathogenesis of necrotising enterocolitis with special reference to intestinal hypersensitivity reactions. *Gut* **21**:265.

Ball, W.S., Jr., Kosloske, A.M., Jewell, P.F., and Bartow, S.A. 1985. Balloon catheter dilatation of focal intestinal strictures following necrotizing enterocolitis. *J. Pediatr. Surg.* **20**:637.

Ballance, W.A., Dahms, B.B., Shenker, N., and Kliegman, R.M. 1990. Pathology of neonatal necrotizing enterocolitis: A ten-year experience. *J. Pediatr. (suppl)* **117**:S6.

Barlow, B., Santulli, T.V., Heird, W.C., Pitt, J., Blanc, W.A., and Schullinger, J.N. 1974. An experimental study of neonatal necrotizing enterocolitis—the importance of breast milk. *J. Pediatr. Surg.* **9**:587.

Bell, M.J., Shackleford, P., and Molleston, J. 1985. Hypothesis: neonatal necrotizing enterocolitis is caused by acquisition of a pathogenic organism by a susceptible host infant. *Surgery* **97**:350.

Blakey, J.L., Lubitz, L., Campbell, N.T., Gillam, G.L., Bishop, R.F., and Barnes, G.L. 1985. Enteric colonization in sporadic neonatal necrotizing enterocolitis. *J. Pediatr. Gastroenterol. Nutrit.* **4**:591.

Book, L.S., Herbst, J.J., Atherton, S.O., and Jung, A.L. 1975. Necrotizing enterocolitis in low-birth-weight infants fed an elemental formula. *J. Pediatr.* **87**:602.

Book, L.S., Herbst, J.J., and Jung, A.L. 1976. Comparison of fast- and slow-feedingrate schedules to the development of necrotizing enterocolitis. *J. Pediatr.* **89**:463.

Book, L.S., Overall, J.C., Jr., Herbst, J.J., Britt, M.R., Epstein, B., and Jung, A.L. 1977. Clustering of necrotizing enterocolitis: interruption by infection-control methods. *N. Engl. J. Med.* **297**:984.

Buescher, E.S., 1994. Host defense mechanisms of human milk and their relations to enteric infections and necrotizing enterocolitis. *Clin. Perinatol.* **21**:247.

Butler, T., Dahms, B., Lindpaintner, K., Islam, M., Azad, M.A.K., and Anton, P. 1987. Segmental necrotising enterocolitis: pathological and clinical features of 22 cases in Bangladesh. *Gut* **28**:1433.

Caplan, M. and MacKendrick, W. 1994. Inflammatory mediators and intestinal injury. *Clin. Perinatol.* **21**:235.

Carrion, V. and Egan, E. 1990. Prevention of neonatal necrotizing enterocolitis. *J. Pediatr. Gastroenterol. Nutr.* **11**:317.

Chany, C., Moscovici, O., Lebon, P., and Rousset, S. 1982. Association of coronavirus infection with neonatal necrotizing enterocolitis. *Pediatrics* **69**:209.

Cooke, R.W.I., Meradji, M., Villeneuve, Y.V.D. 1980. Necrotizing enterocolitis after cardiac catheterization in infants. *Arch. Dis. Child.* **55**:66.

Cooperstock, M.S. and Zedd, A.J. 1983. Intestinal flora of infants, in: *Human Intestinal Microflora in Health and Disease* (D.J. Henges, ed.), pp. 78–93, New York, Academic Press.

Czyrko, C., Del Pin, C.A., O'Neill, J.A., Jr., Peckham, G.J., and Ross, A.J., III. 1991. Maternal cocaine abuse and necrotizing enterocolitis: outcome and survival. *J. Pediatr. Surg.* **26**:414.

de Lemos, R.A., Rogers, J.R. Jr., and McLaughlin, G.W. 1974. Experimental production of necrotizing enterocolitis in newborn goats. *Pediatr. Res.* **8**:380.

Du, X., Liu, Q., Yang, Z., Orazi, A., Rescorla, F.J., Grosfeld, J.L., and Williams, D.A. 1997. Protective effects of interleukin-11 in a murinemodel of bowel ischemia. *Am. J. Physiol.* **272**:G545.

Eibl, M.M., Wolf, H.M., and Furnkranz, H. 1988. Prevention of necrotizing enterocolitis in low-birth-weight infants by IgA-IgG feeding. *N. Engl. J. Med.* **319**:1.

Engel, R.R., Virnig, N.L., Hunt, C.E., and Levitt, M.D. 1973. Origin of mural gas in necrotizing enterocolitis, *Pediatr. Res.* **7**:292.

Gartner, L.M., (Chairperson, American Academy of Pediatrics, Work Group on Breastfeeding). 1997. Breastfeeding and the use of human milk. *Pediatrics* **100**:1035.

Go, L.L., Albanese, C.T., Watkins, S.C., Simmons, R.L., and Rowe, M.I. 1994. Breast milk protects the neonate from bacterial translocation. *J. Pediatr. Surg.* **29**:1059.

Goldman, A.S. 1993. The immune system of human milk: antimicrobial, antiinflammatory and immunomodulating properties. *Pediatr. Infect. Dis. J.* **12**:664.

Graf, J.L., WanderWall, K.J., Adzick, N.S., and Harrison, M.R. 1997. Nitroglycerin attenuates the bowel damage of necrotizing enterocolitis in a rabbit model. *J. Pediatr. Surg.* **32**:283.

Guinan, M., Schaberg, D., Bruhn, F.W., Richardson, C.J., and Fox, W.W. 1979. Epidemic occurrence of neonatal necrotizing enterocolitis. *Am. J. Dis. Child.* **133**:594.

Halac, E., Halac, J., Begue, E.F., Casanas, J.M., Indiveri, D.R., Petit, J.F., Figueroa, M.J., Olmas, J.M., Rodriguez, L.A., Obregon, R.J., Martinez, M.V., Grinblat, D.A., and Vilarrodona, H.O. 1990. Prenatal and postnatal corticosteroid therapy to prevent neonatal necrotizing enterocolitis: a controlled trial. *J. Pediatr.* **117**:132.

Hennekens, C.H. and Buring, J.E. 1987. Intervention studies, in: *Epidemiology in Medicine* (S.L. Mayrent, ed.), pp. 178–204. Little, Brown and Co., Boston.

Holman, R.C., Stehr-Green, J.K., and Zelasky, M.T. 1989. Necrotizing enterocolitis mortality in the United States. *Am. J. Publ. Health* **79**:987.

Jayanthi, S., Seymour, P., Puntis, J.W.L., and Stringer, M.D. 1998. Necrotizingenterocolitis after gastroschisis repair: A preventable complication? *J. Pediatr. Surg.* **33**:705.

Kliegman, R.M., Fanaroff, A.A., Izant, R., and Speck, W.T. 1979. Clostridia as pathogens in neonatal necrotizing enterocolitis. *J. Pediatr.* **95**:287.

Kliegman, R.M. and Fanaroff, A.A. 1984. Necrotizing enterocolitis. *N. Engl. Med.* **310**:1093.

Kosloske, A.M. 1985. Surgery of necrotizing enterocolitis. *World J. Surg.* **9**:277.

Kosloske, A.M. 1990. A unifying hypothesis for pathogenesis and prevention of necrotizing enterocolitis. *J. Pediatr. (suppl.)* **117**:S68.

Kosloske, A.M. 1994. Epidemiology of necrotizing enterocolitis. *Acta Paediatr. Suppl.* **396**:2.

Kosloske, A.M. 1997a. The epidemiology and pathogenesis of necrotizing enterocolitis. *Semin. Neonatol.* **2**:231.

Kosloske, A.M. 1997b. Necrotizing enterocolitis, in: *Surgery of Infants and Children: Scientific Principles and Practice* (K.T. Oldham, P.M. Colombani, and R.P. Foglia, eds.), pp. 1201–1213. Lippincott-Raven, Philadelphia.

Kosloske, A.M., Ulrich, J.A., and Hoffman, H. 1978. Fulminant necrotizing enterocolitis associated with clostridia. *Lancet* **2**:1014.

Kosloske, A.M., Burstein J., and Bartow, S.A. 1980. Intestinal obstruction due to colonic stricture following neonatal necrotizing enterocolitis. *Ann. Surg.* **192**:202.

Kosloske, A.M., Ball, W.S., Jr., Umland, E.T., and Skipper, B. 1985. Clostridial necrotizing enterocolitis. *J. Pediatr. Surg.* **20**:155.

Krissinger, K.D. 1994. Regulation of hemodynamics and oxygenation in developing intestine: insight into the pathogenesis of necrotizing enterocolitis. *Acta Paediatr. Suppl.* **396**:8.

Lawrence, G., Bates, J., and Gaul, A. 1982. Pathogenesis of neonatal necrotizing enterocolitis. *Lancet* **1**:137.

Lawrence, G. and Walker, D.P. 1976. Pathogenesis of enteritis necroticans in Papua New Guinea. *Lancet* **1**:125.

Lawrence, J.P., Brevetti, L., Obiso, R.J., Wilkins, T.D., Kimura, K., and Soper, R. 1997. Effects of epidermal growth factor and *Clostridium difficile* toxin B in a model of mucosal injury. *J. Pediatr. Surg.* **32**:430.

Long, S.S. and Swenson, R.M. 1977. Development of anaerobic fecal flora in healthy newborn infants. *J. Pediatr.* **91**:298.

Leung, M.P., Chau, K.T., Hui, P.W., Tarn, A.Y.C., Chan, F.L., Lai, C.L., and Yeung, C.Y. 1988. Necrotizing enterocolitis in neonates with symptomatic congenital heart disease. *J. Pediatr.* **113**:1044.

Lloyd, J.R. 1969. The etiology of gastrointestinal perforation in the newborn. *J. Pediatr. Surg.* **4**:77.

Lucas, A. and Cole, T.J. 1990. Breast milk and neonatal necrotizing enterocolitis. *Lancet* **336**:1519.

Marchildon, M.B., Buck, B.E., and Abdenour, G. 1982. Necrotizing enterocolitis in the unfed infant. *J. Pediatr. Surg.* **17**:620.

Millar, M.R., Mackay, P., Levene, M., Langdale, V., and Martin, C. 1992. Enterobacteriaceae and neonatal necrotizing enterocolitis. *Arch. Dis. Chil.* 67:53.

Musemeche, C.A., Kosloske, A.M., Bartow, S.A., and Umland, E.T. 1986. Comparative effects of ischemia, bacteria, and substrate on the pathogenesis of intestinal necrosis. *J. Pediatr. Surg.* **21**:536.

Nowicki, P.T. and Miller, C.E. 1992. Effect of increased tissue oxygen uptake on autoregulation in postnatal intestine. *Am. J. Physiol.* **263**:G690.

Okuyama, H., Urao, M. Lee, D., Drongowski, R.A., and Coran, A.G. 1998. The effect of epidermal growth factor on bacterial translocation in newborn rabbits. *J. Pediatr. Surg.* **33**:225.

Oldham, K.T., Coran, A.G., Drongowski, R.A., Baker, P.J., Wesley, J.R., and Polley, T.Z., Jr. 1988. The development of necrotizing enterocolitis following repair of gastroschisis: a surprisingly high incidence. *J. Pediatr. Surg.* **23**:945.

Ostertag, S.G., LaGamma, E.F., Reisen, C.E., and Ferrentino, F.L. 1986. Early enteral feeding does not affect the incidence of necrotizing enterocolitis. *Pediatrics* **77**:275.

Powell, R.W., Dyess, D.L., Collins, J N., Roberts, W.S., Tacchi, E.J., Swafford, A.N., Jr., Ferrara, J.J., and Ardell, J.L. 1999. Regional blood flow response to hypothermia in premature, newborn, and neonatal piglets. *J. Pediatr. Surg.* **34**:193.

Rossier, A., Sarrot S., and Delplanque, J. 1959. L'entercolite ulcero-necrotique dupremature. *Sem. Hop. Paris* **35**:1428.

Rotbart, H.A., Nelson, W.L., Glode, M.P., Triffon, T.C., Kogut, S.J.H., Yolken, R.H., Hernandez, J.A., and Levin, M.J. 1988. Neonatal rotavirus-associated necrotizing enterocolitis: case control study and prospective surveillance during an outbreak. *J. Pediatr.* **112**:87.

Ryder, R.W., Shelton, J.D., and Guinan, M.E. 1980. Necrotizing enterocolitis: a prospective multi-center investigation. *Am. J. Epidemiol.* **112**:113.

Santulli, T.V., Schullinger, J.N., Heird, W.C., Gongaware, R.D., Wigger, J., Barlow, B., Blanc, W.A., and Berdon, W.E. 1975. Acute necrotizing enterocolitis in infancy: a review of 64 cases. *Pediatrics* **55**:376.

Scholander, P.F. 1963. The master switch of life. *Sci. Am.* **209**:92.

Shimura K. 1990. Necrotizing enterocolitis? A Japanese survey. *NICU* **3**:5.

Siebold, J.F. 1825. Gerburtshulfe, Frauenzimmer und Kinderkrankheiten, *Heft. I. Leipzig.* **5**:3.

Speer, M.E., Taber, L.H., Yow, M.D., Rudolph, A.J., Urteaga, J., and Waller, S. 1976. Fulminant neonatal sepsis and necrotizing enterocolitis associated with a "non-enteropathogenic" strain of *Escherichia coli. J. Pediatr.* **89**:91.

Stein, H., Beck, J., Solomon, A., and Schmaman, A. 1972. Gastroenteritis with necrotizing enterocolitis in premature babies. *Br. Med. J.* **2**:616.

Stoll, B.J., Kanto, W.P., Jr., Glass, R.K., Nahmias, A.J., and Brann, A.W., Jr. 1980. Epidemiology of necrotizing enterocolitis. *J. Pediatr.* **96**:447.

Tyson, J.E., deSa, D.J., and Moore, S. 1976. Thromboatheromatous complications ofumbilical arterial catheterization in the newborn period. *Arch. Dis. Child.* **51**:744.

Udall, J.N., Jr. 1990. Gastrointestinal host defense and necrotizing enterocolitis: an update. *J. Pediatr. (suppl.)* **117**:S33.

Williams, A.F. 1997. Role of feeding in the pathogenesis of necrotizing enterocolitis. *Semin. Neona*tol. **2**:221.

Yamataka, A., Yamataka, T., Lane, G.J., Kobayashi, H., Sueyoshi, N., and Miyano, T. 1998. Necrotizing enterocolitis and C-KIT. *J. Pediatr. Surg.* **33**:1682.

Chapter 7

The Protective Properties of Milk and Colostrum in Non-Human Species

Ian Tizard

1. Introduction

Newborn mammals emerge from the clean, stable and sterile uterus into a world where they are immediately exposed to an enormous variety of microorganisms. If they are to survive, newborns therefore must be able to control microbial invasion. The immune system, however, may not be ready for this defensive role. In mammals with a short gestation period such as the marsupials, the immune system may not have developed fully. In mammals with a long gestation period such as the domestic herbivores, although the immune system is structurally complete at birth, it cannot function fully for several weeks. The complete development of immune capability depends on antigenic stimulation. The development of adequate numbers of antigen-sensitive lymphocytes depends on clonal selection and antigen-driven cell multiplication. The first immune responses mounted by a newborn animal must be primary responses with a prolonged lag period and low concentrations of antibodies produced. Thus newborn mammals are highly vulnerable to microbial invasion for the first few weeks of life and unless immunological assistance is provided, they may be killed by microorganisms that present little threat to an adult. This "immunological assistance" is provided by antibodies and other proteins transferred from the mother to her offspring through the placenta and/or by antibodies and lymphocytes transferred through colostrum and milk.

Ian Tizard • Department of Veterinary Pathobiology, Texas A&M University, College Station, Texas 77843

2. Placentation and Passive Immunity

Because of the state of their immune system, all young mammals must acquire maternal antibodies either before or immediately after birth. Whether they do so by transplacental passage or through mammary secretions depends on the structure of their placenta. In the least developed form of placenta (epitheliochorial), the fetal chorionic epithelium comes into direct contact with the intact uterine epithelium (Blüm, 1985). This form of placenta occurs in Pholidota (such as the harbor seal), in primitive primates such as lemurs and in some ungulates such as horses and pigs as well as in their close relatives, the Cetacea such as the bottle-nosed dolphin (Carter *et al.*, 1990). A more developed form of placenta is found in other ungulates such as the ruminants. They have a syndesmochorial placenta in which a uterine epithelium is not present and the chorionic epithelium is in direct contact with uterine tissues. In mammals with these types of placenta, transplacental passage of immunoglobulins cannot occur and the newborn of these species are agammaglobulinemic at birth. These animals, therefore, are totally dependent on antibodies received through mammary secretions for immediate protection. Carnivores, in contrast, have an endotheliochorial placenta in which the chorionic epithelium comes into contact with the endothelium of the maternal capillaries. In these mammals a small fraction (five to ten per cent) of the newborn's IgG may be derived by transplacental transfer but the remainder must be obtained through mammary secretions.

In lagomorphs (rabbits), hystricomorphs (guinea pigs) and rodents there is a yolk sac placenta that permits immunoglobulin transfer. In the advanced primates (including hominoids), the placenta is hemochorial so that maternal blood comes into direct contact with the placental trophoblast. In these mammals, maternal IgG can reach the fetal bloodstream by direct transfer across the chorioallantoic placenta. Because maternal IgG enters the fetal blood stream directly, mammals with this form of placentation are born with circulating IgG levels comparable to those of their mother.

3. Immunoglobulins and Cells in Colostrum and Milk

3.1. Immunoglobulins

In species with placentas that are impermeable to immunoglobulins, newborns must receive both IgG and IgA through mammary secretions. In general they receive IgG from colostrum and IgA from milk (Norcross, 1982).

Colostrum represents the accumulated secretions of the mammary gland over the last few weeks of pregnancy together with proteins actively transported from the bloodstream. Its composition differs among species. In mammals such as the primates with placentas that permit immunoglobulin transfer, colostrum contains predominantly IgA. In mammals with impermeable placentas, IgG must accumulate in colostrum. Very high affinity immunoglobulin receptors (FcR) are expressed on bovine mammary gland acinar cells about a week prior to birth (Sasaki *et al.*, 1977). These transport IgG, and to a lesser extent IgM and IgA, into the mammary gland so that all the IgG, most of the IgM and about half of the IgA in bovine colostrum are derived from the blood. IgG may account for 65 to 90% of the total antibody content of bovine colostrum. Only 30% of the IgG1 and 10% of the IgA in bovine milk are derived from the blood; the rest is produced locally in the udder.

IgG in colostrum provides a source of antibodies that provide systemic immunity in the newborn. However as lactation proceeds, the proportion of locally synthesized IgA increases so that it becomes the predominant immunoglobulin in most milks. One exception to this pattern occurs in ruminants, in which IgG1 is the predominant immunoglobulin isotype in both milk and colostrum. The low level of IgA in ruminant milk, it has been speculated, may permit the earlier development of ruminal flora.

Thatcher and Gershwin (1989) demonstrated that colostral transfer of IgE occurs in the bovine. IgE was transferred to newborn calves, in which maximum levels were reached during the first week of lactation and declined rapidly thereafter. It was suggested that this passive transfer of maternal IgE might be important in providing early protection from intestinal parasites. Likewise, although IgD has not been considered a significant colostral immunoglobulin, it is present in rat milk in extraordinary high concentrations, approximately 100 times those found in the serum of lactating rats (Olson and Leslie, 1982).

Colostrum also contains secretory component both in the free form and bound to IgA. Secretory component may play a role in the transport of IgA within the mammary gland. For the first few weeks of life, while proteolytic digestion is poor, colostral and milk immunoglobulins can be found throughout the intestine and in the feces of young mammals. As the digestive ability of the intestine increases, eventually only secretory IgA molecules remain intact. SIgA is therefore important in protection against enteric infection.

3.2. Cells

It has been recognized for many years that mammary secretions contain many different cells, whose function has been difficult to ascertain (Head and Beer, 1978). They are a mixture of blood leukocytes and mammary epithelial cells. Their numbers and relative proportions vary among species and at different times during lactation. Milk phagocytic cells are loaded with vacuoles containing lipid, casein or cellular debris, and as a result are poorly phagocytic (Crighton, 1984).

Macrophages, neutrophils, and lymphocytes are present in the colostrum and milk of the tammar wallaby (*Macropus eugenii*) (Young *et al.*, 1997). Neutrophils are the predominant cell type in early secretions while macrophages become common later when the young are no longer attached to the teat.

Around two million leukocytes/ml are found in rat, human, pig and cow's milk with lower numbers (8000/ml in mares and water buffalo (Head, 1977; Head and Beer, 1978; Lee *et al.*, 1983). The predominant cell type in humans, rats, cows, sheep and quokkas is the macrophage. Neutrophils may be present in greater numbers in early lactation and in mastitis (Cockson and McNeice, 1980; Outteridge and Lee, 1988).

In the pig, neutrophils predominate in colostrum whereas epithelial cells predominate in milk, although macrophages, eosinophils and lymphocytes are present throughout lactation (Lee *et al.*, 1983). Sow's colostrum contains about 1–2.5 $\times$ 10^6 cells/ml. Ten to 25 percent of these are lymphocytes and 70–90 percent of the lymphocytes are T cells. Their CD4/CD8 ratio is approximately 0.57 which is somewhat lower than their ratio in blood (0.8).

Bovine colostrum contains up to 1 $\times$ 10^6 lymphocytes/ml, about half of which are T cells (Chabaudie *et al.*, 1993). While some milk lymphocytes migrate from the bloodstream others may be derived from the mammary gland itself. They tend to be located close to the alveolar epithelium and in cattle are a mixture of CD4+ and CD8+ T cells as well as IgA+ B cells (Chabaudie *et al.*, 1993).

The functional significance of the cells found in colostrum and milk has been difficult to determine. Nevertheless, the lymphocytes found in mammary secretions are functional and probably play a role in the development of immunity in the newborn. Thus bovine colostral lymphocytes collected in the first two days after calving are responsive to concanavalin A (Con A) (Archambault *et al.*, 1988a). Lymphocytes from the colostrum of cows immunized with a rotavirus vaccine during pregnancy also showed a significant blastogenic response to bovine rotavirus. Colostral lymphocytes showed a higher response than blood lymphocytes, and the response lasted for at least 21 days after parturition (Archambault *et al.*, 1988b).

Riedel-Caspari and Schmidt (1990, 1991a) examined the influence of colostral cells on the concentration of immunoglobulins against *E. coli* in the sera of newborn calves. The calves received either complete colostrum or cell-free colostrum during the first three days of life. The sera of the calves fed complete colostrum had significantly higher concentrations of IgG1 antibodies against *E. coli* than the sera of the calves fed cell-free colostrum. The sera of the calves receiving cells also contained significantly more IgM and slightly more IgA during the first week of life. The groups had equal concentrations of serum IgG. The investigators therefore suggested that the colostral leukocytes enhanced passive immunity in the neonatal calf. They also compared cell-containing and cell-free colostrum for their ability to protect calves against enteropathic *E. coli*. The calves receiving complete colostrum excreted significantly fewer bacteria than the animals receiving cell-free colostrum. The concentration of IgA and IgM-specific antibodies against *E. coli* in the serum of neonatal calves was higher in those that received cells in their colostrum than in those that did not. The calves that received colostrum containing cells had a better response to Con A and an enhanced responsiveness to sheep erythrocytes. Riedel-Caspari and Schmidt (1991b) also examined the effect of ingesting colostral cells on lymphocyte responses in neonatal calves. They found that animals fed cell-free colostrum or a milk substitute showed an enhanced lymphocyte response to mitogens relative to that of calves fed colostrum containing cells. Thus colostral cells may exert immunoregulatory actions as well as immunostimulatory influences.

Leukocytes ingested in mammary secretions can survive within the intestine and migrate to the bloodstream of newborns. Colostral lymphocytes may survive for up 36 hours in the intestine of newborn calves; some penetrate the intestinal wall and reach the lacteal ducts and the mesenteric lymph nodes (Kmetz *et al.*, 1970). Within two hours after receiving colostrum containing labeled cells piglets had maternal lymphocytes in their bloodstream. The cells apparently penetrated between epithelial cells in the duodenum and ileum. Piglets that had received these colostral cells showed enhanced responses to Con A, pokeweed mitogen (PWM) and phytohemagglutinin (PHA) compared to that of control piglets that had been fed a suspension of peripheral blood mononuclear leukocytes. Viable maternal milk lymphocytes can also reach the mesenteric lymph nodes in rats and lambs (Sheldrake and Husband, 1985). In the rat, graft-versus-host disease may result (Head, 1977) but this has not been reported in other species.

4. Absorption of Colostral Immunoglobulins

Young mammals that suckle soon after birth take a large volume of colostrum into their intestinal tract. Naturally suckled calves ingest an average of two liters of colostrum, but individual calves can ingest as much as six liters (Norcross, 1982). In these young mammals, the level of proteolytic activity in the intestine is low and it is further reduced by protease inhibitors in colostrum. Therefore colostral proteins are not degraded and used as a food source but instead remain for long periods within the small intestine. Once within the intestine, colostral IgG probably binds to specialized Fc receptors on the intestinal epithelial cells of newborns (FcRn). Eventually the colostral immunoglobulin reaches the systemic circulation and newborn mammals thus obtain a massive transfusion of maternal IgG. The first detectable IgG in the thoracic duct appears

60–120 minutes after the initial intake of colostrum (Comline *et al.*, 1951; Balfour and Comline, 1962) or in the blood after 180 minutes (Logan *et al.*, 1978, Stott *et al.*, 1979a,b).

Jochims *et al.* (1994) used protein A-gold to investigate the transfer of colostral IgG from the gut lumen to the circulation in newborn calves and found that intracellular micropinocytotic transport was the prime mechanism of transfer throughout the small intestine. Colostral IgG binds to the apical plasma membrane and is taken up in noncoated pits and invaginations, pinocytosed, transported in endocytotic vesicles and accumulated in granules in the apical cytoplasm (Komuves and Heath, 1992). It eventually reaches the lacteals and possibly the intestinal capillaries.

Mammals differ in the selectivity and duration of their intestinal permeability to proteins (Norcross, 1982). For example, in the horse and pig, immunoglobulin absorption is highly selective, so that IgG and IgM are preferentially absorbed while IgA remains mainly in the intestine. In ruminants, however, the intestine is unselectively permeable and all immunoglobulin isotypes are absorbed, although IgA is gradually re-excreted. The lactating mammary gland produces secretory component and secretory IgA (Lejan 1993). As a result, young pigs and probably other young mammals possess large amounts of free secretory component within their intestinal tract. This may inhibit the absorption of both IgM and IgA.

The period for which the intestine is permeable to immunoglobulins also varies among species and among immunoglobulin isotypes. In the ungulates, passage of immunoglobulins across the intestinal wall has largely terminated by 24 hours after birth. In contrast, in suckling rats and mice transfer continues for up to 16–21 days, and in the hedgehog, an insectivore, it lasts for up to 40 days (Manning and Turner, 1976). As discussed below, in marsupials absorption of immunoglobulins occurs for as long as the newborn remains attached to the teat, perhaps for as long as six months. The eventual cessation of immunoglobulin absorption results from the replacement of neonatal Fc receptor (FcRn)-expressing fetal intestinal epithelial cells by more mature cell populations which lack these receptors. In addition, there is an increase in intracellular proteolytic activity by lysosomes. In species such as cattle, pigs and horses, which absorb immunoglobulins immediately after birth, intestinal permeability is highest immediately after birth and begins to decline after six hours. As a rule, absorption of all immunoglobulin isotypes drops to a relatively low level after approximately 24 hours, a phenomenon termed "closure". Feeding colostrum tends to hasten closure whereas a delay in feeding results in a slight delay of closure. The presence of the mother may be associated with increased immunoglobulin absorption. In controlled laboratory studies in which measured amounts of colostrum were fed, there was great variation (±25–35%) in the quantity of immunoglobulin absorbed (Norcross, 1982).

Unsuckled mammals of species with impermeable placentas are effectively lacking in blood immunoglobulins (Halliday, 1978). The absorption of colostral IgG immediately supplies them with blood IgG at a level approaching that found in adults. Peak serum IgG levels are normally reached between 12 and 24 hours after birth. After absorption ceases, these passively acquired antibodies decline through normal catabolic processes. The rate of decline differs among immunoglobulin isotypes, the time taken to decline to unprotective levels depending upon their initial concentration. As intestinal absorption is taking place, a simultaneous proteinuria may occur. This is due to the absorption from the intestine of proteins such as β-lactoglobulin and polypeptides that are sufficiently small to be excreted in the urine. In addition, the glomeruli of newborn mammals are permeable to macromolecules. Thus the urine of neonatal ruminants also contains intact immunoglobulin molecules. This proteinuria ceases spontaneously with the termination of intestinal absorption.

4.1. Fc Receptors

In order to be transported across epithelial barriers, immunoglobulins must first bind to specific receptors. The best characterized of these is the neonatal Fc receptor (FcRn), which has been examined mainly in laboratory rodents. Intestinal epithelial cells of the neonatal rat and mouse express FcRn (Jakoi *et al.*, 1985; Kandil *et al.*, 1995; Raghavan and Bjorkman, 1996; Simister *et al.*, 1997). Although directly involved in immunoglobulin transfer across the intestine, this same receptor is also employed to transfer IgG across the rodent and human placenta.

FcRn is a member of the nonpolymorphic MHC class Ib family. It has an alpha-chain similar to the MHC alpha chain and beta 2-microglobulin (β2M) (Ahouse *et al.*, 1993). The overall exon-intron organization of the FcRn gene is similar to that of classical MHC class I genes. However, regulatory elements normally found in classical MHC class I genes are absent. Sequence analysis of the FcRn gene isolated from eight mouse strains showed that the membrane-distal domain of FcRn has at least three amino acid variants which might be of functional significance. The fact that FcRn is a single copy gene indicates that the same molecule is expressed in both the neonatal intestine and the fetal yolk sac. Phylogenetic analysis suggests that the FcRn gene diverged from MHC class I genes after the emergence of amphibians but before the split of placental and marsupial mammals.

FcRn binds to IgG through a site at the CH2-CH3 domain interface (Medesan *et al.*, 1996). It binds this IgG at the pH of milk in the proximal intestine (pH 6.0–6.5) and releases it at the pH of blood (pH 7.5). Exchange is more than 10 times slower at pH 6.1 than at pH 7.8. These observations imply a pH-dependent conformational change in the FcRn heterodimer that may be related to its physiological function (Raghavan *et al.*, 1993). Like rat and mouse FcRn, human placental FcRn also binds IgG preferentially at low pH, which may imply that IgG binds hFcRn in an acidic intracellular compartment during transport across the placenta (Story *et al.*, 1994).

Neonatal mice homozygous for a targeted disruption of the β2M gene have reduced FcRn expression on their intestinal cells. These mice had much lower serum IgG levels during the first month after birth than littermates with functional FcRn. By fostering mice on mothers with different IgG allotypes, it was shown that none of the IgG in the sera of β2M –/– mice was derived from milk. Thus FcRn is the only transporter of IgG from mother to young in the mouse. There were no differences between the IgG concentrations in the milk of β2M –/– and β2M +/– mice, indicating that FcRn is not involved in the secretion of IgG into milk (Israel *et al.*, 1995).

In addition to transferring immunoglobulins across the intestinal epithelium in rodents, FcRn may be involved in the transmission of IgG from mother to fetus: rat fetal yolk sac, mouse fetal yolk sac and human placental syncytiotrophoblast (Ghetie and Ward, 1997). In addition to being present in these tissues it is also expressed on the surface of mouse hepatocytes (Blumberg *et al.*, 1995). IgG has a much longer half-life in blood than other immunoglobulin isotypes. This selective protection from catabolism is lost in mice that lack a functional FcRn (Junghans and Anderson, 1996). It is likely, therefore, that one function of FcRn is to protect circulating IgG from degradation (Israel *et al.*, 1996).

FcRn has not been characterized in mammals other than laboratory rodents and humans. Nevertheless, it is likely to be equally important in other species. Uncharacterised FcR for IgG1 has been identified on cells in bovine mammary tissues (Barrington *et al.*, 1997).

5. Interference with Active Immunity

The presence of high titered maternal antibodies interferes with the response of newborn mammals to vaccination. Antibodies exert a negative feedback on responding B cells by cross-linking the B cell receptor to a FcR and so sending an inhibitory signal to the B cell (Anderson and Sinclair, 1998). Brar *et al.* (1978) and Bradshaw and Edwards (1996) found that calves could not respond to an infectious bovine rhinotracheitis (IBR) vaccine until maternal antibodies had decreased to undetectable levels. They also showed that, although maternal antibodies inhibited the response to IBR vaccination, priming for a secondary response occurred. Thus, on subsequent vaccination, after maternal antibodies had disappeared, calves responded anamnestically to IBR vaccination. Kitching and Salt (1995) showed that maternally-derived antibody provided immediate protection against infection with foot-and-mouth disease virus, but interfered with the development of active immunity following vaccination. This interference has also been reported in calves for antibody responses to bovine coronavirus (Heckert *et al.*, 1991) and bovine virus diarrhea (Menanteau-Horta *et al.*, 1985). A similar negative feedback is well recognized in companion animals, in which it represents the major cause of vaccination failure in young puppies and kittens. In general this resistance to immunization lasts for eight to twelve weeks after birth. However, for some viral antigens such as canine parvovirus it may persist for as long as 20 weeks (Carmichael, 1983).

6. Specific Mammalian Species

6.1. Marsupials

The marsupials diverged from the eutherian mammals during the Cretaceous period about 145–165 million years ago. Metatheria, the subclass that contains the marsupials, are distinguishable from the Eutheria or placental mammals by differences in their reproductive tract, such as the development of a functional, noninvasive yolk sac placenta. Thus they do not develop a trophoblast or form intimate connections between fetal and maternal tissues. At birth the marsupial, although altricial, has well developed arms and can transport itself to the teat, which in many female marsupials is enclosed in a pouch. It locks onto the teat and suckles for a prolonged period. Marsupial pouches have a diverse bacterial flora and the young therefore require immediate immune protection. Nevertheless, in most marsupials examined, maternal antibodies are not transported across the placenta and immunoglobulins cannot be detected in the embryonic fluids or in the blood of the newborn at birth. Newborn opossums (*Didelphis virginiana* and *Monodelphis domestica*) and the quokka (*Setonyx brachyurus*) do not acquire serum immunoglobulins until they suckle (Yadav *et al.*, 1971, 1973; Hindes and Mizell, 1976; Samples *et al.*, 1986). It is possible that in the tammar wallaby (*Macropus eugenii*) a limited quantity of IgG crosses the placenta since some IgG is detectable two days before birth in fetal serum (Renfree, 1973). Deane *et al.* (1990) also found that IgG increases in concentration in the fetal serum, neonatal serum and colostrum of tammar wallabies. They suggested that both transplacental and intestinal transfer of IgG takes place in this species.

Thus, marsupials secrete colostrum that provides immunoglobulins to the young. Yadav (1971) found that antibodies appeared in the blood of 90-day-old *S. brachyurus* within 15 to 60 minutes of feeding immune serum by stomach tube. In addition, when *S. brachyurus* nursing mothers were immunized with bacteriophage ϕX174 or Salmonella flagellar antigens, specific antibodies could be detected in their colostrum and milk as well as in the serum of pouch young

(Yadav, 1971). Samples *et al.* (1986) immunized lactating *M. domestica* with sheep erythrocytes and found specific antibodies in suckled young 9–10 days later. Antibodies to sheep cells were not found in newborn opossums that had not suckled.

Deane and Cooper (1984) found that pouch young *Macropus robustus* had relatively low levels of serum IgG (20 to 50 mg/dl) up to day 95. However, a steep rise occurred between days 95 and 110 so that concentrations eventually reached 700 mg/dl. This rise coincided with the time that the newborn kangaroo first released from the teat, a time when it is much more likely to ingest harmful microorganisms. The authors suggest that in the first 90–100 days the pouch young are protected largely by passive immunity from the mother, and after this time increasingly mount their own responses (Deane and Cooper, 1984).

Opossum enterocytes express FcR that are detectable up to 52 days of age and can mediate transport of IgG across the intestinal wall (Wild *et al.*, 1994). The disappearance of FcR from marsupial enterocytes appears to correspond to the time of weaning: 7 to 9 weeks in *M. domestica*, 170 to 200 days in *S. brachyurus* and 98 to 145 days in the brush possum (*Trichosurus vulpecula*) (Wild *et al.*, 1994). This long period of FcR expression is in marked contrast to the situation in eutherian mammals. For example, FcRn is expressed for only 21 days in the rat (Griffin and Wild, 1987). However, this pattern of expression corresponds to the suckling period in both species and raises interesting questions regarding cause and effect.

6.2. Eutherian Mammals

The vast majority of mammals belong to the subclass Eutheria. This subclass is characterized by suppression of the yolk sac placenta during early embryonic development and by the development of chorioallantoic placentation, so permitting effective nutrient exchange within the uterus. As pointed out earlier, however, mammalian placentas vary greatly in their structure and in their permeability to IgG.

6.2.1. Primates. Placentations in the more primitive primates such as the lemurs and the Strepsorhines are epitheliochorial and it may be surmised that maternal immunoglobulins can be transferred only through mammary secretions in these species (Blum, 1985; King, 1986). In the Cercopithecidae, Hylobatidae and Pongidae and in humans, placentation is hemochorial and transplacental transfer of immunoglobulins occurs. The main immunological function of mammary secretions in these primates therefore is to protect the newborn intestine by providing IgA. Secretory IgA is present in rhesus macaque milk, though at a lower concentration than in human milk (Kunz and Lonnerdal, 1993). In general, the levels of immunoglobulins in the sera and mammary secretions of baboons and macaques parallel the levels found in humans (Cole and Bowen, 1976). It has proved possible to stimulate an increase in milk antibodies to rotavirus in pregnant baboons by parenteral vaccination (Snodgrass *et al.*, 1995).

6.2.2. Hystricomorphs. The Hystricomorphs (guinea pigs and porcupines) appear to represent a very early branch in the evolution of Eutherian mammals. Many recent studies in rodent taxonomy have focused on the status of the Hystricomorphs, especially guinea pigs and it is clear that they constitute an order distinct from that of other rodents (Derchia *et al.*, 1996). In Hystricomorphs, maternal IgG is transferred in utero across the yolk sac to the fetus (Beer and Billingham, 1976).

6.2.3. Lagomorphs. The Lagomorphs include the hares, rabbits and picas. Rabbits and hares have a unique method of feeding their young. They use an absentee parental care system that requires nursing them only once every 24 hours. Both IgG and IgM antibodies are readily transferred across the yolk sac *in utero* (Beer and Billingham, 1976).

6.2.4. Rodents. The order Rodentia shows an enormous variety of adaptations and considerable difficulty has been experienced in developing a satisfactory phylogenetic classification. Placentation in rodents is hemochorial and it would therefore be anticipated that IgG could be transferred across the placenta. However it is clear that significant quantities of immunoglobulin are also transferred in colostrum and milk (Beer and Billingham, 1976).

In mice, a small but significant transfer of IgG to the fetus occurs by the 15th day of gestation (five days before birth) but the bulk of passively acquired immunoglobulin is derived from the milk after birth (Appleby and Catty, 1983; Barthold *et al.*, 1988). All immunoglobulin acquired *in utero* and later across the intestinal barrier is exclusively IgG, even though the milk contains predominantly IgA. A high level of maternally derived antibody is maintained in the circulation of the young mouse for 24 days or more after gut closure occurs on the 16–20th day postpartum. FcRn on enterocytes binds and transports IgG until the animals are weaned (21 days in mice and 16 days in rats). In addition, IgG binds very rapidly to FcRn on the rat yolk sac membrane and so is efficiently transferred to the fetus (Mucchielli *et al.*, 1983). Up to 60% of mouse milk IgA on the fourth day of lactation is blood derived. Its transport from blood to milk is mediated by the polymeric Ig receptor (pIgR) from which secretory component is derived. Mouse mammary epithelial cells bind IgA through the pIgR and transfer it into the lacteal secretions (Halsey *et al.*, 1980). This process does not occur in the rat where all IgA is produced within the gland (Dahlgren *et al.*, 1981). Early in lactation most of the IgA in mouse milk is transferred via the blood to the mammary gland. However, by day 8 of lactation this transferred IgA is significantly diluted by the influx of IgA synthesized locally within the mammary gland (Halsey *et al.*, 1982).

Maternally-derived passive immunity through colostrum and milk has been shown to protect mice against colibacillosis (Duchet-Suchaux, 1983), mouse hepatitis virus (Homberger, 1992), rotavirus (Gombold and Ramig, 1989), Ross River virus (Milner and Marshall, 1984), *Mycoplasma pneumoniae* (Katsura *et al.*, 1985), *Campylobacter jejuni* (Abimiku and Dolby, 1987) and *Taenia taeniaeformis* (Musoke *et al.*, 1975).

IgD is present in rat's milk in extraordinarily high concentrations, approximately 100 times those found in the blood of lactating rats (Olson and Leslie, 1982). This suggests that IgD might be an important secretory immunoglobulin in this species. It is unclear whether the IgD exerts a protective effect within the mammary gland or within the intestine of the suckling rat. Most of this IgD is synthesized locally within the mammary gland in a manner similar to IgA. Since serum IgD levels do not climb in suckling rats it appears not to be absorbed from the rat intestine. This immunoglobulin may serve several functions, including protection of the mammary gland against bacterial or viral infection, passive transfer of immunity to the intestine of the suckling rat, or as a regulatory signal that influences the normal maturation of the neonatal immune system (Steele and Leslie, 1985).

In cotton rats (*Sigmodon hispidus*), passive maternal immunity protects against respiratory syncytial virus (Prince *et al.*, 1983, 1985). Immune females confer antibody to their young prenatally and postnatally, most of the antibody being transferred via colostrum and milk. Maternally transmitted immunity was found to be more effective in the lungs than in the nose and to betransient in both organs.

6.2.5. Carnivores. Carnivores have an endotheliochorial placentation (Blum, 1985) which is not totally impermeable to immunoglobulins. For all practical purposes, however, young carnivores must ingest colostrum and absorb maternal IgG in order to survive.

Canine colostrum is rich in IgG whereas the predominant immunoglobulin in canine milk is IgA with small quantities of IgM and IgG (Heddle and Rowley, 1975). Maternal antibody is protective against canine parvoviruses (CPV) (Macartney *et al.*, 1988), canine distemper (Krakowka *et al.*, 1978) and canine adenovirus (Carmichael *et al.*, 1962). Antibodies to CPV are transferred from an immune mother to her pups through both placenta and colostrum (Pollock and Carmichael, 1982). Colostral transfer accounts for approximately 90% of the maternally-derived CPV antibody. Likewise, neonatal ferrets (*Mustela erminea*) can be passively immunized following maternal vaccination with killed influenza A vaccine although the relative importance of colostrum and milk in this species has not been examined (Sweet *et al.*, 1987).

Because some maternal immunoglobulins can cross the cat placenta, IgG or IgM can be detected in some fetal and pre colostral kitten sera (Yamada *et al.*, 1991). However, the largest proportion of blood IgG, IgA and IgM is transferred to kittens through colostrum. Casal *et al.* (1996) found that the mean concentration of IgG was higher in colostrum than in maternal serum, implying that there is active transport. Intestinal closure apparently occurs before 16 hours postpartum in kittens since maternal IgG given at or after 16 hours could not be found in any kitten serum.

Passive transfer of immunity has been demonstrated against many different pathogens including feline leukemia virus (Hoover *et al.*, 1977) and *Toxoplasma gondii* (Omata *et al.*, 1994).

In seals (*Phoca vitulina*) the importance of maternal transfer of immunoglobulins through milk is unclear. Neonatal seal pups have a relatively high concentration of some IgG subclasses at birth, suggesting that they either synthesize IgG *in utero* or that it can cross the seal placenta (King, 1986). Ross *et al.* (1994) found that seal pup lymphocytes responded more strongly to mitogens than do the lymphocytes of their mothers. In contrast to newborn cats and dogs, newborn seal pups develop high specific antibody responses after immunization with an inactivated rabies vaccine. Circulating levels of total IgG in newborn seal pups increase rapidly after colostrum intake to 65% of maternal levels after 15 days. Serum IgM levels rise to adult levels by 14 days of age (King, 1986). A similar increase was observed for phocine distemper virus-specific antibodies. The relative immunocompetence of the harbor seal pup at birth may reflect an adaptation to its relatively short nursing period and limited maternal care. In agreement with this view, the IgA concentration in southern elephant seal milk (*Mirounga leonina*) is lower than in other mammals and IgG could not be detected (Marquez *et al.*, 1995).

6.2.6. Ungulata. Since all ungulates appear to have either an epitheliochorial or a syndesmochorial placenta, no significant amount of immunoglobulins can reach the fetus via this route.

Horses have an impermeable epitheliochorial placenta. As a result, foals must obtain their IgG from mare's colostrum. IgG and IgG(T), the latter an immunoglobulin subclass unique to equids, are the predominant immunoglobulins in equine colostrum (MacDougall, 1975) although their concentration varies among breeds. In Arabian mares the concentration of colostral IgG is significantly higher than in thoroughbred mares (Pearson *et al.*, 1984). In suckling foals, immunoglobulin levels climb rapidly for the first 24 hours of life and then fall over the next four weeks to less than half their original values. Maternal antibodies usually disappear completely by six months (Jeffcott, 1975). The presence or absence of maternal IgG significantly affects a foal's chances of developing septicemia but, as might be expected, does not appear to directly affect resistance to surface infections such as diarrheal disease or *Rhodococcus equi* pneumonia (Haas *et al.*, 1996; Raidal, 1996). Robinson *et al.* (1993) found that seven out of eight colostrum deprived foals developed sepsis. Organisms isolated included *Actinobacillus equuli*, *Escherichia coli*, *Pseudomonas* spp. and *Actinomyces pyogenes*. In contrast, none of the colostrum-fed foals became septicemic. The proportion of foals reported to exhibit failure of passive transport (FPT) depends upon the diagnostic criteria used. Perryman and McGuire (1980) and McGuire *et al.* (1977) examined foals with evidence of immunodeficiency (increased susceptibility to bacterial and viral infections) and found that full or partial FPT occurred in 20% of the foals at risk. They suggested that about 10% of foals have FPT as defined by a circulating IgG concentration of <800 mg/dl.

Kohn *et al.* (1989) found that IgG concentrations in foal serum correlated poorly with IgG concentrations in colostrum. This raises the interesting possibility that the high incidence of FPT in foals might indicate a problem with expression of FcRn. Galan *et al.* (1986) examined the passive transfer of immunoglobulins to *Streptococcus equi* in young foals. Shortly after colostral intake, IgG and IgA antibodies were found not only in sera but also in nasal secretions. Intragastrically administered labeled IgA was transported to the nasal mucosa of a newborn foal within a few hours of colostrum uptake. They suggested therefore that passive protection of the foal respiratory tract was derived not only by immunoglobulins directly coating the upper respiratory and oral mucosa but also by secretion of immunoglobulins directly onto the nasopharyngeal mucosa. Passively acquired maternal antibodies have also been shown to protect foals against experimental equine herpesvirus-1 challenge (Kendrick and Stevenson, 1979).

The pig has an impermeable epitheliochorial placenta and newborns are therefore dependent on colostral IgG for immediate protection. Two of the three pig subisotypes (IgGA and IgGB but not IgGC) are preferentially stored in the mammary glands of full-term sows and secreted into colostrum after farrowing (Huang *et al.*, 1992).

The ability of enterocytes of the small intestine of piglets to transport IgG has been assessed by Leary and Lecce (1979). Transport occurred mainly through the enterocytes in the proximal part of the small intestine. When bovine albumin or porcine IgG was presented separately to the piglet's gut, they were transported at about the same low level. However, when mixtures of these proteins were presented IgG was preferentially transported, i.e., albumin enhanced IgG transport. Mature milk is the main source of IgA to the piglet intestine. For instance, a three-week-old piglet may receive 1.6 g/day of IgA from sow's milk (Klobasa *et al.*, 1987).

Passive maternal immunity either in the form of maternal IgG from colostrum or IgA from milk has been shown to protect piglets against transmissible gastroenteritis virus (TGEV) (Stone *et al.*, 1977; Woods and Wesley, 1986; Fu *et al.*, 1990; Lanza *et al.*, 1995; Sestak *et al.*, 1996; Shoup *et al.*, 1997), pseudorabies (McFerran, 1975; Vannier *et al.*, 1995), colibacillosis (Rutter *et al.* 1976; Contrepois *et al.*, 1978; Chidlow and Porter, 1979; Isaacson *et al.*, 1980; Moon, 1981; Furer *et al.*, 1982, 1983; Rijke *et al.*, 1983; Nagy *et al.* 1985), porcine parvovirus (Paul *et al.*, 1980, 1982), *Hemophilus parahaemolyticus* (Nielsen, 1975), eastern equine encephalitis (Elvinger *et al.*, 1996), *Bordetella bronchiseptica* (Smith *et al.*, 1982), and *Pasteurella multocida* toxin (Foged *et al.*, 1989). In general the IgG tends to protect against systemic infections while the prolonged intake of milk IgA protects against enteric disease.

The route by which an antigen enters the body of the sow influences the isotype of immunoglobulin found in colostrum or milk. For example the presence of IgA anti-TGEV antibodies in sow's milk is associated with intestinal infection, whereas IgG antibodies resulted from parenteral antigenic stimulation of the sow (Bohl and Saif, 1975; Bohl *et al.*, 1975; Moxley *et al.*, 1989).

As in cattle, IgG is the predominant immunoglobulin in the milk of the dromedary *Camelus dromedarius* (Azwai *et al.*, 1996). IgM and IgA have also been identified in camel colostrum. Their protective significance has not been ascertained.

Alpacas (*Lama pacos*) lack blood immunoglobulins at birth, but show a 70% increase in total serum proteins within 24 hours because of absorption of gamma globulins from colostrum (Garmendia and McGuire, 1987). IgG is the isotype in highest concentration in colostrum and in serum from 24-hour-old crias. IgM is also absorbed from colostrum but IgG accounts for greater than 85% of the passively transferred protein. Failure of passive transfer of immunoglobulin (IgG $\leq$ 900 mg/dl) is a major risk factor for mortality in newborn alpacas (Garmendia *et al.*, 1987).

6.2.7. Ruminantia. In view of the importance of passive maternal immunity, it is not surprising that in mule deer fawns (*Odocoileus hemionus*) serum gamma globulin concentrations directly affect morbidity and mortality. Parkinson *et al.* (1982) showed that all fawns with serum total protein concentrations of 5 g/dl or less during the first week of life developed diarrhea and died before 17 days of age. Only one of 14 fawns with a serum protein concentration above this level became sick and died (Parkinson *et al.*, 1982). Sams *et al.* (1996) examined a herd of white-tailed deer (*Odocoileus virginianus*) and also showed that increased mortality was associated with lower serum gamma globulin concentrations. This suggests that inadequate absorption of colostrum and partial failure in passive immunity predisposes suckling fawns to increased mortality.

Cattle have a syndesmochorial placenta so that no IgG crosses the bovine placenta. Thus, calves are born agammaglobulinemic and depend on absorption of maternal colostral IgG for survival. There is an inverse correlation between passively acquired serum IgG levels and calf mortality. If failure of passive transport (FPT) occurs then increased neonatal morbidity and mortality occurs. (It may be useful to point out here that the causes of FPT are numerous. They can be classified as; production failure, where insufficient colostrum is produced by the mother—a common feature in multiple births; ingestion failure, where the newborn fails to suckle properly; and absorption failure, where the newborn fails to absorb sufficient immunoglobulin). The prevalence and severity of FPT depends on the definition of the condition. Mortality was significantly increased in calves with serum IgG1 levels <500 mg/dl as a result of FPT (Rea *et al.*, 1996). Calves classified as having FPT had a 9.5 times greater risk of developing infections prior to weaning than calves classified as exhibting sufficient passive transfer of immunoglobulin (Perino *et al.*, 1993). Calves with less than 1000 mg/dl of IgG in their serum by 24–48 hours of age die at more than twice the rate of calves with higher IgG levels. McGuire *et al.* (1976) found that serum IgG1 concentrations in calves less than three weeks old, and dying from infectious disease, were significantly lower than those of clinically normal calves. These low IgG1 concentrations were attributed to FPT. They concluded that "In view of the large numbers of calves that die from neonatal infection each year, it may be assumed that failure in passive transfer, as reflected by low serum immunoglobulin concentrations, is one of the most important factors influencing neonatal calf mortality".

Productivity in calves, as expressed by weight gain, is also affected by FPT (Robison *et al.*, 1988; Virtala *et al.*, 1996). Calves fed colostrum with high immunoglobulin content gained weight from birth to day four while those fed colostrum with low concentrations of immunoglobulins lost weight (Nocek *et al.*, 1984).

Given that mammals transfer IgG to their newborn either through colostrum or via the placenta, it makes sense to enhance resistance to disease by vaccinating pregnant animals. Consequently, antibodies made by the mother will confer immunity on their offspring. Preparturient vaccination of dairy cows induces increases in passive antibody titers in their calves to antigens of *Pasteurella haemolytica* (Hodgkins and Shewen, 1994, 1996), *Salmonella typhimurium* (Jones *et al.*, 1988), rotaviruses (Snodgrass, *et al.*, 1980; Castrucci *et al.*, 1988; Tsunemitsu, *et al.*, 1989) and *E. coli* (Wilson and Jutila, 1976a,b; Valente *et al.*, 1988). Protection by maternal passive immunity has been shown against bovine respiratory syncytial virus (Belknap *et al.*, 1991), bovine rotavirus (Saif *et al.*, 1983; Saif and Smith 1985; Besser *et al.*, 1988), enterotoxigenic *Escherichia coli* (Snodgrass *et al.*, 1982), septicemic *E. coli* (Johnston *et al.*, 1977), infectious bovine rhinotracheitis virus (Mechor *et al.*, 1987), bovine diarrhea virus (Howard *et al.*, 1989; Bolin and Ridpath, 1995), *Salmonella typhimurium* (Staak and Luge, 1995), cryptosporidium (Fayer *et al.*, 1989) *Taenia saginata* (Rickard *et al.*, 1977) and bovine leukemia virus (Lassauzet *et al.*, 1989). There is no doubt that the passive transfer of immunoglobulins through milk and colostrum is absolutely essential for resistance to disease and neonatal health during the first months of life in the major domestic species.

Besser *et al.* (1985) found a negative correlation between the efficiency of absorption and the total amount of immunoglobulin fed to calves. They suggested that there was a physiologic limitation to the amount of immunoglobulin that can be absorbed from a given volume of colostrum. In calves receiving colostrum at six hours after birth, 65.8% of the ingested IgG appeared in the plasma (Matte *et al.*, 1982).

Stott and Menefee (1978) examined the efficiency of absorption of different immunoglobulin isotypes in the colostrum fed to calves. They found that IgG absorption varied from 10 to 46% and IgA absorption from 5 to 50%. The percent of IgM absorbed increased as the amount ingested decreased, whereas, the efficiency of IgA and IgG absorption did not change with intake. Selective transport of IgM, they speculated, may be important for its role as the primary protective immunoglobulin during the first few days of life. Clover and Zarkower (1980) examined the influence of colostrum on calf lymphocyte reactivity. They found that peripheral blood lymphocytes from calves deprived of colostrum during the first 24 hours after delivery were significantly more responsive to mitogens than leukocytes from colostrum-fed calves.

Sheep, like cattle, have a syndesmochorial placenta and lambs consequently require colostral immunoglobulins to survive. Colostrum from ewes contains significantly higher concentrations of immunoglobulins than does their serum, IgG being selectively concentrated over IgM (Sawyer *et al.*, 1977). As in cattle, IgG is the predominant immunoglobulin in sheep milk (Smith *et al.*, 1975).

Absorption of immunoglobulins through the intestinal tract of the lamb appears to be a non-selective process. Concentrations of IgG1 were measured in 590 lambs, among which FPT (less than 600 mg/dl) occurred in twenty (McGuire *et al.*, 1983). Of these, 45% died before three weeks of age, whereas only 5% of the 570 lambs with adequate passive transfer died. Sawyer *et al.* (1977) found that 14% of clinically normal lambs demonstrated some FPT. Protection of lambs through passive transfer of antibodies from colostrum has has been reported for contagious pustular dermatitis (Lejan *et al.*, 1978), K99 colibacillosis (Sojka *et al.*, 1978; Morris *et al.*, 1980; Pugh and Wells, 1985), rotavirus (Snodgrass and Wells, 1978), *Echinococcus granulosus* (Dempster and Harrison, 1995), *Taenia ovis* (Sutton, 1979) and *Taenia hydatigena* (Jacobs *et al.*, 1994).

The IgG concentration in goat colostrum is about 2.5 times greater than in the maternal serum, and 95–98% is IgG1 (Micusan and Borduas, 1977). Given the structure of the goat placenta it is clear that kids are dependent on colostral immunoglobulin for survival.

7. Conclusions

The diversity observed in placental structures testifies to the remarkable nature of mammalian evolution. There appears to be no obvious phylogenetic pattern to placental structure except that less phylogenetically advanced species appear to have less permeable placentas. It can be argued teleologically that continued evolution of the means by which mammals regulate immune responses against fetal antigens permit the evolution of placentas with more efficient transport functions. However, other factors clearly come into play. Less permeable placentas may limit the supply of nutrients to the fetus and so necessitate a longer gestation period. It is also clear that passive transfer of immunoglobulins, and possibly cells, from the mother to the newborn by way of colostrum and milk was an essential step in mammalian evolution. Mammals with impermeable placentas must acquire their immunoglobulins in some other way and milk is the obvious source. It is likely, therefore, that mammalian viviparity and the secretion of milk immunoglobulins evolved at the same time. Once immunoglobulin-permeable placentas evolved, the role of colostrum became less important. On the other hand, placental structure has little bearing on the defense of the newborn's intestine by the continued intake of milk IgA. That seems to be a constant role for milk irrespective of the phylogenetic status of the species.

References

Abimiku, A.G. and Dolby, J.M. 1987. The mechanism of protection of infant mice from intestinal colonization with *Campylobacter jejuni*. *J. Med. Micro.* **23**:339–344.

Ahouse, J.J., Hagerman, C.L., Mittal, P., Gilbert, D.J., Copeland, N.G., Jenkins, N.A., and Simister, N.E. 1993. Mouse MHC class I-like Fc receptor encoded outside the MHC. *J. Immunol.* **151**:6076.

Altmann, K. and Mukkur, T.K. 1983. Passive immunisation of neonatal lambs against infection with enteropathogenic *Escherichia coli* via colostrum of ewes immunised with crude and purified K99 pili. *Res. Vet. Sci.* **35**:234.

Anderson, C.C. and Sinclair, N.R. 1998. FcR-mediated inhibition of cell activation and other forms of coinhibition. *Crit. Rev. Immunol.* **18**:525.

Appleby, P. and Catty, D. 1983. Transmission of immunoglobulin to foetal and neonatal mice. *J. Repro. Immunol.* **5**:203.

Archambault, D., Morin, G., and Elazhary, M. 1988a. Isolation of bovine colostral lymphocytes: in vitro blastogenic responsiveness to concanavalin A and bovine rotavirus. *Ann. Res. Vet.* **19**:169.

Archambault, D., Morin, G., Elazhary, Y., Roy, R.S., and Joncas, J.H. 1988b. Immune response of pregnant heifers and cows to bovine rotavirus inoculation and passive protection to rotavirus infection in newborn calves fed colostral antibodies or colostral lymphocytes. *Am. J. Vet. Res.* **49**:1084.

Azwai, S.M., Carter, S.D., and Woldehiwet, Z. 1996. Immunoglobulins of camel (*Camelus dromedarius*) colostrum. *J. Comp. Path* **114**:273.

Balfour, W.E. and Comline, R.S. 1962. Acceleration of the absorption of unchanged globulins in the newborn calf by factors in colostrum. *J. Physiol.* **160**:234.

Barrington, G.M., Besser, T.E., Davis, W.C., Gay, C.C., Reeves, J.J., and McFadden, T.B. 1997. Expression of immunoglobulin G1 receptors by bovine mammary epithelial cells and mammary leukocytes. *J. Dairy Sci.* **80**:86.

Barthold, S.W., Beck, D.S., and Smith, A.L. 1988. Mouse hepatitis virus and host determinants of vertical transmission and maternally-derived passive immunity in mice. *Arch. Virol.* **100**:171.

Beer, A.E. and Billingham, R.E. 1976. *The Immunology of Mammalian Reproduction*. Prentice Hall, Englewood Cliffs, NJ.

Belknap, E.B., Baker, J.C., Patterson, J.S., Walker, R.D., Haines, D.M., and Clark, E.G. 1991. The role of passive immunity in bovine respiratory syncytial virus-infected calves. *J. Inf. Dis.* **163**:470.

Besser, T.E., Gay, C.C., McGuire, T.C., and Evermann, J.F. 1988. Passive immunity to bovine rotavirus infection associated with transfer of serum antibody into the intestinal lumen. *J. Virol.* **62**:2238.

Besser, T.E., Garmedia, A.E., McGuire, T.C., and Gay, C.C. 1985. Effect of colostral immunoglobulin G1 and immunoglobulin M concentrations on immunoglobulin absorption in calves. *J. Dairy Sci.* **68**:2033.

Blüm, V. 1985. *Vertebrate Reproduction*. Springer-Verlag, Berlin.

Blumberg, R.S., Koss, T., Story, C.M., Barisani, D., Polischuk, J., Lipin, A., Pablo, L., Green, R., and Simister, N.E. 1995. A major histocompatibility complex class I-related Fc receptor for IgG on rat hepatocytes. *J. Clin. Invest.* **95**:2397.

Bohl, E.H., Frederick, T., and Saif, L.J. 1975. Passive immunity in transmissible gastroenteritis of swine: intramuscular injection of pregnant swine with a modified live-virus vaccine. *Amer. J. Vet. Res.* **36**:267.

Bohl, E.H. and Saif, L.J. 1975. Passive immunity in transmissible gastroenteritis of swine: immunoglobulin characteristics of antibodies in milk after inoculating virus by different routes. *Inf. & Imm.* **11**:23.

Bolin, S.R. and Ridpath, J.F. 1995. Assessment of protection from systemic infection or disease afforded by low to intermediate titers of passively acquired neutralizing antibody against bovine viral diarrhea virus in calves. *Amer. J. Vet. Res.* **56**:755.

Bradshaw, B.J. and Edwards, S. 1996. Antibody isotype responses to experimental infection with bovine herpesvirus 1 in calves with colostrally derived antibody. *Vet. Microbiol.* **53**:143.

Brar, J.S., Johnson, D.W., Muscoplat, C.C., Shope, R.E., Jr., and Meiske, J.C. 1978. Maternal immunity to infectious bovine rhinotracheitis and bovine viral diarrhea viruses: duration and effect on vaccination in young calves. *Amer. J. Vet. Res.* **39**:241.

Carmichael, L.E., Robson, D.S., and Barnes, T.D. 1962. Transfer and decline of maternal infectious hepatitis antibody in puppies. *Proc. Soc. Exp. Biol. Med.* **109**:677.

Carmichael, L.E. 1983. Immunization strategies in puppies, why failures? *Comp. Cont. Edu. Pract. Vet.* **5**:1043.

Carter, S.D., Hughes, D.E., and Baker, J.R. 1990. Characterization and measurement of immunoglobulins in the grey seal (*Halichoerus grypus*). *J. Comp. Pathol.* **102**:13.
Casal, M.L., Jezyk, P.F., and Giger, U. 1996. Transfer of colostral antibodies from queens to their kittens. *Amer. J. Vet. Res.* **57**:1653.
Castrucci, G., Frigeri, F., Ferrari, M., Aldrovandi, V., Tassini, F., and Gatti, R. 1988. The protection of newborn calves against experimental rotavirus infection by feeding mammary secretions from vaccinated cows. *Microbiologica* **11**:379.
Chabaudie, N., Leja, C., Oliver, M., and Salmon, H. 1993. Lymphocyte subsets in the mammary gland of sows. *Res. Vet. Sci.* **55**:351.
Chidlow, J.W. and Porter, P. 1979. Intestinal defence of the neonatal pig: interrelationship of gut and mammary function providing surface immunity against colibacillosis. *Vet. Rec.* **104**:496.
Clover, C.K. and Zarkower, A. 1980. Immunologic responses in colostrum-fed and colostrum-deprived calves. *Amer. J. Vet. Res.* **41**:1002.
Cockson, A. and McNeice, R. 1980. Surival in the pouch: The role of macrophages and maternal milk cells. *Comp. Biochem. Physiol.* **66A**:221.
Cole, M.F. and Bowen, W.H. 1976. Immunoglobulins A, G, and M in serum and in some secretions of monkeys (*Macaca fascicularis* syn. irus). *Inf. & Imm.* **13**:1354.
Comline, R.S., Roberts, H.E., and Titchen, D.A. 1951. Route of absorption of colostrum globulin in the newborn animal. *Nature* **167**:561.
Contrepois, M., Girardeau, J.P., Dubourguier, H.C., Gouet, P., and Levieux, D. 1978. Specific protection by colostrum from cows vaccinated with the K99 antigen in newborn calves experimentally infected with E. coli Ent+ K99+. *Ann. Res. Vet.* **9**:385.
Crighton, D.B. (Ed). 1984. *Immunological Aspects of Reproduction in Mammals*. Butterworths London.
Dahlgren, U., Ahlstedt, S., Hedman, I., Wansworth, C., and Hanson, L.A. 1981. Dimeric IgA in the rat is transferred from serum into bile but not into milk. *Scand. J. Immunol.* **14**:95.
Deane, E.M., Cooper, D.W., and Renfree, M.B. 1990. Immunoglobulin G levels in fetal and newborn tammar wallabies (*Macropus eugenii*). *Repro. Fert. & Dev.* **2**:369.
Deane, E.M. and Cooper, D.W. 1984. Immunology of pouch young marsupials. I. Levels of immunoglobulin transferrin and albumin in the blood and milk of euros and wallaroos (hill kangaroos: *Macropus robustus*, marsupialia). *Dev. Comp. Immunol.* **8**:863.
Dempster, R.P. and Harrison, G.B. 1995. Maternal transfer of protection from *Echinococcus granulosus* infection in sheep. *Res. Vet. Sci.* **58**:197.
Derchia, A.M., Gissi, C., Pesole, G., Saccone, C., and Arnason, U. 1996. The guinea-pig is not a rodent. *Nature* **381**:597.
Duchet-Suchaux, M. 1983. Infant mouse model of *E. coli* diarrhoea: clinical protection induced by vaccination of the mothers. *Ann. Res. Vet.* **14**:319.
Elvinger, F., Baldwin, C.A., Liggett, A.D., Tang, K.N., and Dove, C.R. 1996. Protection of pigs by vaccination of pregnant sows against eastern equine encephalomyelitis virus. *Vet. Microbiol.* **51**:229.
Fayer, R., Andrews, C., Ungar, B.L., and Blagburn, B. 1989. Efficacy of hyperimmune bovine colostrum for prophylaxis of cryptosporidiosis in neonatal calves. *J. Parasitol.* **75**:393.
Foged, N.T., Nielsen, J.P., and Jorsal, S.E. 1989. Protection against progressive atrophic rhinitis by vaccination with *Pasteurella multocida* toxin purified by monoclonal antibodies. *Vet. Rec.* **125**:7.
Fu, Z.F., Hampson, D.J., and Wilks, C.R. 1990. Transfer of maternal antibody against group A rotavirus from sows to piglets and serological responses following natural infection. *Res. Vet. Sci.* **48**:365.

Furer, E., Cryz, S.J., Jr., and Germanier, R. 1983. Protection of piglets against neonatal colibacillosis based on antitoxic immunity. *Dev. Biol. Stand.* **53**:161.

Furer, E., Cryz, S.J., Jr., Dorner, F., Nicolet, J., Wanner, M., and Germanier, R. 1982. Protection against colibacillosis in neonatal piglets by immunization of dams with procholeragenoid. *Inf. & Imm.* **35**:887.

Galan, J.E., Timoney, J.F., and Lengemann, F.W. 1986. Passive transfer of mucosal antibody to *Streptococcus equi* in the foal. *Inf. & Imm.* **54**:202.

Garmendia, A.E. and McGuire, T.C. 1987. Mechanism and isotypes involved in passive immunoglobulin transfer to the newborn alpaca (*Lama pacos*). *Amer. J. Vet. Res.* **48**:1465.

Garmendia, A.E., Palmer, G.H., DeMartini, J.C., and McGuire, T.C. 1987. Failure of passive immunoglobulin transfer: a major determinant of mortality in newborn alpacas (*Lama pacos*). *Amer. J. Vet. Res.* **48**:1472.

Ghetie, V. and Ward, E.S. 1997. FcRn: the MHC class I-related receptor that is more than an IgG transporter. *Immunol. Today* **18**:592.

Gombold, J.L. and Ramig, R.F. 1989. Passive immunity modulates genetic reassortment between rotaviruses in mixedly infected mice. *J. Virol.* **63**:4525.

Griffin, P. and Wild, A.E. 1987. Ontogeny of the Fcγ receptor in rat small intestine. *J. Reprod. Immunol.* **11**:287.

Haas, S.D., Bristol, F., and Card, C.E. 1996. Risk factors associated with the incidence of foal mortality in an extensively managed mare herd. *Can. Vet. J.* **37**:91.

Halliday, R. 1978. Immunity and health in young lambs. *Vet. Rec.* **103**:489.

Halsey, J.F., Johnson, B.H., and Cebra, J.J. 1980. Transport of immunoglobulins from serum into colostrum. *J. Exp. Med.* **151**:767.

Halsey, J.F., Mitchell, C., Meyer, R., and Cebra, J.J. 1982. Metabolism of immunoglobulin A in lactating mice: origins of immunoglobulin A in milk. *Eur. J. Immunol.* **12**:107.

Head, J.R. 1977. Immunobiology of lactation. *Sem. Perinatol.* **1**:195.

Head, J.R. and Beer, A.E. 1978. The ImmunologicRole of Viable Leukocytic Cells in Mammary Exosecretions, in: *Lactation, A Comprehensive Treatise*, B.L. Larson, ed. Academic Press, New York, p. 337.

Heckert, R.A., Saif, L.J., Mengel, J.P., and Myers, G.W. 1991. Mucosal and systemic antibody responses to bovine coronavirus structural proteins in experimentally challenge-exposed calves fed low or high amounts of colostral antibodies. *Amer. J. Vet. Res.* **52**:700.

Heddle, R.J. and Rowley, D. 1975. Dog immunoglobulins. I. immunochemical characterization of dog serum, parotid saliva, colostrum, milk and small bowel fluid. *Immunol.* **29**:185.

Hindes, R.D. and Mizell, M. 1976. The origin of immunoglobulins in opossum "embryos". *Dev. Biol.* **53**:49.

Hodgins, D.C. and Shewen, P.E. 1994. Passive immunity to Pasteurella haemolytica A1 in dairy calves: effects of prepartum vaccination of the dams. *Can. J. Vet. Res.* **58**:31.

Hodgins, D.C. and Shewen, P.E. 1996. Preparturient vaccination to enhance passive immunity to the capsular polysaccharide of Pasteurella haemolytica A1. *Vet. Immunol. Immunopathol.* **50**:67.

Homberger, F.R. 1992. Maternally-derived passive immunity to enterotropic mouse hepatitis virus. *Arch. Virol.* **122**:133.

Hoover, E.A., Schaller, J.P., Mathes, L.E., and Olsen, R.G. 1977. Passive immunity to feline leukemia: evaluation of immunity from dams naturally infected and experimentally vaccinated. *Inf. & Imm.***16**:54.

Howard, C.J., Clarke, M.C., and Brownlie, J. 1989. Protection against respiratory infection with bovine virus diarrhoea virus by passively acquired antibody. *Vet. Microbiol.* **19**:195.

Huang, S.C., Hu, Z.L., Hasler-Rapacz, J., and Rapacz, J. 1992. Preferential mammary storage and secretion of immunoglobulin gamma (IgG) subclasses in swine. *J. Reprod. Immunol.* **21**:15.

Isaacson, R.E., Dean, E.A., Morgan, R.L., and Moon, H.W. 1980. Immunization of suckling pigs against enterotoxigenic Escherichia coli-induced diarrheal disease by vaccinating dams with purified K99 or 987P pili: antibody production in response to vaccination. *Inf. & Imm.* **29**:824.

Israel, E.J., Patel, V.K., Taylor, S.F., Marshak-Rothstein, A., and Simister, N.E. 1995. Requirement for a beta 2-microglobulin-associated Fc receptor for acquisition of maternal IgG by fetal and neonatal mice. *J. Immunol.* **154**:6246.

Israel, E.J., Wilsker, D.F., Hayes, K.C., Schoenfeld, D., and Simister, N.E. 1996. Increased clearance of IgG in mice that lack beta 2-microglobulin: possible protective role of FcRn. *Immunol.* **89**:573.

Jacobs, H.J., Moriarty, K.M., Charleston, W.A., and Heath, D.D. 1994. Resistance against *Taenia hydatigena* in sheep after passive transfer of serum or colostrum. *Parasite Immunol.* **16**:351.

Jakoi, E.R., Cambier, J., and Saslow, S. 1985. Transepithelial transport of maternal antibody: purification of IgG receptor from newborn rat intestine. *J. Immunol.* **135**:3360.

Jeffcott, L.B. 1975. The transfer of passive immunity to the foal and its relation to immune status after birth. *J. Reprod. Fert.—Suppl.* **23**:727.

Jochims, K., Kaup, F.-J., Drommer, W., and Pickel, M. 1994. An immunoelectron microscopic investigation of colostral IgG absorption across the intestine of newborn calves. *Res. Vet. Sci.* **57**:75.

Johnston, N.E., Estrella, R.A., and Oxender, W.D. 1977. Resistance of neonatal calves given colostrum diet to oral challenge with a septicemia-producing *Escherichia coli*. *Amer. J. Vet. Res.* **38**:1323.

Jones, P.W., Collins, P., and Aitken, M.M. 1988. Passive protection of calves against experimental infection with *Salmonella typhimurium*. *Vet. Rec.* **21**:536.

Junghans, R.P. and Anderson, C.L. 1996. The protection receptor for IgG catabolism is the beta2-microglobulin-containing neonatal intestinal transport receptor. *Proc. Natl. Acad. Sci. USA* **93**:5512.

Kandil, E., Noguchi, M., Ishibashi, T., and Kasahara, M. 1995. Structural and phylogenetic analysis of the MHC class I-like Fc receptor gene. *J. Immunol.* **154**:5907.

Katsura, T., Kanamori, M., Kitamoto, O., and Ogata, S. 1985. Protective effect of colostrum in *Mycoplasma pneumoniae* infection induced in infant mice. *Microbiol. & Immunol.* **29**:883.

Kendrick, J.W. and Stevenson, W. 1979. Immunity to equine herpesvirus 1 infection in foals during the first year of life. *J. Repro. Fert.—Suppl.* **27**:615.

King, B.F. 1986. Morphology of the Placenta and fetal membranes, in: *Comparative Primate Biology 3: Reproduction and Development*. W.R. Dukelow and J. Erwin, eds. J. Alan R Liss Inc, NY, p. 311.

Kitching, R.P. and Salt, J.S. 1995. The interference by maternally-derived antibody with active immunization of farm animals against foot-and-mouth disease. *Brit. Vet. J.* **151**:379.

Klobasa, F. Werhahn, E., and Butler, J.E. 1987. Composition of sow milk during lactation. *J. Anim. Sci.* **64**:1458.

Kmetz, M., Schultz, R.D., and Dunne, H.W. 1970. Leukocytes as carriers in the transmission of bovine leukemia: in vitro growth of leukocytes from milk and blood of normal and leukemic cattle. *Amer. J. Vet. Res.* **31**:587.

Kohn, C.W., Knight, D., Hueston, W., Jacobs, R., and Reed, S.M. 1989. Colostral and serum IgG, IgA, and IgM concentrations in Standardbred mares and their foals at parturition. *J. Amer. Vet. Med. Assn.* **195**:64.

Komuves, L.G. and Heath, J.P. 1992. Uptake of maternal immunoglobulins in the enterocytes of suckling piglets: improved detection with a streptavidin-biotin bridge gold technique. *J. Histochem. Cytochem.* **40**:1637.

Krakowka, S., Long, D., and Koestner, A. 1978. Influence of transplacentally acquired antibody on neonatal susceptibility to canine distemper virus in gnotobiotic dogs. *J. Inf. Dis.* **137**:605.

Kunz, C. and Lonnerdal, B. 1993. Protein composition of rhesus monkey milk: comparison to human milk. *Comp. Biochem, Physiol.* **4**:793.

Lanza, I., Shoup, D.I., and Saif, L.J. 1995. Lactogenic immunity and milk antibody isotypes to transmissible gastroenteritis virus in sows exposed to porcine respiratory coronavirus during pregnancy. *Amer. J. Vet. Res.* **56**:739.

Lassauzet, M.L., Johnson, W.O., Thurmond, M.C., and Stevens, F. 1989. Protection of colostral antibodies against bovine leukemia virus infection in calves on a California dairy. *Can. J. Vet. Res.* **53**:424.

Leary, H.L., Jr. and Lecce, J.G. 1979. The preferential transport of immunoglobulin G by the small intestine of the neonatal piglet. *J. Nutr.* **109**:458.

Lee, C.S., McCauley, I., and Hartman, P.E. 1983. Light and electron microscopy of cells in pig colostrum, milk and involution secretion. *Acta Anat.* **117**:270.

Lejan, C. 1993. Secretory component and IgA expression by epithelial cells in sow mammary gland and mammary secretions. *Res. Vet. Sci.* **55**:265.

Lejan, C. 1994. A study by flow cytometry of lymphocytes in sow colostrum. *Res. Vet. Sci.* **57**:300.

Lejan, C., L'Haridon, R., Madelaine, M.F., Cornu, C., and Asso, J. 1978. Transfer of antibodies against the CPD virus through colostrum and milk. *Ann. Rech. Vet.* **9**:342.

Logan, E.F., McMurray, C.H., O'Neill, D.G., McParland, P.J., and McRory, F.J. 1978. Absorption of colostral immunoglobulins by the neonatal calf. *Brit. Vet. J.* **134**:258.

Macartney, L., Thompson, H., McCandlish, I.A., and Cornwell, H.J. 1988. Canine parvovirus: interaction between passive immunity and virulent challenge. *Vet. Rec.* **122**:573.

MacDougall, D.F. 1975. Immunoglobulin metabolism in the neonatal foal. *J. Repro. Fert.—Suppl* **23**:739.

Manning, M.J. and Turner, R.J. 1976. *Comparative Immunobiology*, John Wiley and Sons, New York.

Marquez, M.E., Slobodianik, N.H., Ronayne de Ferrer, P.A., Carlini, A.R., Vergani, D.F., and Daneri, G.A. 1995. Immunoglobulin A levels in southern elephant seal (*Mirounga leonina*) milk during the suckling period. *Comp. Biochem. Physiol.* **3**:569.

Matte, J.J., Girard, C.L., Seoane, J.R., and Brisson, G.J. 1982. Absorption of colostral immunoglobulin G in the newborn dairy calf. *J. Dairy Sci.* **65**:1765.

McFerran, J.B. 1975. Studies on immunity to Aujeszky's disease (pseudorabies) virus infection in pigs. *Dev. Biol. Stand.* **28**:563.

McGuire, T.C., Crawford, T.B., Hallowell, A.L., and Macomber, L.E. 1977. Failure of colostral immunoglobulin transfer as an explanation for most infections and deaths of neonatal foals. *J. Amer. Vet. Med. Assn.* **170**:1302.

McGuire, T.C., Pfeiffer, N.E., Weikel, J.M., and Bartsch, R.C. 1976. Failure of colostral immunoglobulin transfer in calves dying from infectious disease. *J. Amer. Vet. Med. Assn.* **169**:713.

McGuire, T.C., Regnier, J., Kellom, T., and Gates, N.L. 1983. Failure in passive transfer of immunoglobulin G1 to lambs: measurement of immunoglobulin G1 in ewe colostrums. *Amer. J. Vet. Res.* **44**:1064.

Mechor, G.D., Rousseaux, C.G., Radostits, O.M., Babiuk, L.A., and Petrie, L. 1987. Protection of newborn calves against fatal multisystemic infectious bovine rhinotracheitis by feeding colostrum from vaccinated cows. *Can. J. Vet. Res.* **51**:452.

Medesan, C., Radu, C., Kim, J.K., Ghetie, V., and Ward, E.S. 1996. Localization of the site of the IgG molecule that regulates maternofetal transmission in mice. *Eur. J. Immunol.* **26**:2533.

Menanteau-Horta, A.M., Ames, T.R., Johnson, D.W., and Meiske, J.C. 1985. Effect of maternal antibody upon vaccination with infectious bovine rhinotracheitis and bovine virus diarrhea vaccines. *Can. J. Comp. Med.* **49**:10.

Micusan, V.V. and Borduas, A.G. 1977. Biological properties of goat immunoglobulins G. *Immunol.* **32**:373.

Milner, A.R. and Marshall, I.D. 1984. The role of colostrum and milk in protection of the neonatal mouse against peripheral infection with Ross River virus. Brief report. *Arch. Virol.* **82**:101.

Moon, H.W. 1981. Protection against enteric colibacillosis in pigs suckling orally vaccinated dams: evidence for pili as protective antigens. *Amer. J. Vet. Res.* **42**:173.

Morris, J.A., Wray, C., and Sojka, W.J. 1980. Passive protection of lambs against enteropathogenic *Escherichia coli*: role of antibodies in serum and colostrum of dams vaccinated with K99 antigen. *J. Med. Microbiol.* **13**:265.

Moxley, R.A., Olson, L.D., and Solorzano, R.F. 1989. Relationship among transmissible gastroenteritis virus antibody titers in serum, colostrum, and milk from vaccinated sows, and protection in their suckling pigs. *Amer. J. Vet. Res.* **50**:119.

Mucchielli, A., Laliberte, F., and Laliberte, M.F. 1983. A new experimental method for the dynamic study of the antibody transfer mechanism from mother to fetus in the rat. *Placenta* **4**:175.

Musoke, A.J., Williams, J.F., Leid, R.W., and Williams, C.S. 1975. The immunological response of the rat to infection with *T. taeniaeformis*. IV. Immunoglobulins involved in passive transfer of resistance from mother to offspring. *Immunol.* **29**:845.

Nagy, L.K., MacKenzie, T., and Painter, K.R. 1985. Protection of the nursing pig against experimentally induced enteric colibacillosis by vaccination of dam with fimbrial antigens of *E coli* (K88, K99 and 987P). *Vet. Rec.* **16**:408.

Nielsen, R. 1975. Colostral transfer of immunity to *Haemophilus parahaemolyticus* in pigs. *Nord. Vetmed.* **27**:319.

Nocek, J.E., Braund, D.G., and Warner, R.G. 1984. Influence of neonatal colostrum administration, immunoglobulin, and continued feeding of colostrum on calf gain, health, and serum protein. *J. Dairy Sci.* **67**:319.

Norcross, N.L. 1982. Secretion and composition of colostrum and milk. *J. Amer. Vet. Med. Assn.* **181**:1057.

Olson, J.C. and Leslie, G.A. 1982. IgD: a component of the secretory immune system?, in: *Immunoglobulin D: Structure and Function*. G.J. Thorbecke and G.A. Leslie, eds. *Ann. N.Y. Acad Sci.* **399**:97.

Omata, Y., Oikawa, H., Kanda, M., Mikazuki, K., Dilorenzo, C., Claveria, F.G., Takahashi, M., Igarashi, I., Saito, A., and Suzuki, N. 1994. Transfer of antibodies to kittens from mother cats chronically infected with Toxoplasma gondii. *Vet. Parasit.* **52**:211.

Outteridge, P.M. and Lee, C.S. 1988. The defense mechanisms of the mammary gland of domestic ruminants. *Progr. Vet. Microbiol. Immunol.* **4**:165.

Parkinson, D.E., Ellis, R.P., and Lewis, L.D. 1982. Colostrum deficiency in mule deer fawns: identification, treatment and influence on neonatal mortality. *J. Wildlife Dis.* **18**:17.

Paul, P.S., Mengeling, W.L., and Brown, T.T., Jr. 1980. Effect of vaccinal and passive immunity on experimental infection of pigs with porcine parvovirus. *Amer. J. Vet. Res.* **41**:1368.

Paul, P.S., Mengeling, W.L., and Pirtle, E.C. 1982. Duration and biological half-life of passively acquired colostral antibodies to porcine parvovirus. *Amer. J. Vet. Res.* **43**:1376.

Pearson, R.C., Hallowell, A.L., Bayly, W.M., Torbeck, R.L., and Perryman, L.E. 1984. Times of appearance and disappearance of colostral IgG in the mare. *Amer. J. Vet. Res.* **45**:186.

Perino, L.J., Sutherland, R.L., and Woollen, N.E. 1993. Serum gamma-glutamyltransferase activity and protein concentration at birth and after suckling in calves with adequate and inadequate passive transfer of immunoglobulin G. *Amer. J. Vet. Res.* **54**:56.

Perryman, L.E. and McGuire, T.C. 1980. Evaluation for immune system failures in horses and ponies. *J. Amer. Vet. Med. Assn.* **176**:1374.

Perryman, L.E., McGuire, T.C., and Torbeck, R.L. 1980. Ontogeny of lymphocyte function in the equine fetus. *Amer. J. Vet. Res.* **41**:1197.

Pollock, R.V. and Carmichael, L.E. 1982. Maternally derived immunity to canine parvovirus infection: transfer, decline, and interference with vaccination. *J. Amer. Vet. Med. Assn.* **180**:37.

Prince, G.A., Horswood, R.L., Camargo, E., Koenig, D., and Chanock, R.M. 1983. Mechanisms of immunity to respiratory syncytial virus in cotton rats. *Inf. & Immun.* **42**:81.

Prince, G.A., Horswood, R.L., and Chanock, R.M. 1985. Quantitative aspects of passive immunity to respiratory syncytial virus infection in infant cotton rats. *J. Virol.* **55**:517.

Pugh, C.A. and Wells, P.W. 1985. Protection of lambs against enteric colibacillosis by vaccination of ewes. *Res. Vet. Sci.* **38**:255.

Raghavan, M. and Bjorkman, P.J. 1996. Fc receptors and their interactions with immunoglobulins. *Ann. Rev. Cell. Dev. Biol.* **12**:181.

Raghavan, M., Gastinel, L.N., and Bjorkman, P.J. 1993. The class I major histocompatibility complex related Fc receptor shows pH-dependent stability differences correlating with immunoglobulin binding and release. *Biochem.* **32**:8654.

Raidal, S.L. 1996. The incidence and consequences of failure of passive transfer of immunity on a thoroughbred breeding farm. *Aust. Vet. J.* **73**:201.

Rea, D.E., Tyler, J.W., Hancock, D.D., Besser, T.E., Wilson, L., Krytenberg, D.S., and Sanders, S.G. 1996. Prediction of calf mortality by use of tests for passive transfer of colostral immunoglobulin. *J. Amer. Vet. Med. Assn.* **208**:2047.

Renfree, M.B. 1973. The composition of fetal fluids of the marsupial *Macropus eugenii*. *Dev. Biol.* **33**:62.

Rickard, M.D., Adolph, A.J., and Arundel, J.H. 1977. Vaccination of calves against *Taenia saginata* infection using antigens collected during in vitro cultivation of larvae: passive protection via colostrum from vaccinated cows and vaccination of calves protected by maternal antibody. *Res. Vet. Sci.* **23**:365.

Riedel-Caspari, G. 1993. The influence of colostral leukocytes on the course of experimental *Escherichia coli* infection and serum antibodies in neonatal calves. *Vet. Immunol. Immunopathol.* **35**:275.

Riedel-Caspari, G. and Schmidt, F.W. 1990. Review article: Colostral leukocytes and their significance for the immune system of newborns. *DTW—Deutsche Tier. Wschrift.* **97**:180.

Riedel-Caspari, G. and Schmidt, F.W. 1991a. The influence of colostral leukocytes on the immune system of the neonatal calf. II. Effects on passive and active immunization. *DTW—Deutsche Tier. Wschrift.* **98**:190.

Riedel-Caspari, G. and Schmidt, F.W. 1991b. The influence of colostral leukocytes on the immune system of the neonatal calf. I. Effects on Lymphocyte responses. *DTW—Deutsche Tier. Wschrift.* **98**:102.

Rijke, E.O., Webster, J., and Baars, J.C. 1983. Vaccination of piglets against post-weaning "*E. coli*" enterotoxicosis. *Dev. Biol. Stand.* **53**:155.

Robinson, J.A., Allen, G.K., Green, E.M., Fales, W.H., Loch, W.E., and Wilkerson, C.G. 1993. A prospective study of septicaemia in colostrum-deprived foals. *Equine Vet. J.* **25**:214.

Robison, J.D., Stott, G.H., and DeNise, S.K. 1988. Effects of passive immunity on growth and survival in the dairy heifer. *J. Dairy Sci.* **71**:1283.

Ross, P.S., de Swart, R.L., Visser, I.K., Vedder, L.J., Murk, W., Bowen, W.D., and Osterhaus, A.D. 1994. Relative immunocompetence of the newborn harbour seal, *Phoca vitulina*. *Vet. Immunol. Immunopathol.* **42**:331.

Rutter, J.M., Jones, G.W., Brown, G.T., Burrows, M.R., and Luther, P.D. 1976. Antibacterial activity in colostrum and milk associated with protection of piglets against enteric disease caused by K88-positive Escherichia coli. *Inf. Immun.* **13**:667.

Saif, L.J., Redman, D.R., Smith, K.L., and Theil, K.W. 1983. Passive immunity to bovine rotavirus in newborn calves fed colostrum supplements from immunized or nonimmunized cows. *Inf. Immun.* **41**:1118.

Saif, L.J. and Smith, K.L. 1985. Enteric viral infections of calves and passive immunity. *J. Dairy Sci.* **68**:206.

Samples, K., Vandenberg, J.L., and Stone, W.H. 1986. Passively acquired immunity in the newborn of a marsupial (*Monodelphis domestica*). *Am. J. Reprod. Immunol. Microbiol.* **11**:94.

Sams, M.G., Lochmiller, R.L., Qualls, C., Jr., Leslie, D., Jr., and Payton, M.E. 1996. Physiological correlates of neonatal mortality in an overpopulated herd of white-tailed deer. *J. Mamm.* **77**:179.

Sasaki, M., Lawsom, B.L., and Nelson, D.R. 1977. Kinetic analysis of the binding of immunoglobulins IgG1 and IgG2 to bovine mammary cells. *Biochim. Biophys. Acta* **497**:160.

Sawyer, M., Willadsen, C.H., Osburn, B.I., and McGuire, T.C. 1977. Passive transfer of colostral immunoglobulins from ewe to lamb and its influence on neonatal lamb mortality. *J. Amer. Vet. Med. Assn.* **171**:1255.

Sestak, K., Lanza, I., Park, S.K., Weilnau, P.A., and Saif, L.J. 1996. Contribution of passive immunity to porcine respiratory coronavirus to protection against transmissible gastroenteritis virus challenge exposure in suckling pigs. *Amer. J. Vet. Res.* **57**:664.

Sheldrake, R.F. and Husband, A.J. 1985. Intestinal uptake of intact maternal lymphocytes by neonatal rats and lambs. *Res. Vet. Sci.* **36**:10.

Shoup, D.I., Jackwood, D.J., and Saif, L.J. 1997. Active and passive immune responses to transmissible gastroenteritis virus (TGEV) in swine inoculated with recombinant baculovirus-expressed TGEV spike glycoprotein vaccines. *Amer. J. Vet. Res.* **58**:242.

Simister, N.E., Jacobowitz Israel, E., Ahouse, J.C., and Story, C.M. 1997. New functions of the MHC class I-related Fc receptor, FcRn. *Biochem. Soc. Trans.* **25**:481.

Smith, I.M., Giles, C.J., and Baskerville, A.J. 1982. Immunisation of pigs against experimental infection with *Bordetella bronchiseptica*. *Vet. Rec.* **110**:488.

Smith, W.D., Dawson, A.M., Wells, P.W., and Burrells, C. 1975. Immunoglobulin concentrations in ovine body fluids. *Res. Vet. Sci.* **19**:189.

Snodgrass, D.R., Campbell, I., Mwenda, J.M., Chege, G., Suleman, M.A., Morein, B., and Hart, C.A. 1995. Stimulation of rotavirus IgA, IgG and neutralizing antibodies in baboon milk by parenteral vaccination. *Vaccine* **13**:408.

Snodgrass, D.R., Fahey, K.J., Wells, P.W., Campbell, I., and Whitelaw, A. 1980. Passive immunity in calf rotavirus infections: maternal vaccination increases and prolongs immunoglobulin G1 antibody secretion in milk. *Inf. Immun.* **28**:344.

Snodgrass, D.R., Nagy, L.K., Sherwood, D., and Campbell, I. 1982. Passive immunity in calf diarrhea: vaccination with K99 antigen of enterotoxigenic *Escherichia coli* and rotavirus. *Inf. Immun.* **37**:586.

Snodgrass, D.R. and Wells, P.W. 1978. The immunoprophylaxis of of rotavirus infections in lambs. *Vet. Rec.* **102**:146.

Sojka, W.J., Wray, C., and Morris, J.A. 1978. Passive protection of lambs against experimental enteric colibacillosis by colostral transfer of antibodies from K99-vaccinated ewes. *J. Med. Micro.* **11**:493.

Staak, C. and Luge, E. 1995. Copro-antibody in calves from dams vaccinated against *Salmonella typhimurium*. *Zblt. Vetmed.—Reihe. B.* **42**:493.

Steele, M.G. and Leslie, G.A. 1985. Immunoglobulin D in rat serum, saliva and milk. *Immunol.* **55**:571.

Stone, S.S., Kemeny, L.J., Woods, R.D., and Jensen, M.T. 1977. Efficacy of isolated colostral IgA, IgG, and IgM(A) to protect neonatal pigs against the coronavirus of transmissible gastroenteritis. *Amer. J. Vet. Res.* **38**:1285.

Story, C.M., Mikulska, J.E., and Simister, N.E. 1994. A major histocompatibility complex class I-like Fc receptor cloned from human placenta: possible role in transfer of immunoglobulin G from mother to fetus. *J. Exp. Med.* **180**:2377.

Stott, G.H., Marx, D.B., Menefee, B.E., and Nightengale, G.T. 1979a Colostral immunoglobulin transfer in calves. IV. Effect of suckling. *J. Dairy Sci.* **62**:1908.

Stott, G.H., Marx, D.B., Menefee, B.E., and Nightengale, G.T. 1979b Colostral immunoglobulin transfer in calves. II. The rate of absorption. *J. Dairy Sci.* **62**:1766.

Stott, G.H., Marx, D.B., Menefee, B.E., and Nightengale, G.T. 1979c. Colostral immunoglobulin transfer. III. Amount of absorption. *J. Dairy Sci.* **62**:1902.

Stott, G.H. and Menefee, B.E. 1978. Selective absorption of immunoglobulin IgM in the newborn calf. *J. Dairy Sci.* **61**:461.

Sutton, R.J. 1979. The passive transfer of immunity to *Taenia ovis* in lambs via colostrum. *Res. Vet. Sci.* **27**:197.

Sweet, C., Bird, R.A., Jakeman, K., Coates, D.M., and Smith, H. 1987. Production of passive immunity in neonatal ferrets following maternal vaccination with killed influenza A virus vaccines. *Immunol.* **60**:83.

Thatcher, E.F. and Gershwin, L.J. 1989. Colostral transfer of bovine immunoglobulin E and dynamics of serum IgE in calves. *Vet. Immunol. Immunopathol.* **20**:325.

Tsunemitsu, H., Shimizu, M., Hirai, T., Yonemichi, H., Kudo, T., Mori, K., and Onoe, S. 1989. Protection against bovine rotaviruses in newborn calves by continuous feeding of immune colostrum. *Jap. J. Vet. Sci.* **51**:300.

Valente, C., Fruganti, G., Tesei, B., Ciorba, A., Cardaras, P., Floris, A., and Bordoni, E. 1988. Vaccination of pregnant cows with K99 antigen of enterotoxigenic *Escherichia coli* and protection by colostrum in newborn calves. *Comp. Immunol. Microbiol. & Inf. Dis.* **11**:189.

Vannier, P., Hutet, E., and Cariolet, R. 1995. Influence of passive immunity on pig immunization with deleted Aujeszky's disease vaccines measured by the amount of wild virus excretion after challenge. *Vet. Microbiol.* **43**:53.

Virtala, A.M., Mechor, G.D., Grohn, Y.T., and Erb, H.N. 1996. The effect of calfhood diseases on growth of female dairy calves during the first 3 months of life in New York State. *J. Dairy Sci.* **79**:1040.

Wild, A.E., Burrows, T.D., and Brand, J. 1994. IgG transport across the gut of the suckling opossum (*Monodelphis domestica*). *Dev. Comp. Immunol.* **18**:75.

Wilson, R.A. and Jutila, J.W. 1976a. Experimental neonatal colibacillosis in cows: serological studies. *Inf. Immun.* **13**:92.

Wilson, R.A. and Jutila, J.W. 1976b. Experimental neonatal colibacillosis in cows: immunoglobin classes involved in protection. *Inf. Immun.* **13**:100.

Woods, R.D. and Wesley, R.D. 1986. Immune response in sows given transmissible gastroenteritis virus or canine coronavirus. *Amer. J. Vet. Res.* **47**:1239.

Yadav, M., Eadie, M., and Stanley, N.F. 1971. The thymus glands of a marsupial, *Setonix brachyurus* (quokka) and their role in immune responses. VI. Maternal transfer of immunoglobulins. *Aust. J. Exp. Biol. Med. Sci.* **49**:322.

Yadav, M., Eadie, M., and Stanley, N.F. 1973. Passage of maternal immunoglobulins to the pouch young of a marsupial *Setonix brachyurus*. *Aust. J. Zool.* **21**:171.

Yamada, T., Nagai, Y., and Matsuda, M. 1991. Changes in serum immunoglobulin values in kittens after ingestion of colostrum. *Amer. J. Vet. Res.* **52**:393.

Young, L., Basden, K., Cooper, D,W, and Deane, E.M. 1997. Cellular components of the milk of the tammar wallaby (*Macropus eugenii*). *Aust. J. Zool.* **45**:423.

Chapter 8

Mammary Gland Defense: The Role of Colostrum, Milk and Involution Secretion

Catherine M. Fetherston, Chee Seong Lee, and Peter E. Hartmann

1. Introduction

The survival of the newborn of all mammals is dependent on an adequate supply of maternal milk that in turn is dependent upon a healthy mammary gland. During lactation, insults to the mammary gland may occur from blockage of ducts, microbial invasion and trauma. These events result in inflammation (mastitis) and occasionally sequelae such as abscess and septicemia. Discussion of protection of the mammary gland against mastitis requires a comprehensive understanding of the process of inflammation and the complex interactions that comprise the defense mechanisms within the mammary gland. One mechanism, not often fully considered in the defense of the mammary gland, is the protective role of milk.

The primary driving force for research on mastitis has come from the dairy industry in which the estimated loss to producers from mastitis is approximately $2 billion dollars in the United States alone (National Mastitis Council, 1987). In contrast the emphasis of research on humans has been on the role played by breast milk components in the protection of the infant against pathogenic microorganisms and disease (Xanthou, 1998; Garofalo and Goldman, 1999).

Catherine M. Fetherston and Peter E. Hartmann • Department of Biochemistry, The University of Western Australia, Nedlands, WA 6907, Australia Chee Seong Lee • Department of Veterinary Science, The University of Melbourne, Parkville Vic 3052, Australia

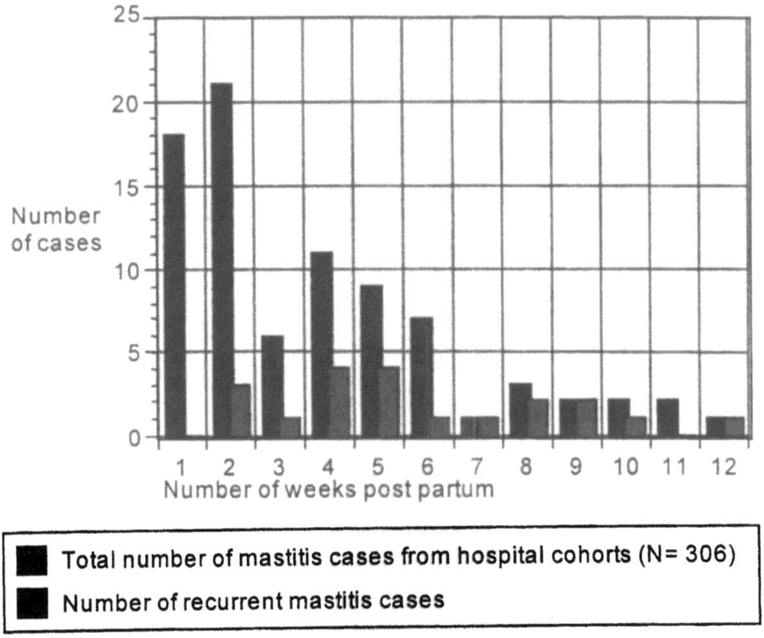

Figure 1. The incidence and recurrence rates for mastitis from the prospective cohort during the first three months post partum (Fetherston, 1997). (Reprinted from Fetherston (1997a) with the permission of Nursing Mothers of Australia).

Women with mastitis experience pyrexia, systemic flu-like symptoms and a localized inflammation of the breast. The pyrexia and flu-like symptoms can be severe and sudden in their onset, and often result in women being debilitated and bedridden. Prospective studies conducted recently in Australia and New Zealand have reported high incidence rates for mastitis. Fetherston (1997a) reported a cumulative incidence rate of 27.1% for mastitis episodes in the first three months post partum as well as a cumulative recurrence rate of 6.5% (Fig. 1). A crude incidence of 20% was estimated by Kinlay *et al.* (1998) for women in the first six months after delivery, and Vogel *et al.* (1999) reported that 23.7% of women suffered one or more episodes of mastitis within the first 12 months post partum. These findings are supported by retrospective, descriptive studies in Sweden and America which identified incidence rates of 24% (Jonsson and Pulkinnen, 1994) and 33% (Riordin and Nichols, 1990). Twenty-one percent of the women in the Australian study who ceased to breast-feed by six weeks did so due to mastitis, representing a major cause of premature cessation in lactation (Fetherston, 1997).

A broad approach to understanding all the epidemiological and physiological risk factors together with the ability of host defense mechanisms to provide protection is required for the successful prevention and appropriate treatment of mastitis. Mammary gland secretions during the lactation cycle are primed to make a major contribution to the protection of the mammary gland. There is a vast wealth of information concerning the role of breast-milk in providing protection to the infant. This, in combination with research into the pathological processes involved in bovine mastitis, provides a sound basis for a review of the role of colostrum, milk and involution secretion in mammary gland defense.

2. Factors Affecting Microbial Invasion of the Mammary Gland

The route of entry for infecting microorganisms in mastitis has been cited as either (i) retrograde, through a lactiferous duct or a traumatized nipple, (ii) from the prelacteal lymphatics or (iii) by hematogenous spread (Lawrence, 1994). Infection of the mammary gland by routes (ii) and (iii) has been reported in cases of tuberculosis and brucellosis in animals (Kennedy and Miller, 1993), but infection by these routes is highly unlikely in women in developed countries due to the low incidence of these diseases. Nevertheless a recent French report of tuberculosis of the breast in an HIV sero-positive patient warns of the possible return of such unusual cases (Leleu *et al.*, 1997). In the absence of systemic pathology, the common portal of entry for infection in women with mastitis is assumed to be through the nipple pores into the lactiferous ducts.

In order for infection to become established the microorganism must not only gain entry through the nipple pore but also then avoid being flushed out as the breast empties during suckling. Indeed, frequent feeding has been shown to be a protective factor against mastitis in women (Fetherston, 1998). It is also of interest that mastitis is less common in beef cattle suckling calves than in dairy cows that are normally only milked twice per day (Hungerford, 1990).

In addition to the flushing action of milk removal, the structure of the gland provides barriers to microorganisms. Despite anatomical and physical differences among species there are common features that protect against microbial penetration, such as the tightness of closure of the sphincter controlling either the teat orifices or nipple pores, protective mechanisms of the keratin lining, and the integrity of the surface epithelium. Studies on the mammary gland of the sheep (Tatarczuch *et al.*, 1997; Lee *et al.*, 1996) have shown that, even during mammary gland involution, the alveolar and ductal epithelia remained intact. It was shown that the gland involuted by apoptosis (Fig. 2), a process that allows deletion of cells without the loss of the basic architecture and integrity of the epithelial lining of the gland.

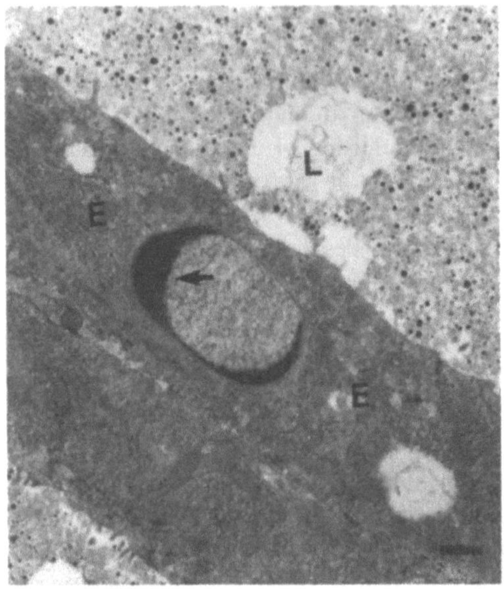

Figure 2. Electron micrograph showing an alveolar epithelial cell (E) of sheep mammary gland undergoing early apoptosis at 2 days after weaning. The chromatin has condensed into crescents (arrow). L, lumen. Bar, 1 μm. (Reprinted from Tatarczuch *et al.*, 1997 with the permission of Cambridge University Press).

The health of the teat or nipple is paramount to the defense functions of this structure. Injury to the nipple or teat has been shown to result in an increased risk for mastitis in women (Jonsson and Pulkinnen, 1994; Fetherston, 1998) and cows (Wadstrom, 1987). Anatomical barriers are often compromised by trauma during early breast-feeding. Attachment difficulties experienced by mothers establishing breast-feeding are commonplace (Fetherston, 1995) and can result, among other things, in a break in the integrity of the surface epithelium of the nipple providing a portal of entry for infection. This often occurs while mothers are still hospitalized and at risk of nosocomial infections, e.g., by organisms such as *S. aureus*. Subsequently the highest proportion of mastitis episodes has been found to occur in the first few weeks post partum (Devereux, 1970; Matheson *et al.*, 1988; Fetherston, 1998). Similar trauma to the teat in cows compromises defense against microbial infiltration of the mammary gland (Sandholm *et al.*, 1991).

Once the anatomical barriers are traversed, adherence is an important step in the process of infection. Adherence of bacteria to epithelial cells enables them to resist being flushed from the gland by the flow of milk. Receptors for binding sites on the fibronectin molecule are implicated in the adherence of bacteria to either host mucous membranes or injured and damaged tissues (Hasty *et al.*, 1989). Fibronectin is a glycoprotein, which mediates adhesion of cells to substrates and has binding sites for many staphylococci (Kuusela, 1978; Maxe *et al.*, 1986) and streptococci (Kuusela *et al.*, 1985). Other matrix proteins such as collagen and fibrinogen, either by themselves or in synergy with fibronectin, also may act as substrates for adhesion (Vercelloti *et al.*, 1985; Nelson *et al.*, 1991; Cifrian *et al.*, 1994). These proteins may present themselves through either micro lesions or blood clots which have resulted from trauma to the teat, nipple or gland, thus providing another way in which mastitis may result from trauma to the mammary gland. Knowledge of the surface properties of bacteria and their interaction with host tissue has lead to the stimulation of anti-adhesive immunity by vaccination. Fibronectin binding protein used as a vaccine component resulted in protection against mastitis in cows and mice challenged with *S. aureus* (Nelson *et al.*, 1991; Mamo *et al.*, 1994).

In human milk there are a number of anti-adhesive molecules such as IgA, glycoproteins, glycolipids, mucins and oligosaccharides that have been shown to function as competitive inhibitors of the adhesion of bacterial pathogens to epithelial receptors (Holmgren *et al.*, 1986; Svanborg *et al.*, 1991; Peterson *et al.*, 1998). The relevance of these molecules is often discussed in relation to protection of the neonate (Newman, 1995; Hamosh, 1998); however, they also may play an important role in protecting the mammary gland.

3. Inflammatory Responses to Microbial Invasion of the Mammary Gland

Once bacteria overcome the physical barriers of the mammary gland the ensuing microbial invasion stimulates an inflammatory response. This response, designed to protect the threatened or injured mammary gland, results from a combination of local, systemic, non-specific, specific, humoral and cellular responses that have been outlined by many authors elsewhere (see: van Miert, 1991; Lees, 1991; Lee *et al.*, 1992; Harmon, 1994; Paape and Capuco, 1996; Peakman and Vergoni, 1997; Sordillo *et al.*, 1997). The inflammatory response is complemented both by the flushing action of the frequent removal of milk and by the protective components in milk. For instance, the cells and the array of immunomodulators and anti-inflammatory factors which can be found in milk (Goldman, 1993) may affect the expression of the inflammatory response within the gland. A model of the inflammatory response within the mammary gland is presented in Figure 3.

These mechanisms are also modified by nutritional and hormonal status, genetic makeup and virulence factors (e.g., bacterial superantigens) of invading microorganisms.

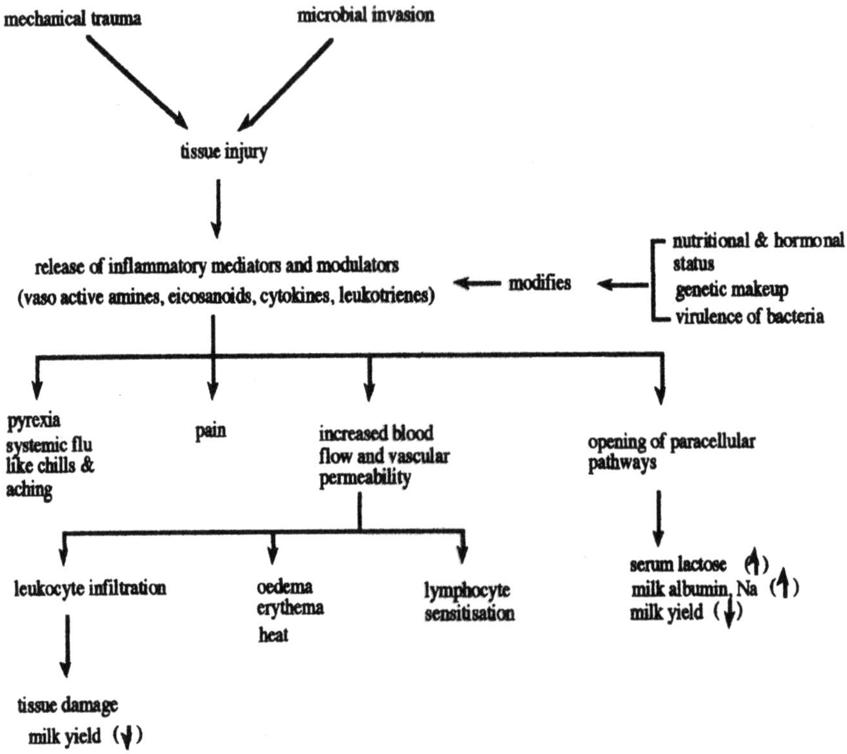

Figure 3. Model for the inflammatory response within the mammary gland.

Changes occurring in the mammary gland during the inflammatory response result in an increase in the permeability of the blood-milk barrier (Stelwagen *et al.*, 1997) leading to edema of the interstitial tissues through the leakage of proteins from both blood and milk. There is also a change in the chemical composition of the milk as a result of the leakage of milk products into the interstitium and the transudation of serum derived products into the milk space. This is reflected in findings of low concentrations of lactose and high concentrations of serum derived factors, such as albumin and sodium, in the milk from glands affected by mastitis. Prentice *et al.* (1985) reported lactose concentrations ranging from 17–54% of normal levels in the milk of women with mastitis, and Conner (1979) reported the sodium level to be 34 times higher in the mastitic breast of one women compared with her unaffected breast. These results are similar to those of other researchers (Prosser and Hartmann, 1983; Golota and Altuev, 1984; Abakada *et al.*, 1990). Abakada *et al.* (1990) also found sodium was sometimes elevated in both symptomatic and asymptomatic breasts. The raised sodium levels in the asymptomatic breast may be suggestive of either a systemic response in which opening of the paracellular pathways occurs in both breasts, or it may reflect the presence of subclinical infection in the asymptomatic breast. Stelwagen *et al.* (1997) found that mammary tight junctions in cows were leaky after approximately 21 hours of milk stasis. A similar response is present in women with elevated concentrations of milk sodium found to occur after 24 hours of milk stasis (Hartmann and Kulski, 1978). Therefore, increased concentrations of sodium in milk would be expected not only with mastitis, but also when either engorgement or blocked segments occur. The increased solute load resulting from opening of the paracellular pathways has been reported to cause breast refusal by infants of mothers with mastitis (Conner, 1979).

Filteau *et al.* (1999a) found that breast milk concentrations of sIgA, lactoferrin, lysozyme and IL-8 correlated positively with the sodium/potassium ratio (Na:K ratio). An elevated ratio was observed in 25% of women (n = 212) at one to three weeks post partum and in 12% of women at 3 months. As IL-8 is an inflammatory cytokine and was strongly associated with a high Na:K ratio, the authors suggested that the increase in the milk Na:K ratio reflected inflammation rather than an opening of paracellular pathways due to other causes. However clinical symptoms of breast inflammation were not observed in the study. Semba *et al.* (1999) found elevated breast milk sodium in 15.6% of a cross sectional sample of 96 lactating women from Malawi at 6 weeks post partum. Elevated sodium concentrations were associated with high concentrations of lactoferrin, secretory leukocyte protease inhibitor (SLPI), IL-8 and the cytokine designated Regulated on Activation Normal T cell Expressed and Secreted. The concentration of sodium correlated more strongly with these immune factors than the Na:K ratio, suggesting that sodium alone may be a more accurate indicator of mastitis.

4. Cellular Defense Mechanisms Present in Milk

4.1. Somatic Cells

The cells present in the milk of a healthy gland include neutrophils, macrophages, lymphocytes and various epithelial cells (Lee *et al.*, 1980; Lee and Outteridge, 1981, Xanthou, 1997). Paape *et al.* (1963) first used the term somatic cells to describe this range of cells found in bovine milk. The term somatic cell count (SCC) is now in common use in the dairy industry, and an increase in the total cell count in milk is used as a measure of inflammation of the mammary gland (Bramley, 1991). The total SCC in the milk of a healthy bovine gland is usually $<10^5$ cells/ml and is $<10^6$ in milk from the healthy human breast. During infection of the gland, the SCC increases to $>10^6$ cells/ml in the milk of both cows and women (Thomsen *et al.*, 1984; Kehrli and Shuster, 1994).

The cellular composition of colostrum, milk and involution secretion appears primed to play a protective role in the mammary gland. Under normal conditions the leukocyte population in milk varies according to stage of lactation. The highest content is present in the early phase of lactation, dropping to low values when lactation is established and then rising again during weaning. It could be postulated that the cell population is highest when the gland is at its most susceptible i.e., at initiation of lactation when the risk of nipple trauma and milk stasis is at its highest and during weaning when the flow of milk through the gland is markedly decreased.

The presence of leukocytes in milk enables the first immunological line of defense, activation of neutrophils, to be enacted promptly once there is microbial penetration of the anatomical barrier. Neutrophil populations in healthy milk vary across species comprising 50–70% of the cell count in bovine milk (Paape and Capuco, 1996), 5–20% in goats' milk (Paape and Capucco, 1996) and 40–50% in human milk (Xanthou, 1997). It is probable that the presence of leukocytes in milk provides the first line of defense against invading bacteria before neutrophils are recruited from the blood stream (Kehrli and Shuster, 1994). However, if the invading bacteria are not overcome by resident leukocytes in milk, they multiply and release chemoattractants that result in the migration of neutrophils, and subsequently macrophages, to the site of invasion. In infected milk, neutrophils comprise more than 95% of the cells present in bovine (Kehrli and Shuster, 1994) and human (Fetherston, unpublished observation) milk.

The mechanism by which bacterial recognition occurs in infection of the mammary gland is not known, but Shuster *et al.* (1996) propose that initial recognition occurs within the milk compartment. These authors investigated the chemotactic activity and upregulation of CD18 expression necessary for neutrophil recruitment during induced intramammary infection with *Escherichia coli*. Multiple inflammatory mediators, including tumour necrosis factor (TNF), several interleukins (IL) and the complement fragment, C5a, were produced in response to the infection, and it was proposed that these mediators were produced locally both because TNF and IL-1 were undetectable in serum and because bacteria and their components would be unable to cross the initially impermeable epithelium to recruit neutrophils directly from the vasculature.

Initial studies (Schalm *et al.*, 1964) provided evidence that a neutrophil count within the range of 4×10^5 and 5×10^6/mL milk is required for protection against experimental challenge with mastitis pathogens in cows. This lead to research into stimulating an artificially high level of milk neutrophils in order to afford better protection against mastitis. Paape *et al.* (1981) inserted polyethylene devices into the udders of cows to aseptically induce leukocytosis in the milk. These devices increased the concentrations of somatic cells, did not damage tissue (as indicated by lysosomal enzyme activity) and depression of milk production was insignificant (Paape and Weinland, 1988). In contrast, Nickerson (1991) found that most devices tested could stimulate leukocytosis, but that the udder quarters fitted with devices exhibited reduced secretory activity. This type of strategy has not been further developed as more sophisticated methods, e.g., cytokine therapy, have been used to increase the numbers and functional activities of neutrophils in the milk of cows with intramammary infection.

The development of recombinant cytokine therapy has resulted in an increase in mononuclear cell populations (Torre *et al.*, 1992; Nickerson *et al.*, 1993) and neutrophil numbers (Nickerson *et al.*, 1989; Daley *et al.*, 1991) and in promotion of neutrophil phagocytic activity (Quiroga *et al.*, 1993a; Sordillo and Babiuk, 1991). A significant elevation in milk neutrophils was observed in cows given a subcutaneous injection of recombinant human granulocyte colony stimulating factor (rhG-CSF) for 15 days as compared with controls that were injected with phosphate buffered saline. It was found that the rate of new infections, as well as the severity of infection, was reduced in cows that had received pre-treatment with rhG-CSF (Nickerson *et al.*, 1989). Kehrli *et al.* (1991) and Sordillo *et al.* (1997) have provided a thorough review and summary of the effects of the administration of various cytokines on mammary immune cells and of the potential application of this intervention with regard to bovine disease prevention. Kehrli *et al.* (1991) suggested that the enhancement of the egression of neutrophils into the mammary gland, through modulation of the immune system by cytokine therapy, will assist in offsetting the inefficiencies of neutrophil action in milk.

4.2. Factors Affecting Neutrophil Function in Milk

Despite the reputation the neutrophil has as the key initial defense element within the milk compartment of the mammary gland (Harmon, 1994) there is evidence to suggest that milk is unable to provide optimal conditions for phagocytic activity. Human milk neutrophils neither respond as quickly to chemotactic agents nor move as rapidly as neutrophils from venous blood (Neville and Jensen, 1995). This observation suggests prior activation of milk neutrophils. Other indicators of activation are an increased expression of CD11b and a decreased expression of L-selectin (Keeney *et al.*, 1993). Indeed in human milk almost all leukocytes are in their activated form (Ozkaragoz *et al.*, 1988; Wirt *et al.*, 1992; Keeney *et al.*, 1993). Membranous material in human milk has been found to activate blood neutrophils *in vivo* and may contribute to the activation of neutrophils in apparently uninfected milk (Keeney *et al.*, 1993). Furthermore Shafer-Weaver *et al.* (1996) suggested that the preferential trafficking of $CD8^+$ suppressor lymphocytes into mammary gland secretions and tissues following parturition in cows, may be responsible for the lower responsiveness of neutrophils in milk compared with the same type of cell from the blood.

The phagocytic activity of the neutrophil may be inhibited by the ingestion of casein, fat and cellular debris (Paape and Wergin, 1977; Russell *et al.*, 1977; Lee *et al.*, 1998). Cellular activity is also likely to be limited by the availability of energy. In milk, the concentration of lactose predominates over that of glucose and, consequently, neutrophils must depend primarily on intracellular glycogen. However, the glycogen content of bovine milk leukocytes has been reported to be only 62% of the level in leukocytes of the blood (Naidu and Newbold, 1973). In addition, the pO_2 in milk is approximately 1% that of the blood so that phagocytes must function under conditions that may compromise the oxygen dependent microbicidal mechanisms of the neutrophil (Targowski, 1983). Other limiting factors with regard to neutrophil activity are the low concentrations of opsonizing immunoglobulins and complement in milk (Kehrli and Shuster, 1994) as well as the presence of a blocking factor for phagocyte Fc receptors that is reported in mammary secretions during colostrogenesis and early involution (Targowski and Niemialtowski, 1996).

Despite numerous factors that appear to limit the activity of neutrophils in milk, both neutropenia (Schalm *et al.*, 1976; Kashiwazaki, 1984) and impaired neutrophil functions (Kehrli and Goff, 1989; Kehrli *et al.*, 1989; Berning *et al.*, 1993; Lee *et al.*, 1998) have been linked to an increased susceptibility to mastitis in cows. Moreover, delayed leukocyte recruitment to the mammary gland in cows with *E. coli* induced mastitis also has been associated with an increased severity of mastitis (Hill, 1981).

Interestingly, a Russian study (Kutushev *et al.*, 1988, English abstract only) reported the use of leukotransfusions following surgery for breast abscess. The abstract did not indicate the study design or specify the type of leukotransfusion, but reported that leukotransfusions arrested mammary inflammation and excluded the development of recurrences and pyo-septic complications. In constrast, Sim and Goldman (1991) reported that granulocyte transfusions have limited value in the treatment of acute mammary infections related to phagocytic dysfunctional diseases and that transfusion was associated with significant complications including production of antibodies against the transfused cells.

4.3. Effects of Neutrophil Action on the Mammary Gland

The high SCC count seen in mastitis in cows correlates negatively with milk production (Fetrow *et al.*, 1988). It is hypothesized that neutrophil recruitment during inflammation may cause impaired milk synthesis through tissue damage to the mammary gland (Akkers and Thompson, 1987). Injury to secretory tissue could occur through release of oxygen radicals and lysosomal proteases during the process of phagocytosis of pathogens (Movat *et al.*, 1987). Studies involving endotoxin infusion into the bovine mammary gland, a procedure which induces inflammation without either bacterial growth or toxin production (Frost *et al.*, 1984), suggest that the effects on the mammary epithelium, and on milk production, could be a result of leukocytic activity (Akkers and Thompson, 1987). In response to this proposition, endotoxin-induced neutrophil migration into the bovine mammary gland was investigated by Waller (1997) who suggested that antagonists to leukotrienes may have potential as therapeutic agents in bovine mastitis. Leukotriene inhibitors, platelet activating factor and IL-1β, were shown to inhibit the accumulation of neutrophils during endotoxin-induced inflammation of the teat cistern.

It has been suggested that the inflammatory effects of neutrophils in human milk may be tempered by the presence of antioxidants. Goldman *et al.* (1990) suggested that milk antioxidants inhibit the generation of superoxide by neutrophils and enhance the degradation of hydrogen peroxide and oxygen radicals. However, it must be noted that many of the anti-inflammatory and antioxidant components present in human milk are either not present, or are only represented in minimal amounts, in cows milk.

4.4. Cell Counts as a Diagnostic Measure for Mastitis

Several factors can affect the normal baseline standards for somatic cells in milk. The percentage of each cell type not only differs across species but also varies according to parity, stage of lactation (Lee *et al.*, 1980; Goldman *et al.*, 1982; Lee and Outteridge, 1981; Sheldrake *et al.*, 1983; Zeng and Escobar, 1995), time of sampling and amount of milk in the mammary gland (Paape and Capuco, 1996) and concomitant disease (Ryan *et al.*, 1993). This presents difficulties in establishing confidence limits for normal and abnormal values (Bramley, 1991). However, variation associated with these factors is considered small compared with the increases seen in the SCC associated with intramammary infection. This, combined with the widespread use of rapid electronic cell counting techniques, makes SCC the most common measurement of bovine milk quality and udder health (Harmon, 1994).

Diagnosis and treatment of lactation mastitis in women presents difficulties as the clinical symptoms of infective and non-infective (due to obstruction or engorgement) forms of mastitis are often very similar. Therefore it is often unclear whether antibiotic treatment is required. In an attempt to address these problems, Thomsen *et al.* (1983) developed diagnostic criteria for women based on the uses of SCC in cows. Using leukocyte counts and microbiological cultivation results of milk from 146 breasts with inflammatory symptoms, the authors grouped cases into three categories: milk stasis (bacteria/ml $<10^3$, leukocytes/ml $>10^6$), noninfective inflammation (bacteria/ml $<10^3$, leukocytes/ml $<10^6$) and infective inflammation (bacteria/ml $>10^6$, leukocytes/ml $>10^6$). A subsequent study (Thomsen *et al.*, 1984) was used to statistically validate these diagnostic categories by observing the duration and outcomes of cases diagnosed according to these criteria. However, Abakada *et al.* (1992) found that low bacterial counts could be recorded even when pathogens were cultured from human milk and, in one case, when a mammary abscess was diagnosed on the basis of a needle aspirate. Thus, in contrast to cow's milk, bacterial and leukocyte counts in human milk cannot be regarded as reliable indices for diagnosing mastitis. In part, this may reflect the difficulties encountered when attempting to collect milk samples exclusively from an infected sector of the human mammary gland. In addition, Matheson *et al.* (1988) have proposed that growth promoting and inhibitory factors present in human milk might interfere with the prognostic value of bacterial counts.

5. Non-Cellular Defense Mechanisms in Milk

Soluble (non cellular) defense factors in milk include specific and non-specific components. They have the capacity to work both independently and synergistically with one other, and with the cellular components in milk, to enact an immune response that will provide protection to the mammary gland.

5.1. Non-Specific Defense

Lactoferrin is a metal binding glycoprotein present in much higher concentrations in milk than in other mucosal fluids. A high concentration of lactoferrin is also found within the specific granules of neutrophils, and the protein is released during an inflammatory response (Lönnerdal and Iyer, 1995). Consequently lactoferrin is potentially able to provide bacteriostatic activity within the milk compartment. In human milk it is present in its unsaturated form and has been shown to have bacteriostatic effects against *Staphylococcus* and *Bacillus* strains (Orem and Reiter, 1968), *E. coli* (Bullen *et al.*, 1972; Rainard, 1986) and *Candida albicans* (Arnold *et al.*, 1977). Lactoferrin contributes to host defense by the sequestering of iron, which is required for microbial growth, and by enhancing the function of phagocytic cells. Bactericidal functions have been shown against Gram-negative bacteria in which lactoferrin directly damages the outer membrane of the cell wall by causing lipopolysaccharide release that results in permeability changes and cell death (Ellison, 1994). It has also been reported that lactoferrin can protect the mammary gland against the effects of inflammation by controlling the release of pro-inflammatory cytokines, such as IL-1, IL-6 and TNF-alpha (Crouch *et al.*, 1992; Hanson *et al.*, 1995), inhibiting the complement system and reducing free radical release (Britigan *et al.*, 1994). A synergistic effect against a milk sensitive strain of *E. coli* has also been demonstrated *in vitro* for lactoferrin when in the presence of IgA (Spik *et al.*, 1978; Akin *et al.*, 1994).

In addition, an antimicrobial peptide derived from the N-terminal region of the lactoferrin molecule, lactoferricin B, has been found to function by a mechanism distinct from iron chelation. A lethal antimicrobial effect was shown for lactoferricin against Gram-negative and Gram-positive bacteria, yeasts and filamentous fungi (Bellamy *et al.*, 1992).

Lactoferrin showed a 30-fold increase in concentration, from a baseline range of 0.2–2.0 × 10^{-3} g/l, in the mammary secretions of cows with experimentally induced *E. coli* infection of the mammary gland (Harmon *et al.*, 1976). This is thought to be a specific response of mammary secretory tissue to inflammatory agents stimulated by the infectious process. Although a much smaller rise, to 2.5 times the normal lactoferrin concentration, was observed in women with mastitis (Prentice *et al.*, 1985), the normal concentrations of this protein in human milk (1.4 – 3.5 g/l) are much greater than the normal concentrations in bovine milk (Lönnerdal and Atkinson, 1995). Prentice *et al.* (1985) reported lower levels of lactoferrin, along with IgA and C3, in the milk of women who later developed mastitis, than in those who did not. Susceptibility to recurrent infections has been previously demonstrated where congenital lactoferrin deficiency exists (Breton-Gorius *et al.*, 1980), a finding which suggests a link between lower levels of lactoferrin in the milk of these women and the development of mastitis.

In cows and ewes involution is associated with increased expression of lactoferrin in the ductal epithelium close to the teat (Molenaar *et al.*, 1996), and in women there is a marked increase in the levels of lactoferrin and immunoglobulins in the mammary secretion expressed after the termination of breast-feeding (Hartmann and Kulski, 1978). Although the anti-microbial proteins in the mammary secretion appear to increase to compensate for the loss of the protective action of milk removal, there is only a small increase during the first 24 hours of milk stasis so that this could be a potential at-risk period for women who wean abruptly.

The enzyme, lysozyme, exerts bactericidal effects through lysis of peptidoglycans in the cell wall of Gram-positive bacteria and in the outer membrane of Gram-negative organisms (Lönnerdal, 1985). In addition, this protein binds complement to sIgA and has bacteriostatic effects against Gram-positive bacteria (Lönnerdal, 1985). Lysozyme is present in human milk at a concentration 300 times that of cows milk (Chandan *et al.*, 1968) and has been shown to increase in concentration around six months post partum (Goldman *et al.*, 1982). The enzyme exhibits a late increase in concentration in response to mastitis in women and the magnitude of the response is reported to correlate with the rise in lactoferrin and IgG levels (Prentice *et al.*, 1985). Finally, cows with low levels of milk lysozyme are also more susceptible to mastitis than cows with higher levels of the enzyme in their milk (Nickerson, 1985).

The lactoperoxidase-thiocyanate-hydrogen peroxide system, which is bacteriostatic for Gram-positive bacteria such as *S. aureus* and streptococci and bactericidal for Gram-negative bacteria such as coliforms (Outteridge and Lee, 1988), is found in high activity in bovine milk. However, the concentration in human milk is low such that this system is and unlikely to exert a strong bacteriostatic effect.

5.2. Specific Defense

Specific immunity within the mammary gland is characterized by the production of specifically sensitized lymphocytes in response to antigen presentation, that in turn leads to immunoglobulin and cytokine production. The total numbers of lymphocytes, including T and B cells, vary considerably across species. Thus, lymphocytes comprise 10–27% of all cells present in bovine milk at mid-lactation (Lee *et al.*, 1980), 10–17% of the cells in ovine milk (Lee and Outteridge, 1981) and 5–10% of the cells in human milk (Xanthou, 1997). The T lymphocytes in bovine milk are functionally competent, represent a subpopulation distinct from that of blood lymphocytes, and their numbers decrease with stage of lactation (Craven and Williams, 1985). In both bovine and human milk, T lymphocytes display characteristics of activation, including phenotypic markers such as the HLA-DR antigen, CD45RO which is characteristic of memory cells, and the high affinity interleukin-2 (IL-2) receptor (Wirt *et al.*, 1992; Park *et al.*, 1992). In cattle and humans the numbers of $CD8^+$ (cytotoxic or suppressor) T lymphocytes exceed those of $CD4^+$ (helper) T cells in both mammary gland tissues and secretions unlike the blood in which the reverse is evident (Wirt *et al.*, 1992; Park *et al.*, 1992). Although the significance of an elevated frequency of $CD8^+$ over $CD4^+$ T lymphocytes is not clear it has been postulated that this phenomenon may play a role in maintaining the integrity of the mucosal epithelium (Taylor *et al.*, 1994). Taylor *et al.* (1997) found that, during naturally acquired bacterial infection of the bovine mammary gland, the subset distribution of T cells changed, with $CD4^+$ lymphocytes representing a greater proportion of the T cells than in normal controls.

The functional significance and activity of lymphocytes in response to mammary gland infections has not been extensively studied compared with other aspects of mammary gland immunity. However, lymphocyte activities have been shown to vary with stage of lactation. Shafer-Weaver and Sordillo (1997) examined the immunoregulatory role of bovine $CD8^+$ lymphocytes in milk during the post partum period. The $CD8^+$ cells present immediately following parturition predominantly expressed suppressor activity, whereas $CD8^+$ lymphocytes obtained during mid-lactation expressed mainly cytotoxic activity. The suppression of lymphocyte activity immediately post partum correlates with a low level of IL-2 activity in colostrum during the final week of gestation (Sordillo *et al.*, 1991). This cytokine enhances the cytotoxic and bactericidal activities of natural killer cell-like lymphocytes in cattle (Sordillo *et al.*, 1991). Thus, the low level of IL-2 and the predominance of suppressor T lymphocytes seen in early lactation appear consistent with the high susceptibility to bovine mastitis experienced at this time.

Suppression of T lymphocyte activities has also been implicated as a factor contributing to the chronic intramammary infection associated with the immunosuppressive activities of bacterial superantigens. Ferens *et al.* (1998) proposed that superantigens, seen in many mastitis isolates, interfere with the development of protective immunity within the bovine mammary gland. In this study, bovine peripheral blood mononuclear cells exposed to staphylococcal enterotoxin C1 (SEC1) were compared with controls exposed to concanavalin A. Relative to concanavalin A-stimulated controls, cultures stimulated with SEC1 generated small numbers of $CD4^+$ $\alpha\beta$Tcells expressing high levels of IL-2Rα, and only a few cells increased to blast size, suggesting that only partial activation had occurred.

$\gamma\delta$T lymphocytes, in both humans and ruminants, have been identified in greater proportions in mammary secretions and parenchyma than in blood (Richie *et al.*, 1982; Shafer-Weaver and Sordillo, 1997). These cells are primarily associated with the protection of epithelial surfaces and, $\gamma\delta$T lymphocytes cultured with IL-2 have been found to recognize and lyse malignant breast carcinoma cells (Miescher *et al.*, 1990). Sordillo *et al.* (1997) suggest that the decrease in numbers of $\gamma\delta$T lymphocytes that occurs within the mammary parenchyma during times of high susceptibility to mastitis indicates that these cells play a role in defense against mammary infection.

The presence of cells expressing the leucocyte common antigen as well as major histocompatability complex (MHC) class II and $CD1^+$ markers, and exhibiting long cytoplasmic processes (Fig. 4) has been reported in the milk of ewes and cows and in the involution secretion of the mammary gland of ewes collected at 15 to 30 days after weaning (Lee *et al.*, 1998). These are probably dendritic cells and, though present in small numbers, they may play an important role in antigen presentation for the purpose of mammary defense.

The percentage of B lymphocytes remains relatively constant between stages of lactation and, in bovine milk, represent a minor population by comparison to peripheral blood (Taylor *et al.*, 1994). A similar situation prevails in sheep, as Lee *et al.* (1989) observed that in the mammary glands of pregnant and non-pregnant sheep the great majority of lymphocytes in the ductal and alveolar epithelium were $CD8^+$ T cells (Fig. 5). Small subpopulations of granular lymphocytes in the epithelium expressed the CD45R antigen but not MHC class II molecules. The CD45R isoform recognized by the monoclonal antibody 20–96

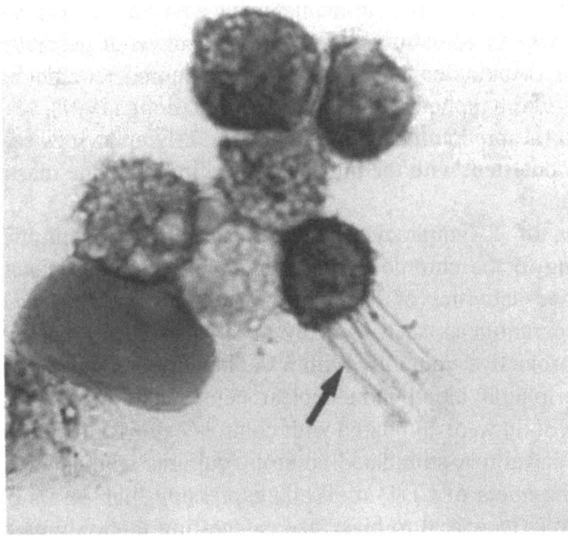

Figure 4. Immunoperoxidase staining of a "dendritic-like" cell from the involution secretion of a sheep mammary gland collected 15 days after weaning. The cell is stained intensely with a monoclonal antibody against MHC class II antigen, and has long cytoplasmic processes (arrows) ×756 (Lee *et al.*, 1992).

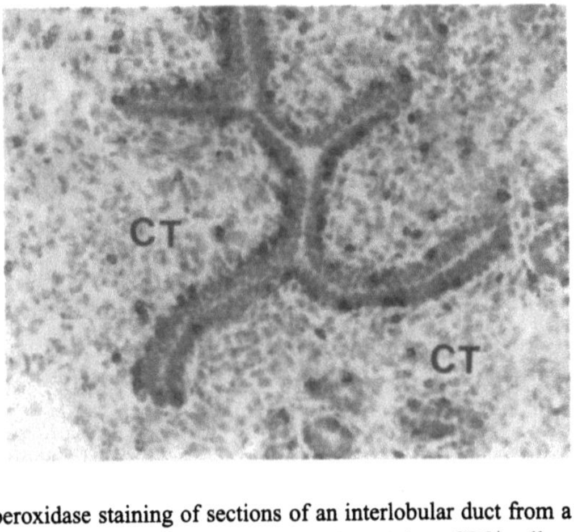

Figure 5. Immunoperoxidase staining of sections of an interlobular duct from a non-pregnant sheep with a monoclonal antibody against the CD8 antigen. Many $CD8^+$ cells are localized within and at the basal regions of the ductal epithelium. CT, connective tissues. ×160. (Reprinted from Lee *et al.* (1989) with the permission of Blackwell Science).

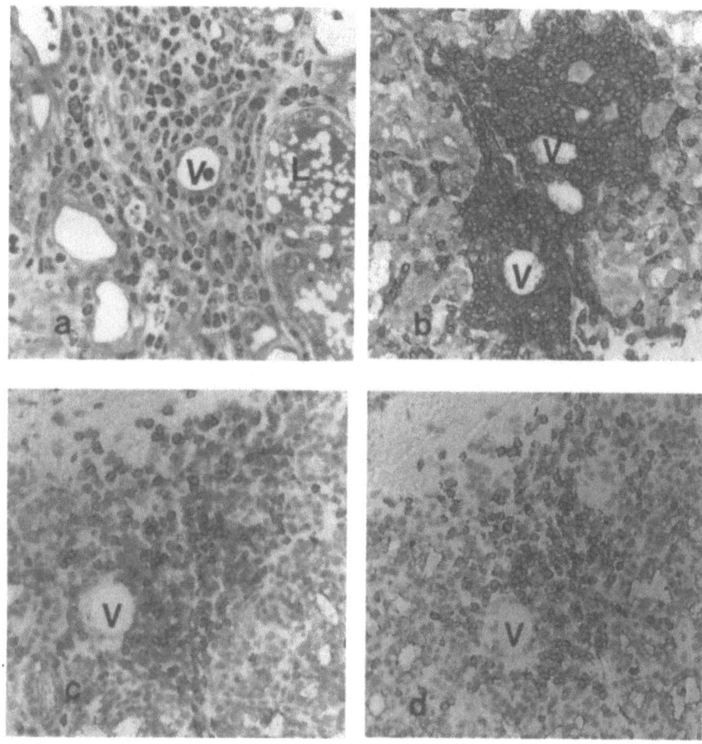

Figure 6 (a–d). Sections from mammary gland of 120-day pregnant sheep. (Reprinted from Lee *et al.* (1989) with the permission of Blackwell Science). (a) Semithin plastic section showing a venule (V) surrounded by a cluster of lymphocytes. The alveolar lumina (L) are filled with secretion. 1% toluidine blue in 1% borax ×400. (b–d) Immunoperoxidase staining of serial sections of a patch of lymphocytes in the interlobular area of the mammary gland with mAbs against (b) CD45, (c) CD5 and (d) CD4 antigens. Note that virtually all the cells around the venule (V) are leucocytes ($CD45^+$) (b); a large proportion of them are $CD5^+$ (c) and $CD4^+$ (d). ×160.

used in these studies has not been identified. However, this isoform is strongly expressed on B cells and a small population of T cells also expresses this antigen (Dutia *et al.*, 1993). Other subpopulations, especially B lymphocytes, were present in much lower numbers and were located in the periductal and intralobular connective tissues. Clusters of lymphocytes were observed around venules (Figs. 6a and b) and, the majority of them were T helper cells, being $CD5^+$ and $CD4^+$ (Figs. 6c and d) with some $CD8^+$ T cells and no B cells ($CD45R^+$, MHC class II^+).

Nevertheless, in a subsequent study Lee *et al.* (1992) demonstrated a striking accumulation of B lymphocytes in the periductal connective tissues

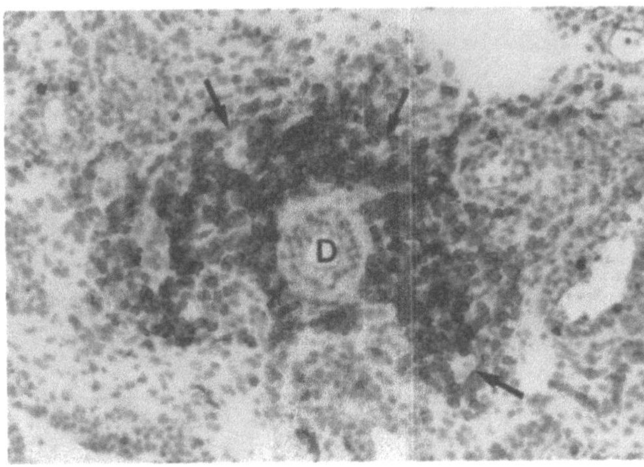

Figure 7. Frozen section of mammary gland tissue from a pregnant sheep 7 days after infusion of killed S. aureus stained with the indirect immunoperoxidase technique using mAb against CD45R* sheep lymphocyte antigen. Numerous B cells ($CD45R^+$)* are seen surrounding the blood capillaries (arrows) in the connective tissue around a small duct (D). Section counterstained with haematoxylin, ×160. (Reprinted from Lee *et al.*, 1992 with permission from Elsevier Science).

when mammary glands of pregnant and non-pregnant sheep were stimulated by infusion of killed *S. aureus* (Fig. 7). These cells were seen to be clustering around blood capillaries with very prominent endothelial cell lining, and some $CD5^+CD4^+$ lymphocytes were scattered among the B cell clusters. Fourteen days after antigen infusion numerous plasma cells were observed, most of them being of the IgA isotype (Fig. 8). It therefore appears that B cells, which are destined to become plasma cells, are prevalent only in infected glands. However this immune response is not sufficient to give the mammary gland protection against mastitis as the bactericidal activity of neutrophils correlated positively only with titres of IgG2, but not with concentrations of IgG1 or IgA antibodies (Watson, 1989).

Protection of the mammary gland by immunoglobulins within the milk compartment occurs by opsonization of bacteria, coverage of bacterial adhesion sites and through the ability of antibodies to induce agglutination of bacteria thereby facilitating their removal from the gland with the passage of milk (Lascelles, 1979). Both the total concentration of immunoglobulin and the relative concentrations of various isotypes vary according to stage of lactation and among species. The IgA system is extremely well developed in the human mammary gland (as is also the case for rabbits, guinea pigs and pigs). Secretory immunoglobulin A (sIgA) is the dominant Ig in human milk with colostral concentrations of approximately 10 g/l which decrease rapidly to 1–2 g/l by the end

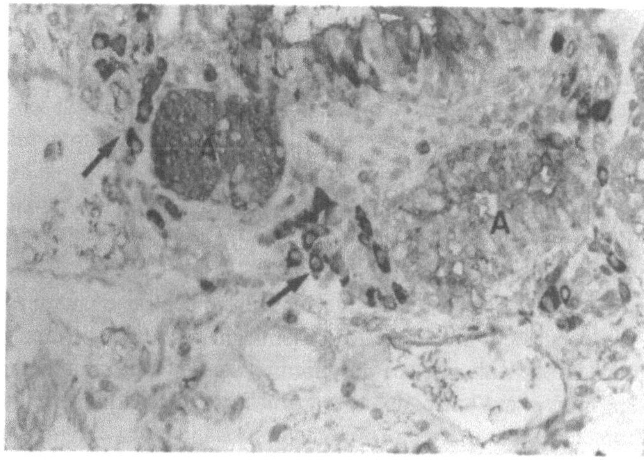

Figure 8. Paraffin section of mammary gland tissue of pregnant sheep 14 days after infusion of killed S. aureus stained with the indirect immunoperoxidase technique for plasma cells of the IgA isotype (arrows). Note the presence of numerous positive cells located in close proximity to the alveoli (A). ×450. (Reprinted from Lee *et al.*, 1992 with permission from Elsevier Science).

of the first week of lactation (Kulski and Hartmann, 1981). Apart from the obvious contribution that these initial high levels make towards protection of the neonate, IgA may also form an important part of the defense of the mammary gland. It has been shown that IgA, in conjunction with lactoferrin and C3, was deficient in the milk of women who developed mastitis relative to other lactating women who did not develop this condition (Prentice *et al.*, 1985). This suggests that a low level of IgA in milk could predispose women to mastitis. Interestingly, IgA deficiency has been cited as the most common form of primary immunodeficiency, with a frequency of 1:300 in the Caucasian population (Islam *et al.*, 1994).

Several physiological variables such as psychological stress, intense endurance exercise and sleep deprivation have been found to result in lowered levels of salivary IgA (Gleeson *et al.*, 1995). During psychological stress the concentration of salivary IgA appears to be inversely related to the level of stress experienced (Graham *et al.*, 1988; Jemmott *et al.*, 1993). Stress has also been found to be a major risk factor for mastitis in multiparous women who are breastfeeding (Fetherston, 1998) with women reporting a significant increase in stress levels in the week prior to developing mastitis. It is possible that there may be a relationship between increased levels of stress, lower levels of IgA and the occurrence of mastitis.

Relative to human milk the levels of IgA in the milk of ruminants is very low. The concentration of IgA in bovine colostrum is 4.4 g/l and this drops to

0.08 g/l in mature milk. This may have implications for the development of mastitis in cows where the lower levels of sIgA could increase the vulnerability of epithelial surfaces to adhesion by invading bacteria. The importance of the IgA system in protection of the mammary gland is implicated in research involving the stimulation of local immunity in a variety of animals with differing milk IgA levels. The production of local antibodies in the mammary glands of animals infused with antigens was much greater and persisted longer in animals that previously had higher levels of milk IgA (Lascelles *et al.*, 1966; McDowell and Lascelles, 1969, 1971; McDowall, 1973).

Since the beginning of this century vaccines have been used in an attempt to reduce the incidence and frequency of mastitis in cattle, sheep and goats. Initially, the intent was to enhance specific immunity by increasing the capacity to synthesize opsonizing antibody on subsequent bacterial challenge. However, vaccine-stimulated increases in antibody concentrations in milk were only effective in lessening the severity of established infections and were unable to prevent the onset of new intramammary infections (Brock *et al.*, 1975). Despite this, substantial progress has been made in recent years in the development of vaccines that interfere with the virulence factors of invading organisms. Sordillo *et al.* (1997) provide an excellent review of current research pertaining to vaccination against bovine mastitis.

In addition to vaccines, amplification of specific immune responses within the bovine udder has also been effected by the use of cytokines. Nickerson *et al.* (1992) found that bovine quarters infused with IL-2 exhibited the greatest increase in concentration of IgG1, IgG2, IgA and IgM in milk compared with quarters infused with other immunostimulants, namely pokeweed mitogen and lipopolysaccharide. Quiroga *et al.* (1993b) also found that treatment of *S. aureus*-infected bovine mammary glands with recombinant IL-2 increased the neutrophil, macrophage, lymphocyte and eosinophil counts in mammary gland secretions. Interleukin-2 has also been used effectively as a vaccine adjuvant in the bovine mammary gland (Pighetti and Sordillo, 1995) and, when used therapeutically as an adjunct to antibiotic therapy, IL-2 and IL-1β effected an increase in recovery rates from mastitis (Dunney and Shuster, 1995; Daley *et al.*, 1992).

6. Nutritional Influences on Mastitis and Immunological Defenses in Milk

During the last 15 years, the ability of the cow to resist mastitis-causing infections has been linked to intake of the micronutrients selenium (Se), vitamin E, vitamin A and β-carotene. Smith *et al.* (1984) found a reduction in the clinical symptoms of mastitis among cows supplemented with Se, vitamin E, or both. The duration of clinical signs was reduced by 46% in the Se-treated group, 44%

in the vitamin E group and 62% in the group given both vitamin E and Se. The effectiveness of vitamin E supplements in protecting against mastitis was investigated further by Weiss *et al.* (1997) in peripartum cows with marginal Se status. Although the animals were fed supplemental Se as recommended by the Food & Drug Authority (1993), their plasma and whole blood concentrations of Se were lower than values generally accepted as adequate. Vitamin E supplementation commenced during the dry period at 60 days prior to anticipated calving and continued through the first 30 days of lactation. Cows receiving a dietary supplement of 1000 IU/d of vitamin E had 30% less clinical mastitis than did cows receiving supplementation of 100 IU/d, and the reduction was 88% in a third group of cows fed 4000 IU/d of vitamin E during the last 14 days of the dry period. Cows with plasma α-tocopherol concentrations of 3 μg/ml, or less, were 9.4 times more likely to have clinical mastitis than cows with concentrations more than 3 μg/ml. The authors recommended that plasma concentrations of α-tocopherol at parturition should be at least 3 μg/ml when Se status is marginal.

Ali-Vehmas *et al.* (1997) found that supplementation of selenium-deficient dairy cows with selenium yeast or selenite improved the recruitment of phagocytes into infected compartments of the udder. Thus, the inflammatory response, as indicated by SCC and lysosomal enzyme activity in milk, was significantly improved in the supplemented cows. In this connection, Ndiweni and Finch (1996) found that *in vitro* supplementation with vitamin E or selenium enhanced both chemotactic and random migration by bovine mammary gland PMN, and increased their production of superoxide. Vitamin E also enhanced the phagocytosis of opsonized *S. aureus* by PMN. These *in vitro* results may indicate mechanisms whereby supplements of vitamin E and selenium enhance the natural resistance of the dairy cow to mastitis. Synergism between the two nutrients was not apparent *in vitro*.

Vitamin A and its precursor, β-carotene have also been implicated in the severity of bovine mastitis. Chew *et al.* (1982) reported a negative correlation between the concentrations of vitamin A and β-carotene and the SCC in bovine milk. Low levels of vitamin A and β-carotene in plasma and milk have been shown to be associated with an increased incidence of mastitis, and supplementation was shown to reduce SCC in subsequent lactations (Chew and Johnson, 1985). This is supported by Tjoelker *et al.* (1986) and Daniel *et al.* (1991) who found that the ability of milk and blood phagocytes to kill *S. aureus in vitro* was enhanced in the presence of β-carotene, retinol and retinoic acid.

Based on evidence from these studies in cows, Filteau *et al.* (1999a) investigated the effects of maternal supplementation with retinol and β-carotene on the amounts of sIgA, lysozyme, lactoferrin, and IL-8 in the breast milk of Bangladeshi women. In this randomized, double blind, placebo controlled trial women received a daily dose of β-carotene, a large single dose of retinol followed by a daily placebo, or a daily placebo until 9 months post partum. There were no significant differences among the three groups in any of the immune factors, although initial vitamin A status in the population was poor and improved with supplementation.

Filteau *et al.* (1999b) also conducted a study designed to assess the impact of dietary vitamin E or vitamin A supplementation on the occurrence of breast inflammation in lactating rural Tanzanian women. The subjects were divided into three groups and supplemented during late pregnancy and throughout lactation either with carotenoid-rich red palm oil or with sunflower oil which is a rich source of α-tocopherol. The third group received dietary advice but no supplement. There were no major differences among the three groups in micronutrient status or breast inflammation at the time of recruitment. Breastmilk concentrations of IL-8 and transforming growth factor β_2, breast milk ratio and plasma concentrations of acute phase proteins (alpha$_1$-acid glycoprotein and C-reactive protein) were used as indicators of breast inflammation. A significant treatment-related effect was seen only for the Na:K ratio. In the group supplemented with sunflower oil, the α-tocopherol concentration in milk increased significantly by three months and women had lower milk Na:K ratios than in the other two groups.

The effects of amount and type of fat in the diet have been investigated in relation to immune cell functions, particularly lymphocyte proliferation in response to mitogens. N-3 polyunsaturated fatty acids are thought to influence gene expression by preventing activation of the transcription factor NF kappa B, thus causing suppression of numerous genes including those for cytokines, cytokine receptors, adhesion molecules and inducible NO synthase (Miles and Calder, 1998). This may have implications for immune activity in milk as research has also shown that the content of n-3 fatty acids in breast milk increases rapidly, in a dose-related manner, in response to an increase in dietary fish oils (Harris *et al.*, 1984) or dietary docosahexaenoic acid (Makrides *et al.*, 1996).

7. Mastitis: Physiology or Pathology?

The question "Mastitis: physiology or pathology?" posed by Bramley formed much of the basis of the 1990 Ghent meeting on the pathogenesis of bovine mastitis. Papers presented at this meeting together with the broad range of subsequent research have brought the understanding of bovine mammary gland immunity to an advanced stage. This is evidenced by the development of experimental vaccines against bovine mastitis and the introduction of recombinant cytokine therapy. There remains, however, a paucity of research into mastitis in lactating women. Therefore "physiology or pathology?" is an important question with repercussions for both prevention and treatment of this condition in women. This is particularly so as the clinical symptoms experienced by women do not always appear to reflect whether mastitis is of an infective or non-infective origin (Brodribb, 1997). The sudden onset of severe flu-like symptoms that can be experienced by women with mastitis is a curious occurrence as these symptoms are usually seen in systemic illnesses such as acute viral influenza and septicemias. Furthermore these extreme symptoms do not occur when a localized inflammation occurs elsewhere in the body. In the breast, however, these symptoms are manifest whether the inflammation occurs as a result of infection or as a result, simply, of ductal obstruction. If non-infective forms of mastitis can produce symptoms similar to those of infective mastitis, this poses many questions that require investigation. Why would severe systemic symptoms be elicited in response to localized non-infectious inflammation within the breast? If bacteria are not involved can the response be explained by an antigen/antibody reaction that occurs when milk leaks from the ducts into breast tissue? Perhaps all mastitis is a result of bacterial proliferation that has been encouraged by the stasis of milk within the gland. Even so, why should such a localized inflammatory event elicit such marked systemic symptoms? Breast permeability increases in engorgement and mastitis so that milk-derived products leak into the interstitium and vasculature. To what extent do the pro-inflammatory cytokines and other phlogistic components present in milk contribute to the symptoms seen in mastitis?

Colostrum, milk and involution secretion undoubtedly contribute to protection of the mammary gland during lactation. However, the extent of this action, especially in women, is not clearly understood. Whether mastitis in lactating women always reflects infection, and what role milk components may play in the manifestation of mastitis symptoms are important areas for future research. A better understanding of the pathogenesis and physiology of mastitis is necessary to facilitate better prevention and treatment of mastitis in lactating women.

References

Abakada, A.O., Hartmann, P.E., and Grubb, W.B. 1990. Sodium and serum albumin in milk: biochemical markers for human lactational mastopathy. *Proceedings of the Australian Biochemical Society* **22**:SP52.

Abakada, A.O., Grubb, W.B., and Hartmann, P.E. 1992. Milk composition from mastopathic breasts with and without a history of mastopathy. *Proceedings of the Australian Biochemical Society* **17**:59.

Akin, D.T., Lu, M.Q., Lu, S.J., Kendall, S., Rundegren, J., and Arnold, R.R. 1994. Bactericidal activity of different forms of lactoferrin, in: *Lactoferrin: Structure and Function: Advances in Experimental Medicine and Biology*, Vol. 357 (T.W. Hutchens, S.V. Rumball, and B. Lönnerdal, eds.), Plenum Press, New York, pp. 61–70.

Akkers, R.M. and Thompson, W. 1987. Effect of induced leukocyte migration on mammary cell morphology and milk component biosynthesis. *J. Dairy Sci.* **70**:1685.

Ali-Vehmas, T., Vikerpur, M., Fanf, W., and Sandholm, M. 1997. Giving selenium supplements to dairy cows strengthens the inflammatory response to intramammary infection and induces a growth suppressing effect on mastitis pathogens in whey. *Zentralblatt Fur Veterinarmedizin—Reihe A.* **44(9–10)**:559.

Arnold, R.R., Cole, M.F., and McGhee, J.R. 1977. A bactericidal effect for human milk lactoferrin. *Science* **197**:263.

Bellamy, W., Takase, M., Wakabayashi, H., Kawase, K., and Tomita, M. 1992. Antibacterial spectrum of lactoferricin B, a potent bactericidal peptide derived from the N-terminal region of bovine lactoferrin. *J. Appl. Bact.* **73(6)**:472.

Berning, L.M., Paape, M.J., and Peters, R.R. 1993. Functional variation in endogenous and exogenous immunoglobulin binding to bovine neutrophils relative to parturition. *Am. J. Vet. Res.* **54**:1145.

Bramley, A.J. 1991. Mastitis: physiology or pathology?, in: *New Insights into the Pathogenesis of Mastitis* (C. Burvenich, G. Vandeputte-Van Messon, A.W. Hill, eds.), Rijksuniversiteit Gent, Belgium, pp. 3–8.

Breton-Gorius, J., Mason, D.Y., Buriot, D., Vilde, J., and Griscelli, C. 1980. Lactoferrin deficiency as a consequence of a lack of specific granules in neutrophils from a patient with recurrent infections. *Am. J. Pathol.* **99(2)**:413.

Britigan, B.E., Serody, J.S., and Cohen, M.S. 1994. The role of lactoferrin as an anti-inflammatory molecule, in: *Lactoferrin: Structure and Function: Advances in Experimental Medicine and Biology*, Vol. 357 (T.W. Hutchens, S.V. Rumball, and B. Lönnerdal, eds.), Plenum Press, New York, pp. 143–156.

Brock, J.H., Steel, E.D., and Reiter, B. 1975. The effect of intramuscular and intramammary vaccination of cows on antibody levels and resistance to intramammary infections by *S. aureus*. *Res. Vet. Sci.* **19**:152.

Brodribb, W. (ed.). 1997. *Breastfeeding Management in Australia: A Reference and Study Guide.* Nursing Mothers' Association of Australia, Vic., Australia.

Bullen, J.J., Rodgers, H.J., and Leigh, L. 1972. Iron binding proteins in milk and resistance to Escherichia coli infection in infants. *Br. Med. J.* **1**:69.

Chandan, R.C., Parry, R.M., and Shahani, K.M. 1968. Lysosyme, lipase and ribonuclease in milk of various species. *J. Dairy Sci.* **51**:606.

Chew, B.P., Hollen, L.L., Hillers, J.K., and Hertugson, M.L. 1982. Relationship between vitamin A and beta catotene in blood plasma and milk and mastitis in Holsteins. *J. Dairy Sci.* **65(11)**:2111.

Chew, B.P. and Johnston, L.A. 1985. Effects of supplemental vitamin A and beta carotene on mastitis in dairy cows. *J. Dairy Sci.* **68**(suppl. 1):191.

Cifrian, E., Guidry, A.J., O'Brien, C.N., and Nickerson, S.C. 1994. Adherence of Staphylococcus aureus to cultured bovine mammary epithelial cells. *J. Dairy Sci.* **77**:970.

Conner, A.E. 1979. Elevated levels of sodium and chloride in milk from mastitic breast. *Paediatr.* **63(6)**:910.

Craven, N. and Williams, M.R. 1985. Defences of the bovine mammary gland against infection and prospects for their enhancement. *Vet. Immun. Immunopath.* **10**:71.

Crouch, S.P.M., Slater, K.J., and Fletcher, J. 1992. Regulation of cytokine release from mononuclear cells by iron binding protein lactoferrin. *Blood* **80**:235.

Daley, M.J., Williams, T., Dougherty, R., Coyle, P., Furda, G., and Hayes, P. 1991. *Staphylococcus aureus* mastitis: pathogenesis and treatment with bovine interleukin-1 and interleukin-2. *J. Dairy Sci.* **74**:4413.

Daley, M.J., Furda, G., Dougherty, R., Coyle, P., Williams, T., and Johnston, P. 1992. Potentiation of antibiotic therapy of *Staphylococcus aureus* mastitis by recombinant bovine interluekin-2. *J. Dairy Sci.* **75**:3330.

Daniel, L.R., Chew, B.P., Tanaka, T.S., and Tjoelker, L.W. 1991. B-carotene and vitamin A effects on bovine phagocyte function in vitro during the peripartum period. *J. Dairy Sci.* **74**:124.

Devereux, W.P. 1970. Acute puerperal mastitis. *Am. J. Obstet. Gynaecol.* **108(1)**:78.

Dunney, S. and Shuster, D. 1995. Recombinant bovine interluekin-1B as an adjunct therapuetic for Staphylococcus aureus mastitis in lactating dairy cows. *Proceedings of the 4th International Veterinary Immunology Symposium*. Davis, California, USA, 16–21 July, p. 87.

Dutia, B.M., Ross, A.J., and Hopkins, J. 1993. Comparison of workshop on CD45R monoclonal antibodies with OvCD45R monoclonal antibodies in sheep. *Vet. Immunol. Immunopath.* **39**:121.

Ellison III, R.T. 1994. The effects of lactoferrin on gram-negative bacteria, in: *Lactoferrin: Structure and Function. Advances in Experimental Medicine and Biology*, Vol. 357 (T.W. Hutchens, S.V. Rumball, and B. Lönnerdal, eds.), Plenum Press, New York, pp. 61–70.

Ferens, W.A. Davis, W.C. Hamilton, M.J. Park, Y.H. Deobald, C.F. Fox, L., and Bohach, G. 1998. Activation of bovine lymphocyte subpopulations by staphylococcal entertoxin C. *Infect. Immun.* **66(2)**:573.

Fetherston, C. 1995. Factors influencing breastfeeding initiation and duration in a private Western Australian maternity hospital. *Breastfeeding Review* **3(1)**:9.

Fetherston, C. 1997. Characteristics of lactation mastitis in a Western Australian cohort. *Breastfeeding Review* **5(2)**:5.

Fetherston, C. 1998. Risk factors for lactation mastitis. *J. Hum. Lact.* **14**:101.

Fetrow, J., Anderson, K., Sexton, S., and Butcher, K. 1988. Herd composite somatic cell counts: average linear score and weighted average somatic cell count score and milk production. *J. Dairy Sci.* **68**:2100.

Filteau, S.M., Rive, A.M., Ball, J.J., Chakrborty, J., Stoltzfus, R., and de Francisco, A. 1999a. Breastmilk immune factors in Bangladeshi women supplemented postpartum with retinol or B-carotene. *Am. J. Clin. Nutr.* **69**:953.

Filteau, S.M., Lietz, G., Mulokozi, G., Bilotta, S., Henry, C.J.K., and Tomkins, A.M. 1999b. Milk cytokines and subclinical breast inflammation in Tanzanian women: effects of dietary red palm oil or sunflower oil supplementation. *Immunology* **7**:595.

Frost, A.J., Brooker, B.E., and Hill, A.W. 1984. The effect of Escherichia coli endotoxin and culture filtrate on the lactating bovine mammary gland. *Aust. Vet. J.* 61:77.

Garofalo, R.P. and Goldman, A.S. 1999. Expression of functional immunomodulatory and anti-inflammatory factors in human milk. *Clinics in Perinatology* **26(2)**:361.

Gleeson, M., Cripps, A.W., and Clancy, R.L. 1995. Modifiers of the human mucosal immune system. *Immunol. Cell Biol.* **73**:397.

Goldman, A.S. 1993. The immune system of human milk: antimicrobial, antiinflammatory and immunomodulating properties. *Pediatr. Infect. Dis. J.* **12**:664.

Goldman, A.S., Garza, C.S., Nichols, B.L., and Goldblum, R.M. 1982. Immunological factors in human milk during the first year of lactation. *J. Paediatr.* **100**:563.

Goldman, A.S., Goldblum, R.M., and Hanson, L.A. 1990. Anti-inflammatory systems in human milk. *Adv. Exp. Med. Biol.* **262**:69.

Golota, V.Y. and Altuev, G.N. 1984. Composition of milk during mastitis. *Akusherstvo I Ginekologia, Moscow* **9**:48.

Graham, N.M.H., Chiron, R., Barlholomeusz, A., Taboonpong, N., and LaBrooy, J.T. 1988. Does anxiety reduce the secretion rate of secretory IgA in saliva? *Med. J. Aust.* **148**:131.

Hamosh, M. 1998. Protective functions of proteins and lipids in milk. *Biol. Neonate* **74**:163.

Hanson, L.A., Mattsby-Baitzer, I., Engberg, I., Roseneanu, A., Elverfors, J., and Motas, C. 1995. Anti-inflammatory capacities of human milk: lactoferrin and secretory IgA inhibit endotoxin-induced cytokine release, in: *Advances in Mucosal Immunology* (J. Mestecky, ed.), Plenum Press, New York, pp. 669–672.

Harmon, R.J. 1994. Physiology of mastitis and factors affecting somatic cell counts. *J. Dairy Sci.* **77**:2103.

Harmon, R.J., Schanbacher, F.L., Ferguson, L.C., and Smith, K.L. 1976. Changes in lactoferrin, immunoglobulin G, bovine serum albumin, and alpha lactalbumin during acute experimental and natural coliform mastitis in cows. *Infect Immunol.* **13(2)**:533.

Hartmann, P.E. and Kulski, J.K. 1978. Changes in the composition of the mammary secretion of women after abrupt termination of breastfeeding. *J. Physiol. (Lond).* **275**:1.

Harris, W.S., Conner, W.E., and Lindsey, S. 1984. Will dietary n-3 fatty acids change the composition of human milk? *Am. J. Clin. Nutr.* **40**:780.

Hasty, D.L., Beachey, E.H., Courtney, H.S., and Simpson, W.A. 1989. Interactions between fibronectin and bacteria, in: *Fibronectin in Health and Disease* (S. Carsons, ed.), CRC Press, Florida, pp. 89–112.

Hill, A.W. 1981. Factors influencing the outcome of *E coli* mastitis in cows. *Res. Vet. Sci.* **31**:107.

Holmgren, J., Svennerholm, A.M., Lindbald, M., and Strecker, G. 1986. Inhibition of bacterial adhesion and toxin binding by gycoconjugate and oligosaccharide receptor analogues in human milk, in: *Lactation III: The Effects of Human Milk Upon the Recipient Infant* (A.S. Goldman, S.A. Atkinson, and L.A. Hanson, eds.), Plenum Press, New York, pp. 251–259.

Hungerford, T.G. 1990. *Diseases of Livestock.* McGraw Hill. NSW, Australia.

Islam, K.B., Baskin, B., Nilsson, L., Hammerstrom, L., Sideras, P., and Smith, C.I. 1994. Molecular analysis of IgA deficiency: evidence for impaired switching to IgA. *J. Immun.* **152**:1442.

Jemmott, J.B., Borysenko, J.Z., Borysenko, M., McClelland, D.C., Chapman, R., Meyer, D., and Benson, H. 1993. Academic stress, power, motivation and decrease in secretion rate of salivary secretory immunoglobulin A. *Lancet* **1**:1400.

Jonsson, S. and Pulkinnen, M.O. 1994. Mastitis today: incidence, prevention and treatment. *Annales Chirugiae et Gynaecologiae* **83**:84.

Kashiwazaki, Y. 1984. Lymphocyte activities in dairy cows with special reference to outbreaks of mastitis in pre and post partus. *Jpn. J. Vet. Res.* **32**:101.

Keeney, S.E., Schmalstieg, F.C., Palowetz, K.H., Rudolf, H.E., Le, B.M., and Goldman, A.S. 1993. Activated neutrophils and neutrophil activators in human milk: increased expression of CD11b and decreased expression of L-selectin. *J. Leuk. Biol.* **54(2)**:97.

Kehrli, M.E. Jr and Goff, J.P. 1989. Periparturient hypocalcaemia in cows: effects on peripheral blood neutrophil and lymphocyte function. *J. Dairy Sci.* **72**:1188.

Kehrli, M.E. and Shuster, D.E. 1994. Factors affecting milk somatic cells and their role in health of the bovine mammary gland. *J. Dairy Sci.* **77**:619.

Kehrli, M.E., Cullor, J.S., and Nickerson, S.C. 1991. Immunobiology of hematopoeitic colonoy stimulating factors: Potential appliation to disease prevention in the bovine. *J. Dairy Sci.* **74**:4399.

Kehrli, M.R. Jr., Nonnecke, B.J., and Roth, J.A. 1989. Alterations in bovine peripheral blood neutrophil function during the periparturient period. *Am. J. Vet. Res.* **50**:207.

Kennedy, P.C. and Miller, R.B. 1993. The mammary gland, in: *Pathology of Domestic Animals*, 4th ed., Vol 3 (K.V.F. Jubb, P.C. Kennedy, and N. Palmer, eds.), Academic Press, New York, pp. 454–470.

Kinlay, J., O'Connell, D.L., and Kinlay, S. 1998. Incidence of mastitis in breastfeeding women during the six months after delivery: a prospective study. *Med. J. Aust.* **169**:310.

Kulski, J.K. and Hartmann, P.E. 1981. Changes in human milk composition during the initiation of lactation. *Aust. J. Exp. Biol. Medical Sc.* **59**:769.

Kutushev, F.K., Spesivtsev, I.A., Andreev, A.V., Mel'nikova, V.N., and Volkova, S.D. 1988. Use of the leukocyte mass in treatment of patients with suppurative lactation mastitis (abstract only). *Vestn. Khir.* **141(8)**:95.

Kuusela, P. 1978. Fibronectin binds to Staphylococcus aureus. *Nature* **276**:718.

Kuusela, P., Vartio, T., Vuento, M., and Myhre, E.B. 1985. Attachment of staphylococci and streptococci on fibronectin, fibronectin fragments and fibrinogen bound to a solid phase. *Infect. Immun.* **50**:77.

Lascelles, A.K. 1979. The immune system of the ruminant mammary gland and its role in the control of mastitis. *J. Dairy Sci.* **62**:154.

Lascelles, A.K., Outteridge, P.M., and Mackenzie, D.S.S. 1966. Local production of antibody by the lactating mammary gland following antigenic stimulation. *Aust. J. Exp. Biol. Med. Sci.* **44**:169.

Lawrence, R. 1994. *Breastfeeding: A Guide for the Medical Profession*, 4th Ed., Mosby, St. Louis.

Lee, C.S. and Outteridge, P.M. 1981. Leukocytes of sheep colostrum, milk and involution secretion with particular reference to ultrastructure and lymphocyte subpopulations. *J. Dairy Res.* **48**:225.

Lee, C.S., Wooding, F.B.P., and Kemp, P. 1980. Identification, properties and differential counts of cell populations using electron microscopy of dry cow secretions, colostrum and milk from normal cows. *J. Dairy Res.* **47**:39.

Lee, C.S., Meeusen, E., and Brandon, M.R. 1989. Subpopulations of lymphocytes in the mammary gland of sheep. *Immunol.* **66**:388.

Lee, C.S., Meeusen, E., and Brandon, M.R. 1992. Local immunity in the mammary gland. *J. Vet. Immunopathol.* **32**:1.

Lee, C.S., Tatarczuch, L., and Philip, C. 1996. Mechanism of mammary gland involution in the sheep. *Proc. Anat. Soc. Aust. N.Z.* **33**:12.

Lee, C.S., Tatarczuch, L., Philip, C., and Dischof, R. 1998. Dendritic cells in involuting mammary gland secretions. *Australasian Society for Immunology: 28th Annual Conference.*

Lees, P. 1991. General aspects of inflammation, in: *New Insights into the Pathogenesis of Mastitis* (C. Burvenich, G. Vandeputte-Van Messon, and A.W. Hill, eds.), Rijksuniversiteit Gent, Belgium, pp. 43–49.

Leleu, O., Aubrey, P., Verhoest, P. Jounieaux, V. 1997. Tuberculosis of the breast. *Revue des Maladies Respiratoires* **14(5)**:401.

Lönnerdal, B. 1985. Biochemistry and physiological function of human milk proteins. *Am. J. Clin. Nutr.* **42**:1229.

Lönnerdal, B. and Atkinson, S. 1995. Nitrogenous Components of Milk: Human Milk Proteins, in: *Handbook of Milk Composition* (R.G. Jensen, ed.), Academic Press, California, pp. 351–364.

Lönnerdal, B. and Iyer, S. 1995. Lactoferrin: molecular structure and biological function. *Ann. Rev. Nutr.* **15**:93.

Makrides, M., Neumann, M.A., and Gibson, R.A. 1996. Effect of maternal docashexanaeoic acid (DHA) supplementation on breast milk composition. *Eur. J. Clin. Nutr.* **50**:352.

Mamo, W., Jonsson, P., Flock, J.I., Lindberg, M., Muller, H.P., Wadstrom, T., and Nelson, L. 1994. Vaccination against Staphylococcus aureus mastitis: immunological response of mice vaccinated with fibronectin-binding protein (FnBP-A) to challenge with *S. aureus*. *Vaccine* **12(11)**:988.

Matheson, I, Aursnes, I., Horgen, M., Aabo, O., and Melby, K. 1988. Bacteriological findings and clinical symptoms in relation to clinical outcome in puerperal mastitis. *Acta Obstet. Gynaecol. Scand.* **67**:723.

Maxe, I., Ryden, C., Wadstrom, T., and Rubin, K. 1986. Specific attachment of *S. aureus* to immobilised fibronectin. *Inf. and Immun.* **54(3)**:695.

McDowell, G.H. 1973. Local antigenic stimulation of guinea pig mammary gland. *Aust. J. Exp. Biol. Med. Sci.* **51**:237.

McDowell, G.H. and Lascelles, A.K. 1969. Local production of antibody by ovine mammary glands infused with Salmonella flagellar antigens. *Aust. J. Exp. Biol. Med. Sci.* **47**:669.

McDowell, G.H. and Lascelles, A.K. 1971. Local production of antibody by the lactating mamary gland of the ewe and the effect of systemic immunisation. *Res. Vet. Sci.* **12**:113.

Miescher, S.M., Schreyer, C., Baras, C., Capasso. C., and Von Fleidner. 1990. Sparse distributions of gamma delta T lymphocytes around epithelial tumours predominately infiltrated by primed memory T cells. *Cancer Immunol. Immunotherapy* **32**:81.

Miles, E.A. and Calder, P.C. 1998. Modulation of immune function by dietary fatty acids. *Proceedings of the Nutrition Society* **57**:277.

Molenaar, A.J., Kuys, Y.M., Davis, S.R., Wilkins, R.J., Mead, P.E., and Tweedie, J.W. 1996. Elevation of lactoferrin gene expression in developing, ductal, resting and regressing parenchymal epithelium of the ruminant mammary gland. *J. Dairy Sci.* **79**:1198.

Movat, H.Z., Cybulsky, M.I., Colditz, I.G., Chan, M.K., and Dinarello, C.A. 1987. Acute inflammation in gram-negative infection: endotoxin, interleukin 1, tumor necrosis factor, and neutrophils. *Fed. Proc.* **46(1)**:97.

Naidu, T.G. and Newbold, F.H.S. 1973. Glycogen in leukocytes from bovine milk and blood. *Can. J. Comp. Med.* **37**:47.

National Mastitis Council. 1987. *Current Concepts of Bovine Mastitis*. 3rd Ed. National Research Council, Arlington, Virginia.

Ndiweni, N. and Finch, J.M. 1996. Effects of *in vitro* supplementation with alpha-tocopherol and selenium on bovine neutrophil functions: implications for resistance to mastitis. *Vet. Immunol. Immunopath.* **51(1–2)**:67.

Nelson, L., Flock, J.I., Hook, M., Linberg, M., Muller, H.P., and Wadstrom, T. 1991. Adhesins in staphylococcal mastitis as vaccine components, in: *New Insights into the Pathogenesis of Mastitis* (C. Burvenich, G. Vandeputte-Van Messon, and A.W. Hill, eds.), Rijksuniversiteit Gent, Belgium, pp. 111–125.

Neville, M.C. and Jensen, R.G. 1995. The physical properties of human and bovine milks, in: *Handbook of Milk Composition* (R.G. Jensen, ed.), Academic Press, California, pp. 81–85.

Newman, J. 1995. How breastmilk protects newborns. *Sci. Amer.* **273(6)**:58.

Nickerson, S.C. 1985. Immune mechanisms of the bovine udder: An overview. *JAVMA* **187**:41.
Nickerson, S.C. 1991. Effects of cytokines on bovine mammary gland immunity. *SAAS Bulletin, Biochem. and Biotech.* **4**:60.
Nickerson, S.C., Owens, W.E., and Watts, J.L. 1989. Effects of recombinant granulocyte colony stimulating factor on Staphylococcus aureus mastitis in lactating dairy cows. *J. Dairy Sci.* **72**:3286.
Nickerson, S.C., Owens, W.E., Boddie, R.L., and Boddie, N.T. 1992. The effect of chronic immunostimulation of the of the nonlactating bovine mammary galnd with interleukin-2, pokeweed mitogen and lipopolysaccaide. *J. Dairy Sci.* **75(12)**:3339.
Nickerson, S.C., Owens, W.E., Rejman, J.J., and Oliver, S.P. 1993. Effects of interleukin-1 and interleukin-2 on mammary gland leukocyte populations and histology during the early non lactating period. *Zentralblatt Fur Veterinarmedizin—Reihe B.* **40**:621.
Orem, J.D. and Reiter, B. 1968. Inhibition of bacteria by lactoferrin and other iron chelating agents. *Biochem. Biophys. Acta* **170**:351.
Outterridge, P.M. and Lee, C.S. 1988. The defence mechanisms of the mammary gland of domestic ruminants. *Prog. Vet. Microbiol. Immun.* **4**:165.
Ozkaragoz, F., Rudoloff, H.E., Rajaraman, S., Mushtaha, A.A., Schmalsteig, F.C., and Goldman, A.S. 1988. The motility of human milk macrophages in collagen gels. *Paediatr. Res.* **23**:449.
Paape, M.J. and Capuco, A.V. 1996. Cellular defence mechanisms in the udder and lactation of goats. *J. Anim. Sci.* **75**:556.
Paape, M.J. and Weinland, B.T. 1988. Effect of abraded intramammary device on milk yield, tissue damage and cellular composition. *J. Dairy Sci.* **71**:250.
Paape, M.J. and Wergin, W.P. 1977. The leukocyte as a defence mechanism. *J. Am. Vet. Med. Assoc.* **36**:377.
Paape, M.J., Hafs, H.D., and Snyder, W.W. 1963. Variation of estimated numbers of milk somatic cells stained with Wrightís stain or pyronin-y-methyl green stain. *J. Dairy Sci.* **46**:1211.
Paape, M.J., Schultze, A.J., Guidry, A.J., Kortum, W.M., and Weinland, B.T. 1981. Effect of an intramammary polyethelene device on the concentration of leukocytes and immunoglobulins in milk and on the leukocyte response to *E. coli* endotoxin and challenge exposure with S. aureus. *Am. J. Vet. Res.* **42**:774.
Park, Y.H., Fox, L.K. Hamilton, M.J., and Davis, W.C. 1992. Bovine mononuclear leukocyte populationsin peripheral blood and mammary gland secretions. *J. Dairy Sci.* **75**:998.
Peakman, M. and Vergoni, D. 1997. *Basic and Clinical Immunology*. New York, Churchill Livingstone.
Peterson, J.A., Patton, S., and Hamosh, M. 1998. Glycoproteins of the human milk fat glocbule in the protection of the breastfed infant against infections. *Biol. Neonate* **74**:143.
Pighetti, G.M. and Sordillo. 1995. Enhanced mammary immunity following primary immunisation with interferon-gamma. *J. Dairy Sci.* **78**:528.
Prentice, A., Prentice, A.M., and Lamb, W.H. 1985. Mastitis in rural Gambian mothers and the protection of the breast by milk antimicrobial factors. *Trans. Royal Soc. Trop. Med. Hyg.* **79**:90.
Prosser, C.G. and Hartmann, P.E. 1983. Comparison of mammary gland function during the ovulatory menstrual cycle and acute breast inflammation in women. *Aust. J. Exp. Biol. Med. Sci.* **61(Pt 3)**:277.
Quiroga, G.H., Nickerson, S.C., and Adkinson, R.W. 1993a. Histologic response of the heifer mammary gland to intramammary infusion of interluekin-2 or interferon-gamma. *J. Dairy Sci.* **76(10)**:2913.

Quiroga, G.H., Sordillo, L.M., Adkinson, R.W., and Nickerson, S.C. 1993b. Cytological responses of Staphylococcus aureus infected mammary glands of heifers to interferon gamma and interluekin-2 treatment. *Am. J. Vet. Res.* **54(11)**:1894.
Rainard, P. 1986. Bacteriostatic activity of bovine milk lactoferrin against mastitic bacteria. *Vet. Microbiol.* **11**:387.
Richie, E.R., Bass, R., Meistrich, M.L., and Demmison, D.K. 1982. Distribution of T-lymphocyte subsets in human colostrum. *J. Immunol.* **129**:1116.
Riordin, J.M. and Nichols, F.H. 1990. A descriptive study of lactation mastitis in long term breastfeeding women. *J. Hum. Lact.* **6(2)**:53.
Russell, M.W., Brooker, B.E., and Reiter, B. 1977. Electron microscopic observations of the interactions of casein micelles and milk fat globules with bovine polymorphonuclear leukocytes during the phagocytosis of staphylococci in milk. *J. Comp. Path.* **87**:43.
Ryan, D.P., Greenwood, P.L., and Nichols, P.J. 1993. Effect of caprine arthritis-encephalitis virus infection on milk cell count and n-Acetyl-B-glucosaminidase activity in bovine neutrophils. *Am. J. Vet. Res.* **52**:1201.
Sandholm, M., Ali-Vehmas, T., Nyholm, K., Honkanen-Buzalski, T., and Louhi, M. 1991. Failure mechanisms in lactational therapy of staphylococcal mastitis, in: *New Insights into the Pathogenesis of Mastitis* (C. Burvenich, G. Vandeputte-Van Messon, and A.W. Hill, eds.), Rijksuniversiteit Gent, Belgium, pp. 171–178.
Schalm, O.W., Lasmanis, J., and Jain, N.C. 1976. Conversion of chronic staphylococcal mastitis to acute gangrenous mastitis after neutropenia in blood and bone marrow produced by equine anti-bovine leukocyte serum. *Am. J. Vet. Res.* **37**:885.
Schalm, O.W., Lasmanis, J., and Carroll, E.J. 1964. Effects of pre existing leukocytosis on experimental coliform (*Aeobacter aerogenes*) mastitis in cattle. *Am. J. Vet. Res.* **27**:1537.
Semba, R.D., Kumwenda, N., Taha, T.E., Hoover, D.R., Lan, Y., Eisinger, W., Mtimavalye, L., Broadhead, R., Miotti, P.G., van der Hoeven, L., and Chiphangwi, J.D. 1999. Mastitis and immunological factors in breastmilk of lactating women in Malawi. *Clin. Diag. Lab. Immunol.* **6(5)**:671.
Shafer-Weaver, K.A., and Sordillo, L.M. 1997. Bovine CD8+ suppressor lymphocytes alter immune responsiveness during the post partum period. *Vet. Immunol. Immunopath.* **56(1–2)**:53.
Shafer-Weaver, K.A., Pighetti, G.M., and Sordillo, L.M. 1996. Diminished mammary gland lymphocyte functions parallel shifts in trafficking patterns during the post partum period. *Proc. Soc. Exp. Biol. Med.* **212(3)**:271.
Sheldrake, R.F., McGregor, G.D., and Hoare, R.J.T. 1983. Somatic cell count, electrical conductivity and serum albumin concentration for detecting bovine mastitis. *J. Dairy Sci.* **66**:548.
Shuster, D.E., Lee, E.K., and Kehrli, M.E., Jr. 1996. Bacterial growth, inflammatory cytokine production, and neutrophil recruitment during coliform mastitis in cows ten days after calving compared with cows midlactation. *Am. J. Vet. Res.* **57(11)**:1569.
Sim, T.C. and Goldman, A.S. 1991. Defects in neutrophils. *Allergy Proc.* **91**:131.
Smith, K.L., Harrison, J.H., Hancock, D.D., Todhunter, D.A., and Conrad, H.R. 1984. Effect of vitamin E and selenium supplementation on incidence of clinical mastitis and duration of clinical symptoms. *J. Dairy Sci.* **67**:1293.
Sordillo, L.M. and Babiuk, L.A. 1991. Modulation of mammary neutrophil function during the periparturient period following *in vitro* exposure to recombinant bovine interferon gamma. *Vet. Immunol. Immunopath.* **27**:393.

Sordillo, L.M., Campos, M., and Babiuk, L.A. 1991. Anti-bacterial activity of the bovine mammary gland lymphocytes following treatment with interleukin-2. *J. Dairy Sci.* **74**:3370.

Sordillo, L.M., Shafer-Weaver, K., and DeRosa, D. 1997. Symposium: bovine immunology. Immunobiology of the mammary gland. *J. Dairy Sci.* **80**:1851.

Spik, G., Cheron, A., Montreuil, J., and Dolby, J.M. 1978. Bacteriostasis of a milk sensitive strain of Escherichia coli by immunoglobulins and iron binding proteins in association. *Immunology* **35**:663.

Stelwagen, K., Farr, V.C., McFadden, H.A., Prosser, C.G., and Davis, S.R. 1997. Time course of milk accumulation-induced opening of mammary tight junctions, and blood clearance of milk components. *Am. J. Physiol.* **273** (*Regulatory Integrative Comp. Physiol.* **42**):R379.

Svanborg, C., Aniansson, G., Mestecky, J., Sabharwal, H., and Wald, A. 1991. Antiadhesive molecules in milk. *Adv. Exp. Med. Biol.* **310**:167.

Targowski, S.P. 1983. Role of immune factors in protection of the mammary gland. *J. Dairy Sci.* **66(8)**:1781.

Targowski, S.P. and Niemialtowski, M. 1996. Inhibition of lacteal leukocyte phagocytosis by colostrum, non lactating secretion and mastitis milk. *Am. J. Vet. Res.* **47**:1940.

Tatarczuch, L., Philip, C., and Lee, C.S. 1997. Involution of the sheep mammary gland. *J. Anatomy* **190**:405.

Taylor, B.C., Dellinger, J.D., Cullor, J.S., and Stott, J.L. 1994. Bovine milk lymphocytes display the phenotype of memory T cells and are predominately CD8$^+$. *Cell Immunol.* **156**:245.

Taylor, B.C., Keefe, R.G., Dellinger, J.D., Nakamura, Y., Cullor, J.S., and Stott, J.L. 1997. T cell populations and cytokine expression in milk derived from normal and bacteria-infected bovine mammary glands. *Cell Immunol.* **182(1)**:68.

Thomsen, A.C., Hansen, K.B., and Moller, B.R. 1983. Leukocyte counts and microbiological cultivation in the diagnosis of puerperal mastitis. *Am. J. Obs. Gynae.* **146(8)**:938.

Thomsen, A.C., Esperson, T., and Maigarrd, S. 1984. Course and treatment of milk stasis, non infectious inflammation of the breast, and infectious mastitis in nursing women. *Am. J. Obs. Gynae.* **149(5)**:492.

Tjoelker, L.W., Chew, B.P., Tanaka, T.S., and Daniel, L.R. 1986. Effects of vitamin A and β-carotene on phagocytosis and killing by bovine mammary neutrophils *in vitro*. *J. Dairy Sci.* **69**(Supplement):386.

Torre, P.M., Konur, P.K., and Oliver, S.P. 1992. Proliferative response of mammary gland mononuclear cells to recombinant bovine interluekin-2. *Vet. Immunol. Immunopath.* **32**:351.

Van Miert, A.S.J.P.A.M. 1991. Acute phase response and non cellular defence mechanisms, in: *New Insights into the Pathogenesis of Mastitis* (C. Burvenich, G. Vandeputte-Van Messon, and A.W. Hill, eds.), Rijksuniversiteit Gent, Belgium, pp. 69–87.

Vercellotti, G.M., McCarthy, J.B., Lindholm, P., Peterson, P.K., Jacob, H.S., and Furcht, L. 1985. Extracellular matrix proteins (fibronectin, laminin and type IV collagen) bind and aggregate bacteria. *Am. J. Pathol.* **120**:13.

Vogel, A., Hutchison, B.L., and Mitchell, E.A. 1999. Mastitis in the first year post partum. *Birth* **26(4)**:218.

Wadstrom, T. 1987. Molecular aspects on pathogenesis of wound and foreign body infections due to staphylococci. *Zbl. Bact. Hyg. A.* **226**:191.

Waller, K.P. 1997. Modulation of endotoxin-induced inflammation in the bovine teat using antagonists/inhibitors to leukotrienes, platelet activating factor and interluekin-1 beta. *Vet. Immunol. Immunopath.* **57**:239.

Watson, E.D. 1989. Specific antibody in milk whey and phagocytosis of Actinomyces pyogenes by neutrophils *in vivo*. *Res. Vet. Sci.* **47**:253.

Weiss, W.P., Hogan, J.S., Todhunter, D.A., and Smith, K.L. 1997. Effect of vitamin E supplementation in diets with a low concentration of selenium on mammary gland health of dairy cows. *J. Dairy Sci.* **80**:1728.

Wirt, D., Adkins, L.T., Palkowetz, K.H., Schmalsteig, F.C., and Goldman, A.S. 1992. Activated memory T cells in human milk. *Cytometry* **13**:282.

Xanthou, M. 1997. Human milk cells. *Acta Paediatr.* **86**:1288.

Xanthou, M. 1998. Immune Protection of Human Milk. *Biology of the Neonate* **74**:121.

Zeng, S.S. and Escobar, E.N. 1995. Factors affecting somatic cell count of goat milk: Breed and farm, in: *Proc. Of the National Mastitis Council Inc.*, Orlando, VA, p. 168.

Part III

Passively Transferred Anti-Infectious Agents of Milk

Chapter 9

Colostrum and Milk in the Treatment of Disease

Saeed Hosseini, Paula Inserra,
Mohsen Araghi-Niknam, and Ronald R. Watson

1. Introduction

Although pharmaceutical companies have come close to duplicating breast milk, none has a formula that exactly mimics the effects of mother's milk on the infant. These formulas lack the hormones, the growth factors and the passive and active protection of the neonate's intestinal tract against microorganisms and antigens provided by breast milk. The frequency of many infections is lower in breast-fed neonates than in formula-fed babies (Goldblum and Goldman, 1994; Carlsson and Hanson, 1994). In addition, breast-fed infants are less susceptible than formula-fed infants to insulin-dependent diabetes mellitus (IDDM) (Mayer *et al.*, 1988) as well as Crohn's Disease (Koletzko *et al.*, 1989) and they have a much reduced risk of allergic disease (Chandra, 1979). In addition to providing nutritional benefits and passive immunity against microbial infection, breast milk also influences development of the active immune system in the newborn (Mandalapu *et al.*, 1995). Commercial formulas can meet the nutritional demands of the neonate; however, the immunological properties of breast milk are unique.

Saeed Hosseini, Paula Inserra, Mohsen Araghi-Niknam, and Ronald R. Watson • Arizona Prevention Center, P.O. Box 245155, 1501 N. Campbell Ave, University of Arizona, Tucson, AZ 85724

2. Protective Factors in Milk

Colostrum is produced in humans during the first two to four days postpartum. It contains immunoglobulins (Ig), components of the complement system, carrier proteins, bifidus factors, epithelial growth factors, enzymes and hormone-like substances. It also contains B and T lymphocytes, particularly memory T-cells (Bertotto *et al.*, 1990), monocytes, macrophages, polymorphonuclear leukocytes and epithelial cells. The immunoglobulins of colostrum include IgA and IgM antibodies to autoantigens such as native DNA, transferrin, thyroglobulin, actin, myosin, myoglobin and laminin (Vassilev and Veleva, 1996).

The perinatal period is critical for the maturation of the immune repertoire of the child. In the last weeks prior to delivery maternal IgG enters the circulation of the unborn child by transplacental passage (Carlsson and Hanson, 1994). Thus, the immune system of full-term, breast-fed neonates develops in the presence of Ig from the maternal blood and milk. Preterm, low birth weight infants are deficient in maternal IgG. Therefore breast-feeding ensures passive immunization with secretory Ig specific to foreign and self-antigens. A formula-fed infant is left without these protective factors (Goldblum and Goldman, 1994).

Human and bovine colostrum contains colostral inhibitory factor, (CIF) which depresses IL-2 production *in vitro*. CIF does not belong to the transforming growth factor (TGF) family and its mode of immunodepressive action differs from that of TGF-β (Mandalapu *et al.*, 1995). CIF may play a role in the prevention of inappropriate cellular immune responses in the neonate that give rise to conditions such as IDDM (Mayer *et al.*, 1988) and Crohn's disease (Koletzko *et al.*, 1989). Thus, it could aid in the proper development of the immune system. CIF no longer appears in milk after several weeks of lactation (Mandalapu *et al.*, 1995), whereas TGF-β, both B1 and B2, is found, at least in bovine milk, at all stages of lactation (Jin *et al.*, 1991).

Colostrum also contains 10^5-10^7 white blood cells per mL, a level that decreases five-fold to ten-fold by the eighth day postpartum. The macrophages of human milk are more motile than blood monocytes *in vitro* (Ozkaragoz *et al.*, 1988) and they produce several host resistance factors, including lysozyme, prostaglandin, C3 and C4 components of complement and lactoferrin. Prostaglandin E2 is present in large quantities in human milk and exerts numerous influences on the gastrointestinal tract including promotion of mucosal integrity, stimulation of intestinal peristalsis and release of brush-border enzymes (Mirhom *et al.*, 1977).

The T lymphocytes in human milk include both CD4+ (helper/inducer) cells and CD8+ (cytotoxic/suppresser) cells. The ratio of CD4+ cells to CD8+ cells is similar to, or lower than, that found in human peripheral blood (Richie *et al.*, 1982; Keller *et al.*, 1986). T cells in milk are highly reactive to organisms invading the gastrointestinal tract. Breast milk antibodies are directed against gastrointestinal or respiratory tract pathogens with which the mother has come into contact. In humans, T and B lymphocytes sensitized by gastrointestinal or respiratory tract pathogens are transported from the maternal gastrointestinal or bronco-tracheal lymphatic tissues to the mammary gland, where the B cells differentiate to become plasma cells and secrete antibodies (Orlando, 1995). Therefore, the antibodies found in milk result from maternal exposure to microorganisms that should be similar to the flora with which the infant will come in contact.

Secretory IgA constitutes 90% of the Ig found in human milk (Goldman *et al.*, 1982), and it is widely accepted as being important in reducing the incidence and severity of enteric infections among breast-fed infants. IgA is probably the main anti-microbial factor found in human milk. Its major action in the infant's intestinal tract is to block the adhesion of enteropathogenic bacteria, viruses and bacterial enterotoxins to epithelial cells (Cravito *et al.*, 1991). As the volume of milk produced increases in the early stages of lactation, the concentration of IgA declines to a plateau level that remains higher than maternal serum IgA levels.

3. Anti-Bacterial, Anti-Parasitic, and Anti-Viral Effects of Milk

In many developing countries, enteropathogenic *E. coli* (EPEC) is the main cause of acute diarrhea in infants under one year of age (Levine, 1987). EPEC organisms adhere to cup-like projections of the apical enterocyte membrane and cause localized destruction of brush border microvilli and development of lesions. *In vitro*, EPEC strains can attach to HeLa or Hep-2 cells (Scaletsky *et al.*, 1984). Camara *et al.* (1995) showed that delipidated colostrum can strongly inhibit the adherence of EPEC strains to HeLa cells but that colostrum without IgA does not inhibit bacterial adherence. A 94KDa EPEC surface protein appears to be recognized by IgA antibodies found in milk (Camara *et al.*, 1995). Therefore, it was suggested that milk IgA likewise inhibits EPEC adherence to epithelial cells. Since colostrum heated to 56°C still offered protection against EPEC organisms, heat-labile factors such as complement proteins appeared not to be involved (Camara *et al.*, 1995). Apart from IgA, carbohydrate residues in milk may affect attachment of some diarrheagenic *E. coli* to the intestinal mucosa (Ashkenazi, 1994). Attachment occurs between bacterial surface proteins and sugar residues of intestinal cell receptors (Ofek and Beachey, 1980), and some milk glycoconjugates appear to act as receptor analogues that interfere with enteric bacterial binding to epithelial cells of the intestine (Ashkenazi and Mirelman, 1987;

Ashkenazi *et al.*, 1991). Similar mechanisms might be used against other bacteria (Anderson *et al.*, 1986) or toxins (Otnaess *et al.*, 1983).

Rotavirus infection is a well-known cause of acute diarrhea in infants and children throughout the world. The infection is characterized by vomiting, watery diarrhea and fever. Careful evaluation has demonstrated that breast-fed infants exhibit milder symptoms of rotavirus infection than formula-fed infants (Duffy *et al.*, 1986). Polymeric IgA in human milk can act in harmony with other antiviral milk components. Bell *et al.* (1988) showed that rotavirus-specific IgA and IgG antibodies remain at high levels for many months postpartum. However, the prevention of rotavirus infection during the early neonatal period is reported to depend on the combined presence of anti-viral IgA and trypsin inhibitors in human milk (Jayashree *et al.*, 1988). IgA of human colostrum and breast milk has also been shown to inhibit seroconversion by breast-fed infants in response to oral poliovirus vaccine (Plotkin *et al.*, 1966). Consequently, cessation of breast-feeding for several hours before and after rotavirus immunization may be needed, as is the practice for immunization against the poliovirus.

Breast-feeding also provides protection against at least some respiratory viruses including *Respiratory syncytial virus*, a significant cause of infection in early infancy. In this connection, polymeric IgA antibodies specific to a surface protein of this virus appeared to suppress viral replication in the upper respiratory tract of suckling infants (Tsutsumi *et al.*, 1989). In contrast, milk secretory IgA against *rubella* (Losonsky *et al.*, 1982) and *Cytomegalovirus* (Stango *et al.*, 1980; Dworsky *et al.*, 1983) exerted no neutralizing or other protective effects. Despite the presence of specific antibodies in human milk against *Herpes simplex* virus, the importance of this antibody is unknown (Lin *et al.*, 1993).

4. Milk Produced by Hyper-Immunized Dairy Cows

Repetitive injection of antigen achieves a heightened state of immunity that can be termed hyper-immunization. Dairy cows hyper-immunized with a polyvalent bacterial vaccine secrete a variety of biologically active substances into their milk, including a potent hyper-immune milk anti-inflammatory factor (HIMF) (Beck, 1990). HIMF has been shown to suppress inflammation by inhibiting neutrophil emigration from the vasculature (Ormrod and Miller, 1991, 1992). Thus, this factor may be useful for the inhibition of tissue destructive infectious processes and in situations requiring suppression of lymphocyte functions (Ormrod and Miller, 1992).

Milk obtained from cows hyper-immunized with a variety of human gut bacteria (*E. coli, S. dysenteriae, S. typhimurium, S. pyogenes, A. aerogenes, K. pneumoniae, H. influenzae, P. vulgaris, S. mittis* and others) exhibited high antibody activities specific for intestinal bacteria (Golay *et al.*, 1990) and protected mice from radiation-induced lethality when administered orally (Ishida *et al.*, 1992). Although the mechanism of this protective action was not elucidated, other studies suggest it is due to an inhibitory effect of immunologically-enriched milk on the growth of intestinal *Enterobacteriaceae* and on the establishment of EPEC bacterial infection (Tacket *et al.*, 1988). Two hypotheses may be proposed regarding the mechanisms whereby the invasion of enteric *E. coli* was prevented. Firstly, milk antibodies may have activated complement components to lyse the bacteria or to promote phagocytosis by PMN and/or macrophages in the intestinal lumen. Secondly, milk antibodies may have prevented the bacteria from adhering to and invading the intestinal mucosal membrane (Janeway, 1988; Fergusen, 1990). Immunologically-enriched milk may also reduce the number of bacteria translocating from the gut and augment the effectiveness of gut-associated lymphoid tissues against the invasion of intestinal bacteria, thereby protecting against radiation-induced mortality (Ishida *et al.*, 1992). Thus, immunologically-enriched milk may be useful prophylactically to control invasion by intestinal microflora during radiotherapy and chemotherapy (Ishida *et al.*, 1992).

Enterohemorrhagic *Escherichia coli* (EHEC) is a pathogen of the human gastrointestinal tract. As many as 10% of children with EHEC- associated diarrhea develop extra-intestinal sequelae, among which the hemolytic uremic syndrome is the most common (Tarr, 1995). Severe complications such as long-term renal or cerebral damage can occur in up to 30% of patients with the hemolytic uremic syndrome (Siegler *et al.*, 1991). Cows are a natural reservoir for EHEC; there is a high prevalence of this type of organism in apparently healthy animals (Chapman *et al.*, 1994). The transfer of SLT-specific Ig via colostrum is an important aspect of the early passive immunization of the newborn calf against EHEC infections (Pirro *et al.*, 1995). Inasmuch as bovine colostrum preparations are safe and effective agents for the prevention and treatment of diarrhea in humans caused by rotavirus (Brussow *et al.*, 1987), enterotoxigenic *E. coli* (Tacket *et al.*, 1988) and the protozoan parasite *Cryptosporidium parvum* (Nord *et al.*, 1990), bovine colostrum appears to have potential in the treatment of EHEC infections also.

Shiga-like toxins (SLTs) are associated with the development of hemorrhagic colitis as well as systemic complications in EHEC infections. However, some symptoms of these infections cannot be explained by the action of SLT alone (Boyce *et al.*, 1995). In addition to SLT, most EHEC strains express the protein EHEC-hemolysin (EHEC-Hly) (Schmidt *et al.*, 1995). Belonging to the pore-forming branch of Repeats in ToXin (RTX) exo-proteins, EHEC-Hly is cytolytic for erythrocytes and for cultures of human and animal cells (Bauer and Welch, 1996). Antibodies that neutralize SLTs are found in bovine colostrum (Pirro *et al.*, 1995). Lissner *et al.* (1996) investigated a commercial product designated Lactobin®, which is prepared from hyper-immune bovine colostrum that has high anticryptosporidial activity and antibody titers against EHEC-Hly and SLTs according to *in vitro* assays for cytolytic activity. Immunoblotting showed high IgG reactivity against potentially pathogenic factors associated with EHEC infections. Moreover, Lactobin® neutralized EHEC SLTs. These findings have therapeutic implications because the oral administration of Lactobin® not only may modify the severity of diarrheal illness in humans but also prevent the hemolytic uremic syndrome in EHEC-infected patients (Lissner *et al.*, 1996).

Cryptosporidium parvum is an intracellular protozoan parasite that infects a wide range of mammals including humans. The organism causes transient, mostly asymptomatic and usually self-limited infection in immunocompetent individuals (DuPont *et al.*, 1995). In contrast, immunocompromised individuals can develop persistent, severe diarrhea that leads to volume depletion, malabsorption, weight loss and possibly death. For example, *C. parvum* is an extremely common cause of diarrhea in AIDS patients. As there is no antibiotic therapy for this parasite, its clearance by the immune system is critical, and both humoral and cellular immunity are involved (Mead *et al.*, 1991). *C. parvum* is a common bovine infection against which cows are naturally immunized. Thus, oral passive immunotherapy with bovine colostral-derived Ig to *C. parvum* has potential as a therapeutic strategy to address a difficult disease management problem.

In vitro and animal studies using either neonatal or genetically immune deficient mice suggest that colostrum from cows immunized with *C. parvum* oocysts contains high antibody titers that are effective in combatting *C. parvum* infection (Fayer *et al.*, 1990; Perryman *et al.*, 1990; Flanigan *et al.*, 1991; Riggs *et al.*, 1994). Greenberg and Cello (1996) administered a bovine Ig concentrate to *C. parvum*-infected AIDS victims with chronic diarrhea. The Ig concentrate was prepared from colostrum obtained from the first six postpartum milkings of cows immunized with *C. parvum* oocysts. It had an anti-*C. parvum* antibody titre 16- to 20-fold greater than the activity found in a control product that was not derived from hyper-immune colostrum. The colostrum-derived product produced a significant reduction in stool frequency and weight, and promoted stabilization of body weight among the patients.

Despite the successful use of hyper-immunized bovine colostrum under experimental immunoprophylactic and immunotherapeutic conditions, the biologically active substances in hyper-immunized bovine colostrum have not been determined. IgA and IgG1 in hyper-immunized bovine colostrum have been reported to be more effective than IgG2 or IgM (Fayer *et al.*, 1990). Other work has shown that bovine colostrum with undetectable levels of anti-Cryptosporidium antibodies can reduce Cryptosoporidial colonization of mice immunocompromised with murine AIDS (Watzl *et al.*, 1993). In this investigation, immunocompromised mice shed higher numbers of fecal oocysts and developed a persistently higher intestinal parasite load than immunocompetent controls, which were essentially refractory to cryptosporidiosis. Feeding non-immune bovine colostrum for five days reduced fecal oocyst shedding and cryptosporidial colonization of the intestine in the immunocompromised mice. Other workers have reported that colostrum from unimmunized cows offers protection against cryptosporidiosis (Fayer *et al.*, 1990; Perryman *et al.*, 1990). Colostrum from non-immunized cows may have a low level of anti-Cryptosporidial antibodies from natural exposure. Prophylactic efficacy against *Cryptosporidium* may be augmented by the presence of cytokines such as INF-γ and IL-1 (Svedberg *et al.*, 1985; Soder, 1987).

Although the focus of many passive immunization studies has been on enteric infections, attention recently has been directed toward passive immunization against oral bacterial infections. For example, monoclonal antibodies directed against surface molecules of *Streptococcus* reduced the level of this cariogenic bacterium in the mouths of rhesus monkeys (Lehner *et al.*, 1985) and humans (Ma *et al.*, 1987). Likewise, hens' egg yolk antibodies from chickens immunized with *S. mutans* reduced caries formation in specific pathogen-free rats carrying the *S. mutans* bacterium (Otake *et al.*, 1991). Furthermore, rabbit antiserum directed against the periodontopathic organism *Porphyromonas gingivalis* prevented colonization of the oral cavities of hamsters by this pathogen (Okuda *et al.*, 1988). Similarly, antibodies prepared from the milk of cows immunized with *S. mutans* reduced oral *S. mutans* infection and subsequent caries formation in rats (Michalek *et al.*, 1987), and when used in an oral rinse reduced the numbers of *S. mutans* organisms in human dental plaque (Filler *et al.*, 1986, 1991). Preliminary evidence indicates that bovine antibodies exert their protective influence at least partly through inhibition of enzyme activities in the infecting bacteria, e.g., glucosyltransferase and lactate dehydrogenase (Filler *et al.*, 1991), although blocking and agglutination activities probably also contribute. Cows can be immunized with oral bacteria to yield milk preparations with high titers of IgG2, IgA and IgM, and with particularly high concentrations of IgG1, directed against the immunizing microbial strains. These antibodies exhibit minimal cross-reaction with other oral organisms (Takahashi *et al.*, 1992) and large quantities can be prepared against selected members of the oral flora.

5. Conclusion

More than twenty years ago, Hilpert *et al.* (1977) pointed out that colostrum produced by hyper-immunized cows could open the way to new preventive and therapeutic modalities for infectious disease. The utilization of bovine antibodies for passive immunization of humans has several advantages. Large quantities of antibody can be obtained readily and the specificity of the preparations can be controlled. In addition, the preparations are robust during processing, can be stored for long periods of time without loss of activity and the products can be ingested safely by humans. Evidence is now available from studies on humans as well as animals that bovine colostral products can be useful tools in the management of refractory infections even in immunocompromised subjects.

References

Anderson, B., Porras, O., Hanson L.A., Lagergard, T., and Svanborg-Eden, C. 1986. Inhibition of attachment of Streptococus pneumoniaeand Haemophilus influenza by human milk. *J. Infect.* Dis. **153**:232.

Ashkenazi, S. 1994. A review of the effect of human milk fractions on the adherence of diarrheogenic Escherichia coli to the gut in animal model. *Isr. J. Med. Sci.* **30**:335.

Ashkenazi, S. and Mirelman, D. 1987. Non immunoglobulin fractions of human milk inhibits the adherence of certain enterotoxigenic Escherichia coli strains to guinea pig intetinal cells. *Pediatr. Res.* **22**:130.

Ashkenazi, S., Newburg, D.S., and Cleary, T.G. 1991. The effect of human milk on the adherence of enterohemorrhagic E. coli to rabbit intestinal cells. *Adv. Exp. Med. Bio.* **310**:173.

Bauer, M.E. and Welch, R.A. 1996. Characterization of an RTX toxin from enterohemorrhagic Escherichia coli O157: H 7. *Infect. Immunol.* **64**:167.

Beck, L.R. 1990. Method of treating inflammation using bovine milk. United States patent number 4,956,349.

Bell, L.M., Clark, H.F., Offit, P.A., Slifht, P.H., Arbeter, A.M., and Plotkin, S.A. 1988. Rotavirus serotype-specific neutralizing activity in human milk. *Am. J. Dis. Child.* **142(3)**:275.

Bertotto, A., Gerli, R., Fabietti, G., Crupi, S., Arcangeli, C., Scalise, F., and Vaccaro, R. 1990. Human breast milk T lymphocytes display the phenotypeand functional characteristics of memory T cells. *Eur. J. Immunol.* **20**:1877.

Boyce, T.G., Swerdlow, D.L., and Griffin, P.M. 1995. Escherichia coli O157: H7and the hemolytic-uremic syndromr. *N. Engl. J. Med.* **333**:364.

Brussow, H., Hilpert, H., Walther, I., Sidoti, J., Mietens, C., and Bachmann, P. 1987. Bovine milk immunoglobulins for passive immunity to infantile rotavirus gasteroenteritis. *J. Clin. Microbiol.* **25**:982.

Camara, L.M., Carbonare, S.B., Scaletsky, I.C.A., Lourdes Monterio da Silva, M., and Carneiro-Sampaio, M.M.S. 1995. Inhibition of enteropathogenic E. coli (EPEC) adherence to HeLa cells by human colostrum. Detection of specific sIgA related to EPEC outer-membrane proteins, in: *Advances in Mucosal Immunology* (J. Mestecky, S. Russel, S. Jakson, S.M. Michalek, H. Tlaskalova, and J. Sterzl, eds.), pp. 673–676, Plenum Press, New York.

Carlsson, B. and Hanson, L.A. 1994. Immunologic effects of breast feeding on the infant, in: *Handbook of Mucosal Immunology* (P.L. Ogra, W. Strober, G.R. McGhee, M.E. Lamm, and J. Bienenstock, eds.), pp. 653–660, Academic Press, San Diego.

Chandra, R.K. 1979. Prospective studies of the effect of breast feeding on incidence of infectionand allergy. *Acta Paediatr. Scand.* **68**:691.

Chapman, P.A., Wright, D.J., and Siddons, C.A. 1994. A comparison of immunomagnetic separationand direct culture for the isolation of verocytotoxin-producing Escherichia coli O157 from bovine feces. *J. Med. Microbiol.* **40**:424.

Cravito, A., Tello, A., Villafan, H., Ruiz, J., del Vedovo, S., and Neeser, J.R. 1991. Inhibition of localized adhesion of enteropathogenic E. coli to HEP-2 cells by immunoglobolinand oligosaccharide fractions of human colostrumand breast milk. *J. Infect. Dis.* **163**:1247.

Duffy, L.C., Byers, T.E., Reipenhoff-Talty, M., La Scolea, O.L.J., Zielezny, M., and Ogra, P.L. 1986. The effects of infant feeding on rotavirus-induced gasteroentritis. A prospective study. *Am. J. Pub. Health* **76**:259.

DuPont, H.L., Chappell, C.L., Sterling, C.R., Okhuysen, P.C., Rose, J.B., and Jakuboswki, W. 1995. The infectivity of Cryptosporidium parvum in healthy volunteers. *N. Engl. J. Med.* **325**:855.

Dworsky, M., Yow, M., Stango, S., Pass, R.F., and Alford, C. 1983. Cytomegalovirus infection of breast milkand transmission in infancy. *Pediatrics* **72**:295–299.

Fayer, R., Guidry, A., and Blagburn, B.L. 1990. Immunotherapeutic efficacy of bovine colostral immunoglobulins from a hyperimmunized cow against cryptosporidiosis in neonatal mice. *Infect. Immun.* **58**:2962.

Fergusen, A. 1990. Mucosal immunology. *Immunol. Today* **11**:1.

Filler, S.J., Gregory, R.L., Michalek, S.M., Katz, J., and McGhee, J.R. 1986. Effect of bovine milk rinse containing antibodies to Streptococcus mutans on human plaque, in: *Molecular Microbiologyand Immunobiology of Streptococcus mutans* (S. Hamada, S. Michalek, H. Kiyono, L. Menaker, and J. McGhee, eds.), pp. 475–483. Elsevier Science Publishers, New York.

Filler, S.J., Gregory, R.L., Michalek, S.M., Katz, J., and McGhee, J.R. 1991. Effect of immune bovine milk on Streptococcus mutans in human dental plaque. *Arch. Oral Biol.* **36**:41.

Flanigan, T., Marshall, R., Redman, D., Kaetzel, C., and Ungar, B. 1991. In vitro screening of therapeutic agents against Cryptosporidium: hyperimmune cow colostrum is highly inhibitory. *J. Protozool.* **38**:225S.

Golay, A., Ferrara, J.M., Felber, J.P., and Schneider, H. 1990. Cholestrol-lowering effect of skim milk from immunized cows in hypercholestrolemic patients. *Am. J. Clin. Nutr.* **52**: 1014.

Goldblum, R.M. and Goldman, A.S. 1994. Immunological components of milk: formationand function, in: *Handbook of Mucosal Immunology* (P.L. Ogra, W. Strober, G.R. McGhee, M.E. Lamm, and *J. Bienenstock*, eds.), pp. 643–652. Academic Press, San Diego.

Goldman, A.S., Garza, C., Nichols, B.L., and Goldblum, R.M. 1982. Immunologic factors in human milk during the first year of lactation. *J. Pediatr.* **100(4)**:563.

Greenberg, P.D. and Cello, J.P. 1996. Treatment of severe diarrhea caused by *Cryptosporidium parvum* with oral bovine immunoglobulin concentrate in patients with AIDS. *J. Acquired Immune Deficiency Syndromesand Human Retrovirology* **13**:348.

Hawes, C.S. and Jones, W.R. 1985. Human milk cell migrationand production of monocyte chemotactic factor. Lack of activity. *Pediatr. Res.* **19**:996.

Hilpert, H., Gerber, H., Amster, H., Pahud, J.J., Ballabriga, A., Arcalis, L., Farriaux, F., Peyer, E., and Nussle, D. 1977. Bovine milk Ig, their possible utilization in industrially prepared infant's formulae, in: *Foodand Immunology. Proceedings of a Symposium Co-sponsored by the Swedish Medical Research* Council, 1975, June 9–11 (L. Hambraeusand H. McFarlane, eds.), pp. 182–196. Almqvist & Wiskell, Stockholm.

Ishida, A., Yoshikai, Y., Murosaki, S., Hidaka, Y., and Nomoto, K. 1992. Administration of milk from cows immunized with intestinal bacteria protects mice from radiation-induced lethality. *Biotherapy* **5**:215.

Janeway, C.A. 1988. Frontiers of the immune system. *Nature* **333**:804.

Jayashree, S., Bhan, M.K., Kumar, R., Bhandari, N., and Sazawal, S. 1988. Protection against neonatal rotavirus infection by breast milk antibodiesand trypsin inhibitors. *J. Med. Virol.* **26**:333.

Jin, Y., Cox, D.A., Knecht, R., Raschdorf, F., and Cerletti, N.J. 1991. *Protein Chem.* **10**:565.

Keller, M.A., Faust, J., Rolewic, L.J., and Stewart, D.D. 1986. T-cell subsets in human colostrum. *J. Pediatr. Gasteroentrol. Nutr.* **5**:439.

Koletzko, S., Sherman, P., Corey, M., Griffiths, A., and Smith, C. 1989. Role of infant feeding practices in development of Crohns disease in childhood. *Br. Med. J.* **298**:1617.

Lehner, T., Caldwell, J., and Smith, R. 1985. Local passive immunization by monoclonal antibodies against streptococcal antigen I/II in the prevention of dental caries. *Infect. Immun.* **50**:796.

Levine, M.M. 1987. Escherichia coli that cause diarrhea: entrotoxigenic, entropathogenics, entroinvasive, entrohemorrhagicand entroadherent. *J. Infect. Dis.* **155**:377.

Lissner, R., Schmidt, H., and Karch, H. 1996. A standard immunoglobulin preparation produced from bovine colostra shows antibody reactivityand neutralization activity against Shiga-like toxinsand EHEC-hemolysin of Escherichia coli O157: H7. *Infection* **24**:378.

Liu, A., Sancher-Prescador, S., Ruiz-Palacios, G., Morrow, A.L., Pickering, L.K., and Kohl, S. 1993. A companion of herpes simplex virus specific antibodies in human milkand serum. *Pediatr. Res.* **31**:591.

Losonsky, G.A., Fishaut, J.M., Strussenberg, J., and Ogra, P.L. 1982. Effect of immunization against rubella on lactation products. I. Developmentand characterisation of specific immunologic reactivity in breast milk. *J. Infect. Dis.* **145**:654.

Ma, JK-C., Smith, R., and Lehner, T. 1987. Use of monoclonal antibodies in local passive immunization to prevent colonization of human teeth by Streptococcus mutans. *Infect. Immun.* **55**:1274.

Mandalapu, P., Pabst, H.F., and Paetkau, V. 1995. A novel immunosuppressive factor in human colostrum. *Cell Immunol.* **162(2)**:178.

Mayer, E.J., Hamman, R.F., Gay, E.C., Lezotte, D.C., Savitz, D.A., and Klingensmith, G.J. 1988. Reduced risk of IDDM among breast-fed children. The Colorado IDDM Registry. *Diabetes* **37**:1625.

Mead, J.R., Arrowood, M.J., Sidewell, R.W., and Healey, M.C. 1991. Chronic Cryptosporidium parvum infections in congenitally immunodeficient SCIDand nude mice. *J. Infect. Dis.* **163**:1297.

Michalek, S.M., Gregory, R.L., Harmon, C.C., Katz, J., Richardson, G.J., Hilton, T., Filler, S.J., and McGhee, J.R. 1987. Protection of gnotobiotic rats against dental caries by passive immunization with bovine milk antibodies to Streptococcus mutans. *Infec. Immun.* **55(10)**:2341.

Mirhom, R., Eloy, R., Vaultier, J.P., Ananna, A., Clendinnen, G., and Grenier, J.F. 1977. Prostaglandin E1and E2 stimulate release of intestinal brush border enzymes. *Prostaglandins* **14(3)**: 463.

Nord, J., Ma, P., DiJohn, D., Tzipori, S., and Tacket, C.O. 1990. Treatment with bovine hyperimmune colostrum of cryptosporidial diarrhea in AIDS patients. *AIDS* **4**:581.

Ofek, I. and Beachey, E.H. 1980. Bacterial adherence. *Adv. Intern. Med.* **25**:503.

Okuda, K., Kato, T., Naito, Y., Takazoe, I., Kikuchi, Y., Nakamura, T., Kiyoshige, T., and Sasaki, S. 1988. Protective efficacy of activeand passive immunizations against experimental infection with Bacteroides gingivalis in ligated hamsters. *J. Dent. Res.* **67(5)**:807.

Orlando, S. 1995. The immunologic significance of breast milk. *J. Obstetrics, Gynecologyand Neonatal Nursing* **24**:678.

Ormrod, D.J. and Miller, T.E. 1991. The anti-inflammatory of a low molecular weight component derived from the milk of hyperimmunized cows. *Agentsand Actions* **32**: 160.

Ormrod, D.J. and Miller, T.E. 1992. The anti-inflammatory of a low molecular weight component derived from the milk of hyperimmunized cows suppress inflammation by inhibiting neutrophil emigration. *Agentsand Actions* **37**:70.

Otake, S., Nishihara, Y., Makimura, M., Hatta, H., Kim, M., Yamamoto, T., and Hirasawa, M. 1991. Protection of rats against dental caries by passive immunization with hen-egg-yolk antibody (IgY). *J. Dent. Res.* **70(3)**:162.

Otnaess, A.B., Laegreid, A., and Ertresvag, K. 1983. Inhibition of enterotoxin from Escherichia coliand Vibrio cholerae by gangliosides from human milk. *Infect. Immun.* **40**:563.

Ozkaragoz, F., Rudloff, H.B., Rajaraman, S., Mushtaha, A.A., Schmalstieg, F.C. and Goldman, A.S. 1988. The motility of human milk macrophages in collagen gels. *Pediatr. Res.* **23(5)**:449.

Perryman, L E., Riggs, M.W., Mason, P.H., and Fayer, R. 1990. Kinetics of Cryptosporidium parvum sporozoite neutralization by monoclonal antibodies, immune bovine serumand immune bovine colostrum. *Infect. Immun.* **58**:257.

Pirro, F., Wieler, L.H., Failing, K., Bauerfeind, R., and Baljer, G. 1995. Neutralizing antibodies against Shiga-like toxins from Escherichia coli in colostraand sera of cattle. *Vet. Microbiol.* **43**:131.

Plotkin, S.A., Katz, M., Brown, R.E., and Pagano, J.S. 1966. Oral poliovirus vaccination in newborn African infants. *Am. J. Dis. Child.* **111(1)**:27.

Richie, E.R., Bass, R., Meistrich, M.L., and Dennison, D.K. 1982. Distribution of T-lymphocyte subsets in human colostrum. *J. Immunol.* **129(3)**:1116.

Riggs, M.W., Cama, V.A., Leary, H.L., and Sterling, C.R. 1994. Bovine antibody against Cryptosporidium parvum elicits a circumsporozite precipitate-like reactionand has immunotherapeutic effect against persistent cryptosporidiosis in SCID mice. *Infect. Immun.* **62**:1927.

Scaletsky, I.C., Silva, M.L., and Trabulsi, L.R. 1984. Distinctive patterns of adherence of enteropathogenic Escherichia coli to HeLa cells. *Infec. Immun.* **45**:534.

Schmidt, H., Beutin, L., and Karch, H. 1995. Molecular analysis of the plasmid-encoded hemolysin of Escherichia coli O157:H7 strain EDL 933. *Infect. Immunol.* **63**:1055.

Siegler, R.L., Milligan, M.K., Burningham, T.H., Christoferson, R.D., Chang, S.Y., and Jorde, L.B. 1991. Long-term outcomeand prognostic indicators in the hemolytic-uremic syndrom. *J. Pediatr.* **118**:195.

Soder, O. 1987. Isolation of interleukin-1 from human milk. *Int. Arch. Allergy Appl. Immunol.* **83**:19.

Stango, S., Reynolds, D.W., Pass, R.F., and Alford, C. 1980. Breast milkand the risk of cytomegalovirus infection. *New Engl. J. Med.* **302**:1073.

Svedberg, J., Deltaas, J., Leimenstoll, G., Paul, F., and Teschemacher, H. 1985. Demonstration of beta-casomorphin immunoreactive materials in vitro digests of bovine milkand in small intestinal contents of bovine milk ingestion in adult humans. *Peptides* **6**:825.

Tacket, C.O., Losonsky, G., Link, H., Hoang, Y., Guesry, P., Hilpert, H., and Levine, M.M. 1988. Protection by milk immunoglobulin concentrate against oral challenge with enterotoxigenic Escherichia coli. *N. Engl. J. Med.* **318**:1240.

Takahashi, N., Eisenhuth, G., Lee, I., Laible, N., Binion, S., and Schachtele, C. 1992. Immunoglobulins in milk from cows immunized with oral strains of Actinomyces, Prevotella, Porphyromonasand Fusobacterium. J. Dent. Res. **71**:1509.

Tarr, P.I. 1995. Escherichia coli O157:H7, clinical, diagnosticand epidemiological aspects of human infection. *Clin. Infec. Dis.* **20**:1.

Tsutsumi, H., Honjo, T., Nagai, K., Chiba, Y., Chiba, S., and Tsugawa, S., 1989. Immunoglobulin A antibody response to respiratory syncytial virus structural proteins in colostrumand milk. *J. Clin. Microbiol.* **27(9)**:1949.

Vassilev, T.L. and Veleva, K.V. 1996. Natural polyreactive IgAand IgM autoantibodies in human colostrum. *Scand. J. Immunol.* **44**:535.

Watzl, B., Huang, D.S., Alak, J., Darban, H., Jenkins, E.M., and Watson, R.R. 1993. Enhancement of resistance to Cryptosporidium parvum by pooled bovine colostrum during murine retroviral infection. *Am. J. Trop. Med. Hyg.* **48**:519.

Chapter 10

The Role of Human Milk Secretory IgA in Protecting Infants from Bacterial Enteritis

Marita Noguera-Obenza and Thomas G. Cleary

1. Introduction

In most of the world diarrhea is the leading cause of death in infancy. The risks of bacterial enteritis include acute life-threatening fluid losses, renal failure, and impaired growth. Exposure to bacterial enteropathogens varies geographically. In areas where sanitation is poor a variety of pathogens occur at high frequency. In more developed countries a limited range of pathogens is found with only organisms easily spread by direct contact or through contaminated food being common.

Human milk is generally believed to provide the suckling infant with protection from bacterial enteritis (Feacham and Koblinsky, 1984; Victora *et al.*, 1989; Hanson *et al.*, 1991). The basis for this defense mechanism is complex. Breast-feeding serves as an uncontaminated source of food so that there is decreased exposure to enteropathogens. It contains pathogen-specific secretory IgA (sIgA) and non-specific protective factors including cell receptor analogs (glycolipids, glycoproteins, oligosaccharides), and anti-infectives such as lactoferrin and lysozyme (Pickering and Kohl, 1986). Breast-feeding lowers stool pH so that acid-intolerant enteropathogens fail to thrive in the intestinal milieu. The best studied of the milk factors is sIgA.

Marita Noguera-Obenza and Thomas G. Cleary • University of Texas-Houston Health Science Center, Department of Pediatrics, Infectious Diseases Division, 6431 Fannin, JFB 1.739, Houston, Texas 77030

sIgA in human milk is produced as a result of prior maternal experience with infectious agents. Bacterial virulence factors presented to maternal Peyer's patches and lymphoid follicles are taken up by M cells, processed, and presented to lymphocytes. These primed sIgA-secreting cells then migrate transiently through the blood and "home" to sites along the intestine where pathogens may again present themselves. In addition, there is homing of these B-lymphocytes to the mammary gland where, during lactation, they produce pathogen-specific sIgA (the enteromammary axis of the common mucosal immune system) (Roux *et al.*, 1977; Qadri *et al.*, 1998). Thus, the infant's feedings are rich in sIgA antibodies to the broad range of virulence antigens to which the mother has had exposure. Much of this antibody survives passage through the infant's gut intact despite digestive enzymes, stomach acid, and bacterial IgA proteases.

The focus of this chapter is on pathogen-specific sIgA found in human milk. For each of the major bacterial enteropathogens, there are several possible levels of understanding of sIgA. For many pathogens there are descriptive data regarding the presence or absence of antibodies to specific virulence antigens. In some cases, there are tissue culture and animal model studies demonstrating the potential of antigen-specific sIgA to provide protection. For a few bacterial enteropathogens there are epidemiologic data showing that specific antibodies are related to infant protection. Finally, there are scattered reports of attempts to manipulate maternal immunity through vaccination to better protect infants. For several major bacterial enteropathogens (*Clostridium difficile, Yersinia enterocolitica*) there are few data regarding the role of human milk secretory IgA in infant health.

2. *Vibrio cholerae*

Cholera is a clinical syndrome caused by intestinal infection with specific serotypes of *Vibrio cholerae* in which severe, often life-threatening, water and electrolyte losses occur in stool. In much of the world cholera is a major cause of childhood death. The organism has the ability to cause watery diarrhea because it adheres to intestinal epithelial cells and delivers a toxin, cholera toxin, that causes ADP ribosylation of an adenylate cyclase regulatory protein (Sears and Kaper, 1996). Although other toxic factors are produced, it is generally thought that cholera toxin is this organism's key virulence factor. The role of human milk in protection has been better studied for cholera than for many other diarrheal diseases. Descriptive, epidemiologic, and immune manipulation data, as well as tissue culture and animal model data on sIgA activity exist.

Human milk has been shown to contain sIgA to cholera toxin (Simhon *et al.*, 1979) and to *V. cholerae* O antigen (Carlsson *et al.*, 1976). The O antigen (also called somatic antigen or lipopolysaccharide (LPS)) has repeating carbohydrate units linked to lipids that form part of the outer membrane lipopolysaccharide of Gram negative bacteria. Despite the lack of cholera in Central America prior to 1991 (Ries *et al.*, 1992), separate pools of colostral and 15- to 30-day post-delivery milk samples collected from women living in Guatemala in the 1970s contained antibodies capable of neutralizing cholera toxin when injected into ligated loops of rabbit ileum [rabbit ileal loop assay]. The amount of protective activity directly correlated with the amount of sIgA present in the milks (Stoliar *et al.*, 1976). The protective activity in these milks was probably related to the frequency of enterotoxigenic *E. coli* [ETEC] in this setting since many ETEC produce heat-labile enterotoxin (LT), a toxin closely related to cholera toxin (Stoliar *et al.*, 1976). LT and cholera toxin share about 80% protein sequence identity (Sears and Kaper, 1996). A key concept in immunity to bacterial enteropathogens is that in the fecal environment virulence traits have been shared by multiple pathogens over time. Cross protective immunity to pathogens that the host has never previously experienced is therefore possible.

There are population-related differences in frequency and amount of cholera toxin neutralizing antibodies in human milk. Holmgren *et al.* (1976) found that severely undernourished Pakistani women had a higher frequency than Swedish women of milks positive for sIgA against cholera toxin detected by enzyme-linked immunoassay (ELISA). However, the Y1 adrenal cell assay to assess enterotoxin neutralizing activity showed only one of eighteen samples from Pakistani women with activity against cholera toxin whereas twelve of eighteen milk samples from Pakistani women showed activity against ETEC LT. ETEC-LT was speculated to be the primary stimulus for production of these antibodies to cholera toxin rather than infection with *V. cholerae*. This may account for the difference in activity against the two toxins in the above assay. Both anti-cholera LPS and anti-cholera toxin antibodies have been demonstrated in colostrum from Indian women at higher titer than in Swedish women. Purified sIgA (separated by gel filtration) from pooled colostrum as well as unprocessed pooled colostrum of Indian women protected in the rabbit ileal loop assay (Majumdar and Ghose, 1982).

There are several epidemiologic studies that have suggested that human milk feedings protect against cholera. In endemic areas, breast-feeding has been associated with a 70% reduction in the risk of developing severe cholera (Clemens *et al.*, 1990a). A retrospective case-control study of an outbreak of El Tor cholera that occurred in Bahrain showed that more bottle-fed than breast-fed infants became ill. However, it was unclear whether protection was due to decreased exposure to the pathogen or to ingestion of a protective factor in the milk (Gunn *et al.*, 1979). This issue was addressed subsequently in a prospective study in which families of patients with cholera were followed with regular stool cultures for *V. cholerae*. The amount of breast milk antibodies to cholera toxin and *V. cholerae* LPS consumed by the infants correlated with presence or absence of symptoms. In those infants who were asymptomatic, *V. cholerae* were still isolated; thus, the antibodies did not prevent colonization of these infants (Glass *et al.*, 1983).

Immunization of Pakistani women with a single dose of cholera vaccine induced a rise in both serum and milk secretory antibodies to *V. cholerae* LPS in the majority of women (70%). Thus, parenteral immunization boosts secretory immunity in women previously primed by natural infection (Svennerholm *et al.*, 1977). Similar data have been found in studies of Pakistani women (Hahn-Zoric *et al.*, 1989). In contrast, a non-immune population of Swedish women failed to develop secretory antibodies in their milk even when immunized twice with parenteral cholera vaccine even though serum IgG antibodies did develop. These studies showed that the secretory immune response to systemic immunization in naturally primed women was different from that of immunologically naive individuals (Svennerholm *et al.*, 1980). Surprisingly, oral immunization with cholera B subunit-whole cell vaccine resulted in little or no sIgA antitoxin responses in milk of Bangladeshi women compared to that which developed following natural infection (29% vs. 90%) (Jertborn *et al.*, 1986; Clemens *et al.*, 1990b).

3. *Shigella spp.*

Shigella spp. infections are among the most serious that occur in infants and toddlers. These infections are characterized by high fever, watery or bloody diarrhea, and toxicity. The clinical features reflect the fact that the fundamental virulence trait of these organisms is their ability to invade the intestinal epithelium.

There is evidence that breast-feeding may decrease the risk and severity of illness due to *Shigella*. Clemens *et al.* (1986) related breast-feeding to illness severity and showed that breast-feeding was protective with an odds ratio of 0.49 (95% CI = 0.28 – 0.86). Protection was evident up to 35 months of age. A community-based study conducted in a rural area of Bangladesh (Ahmed *et al.*, 1992) showed a protective association with an odds ratio relating breast-feeding to confirmed or presumptive shigellosis of 0.48 (95% CI = 0.32 to 0.72). The protection related to breast-feeding decreased with age but was still significant through the third year of life.

Two groups of antigens have been shown to be involved in mucosal immunity against *Shigella*: (1) LPS somatic antigens, and (2) invasion plasmid coded antigens. Secretory IgA against each has been characterized in breast milk.

The concentration of anti-LPS sIgA is highest in colostrum and decreases in the first 6 days of lactation, reaching stable levels thereafter. Although a decline in anti-LPS sIgA titers is reported, they remain detectable as late as the 247th day of lactation in mothers without symptoms of infection (Hayani *et al.*, 1991). This study was done on women living in Houston who are infrequently exposed to *Shigella* and who commonly have anti-*Shigella* LPS antibodies. This presumably reflects a long-term memory of the mucosal immune system rather than repeated antigenic stimulation. Cruz and Arevalo (1985) observed fluctuations in anti-*Shigella* LPS sIgA antibodies independent of total sIgA concentration in a non-uniform and unpredictable manner. In a subsequent study (Cruz *et al.*, 1991) in which they followed levels of sIgA concomitant with routinely collected maternal fecal samples to detect infection, they observed a fall in antibody levels in 7 instances among 6 individuals typically following infections by *Shigella spp.* They postulated that intestinal infections in lactating mothers may account for transient decreases in the concentration of specific sIgA. A fall in specific sIgA during maternal intestinal infection may decrease protection for the suckling infant at a time when exposure may be present.

Titers of class-specific antibodies against *Shigella* antigens appear to reflect exposure of mothers to the prevalent organisms in a particular area. Higher titers of sIgA against LPS have been found in areas with poor socioeconomic conditions where incidence of shigellosis and infections with other bacterial enteropathogens are higher. Achi *et al.* (1992) compared titers of sIgA against LPS antigens of *Shigella dysenteriae* type 1, *Plesiomonas shigelloides* (which has the same O-antigen as *Shigella sonnei*), *Shigella flexneri* type Y, and *Salmonella* serogroup B in colostrum and breast milk of Swedish, Vietnamese and Costa Rican mothers living in various socioeconomic conditions. Their data showed high IgA antibody titers found in Vietnamese mothers particularly against *S. flexneri* type Y. Costa Rican mothers from low and middle socioeconomic classes had a similar pattern. IgA antibodies were low in both Swedish mothers and Costa Rican mothers of high socioeconomic level. Thus, exposure was presumably related to socioeconomic conditions.

Although the presence of sIgA against *Shigella* LPS antigen has been shown in human milk, there are no animal or human models which show its direct effect in curtailing infection or modifying disease. Hyper-immune bovine milk with a high titer to *S. flexneri* 2a strain 2457T LPS orally administered to healthy adult volunteers prevented infection (Tacket *et al.*, 1992).

Virulence plasmids of *Shigella spp.* and enteroinvasive *E. coli* are related. The major antigens encoded by these plasmids have been called invasion plasmid antigens A–D (IpaA–D). Human milk contains sIgA against invasion plasmid-coded antigens (Cleary *et al.*, 1989). This could account for protection across *Shigella* serotypes. The concentration of these antibodies was highest in colostrum. The titers dropped in the first 2 weeks postpartum. Levels stabilized by the first month at a low but constant level, persisting to 665 days of lactation. Titers of sIgA against Ipa antigens also varied in different population groups and geographic areas, presumably as a reflection of exposure. Milk samples from

Mexican mothers contained significantly more sIgA to virulence plasmid antigens than those of women from Houston where *Shigella* infection is less common (Cleary *et al.*, 1991). A similar trend was seen between Vietnamese and Swedish mothers with higher antibody titers in the former. Similar to anti-*shigella* LPS sIgA, higher sIgA antibody titers against invasion plasmid antigens were seen in colostrum of Costa Rican mothers in the low income group compared to those from the high income group (Cam *et al.*, 1992).

The protective effect of antibodies to the virulence plasmid invasion antigens has been evaluated. A prospective, community based study by Hayani *et al.* (1992) followed 316 infants from birth to 24 months. Stool cultures were performed weekly on the infants and breast milk samples collected monthly. Asymptomatic *Shigella* infection in suckling infants was related to higher titer of sIgA against Ipa antigens in the milk collected in the month prior to *Shigella* infection compared to those with symptomatic infection. The infecting serogroups included *S. flexneri*, *S. sonnei* and *S boydii* 2 isolated in both symptomatic and asymptomatic infection. The LPS titers did not differ significantly between those who were symptomatic and those who were asymptomatic.

4. *Escherichia coli*

Five groups of diarrheagenic *Escherichia coli* have been identified based on virulence traits and clinical manifestations (Nataro and Kaper, 1998). The five groups are: enterotoxigenic *E. coli* (ETEC), enteropathogenic *E. coli* (EPEC), enteroinvasive *E. coli* (EIEC), shigatoxin-producing *E. coli* (STEC) (also called enterohemorrhagic *E. coli* (EHEC) or verotoxin-producing *E. coli* (VTEC)) and enteroaggregative *E. coli* (EAggEC). Certain serotypes, as defined by somatic (O) and flagellar (H) antigens, are restricted to one particular pathogenic group. Each group has characteristic virulence factors that mediate adherence (proteins known as adhesins) or invasion, and/or toxin production. These virulence traits confer characteristic clinical manifestations (see Table 1). Studies on specific immunity in human milk have involved antibodies against some of these adhesins and toxins. The studies related to breast milk antibodies against the various diarrheagenic *E. coli* are sparse; this is due to both changing definitions and recognition of new mechanisms of pathogenesis. The repertoire of relevant antigens that need to be studied to characterize immune responses continues to grow. Most of the studies on breast milk sIgA have involved enterotoxigenic *E. coli*; there are no data regarding enteroaggregative *E. coli* where the mechanism of pathogenesis is least defined.

ETEC synthesize heat-labile toxin (LT) and/or heat stable toxin (ST). Heat-labile enterotoxins are high molecular weight proteins composed of a single enzymatic A subunit combined with five B subunits which are involved in receptor

Table 1. Comparison of Diarrhea-Causing *Escherichia coli*

	Virulence factors*	Pathogenic Phenotype	Distinctive clinical manifestation
EPEC	Bundle-forming pilus Secreted proteins: Intimin, tir, espA, espB	Localized adherence Attaching and effacing lesion	Acute watery diarrhea Chronic diarrhea
STEC	Secreted proteins: Intimin, tir, espA, espB Stx 1, Stx 2 Hemolysin	Attaching and effacing lesion	Hemorrhagic colitis Hemolytic uremic syndrome**
ETEC	Heat-stable toxin (ST) Heat-labile toxin (LT) Colonization factor antigens	Enterotoxin production	Acute watery diarrhea
EIEC	Invasion plasmid antigens (IpA)	Invastion	Dysentery
EaggEC	Aggregative adherence fimbria	Aggregative adherence⁺	Acute or persistent watery diarrhea

* List of virulence factors is not exhaustive.
** Syndrome which includes acute renal failure, thrombocytopenia, and hemolytic anemia.
⁺ Stacked brick pattern of adherence on light microscopy.

recognition. LT is structurally and functionally related to cholera toxin. Heat stable enterotoxins (ST) are small non-immunogenic peptides (Sears and Kaper, 1996). Breast milk has been shown to contain antibodies against *E. coli* LT enterotoxin. Using the adrenal cell assay, Holmgren *et al.* (1976) demonstrated ETEC enterotoxin-neutralizing activity in all breast milk samples from a group of malnourished Pakistani women. In contrast, only one sample from a group of Swedish mothers contained neutralizing activity. This may reflect differences in exposure to ETEC between the two groups. *In vivo* protection of breast milk against *E. coli* heat labile enterotoxin was demonstrated by Stoliar *et al.* (1976) using a rabbit ileal loop model. Incubation of either *V. cholerae* toxin or *E. coli* toxin with varying dilutions of colostrum or breast milk from Guatemalan women inhibited fluid accumulation in rabbit ileal loops. There was a linear relationship between the quantity of colostrum used and degree of protection seen. The IgA fraction from colostrum was shown to be the component responsible for this protection. Protective activity persisted for at least the first month of lactation.

Breast-feeding has been correlated with protection against diarrhea caused by ETEC. A community-based prospective study by Long *et al.* (1994) in urban Mexican children revealed that strictly formula-fed infants colonized by ETEC-LT were symptomatic with diarrhea nearly three times as often as strictly breast-fed infants and twice as often as infants receiving mixed diet. The protective effect of breast milk against ETEC-LT diarrhea as related to its sIgA content has been shown in epidemiologic studies. In a prospective study by Cruz *et al.* (1988) of 130 children in Guatemala, twenty infections by heat-labile toxin-producing *Escherichia coli* were detected among breast-fed infants in a three- to nine-month follow-up. Among the twenty cases analyzed, only one of eight infants ingesting high-titered milk for ETEC-LT became symptomatic while 8 out of 12 infants ingesting low-titered milk became symptomatic.

In addition to secretion of enterotoxins, ETEC have colonization factor antigens (CFAs) which enable the bacteria to adhere to enterocytes. Many CFAs have been identified. The predominant CFA produced by ETEC strains varies geographically (McConnell *et al.*, 1991). In Burma, strains were mainly positive for CFA I (38%), while CFA IV positive strains were predominantly in Central Africa (33%) and in Peru (17%). This difference is important with respect to specific immunity and the extent of protective effect of breast milk antibodies across strains. Because different CFAs are known to agglutinate erythrocytes of specific animal species, Kletter *et al.* (1988) used hemagglutination inhibition tests of different animal erythrocytes to study colostrum. Colostrum inhibited hemagglutination of human erythrocytes in the presence of *E. coli* known to have CFA I and inhibited bovine erythrocyte agglutination in the presence of *E. coli* known to have CFA II. Using ELISA, Nathavitharana *et al.* (1995) demonstrated the presence of specific sIgA antibodies against CFA I in breast milk. 37.5% of Sri Lankan women and 25% of immigrant Asian women had CFA I specific sIgA in their breast milk. Caucasian women controls lacked these antibodies. Evidence is lacking with regard to the protective effect conferred by the presence of these antibodies in human milk but a hyperimmune bovine milk antibody product with specific activity against purified CFA I has been shown to be effective against oral challenge with a CFA I-bearing ETEC (Freedman *et al.*, 1998).

The definition of EPEC has undergone multiple revisions over the last 20 years as our understanding of bacterial virulence factors has evolved. Until recently, EPEC were defined on the basis of O and H antigens and on epidemiologic data suggesting that certain serotypes were associated with outbreaks of infantile diarrhea. As currently understood, EPEC cause disease through their ability to adhere to intestinal cells and distort cellular architecture and function. This trait can be demonstrated *in vitro* by localized adherence of the organism to Hep-2 cells and *in vivo* by the presence of the attaching and effacing lesion (A/E) in the intestine. The A/E change is characterized by loss and blunting of microvilli with formation of an enterocyte surface protrusion or pedestal to which the bacteria are adherent. A 94 kDa protein (intimin) is associated with the A/E lesion, whereas the adhesin involved in localized adherence is a fimbria called bundle forming pilus (BFP). The A/E lesion is mediated by proteins secreted by the bacteria in a complex delivery machinery. EPEC export a protein called intimin, which then mediates attachment to the host cell via a specific receptor called tir (translocated intimin receptor); tir is exported by EPEC into host cells and is inserted into the host cell plasma membrane (Frankel *et al.*, 1998).

Many of the early studies of EPEC (prior to 1980) are now largely uninterpretable because it is no longer clear that the organisms studied would still be considered pathogens. Some of these data are provocative. Human colostrum was successfully used to cure a group of infants who had chronic diarrhea due to organisms classified at that time as EPEC (Coello-Ramirez *et al.*, 1977). Human milk was shown to have a bacteriostatic effect on EPEC when both lactoferrin and anti-O secretory antibodies were present; paradoxically, there was no growth inhibition of normal intestinal flora (Rogers and Synge, 1978; Stephens *et al.*, 1980; Dolby and Stephens, 1983). Differences in frequency of anti-O antibodies in milk have been described that are population and exposure related. Women from Asia are far more likely to have antibodies to EPEC O antigens in their milks than are women in the US (Nathavitharana *et al.*, 1994). Even those who have resided for many years outside of Asia still secrete these antibodies, suggesting that there is a long term memory for sIgA secretion in milk.

By Western blot analysis, Cravioto *et al.* (1991) demonstrated the presence of breast milk sIgA against a 94 kDa outer membrane protein found in *E. coli* with a localized type of adhesion pattern. A purified IgA fraction of the breast milk samples inhibited adherence of EPEC to Hep-2 cells or HeLa cells (Cravioto *et al.*, 1991, Camara *et al.*,1994). Since intimin is a 94 kDa protein, it is possible that these antibodies are directed against it. Costa-Carvalho *et al.* (1994) also demonstrated the presence of specific sIgA against a 94 kDa outer membrane protein. In addition, they found sIgA against a 32-kDa outer membrane protein. A 32-kDa protein was produced after protease digestion of the 94-kDa protein, suggesting that this smaller protein may be a component of the 94-Da outer membrane protein. All samples of colostrum contained sIgA against the 32 kDa OMP (22/22) compared to only 59% for sIgA against the 94 kDa OMP. The partially digested protein may contain cryptic immunogenic epitopes. Delneri *et al.* (1997) in Brazil showed the presence of sIgA against a 94 kDa protein similar to that seen by other investigators but also observed sIgA against other protein components including an 18.4 kDa protein consistent with the size of BFP. Milk from mothers who delivered preterm also showed this activity. Using purified virulence proteins and nonpathogenic *E. coli* overexpressing EPEC virulence factors as antigens in a Western blot, sIgA against the EPEC secreted proteins involved in the formation of the attaching and effacing lesion, namely espA and espB as well as intimin, was shown to be present in pooled colostrum and in purified colostral sIgA from mothers in Sao Paolo, Brazil (Loureiro *et al.*, 1998). Opsonic IgA from pooled colostrum and purified colostral IgA enhanced phagocytosis by milk mononuclear cells (Honorio-Franca *et al.*, 1997). The protective effect of breast milk against EPEC has been inferred from studies showing inhibition of adhesion to Hep-2 cells in assays using colostrum or purified sIgA (Cravioto *et al.*, 1991; Camara *et al.*, 1994; Carbonare *et al.*, 1997). The contribution of antibodies directed against specific secreted proteins in the prevention of adherence is not known. *In vivo* studies are needed to demonstrate the specific protective effect of breast milk against EPEC.

The pathogenesis of EIEC is identical to the invasive mechanism of *Shigella*. A 140 megadalton plasmid present in both bacteria codes for invasion plasmid antigens (IpaA–D). These organisms cause dysentery. Carbonare *et al.* (1995) demonstrated the ability of human colostrum to inhibit invasion of Hep-2 cells by EIEC. 95.9% of colostrum samples shown to contain high levels of IgA by ELISA prevented invasion of Hep-2 cells. By immunoblotting, antibodies to proteins of a size consistent with IpaA–D were shown. Although protection *in vivo* by these antibodies specific to EIEC invasion plasmid antigens is yet to be demonstrated, sIgA against Ipa antigens of *Shigella* appears to be protective (Hayani *et al.*, 1992).

STEC pathogenesis involves the production of attaching and effacing lesions as seen in EPEC and secretion of shigatoxin 1 and 2 (Stx 1, Stx 2). However, STEC does not have BFP. There are no published studies involving sIgA in human milk against Stx 1/Stx 2 or STEC proteins. It is likely that antibodies similar to those found in EPEC are present. Colostra of non-immunized cows have been shown to contain specific antibodies against Stx and STEC hemolysin (Lissner *et al.*, 1996). The colostral immunoglobulin fraction can block the cytotoxic effect of Stx 1and Stx 2 on Vero cell monolayers. It is possible that similar activity exists in human milk.

5. *Salmonella* and Non-Typhoidal *Salmonella*

Infection with the various *Salmonella spp.* is associated with a variety of intestinal and extraintestinal manifestations including typhoid fever, gastroenteritis, sepsis, and focal infections. *Salmonella spp.* share the ability to invade through the intestinal tract but differ in their ability to cause extra-intestinal disease. Although the peak risk for non-typhoidal salmonellosis is in the first few months of life, the data regarding the role of milk sIgA in prevention are not extensive.

Several studies have suggested that breast-feeding protects infants from *Salmonella* infections. Retrospectively, France *et al.* (1980) found that of 253 infants less than 12 months of age who were infected with *Salmonella* (based on microbiology records of three institutions) only one was breast-fed at the time of study and only twelve had ever been breast-fed. In separate groups of infants classified retrospectively to a breast-feeding group or a bottle-feeding group, the incidence of *Salmonella* in the latter was significantly higher (France *et al.*, 1980).

Likewise, a case control study in Italy suggested that breast-feeding was the single most important risk factor for non-typhoid *Salmonella* infection; it was associated with a five-fold reduction in frequency of disease (Borgnolo *et al.*, 1996).

Opsonizing antibodies have been demonstrated in colostrum and milk (France *et al.*, 1980). As with many enteropathogens, there are population-related differences in concentrations of milk sIgA antibodies against *Salmonella spp.* lipopolysaccharides that may relate to the frequency of exposure to these organisms (Cruz *et al.*, 1985; Achi *et al.*, 1992). These differences are both socioeconomic and geographical. However, in areas where exposure is common, the immune system seems to put special value on preserving these antibodies. Studies in Indian women suggest that nutritional status does not affect the development of agglutinins specific for *S. typhi* in milk (Garg *et al.*, 1989). These data underscore the importance of the survival advantage that is associated with these antibodies.

Whether vaccination of women as a strategy for preventing Salmonellosis in suckling infants may be useful remains to be seen. Live oral typhoid vaccine of immune (Pakistani) women was followed by erratic and unpredictable sIgA booster responses in milk with some women having a rise in titers and others having a fall (Hahn-Zoric *et al.*, 1989). The results obtained with the live vaccine contrasted with data based on the use of a parenteral *Vibrio cholerae* whole cell (killed) vaccine which showed enhanced serum antibody and milk sIgA responses.

6. *Campylobacter jejuni* and *C. coli*

Campylobacter spp. are important intestinal pathogens particularly in the first four years of life. They cause both watery diarrhea and bloody diarrhea. In many places they are among the most common causes of bacterial diarrhea. Some strains are invasive and others produce a heat labile toxin related to cholera toxin. Several lines of evidence suggest that human milk may play an important role in protecting infants from these organisms.

The studies of Ruiz-Palacios in Mexico City convincingly demonstrated that human milk protected infants from *Campylobacter* gastroenteritis (Ruiz-Palacios *et al.*, 1990). In these studies a cohort of infants was followed from birth to 2 years of age with weekly stool cultures and monthly collections of mother's milk. Breast-feeding lowered the risk of symptomatic *Campylobacter* infection by 2.3-fold compared to non-breast-fed infants and delayed the time of first symptomatic infection. Human milk consumed by infants who developed *Campylobacter* diarrhea lacked sIgA antibodies to certain antigens that are extracted by acid glycine. Milk of women living in both France and Central Africa has sIgA against the flagellae of *Campylobacter jejuni* (Renom *et al.*, 1992). Such antibodies may be relevant to infant protection. Mexican infants have been shown to be infected less often when fed breast milk containing anti-flagellin IgA antibodies (1.2 infections/child/yr vs. 3.3 infections/child/yr) (Nachamkin *et al.*, 1994). Prospective data from Central America, derived from Western blot analysis of individual samples of breast milk and infant stools, showed that ingestion of milk containing antibodies to three high molecular weight surface antigens [95, 110 and 185 kDa] of *Campylobacter jejuni* was correlated with absence of *C. jejuni* diarrhea (Torres and Cruz, 1993).

7. Conclusions

There are several general principles that can be drawn from review of the sIgA data relevant to bacterial enteropathogens (Table 2). The frequency of antibodies to infrequent pathogens (e.g., antibodies to *Shigella* lipopolysaccharides in Houston women's milk) strongly suggests that the recruitment of sIgA cells from the gut reflects more than just recent antigenic stimulation. The population related differences in antibodies secreted undoubtedly reflect exposure but because of the promiscuity of bacterial enteropathogens, virulence factors are often closely related and cross protective antibodies are made against pathogens that the mother's immune system has never seen (e.g., cholera toxin). A few studies suggest that antibody persistence is erratic but most suggest that antibodies secreted early in lactation continue to be secreted at lower levels

Table 2. General Principles Regarding the Role of Human Milk Secretory IgA Directed toward Virulence Determinants of Bacterial Enteropathogens

1. The memory of the secretory immune system seems to be long lasting
2. Although antibody levels are highest in colostrum, they persist at relatively constant levels throughout lactation
3. Premature as well as term delivery is followed by similar antibody secretion during lactation
4. Although the antibodies present in milk are related to prior antigenic stimulation, the close relationship (shared antigen determinants) between virulence characteristics of many bacterial enteropathogens often results in cross-protective immunity
5. Protection in tissue culture or animal models typically predicts protection of infants when appropriately controlled studies are performed
6. Parenteral re-stimulation of an immune mother can boost milk antibodies but oral re-stimulation is less predictable in its ability to boost milk sIgA
7. Poor maternal nutritional status does not impair production of milk sIgA
8. The sIgA antibodies to bacterial enteropathogens seem to have more impact on disease than colonization

throughout lactation. At least in humans, Mother Nature puts a high priority on preservation of milk sIgA delivery so that neither maternal nutritional status nor prematurity adversely impact on sIgA production. Although these antibodies appear to prevent disease in the infant, they have little impact on asymptomatic colonization of the infants' gut. Finally, although only a few of the specific antibodies have been clearly related to protection from disease, the data strongly suggest that many of the antibodies detected are likely to be biologically relevant.

References

Achi, R., Dac Cam, P., Forsum, U., Karlsson, K., Saenz, P., Mata, L., and Lindberg, A.A. 1992. Titres of class-specific antibodies against *Shigella* and *Salmonella* lipopolysaccharide antigens in colostrum and breast milk of Costa Rican, Swedish and Vietnamese mothers. *J. Infect.* **25**:89.

Ahmed, F., Clemens, J.D,. Rao, M.R., Sack, D.A., Khan, M.R., and Haque, E. 1992. Community-based evaluation of the effect of breast-feeding on the risk of microbiologically confirmed or clinically presumptive shigellosis in Bangladeshi children. *Pediatr.* **90**:406.

Borgnolo, G., Barbone, F., Scornavacca, G., Franco, D., Vinci, A., and Iuculano, F., 1996. A case-control study of *Salmonella* gastrointestinal infection in Italian children. *Acta Paediatr* **85**:804.

Cam, P.D., Achi, R., Lindberg, A.A., and Pal, T. 1992. Antibodies against invasion plasmid coded antigens of *shigellae* in human colostrum and milk. *Acta Microbiol. Hung.* **39**:263.

Camara, L.M., Carbonare, S.B., Silva, M.L., and Carneiro-Sampaio, M.M. 1994. Inhibition of enteropathogenic *Escherichia coli* (EPEC) adhesion to HeLa cells by human colostrum: detection of specific sIgA related to EPEC outer-membrane proteins. *Int. Arch. Allergy Immunol.* **103**:307.

Carbonare, S.B., Silva, M.L., Palmeira, P., and Carneiro-Sampaio, M.M. 1997. Human colostrum IgA antibodies reacting to enteropathogenic *Escherichia coli* antigens and their persistence in the faeces of a breastfed infant. *J. Diarrhoeal Dis. Res.* **15**:53.

Carbonare, S.B., Silva, M.L., Trabulsi, L.R., and Carneiro-Sampaio, M.M. 1995. Inhibition of HEp-2 cell invasion by enteroinvasive *Escherichia coli* by human colostrum IgA. *Int. Arch. Allerg. Immunol.* **108**:113.

Carlsson, B., Gothefors, L., Ahlstedt, S., Hanson, L.A., and Winberg, J. 1976. Studies of *Escherichia coli* O antigen specific antibodies in human milk, maternal serum and cord blood. *Acta Paediatr. Scand.* **65**:216.

Cleary, T.G., West, M.S., Ruiz-Palacios, G., Winsor, D.K., Calva, J.J., Guerrero, M.L., and Van, R. 1991. Human milk secretory immunoglobulin A to *Shigella* virulence plasmid-coded antigens. *J. Pediatr.* **118**:34.

Cleary, T.G., Winsor, D.K., Reich, D., Ruiz-Palacios, G., and Calva, J.J. 1989. Human milk immunoglobulin A antibodies to *Shigella* virulence determinants. *Infect. Immun.* **57**:1675.

Clemens, J.D., Sack, D.A., Harris, J.R., Khan, M.R., Chakraborty, J., Chowdhury, S., Rao, M.R., van Loon, F.P., Stanton, B.F., Yunus, M., Ali, M., Ansaruzzaman, M., Svennerholm, A.M., and Holmgren, J. 1990a. Breast feeding and the risk of severe cholera in rural Bangladeshi children. *Am. J. Epidemiol.* **131**:400.

Clemens, J.D., Sack, D.A., Chakraborty, J., Rao, M.R., Ahmed, F., Harris, J.R., van Loon, F., Khan, M.R., Yunis, M., Huda, S., Kay, B.A., Svennerholm A.M., and Holmgren, J. 1990b. Field trial of oral cholera vaccines in Bangladesh: evaluation of anti-bacterial and anti-toxic breast-milk immunity in response to ingestion of the vaccines. *Vaccine* **8**:469.

Clemens, J.D., Stanton, B., Stoll, B., Shahid, N.S., Banu, H., and Alauddin Chowdhury, A.D.M. 1986. Breast feeding as a determinant of severity in shigellosis. Evidence for protection throughout the first three years of life. *Am. J. Epidemiol.* **122**:710.

Coello-Ramirez, P., de J Garcia-Cortes, M., Diaz-Bensussen, S., Dominguez-Camacho, C., and Zuniga, V. 1977. Treatment with human colostrum of children with prolonged infectious gastroenteritis. *Bol. Med. Hosp. Infant Mex.* **34**:487.

Costa-Carvalho, B.T., Bertipaglia, A., Sole, D., Naspitz, C.K., and Scaletsky, I.C. 1994. Detection of immunoglobulin (IgG and IgA) anti-outer-membrane proteins of enteropathogenic *Escherichia coli* (EPEC) in saliva, colostrum, breast milk, serum, cord blood and amniotic fluid. Study of inhibition of localized adherence of EPEC to HeLa cells. *Acta Paediatr.* **83**:870.

Cravioto, A., Tello, A., Villafan, H., Ruiz, J., del Vedovo, S., and Neeser, J.R. 1991. Inhibition of localized adhesion of enteropathogenic *Escherichia coli* to HEp-2 cells by immunoglobulin and oligosaccharide fractions of human colostrum and breast milk. *J. Infect. Dis.* **163**:1247.

Cruz, J.R., Cano, F., and Caceres, P. 1991. Association of human milk sIgA antibodies with maternal intestinal exposure to microbial antigens. *Adv. Exper. Med. Biol.* **310**:193.

Cruz, J.R. and Arevalo, C. 1985. Fluctuation of specific IgA antibodies in human milk. *Acta Paediatr. Scand.* 74:897.

Cruz, J.R., Gil, L., Cano, F., Caceres, P., and Pareja, G. 1988. Breast milk anti-*Escherichia coli* heat-labile toxin IgA antibodies protect against toxin-induced infantile diarrhea. *Acta Paediatr. Scand.* **77**:658.

Delneri, M.T., Carbonare, S.B., Silva, M.L., Palmeira, P., and Carneiro-Sampaio, M.M. 1997. Inhibition of enteropathogenic *Escherichia coli* adhesion to Hep-2 cells by colostrum and milk from mothers delivering low-birth-weight neonates. *Eur. J. Pediatr.* **156**:493.

Dolby, J.M. and Stephens, S. 1983. Antibodies to *Escherichia coli* O antigens and the in-vitro bacteriostatic properties of human milk and its IgA. *Acta Paediatr. Scand.* **72**:577.

Feachem, R.G. and Koblinsky, M.A. 1984. Interventions for the control of diarrhoeal diseases among young children: promotion of breast-feeding. *Bull. World Health Org.* **62**:271.

France, G.L., Marmer, D.J., and Steele, R.W. 1980. Breast-feeding and *Salmonella* infection. *Am. J. Dis. Child.* **134**:147.

Frankel, G., Phillips, A.D., Rosenshine, I., Dougan, G., Kaper, J.B., and Knotton, S. 1998. Enteropathogenic and enterohaemorrhagic Escherichia coli: more subversive elements. *Molec. Microbiol.* **30**:911.

Freedman, D.J., Tacket, C.O., Delehanty, A., Maneval, D.R., Nataro, J., and Crabb, J.H. 1998. Milk immunoglobulin with specific activity against purified colonization factor antigens can protect against oral challenge with enterotoxigenic *Escherichia coli*. *J. Infect. Dis.* **177**:662.

Garg, M., Thirupuram, S., Prasad, A.K., and Saha, K. 1989. Specific antibodies in colostrum of north Indian mothers with special reference to their nutritional status. *J. Trop. Pediatr.* **35**:60.

Glass, R.I., Svennerholm, A.M., Stoll, B.J., Khan, M.R., Hossain, K.M., Huq, M.I., and Holmgren, J. 1983. Protection against cholera in breast-fed children by antibodies in breast milk. *N. Engl. J. Med.* **308**:1389.

Gunn, R.A., Kimball, A.M., Pollard, R.A., Feeley, J.C., Feldman, R.A., Dutta, S.R., Matthew, P.P., Mahmood, R.A., and Levine, M.M. 1979. Bottle feeding as a risk factor for cholera in infants. *Lancet* **2**:730.

Hahn-Zoric, M., Carlsson, B., Jalil, F., Mellander, L., Germanier, R., and Hanson L.A. 1989. The influence on the secretory IgA antibody levels in lactating women of oral typhoid and parenteral cholera vaccines given alone or in combination. *Scand. J. Infect. Dis.* **21**:421.

Hanson, L.A., Jalil F., Ashraf, R., Bernini, S., Carlsson, B., Cruz, J.R., Gonzalez, T., Hahn-Zoric, M., Mellander, L., Minoli, Y., Moro, G., Nave, F., Zaman, S., Mata, L., Karlberg, J., and Lindblad, B.S. 1991. Characterisitics of human milk antibodies and their effect in relation to the epidemiology of breastfeeding and infections in a developing country. *Adv. Exp. Med. Biol.* **310**:1.

Hayani, K,C., Guerrero, M.L., Ruiz-Palacios, G.M., Gomez, H.F., and Cleary, T.G. 1991. Evidence for long-term memory of the mucosal immune system: milk secretory immunoglobulin A against *Shigella* lipopolysaccharides. *J. Clin. Microbiol.* **29**:2599.

Hayani, K.C., Guerrero, M.L., Morrow, A.L., Gomez, H.F., Winsor, D.K., Ruiz-Palacios, G.M., and Cleary, T.G. 1992. Concentration of milk secretory immunoglobulin A against *Shigella* virulence plasmid-associated antigens as a predictor of symptom status in *Shigella*-infected breast-fed infants. *J. Pediatr.* **121**:852.

Holmgren, J., Hanson, L.A., Carlson, B., Lindblad, B.S., and Rahimtoola, J. 1976. Neutralizing antibodies against *Escherichia coli* and *Vibrio cholerae* enterotoxins in human milk from a developing country. *Scand. J. Immunol.* **5**:867.

Honorio-Franca, A.C., Carvalho, M.P., Isaac, L., Trabulsi, L.R., and Carneiro-Sampaio, M.M. 1997. Colostral mononuclear phagocytes are able to kill enteropathogenic *Escherichia coli* opsonized with colostral IgA. *Scand. J. Immunol.* **46**:59.

Jertborn, M., Svennerholm, A.M., and Holmgren, J. 1986. Saliva, breast milk, and serum antibody responses as indirect measures of intestinal immunity after oral cholera vaccination or natural disease. *J. Clin. Microbiol.* **24**:203.

Kletter, Y., Goldhar, J., Gutman, R., Fried, D., and Zakut, H. 1988. Antibodies against enterotoxigenic *Escherichia coli* in the colostrum isolated from infants with diarrhea. *Gyn. Obstet. Invest.* **25**:31.

Lissner, R., Schmidit, H., and Karch, H. 1996. A standard immunoglobulin preparation produced from bovine colostra shows antibody reactivity and neutralization activity against Shiga-like toxins and EHEC-hemolysin of *Escherichia coli* O157:H7. *Infection* **24**:378.

Long, K.Z., Wood, J.W., Vasquez Gariby, E., Weiss, K.M., Mathewson, J.J., de la Cabada, F.J., DuPont, H.L., and Wilson, R.A. 1994. Proportional hazards analysis of diarrhea due to enterotoxigenic *Escherichia coli* and breast feeding in a cohort of urban Mexican children. *Am. J. Epidemiol.* **139**:93.

Loureiro, I., Frankel, G., Adu-Bobie, J., Dougan, G., Trabulsi, L.R., and Carneiro-Sampaio, M.M. 1998. Human colostrum contains IgA antibodies reactive to enteropathogenic *Escherichia coli* virulence-associated proteins: intimin, BfpA, EspA, and EspB. *J. Pediatr. Gastroenterol. Nutr.* **27**:166.

Majumdar, A.S., and Ghose, A.C. 1982. Protective properties of anti-cholera antibodies in human colostrum. *Infect. Immun.* **36**:962.

McConnell, M.M., Hibberd, M.L., Penny, M.E., Scotland, S.M., Cheasty, T., and Rowe, B. 1991. Surveys of human enterotoxigenic *Escherichia coli* from three different geographical areas for possible colonization factors. *Epidemiol. Infect.* **106**:477.

Nachamkin, I., Fischer, S.H., Yang, X.H., Benitez, O., and Cravioto, A. 1994. Immunoglobulin A antibodies directed against *Campylobacter jejuni* flagellin present in breast-milk. *Epidemiol. Infect.* **112**:359.

Nataro, J.P., and Kaper, J.B. 1998. Diarrheagenic *Escherichia coli. Clin. Microbiol. Rev.* **11**:142.

Nathavitharana, K.A., Catty, D., and McNeish, A.S. 1994. IgA antibodies in human milk: epidemiological markers of previous infections? *Arch. Dis. Child. Fetal Neonat. Ed.* **71**:F192.

Nathavitharana, K.A., Catty, D., Raykundalia, C., and McNeish, A.S. 1995. Presence of secretory IgA antibodies to an enteric bacterial pathogen in human milk and saliva. *Arch. Dis. Child. Fetal Neonat. Ed.* **72**:F102

Pickering, L.K. and Kohl, S. 1986. Human milk humoral immunity and infant defense mechanisms. In: *Human Milk in Infant Nutrition and Health* (Howell, R.R., Morriss Jr., F.H., and Pickering, L.R. eds.), pp. 123–127. Charles C. Thomas Publisher, Springfield, Illinois.

Qadri, F.Q., Makela, P.H., Holmgren, J., Albert, M.J., Mannoor, K., Kantele, A., Saha, D., Salam M.A., and Kantele, J.M. 1998. Enteric infections in an endemic area induce a circulating antibody-secreting cell response with homing potentials to both mucosal and systemic tissues. *J. Infect. Dis.* **177**:1594.

Renom, G., Kirimat, M., Georges, A.J., Philippe, J.C., and Martin, P.M. 1992. High levels of anti-*Campylobacter*-flagellin IgA antibodies in breast milk. *Res. Microbiol.* **143**:93.

Ries, A.A., Vugia D.J., Beingolea L., Palacios A.M., Vasquez, E., Wells J.G., Baca, N.G., Swerdlow, D.L., Pollack, M., Bean N.H., Seminario, L., and Tauxe, R.V. 1992. Cholera in Piura, Peru: a modern urban epidemic. *J. Infect. Dis.* **166**:1429.

Rogers, H.J. and Synge, C. 1978. Bacteriostatic effect of human milk on *Escherichia coli*: the role of IgA. *Immunology* **34**:19.

Roux, M.E., McWilliams, M.G., Phillips-Quagliata, J.M., Weisz-Carrington P., and Lamm, M.E. 1977. Origin of IgA-secreting plasma cells in the mammary gland. *J. Exp. Med.* 146:1311.

Ruiz-Palacios, G.M., Calva, J.J., Pickering, L.K., Lopez-Vidal, Y., Volkow, P., Pezzarossi, H., and West, M.S. 1990. Protection of breast-fed infants against *Campylobacter* diarrhea by antibodies in human milk. *J. Pediatr.* **116**:707.

Sears, C.L. and Kaper, J.B. 1996. Enteric bacterial toxins: mechanism of action and linkage to intestinal secretion. *Microbiol. Rev.* **60**:167.

Simhon, A., Yolken, R.H., and Mata, L. 1979. S-IgA cholera toxin and rotavirus antibody in human colostrum. *Acta Paediatr. Scand.* **68**:161.

Stephens, S., Dolby, J.M., Montreuil, J., and Spik, G. 1980. Differences in inhibition of the growth of commensal and enteropathogenic strains of *Escherichia coli* by lactotransferrin and secretory immunoglobulin A isolated from human milk. *Immunol.* **41**:597.

Stoliar, O.A., Pelley, R.P., Kaniecki-Green, E., Klaus, M.H., and Carpenter, C.C. 1976. Secretory IgA against enterotoxins in breast-milk. *Lancet* **1**:1258.

Svennerholm, A.M., Holmgren, J., Hanson, L.A., Lindblad, B.S., Quereshi, F., and Rahimtoola, R.J. 1977. Boosting of secretory IgA antibody responses in man by parenteral cholera vaccination. *Scand. J. Immunol.* **6**:1345.

Svennerholm, A.M., Hanson, L.A., Holmgren, J., Lindblad, B.S., Nilsson, B., and Quereshi, F. 1980. Different secretory immunoglobulin A antibody responses to cholera vaccination in Swedish and Pakistani women. *Infect. Immun.* **30**:427.

Tacket, C.O., Binion, S.B., Bostwick, E., Losonsky, G., Roy, M.J., and Edelman, R. 1992. Efficacy of bovine milk immunoglobulin concentrate in preventing illness after *Shigella flexneri* challenge. *Am. J. Trop. Med. Hyg.* **47**:276.

Torres, O. and Cruz, J.R. 1993. Protection against *Campylobacter* diarrhea: role of milk IgA antibodies against bacterial surface antigens. *Acta Paediatr.* **82**:835.

Victora, C.G., Smith P.G., Vaughan, J.P., Nobre, L.C., Lombardi C., Teixeira, A.M.B., Fuchs, S.C., Moreira L.B., Gigante, L.P., and Barros, F.C. 1989. Infant feeding and deaths due to diarrhea. *Am. J. Epidemiol.* **129**:1032.

Chapter 11

Chemistry of Milk Mucins and Their Anti-Microbial Action

H. Schroten

1. Introduction

Mucins are high molecular weight glycoproteins. They are the major component of all mucus secreted from epithelial cells. As a visco-elastic layer they line the surfaces of various body cavities, including the respiratory, gastrointestinal and urogenital tracts. Depending on their site or origin, mucin monomers have molecular weights between 250 and 600 kDa, and are thus significantly larger than most other glycoproteins. At least 50% of the total molecular mass of the mucin molecule is made up of the carbohydrate chains (Feizi *et al.*, 1984; Neutra and Forstner, 1987).

Polymorphic epithelial mucin (PEM) is the best characterized human mucin molecule (Gendler *et al.*, 1990). It is identical to the mucin that was first isolated from human milk fat globules by Shimizu and Yamauchi (1982). With a view to its modest electrophoretic mobility, they named it Periodate acid Schiff (PAS-O). According to the excellent review by Patton *et al.* (1995), this primarily membrane-bound human milk mucin is nowadays referred to as MUC 1. Shimizu *et al.* (1986) described another, significantly more glycosylated mucin entitled Component A, in addition to the well established PAS-O, which they separated into two components, B and C, the former apparently a dimer of the latter. This mucin is called MUC X in the classification of Patton *et al.* (1995) since its relationship to the known members of the MUC family is still unclear.

H. Schroten • University Children's Hospital, Moorenstr. 5, 40225 Düsseldorf, Germany

2. Human MUC 1

2.1. Biochemical, Structural, and Genetic Characteristics

MUC 1 mucin corresponds to the first human mucin gene that has been sequenced among a series of 9 MUC genes (MUC 1, MUC 2, MUC 3, MUC 4, MUC 5A, MUC 5B, MUC 6, MUC 7, MUC 8). Dependent on its source it has previously been designated PEM, PUM, PAS-O, etc. Owing to its high molecular weight (250–450 kD) and its high carbohydrate content, MUC 1 does not migrate during SDS polyacrylamide gel electrophoresis. It stains positive with PAS and is found at the top of the gel. The mucin does not stain well with Coomasie blue (Shimizu and Yamauchi, 1982). MUC 1 from human milk can be easily enriched by first removing the non-glycosylated proteins using phenol extraction and subsequently removing the non-mucin glycoproteins by gel exclusion chromatography with Sephacryl S 300 (Hanisch *et al.*, 1985). Alternatively it was first isolated by urea/deoxycholate extraction of milk fat globule membranes followed by size exclusion on Sepharose CL 4B (Shimizu and Yamauchi, 1982).

Sequence analysis of the MUC 1 c-DNA revealed multiple repetitions (tandem repeats) of a short peptide motif of 20 amino acids (Gendler *et al.*, 1987, 1990). Repetitive peptide motifs are a common feature of proteins involved in protective functions (see section 5). They seem to be particularly characteristic of mucin core proteins. In these tandem repeats, threonine, serine, proline, alanine and glycine comprise >80% of all amino acids. The central domain, making up between 65% and 80% of the total protein, has between 32 and 120 tandem repeats. At both ends of the central domain there are the amino-terminal and the carboxy-terminal regions, each containing degenerated tandem repeats (Figure 1) (Gendler *et al.*, 1987, 1990). Recent evidence suggests that the tandem repeat domain is also polymorphic with respect to the repetitive peptide sequence. At least three amino acid replacements could be defined with frequencies exceeding 50% (Müller *et al.*, 1999).

As a membrane protein, MUC 1 contains a transmembrane segment of 31 hydrophobic amino acids and a cytoplasmic tail (Figure 2). This transmembrane portion shows high homology through different species (Spicer *et al.*, 1991; Patton *et al.*, 1995). The cytoplasmatic tail of the mucin molecule is connected to actin filaments within the mammary cell. It may be a regulating factor for the size of milk fat globules because, for expansion of the droplet, the membrane has to detach from the cytoskeleton (Patton *et al.*, 1995).

A minor portion of the protein-bound glycans located within the C-terminal flanking regions are N-glycosidically linked to asparagine (Hilkens and Buijs, 1988), but the majority are O-glycosidically bound to threonine or serine via N-acetylgalactosamine (GalNAc). Five potential sites are located within the

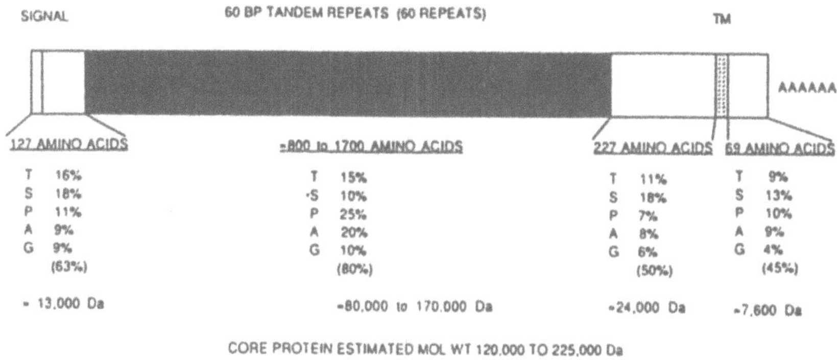

Figure 1. MUC 1 cDNA—various segments of MUC 1 protein. Indicated are the respective amounts of threonine, serine, proline, alanine, and glycine. TM transmembrane. Reproduced from Gendler *et al.* (1990).

20 amino acid repeat: AHGVTSAPDTRPAPGST-APP. On analyses of lactation associated MUC 1 for glycolysation status, an average of only 2.6 of these putative O-glycosiation sites were substituted with a glycan (Muller *et al.*, 1997). MUC 1 contains approximately 50% carbohydrates; more than 10% of these consist of terminal sialic acid. This sialic acid is responsible for a negative surface charge which may prevent clumping of milk fat globules (Patton *et al.*, 1995).

Human MUC 1 contains only small amounts of mannose. The glycans exhibit an extraordinary degree of structural diversity with as many as 50 different carbohydrate side chains between one and 20 monosaccharide units in length (Snow *et al.*, 1977; Hanisch, 1993). Hanisch *et al.* (1989, 1991) have characterized the molecular structure of major MUC 1 oligosaccharides by a combination of mass spectometry, nuclear magnetic resonance spectroscopy and methylation analysis (Figure 3). They found that both neutral and sialic acid containing oligosaccharides are made up of a mixture of branched and unbranched polylactosaminoglycans which range in size from 1 to 16 monosaccharides, their average length being 6–7 monosaccharides.

As MUC 1 is highly immunogenic, various monoclonal antibodies against this mucin have been produced which have helped to characterize and identify it in various tissues (Kufe *et al.*, 1984; Hilkens *et al.*, 1984). MUC 1 is probably only weakly expressed in non-lactating mammary tissue, but highly so in breast cancer. In fact, in patients with breast cancer MUC 1 is even found in the serum. It is for this reason that oncologic researchers in particular have pursued the development of antibodies against various epitopes of this glycoprotein (Hilkens *et al.*, 1984).

Highly glycosylated glycoproteins also can be characterized on the basis of their lectin binding patterns. MUC 1, for instance, is positive for peanut (*Arachis*

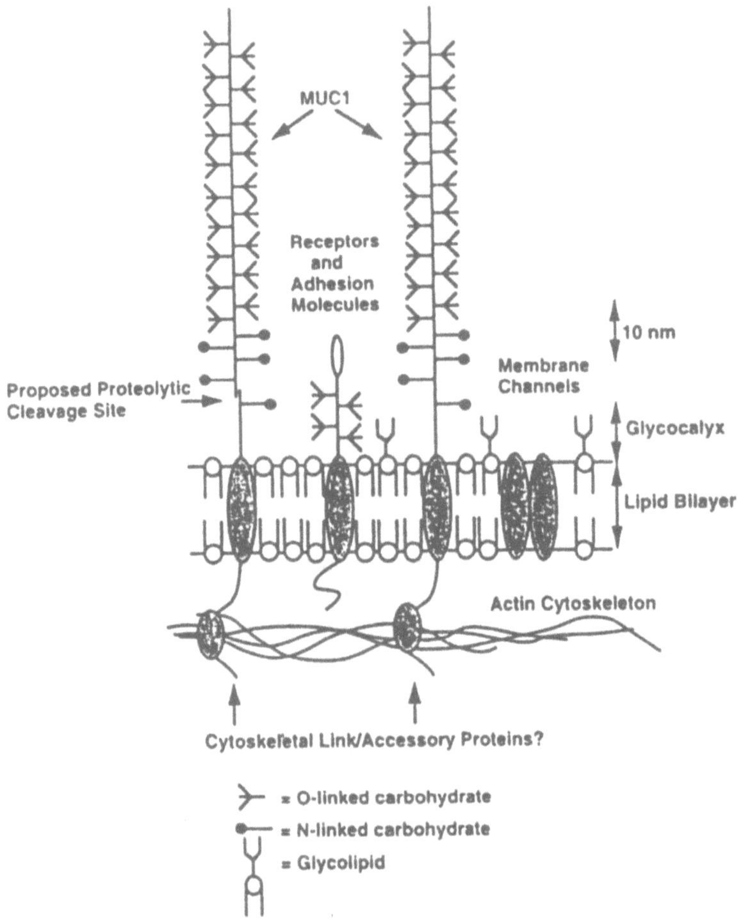

Figure 2. Scheme of the membrane mucin MUC 1. Depicted are the transmembrane configuration and the connection to the cytoskeleton. The location of a proposed proteolytic cleavage site is shown on the left. The scheme does not show the entire extracellular projection of human MUC 1 which, on the basis of a 0.25 nm per amino acid, may range between 0.2 and 0.7 μm. Reproduced from Patton *et al.* (1995).

hypogaea) agglutinin (PNA) binding. This lectin recognizes the terminal Galβ(1–3)GalNAcα sequence, the so-called Thomsen-Friedenreich antigen. As to be expected because of its high sialic acid content, MUC 1 is also positive for wheat germ (*Tritium vulgaris*) agglutinin (WGA) binding (Fischer *et al.*, 1984; Shimizu *et al.*, 1986). There is some evidence that other blood group antigens may be present on MUC 1 (Dion *et al.*, 1990).

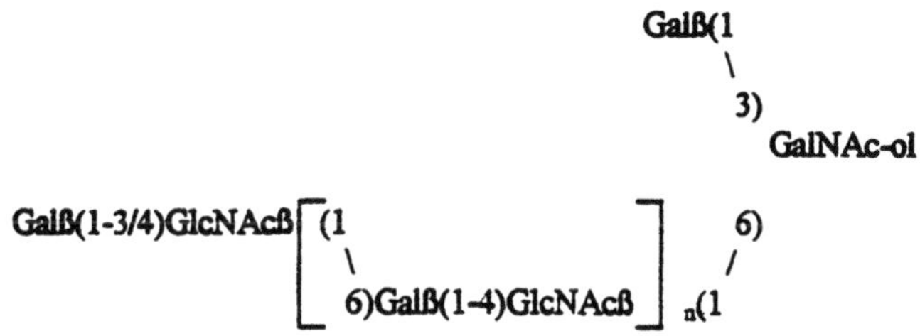

Figure 3. Representative structural model of major MUC 1 oligosaccharides.

The human MUC 1 gene locus has been mapped to chromosome 1 region q21 (Swallow *et al.*, 1987; Lancaster *et al.*, 1990). The gene contains seven exons and its sequence recently has been described (Gendler *et al.*, 1990). The MUC 1 gene is highly polymorphic with respect to the number of tandem repeats and is transmitted in a co-dominant fashion. Each individual expresses two alleles of MUC 1 with a distinct number of repeats and differing molecular masses. By SDS-PAGE, the polymorphisms usually can be separated into two distinct bands (Patton and Huston, 1987).

2.2. Biosynthesis and Cellular Origin

The neutral mucin polypeptide chain is synthesized on membrane-bound polysomes of the rough endoplasmic reticulum (Neutra and Forstner, 1987). The extensive glycosylation takes place subsequently inside the vesicles of the Golgi apparatus, where membrane-bound glycosyl transferases advance the stepwise binding of monosaccharides. This latter process has been impressively demonstrated with the use of radioactively labeled monosaccharides (Kramer *et al.*, 1985) and immunochemical proof of the glycosyl transferases (Roth, 1984). Furthermore, immunohistochemical investigations have demonstrated that the mucins are located in the apical plasma membrane of the lactating mammary glandular cell (Parry *et al.*, 1990). Thus, it is hardly surprising that MUC 1 also can be demonstrated in the milk fat globule membrane (Shimizu and Yamauchi, 1982), since the milk fat globule is nothing other than the product of apocrine extrusion from the mammary gland (Patton and Keenan, 1975; Moyer-Mileur and Chan, 1989). Membrane bound MUC 1 seems to be connected to microfilaments of the cytoskeleton (Parry *et al.*, 1990) (Figure 2). An electron micrograph of the human milk fat globule membrane surface makes the numerous, mostly radially oriented filaments visible. A major part of these consist of MUC 1 (Figure 4) (Buchheim *et al.*, 1988).

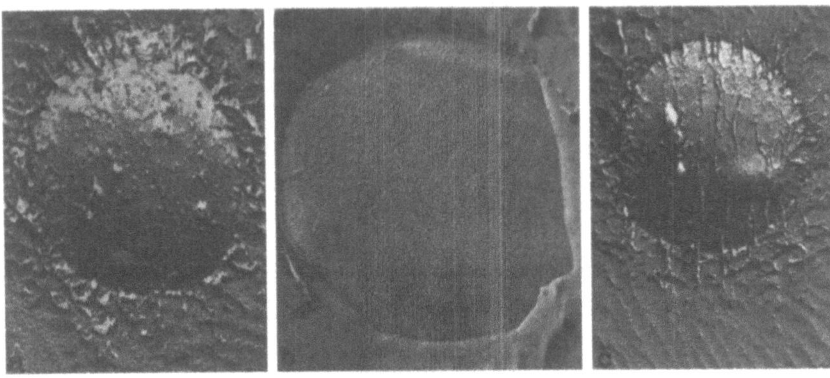

Figure 4. Freeze-etching preparation showing the surface of human (a; x 19000), cow (b; x 28300) and horse (c; x 18600) milk fat globules; radially oriented filaments are marked with arrows. Reproduced from Welsch *et al.* (1988).

2.3. Secreted Milk Mucins

Most milk mucins are membrane-bound, but MUC 1 is also found in the casein fraction of human skim milk (Shimizu *et al.*, 1986; Hanisch *et al.*, 1991). It remains uncertain whether these mucins are secretory entities, membranous in origin and produced by proteolytic cleavage of the extramembranous domain of MUC 1 (Figure 2), or released under specific conditions during preparation (Peterson *et al.*, 1998). The fact that MUC 1 is released into the serum of breast cancer patients could support the notion of non-membrane bound, soluble mucins as a seperate entity (Burchell *et al.*, 1984). Cultured breast cancer cells secrete MUC 1 and it is assumed that secretory epithelial cells also can secrete an MUC 1 isoform (Smorodinsky *et al.*, 1996).

2.4. Quantitation During Lactation

There have been few solid quantitative investigations on MUC 1 in human milk. Between 70 and 100 mg/l mucin (MUC 1) were found in skimmed human milk (Hanisch *et al.*, 1989; Hanisch, 1993). If soluble mucins make up at most one third of total mucins, about 200 mg/l mucin should be found in human milk. Recent data indicated a mucin concentration of 500–900 mg/l (Peterson *et al.*, 1998).

Putative changes in mucin secretion patterns throughout the lactation period can be assessed indirectly. Over the course of lactation there is a phase dependent change in the overall glycosylation of MUC 1. The average chain length of the oligosaccharides and the proportion of sialic acid versus fucose decreases significantly over the first three months (Schroten and Hanisch, unpublished observation). Skim milk from later phases of lactation has been demonstrated to be deficient as to its ability to neutralize the adhesion of S-fimbriae carrying *E. coli* to human buccal epithelial cells. As mucins were shown to be the relevant component for inhibition of adhesion, this could point to a reduction of the mucin concentration over the course of the lactation period (Schroten *et al.*, 1993). The 50% decrease in membrane material in mature human milk relative to the concentration found in colostrum and transitional milk and the well established reduction over time in total sialic acid content point in the same direction (Moyer-Mileur and Chan, 1989; Carlson, 1985). Recently a decline of total milk mucin concentration during lactation was proved (Peterson *et al.*, 1998). Milk from mothers of preterm infants seems to contain more mucins than milk from those of full term infants (Peterson *et al.*, 1998). The extent to which this apparent decline in mucin concentration and the possibly ensuing loss of protective activity from milk mucin is neutralized by the increased total volume of milk remains uncertain.

2.5. Component A, Alias MUC X

Shimizu *et al.* (1986) demonstrated three distinct high molecular weight bands using gradient SDS PAGE of human milk and fat globule membranes. They called the component that least penetrated the gel Component A. Patton *et al.* (1995) refer to this fraction as MUC X. The two other components, B and C, were derived from the previously described PAS-0 (Shimizu and Yamauchi, 1982), B presumably representing a dimeric form of C. Component A reacted with soybean (*Glycine max*) agglutinin, PHA and WGA. Like MUC 1, Component A/MUC X is highly polymorphic (Patton *et al.*, 1989) and contains 80% carbohydrates. A number of findings support the notion that component A is not solely a differently glycosylated variant of MUC 1: a) Component A/MUC X contains significantly more threonine and less glutamine, glycine, and alanine (Shimizu *et al.*, 1986); b) some antibodies against MUC 1, for instance HMFG 1 and 2, do not recognize Component A/MUC X (Shimizu *et al.*, 1986); c) human milks show interindividual differences in the number of mucin bands (some individuals having two MUC 1 and one MUC X band, others only one MUC 1 but two MUC X bands) (Patton *et al.*, 1989); d) during gel electrophoresis, MUC 1 and MUC X clearly separate into two distinct bands.

2.6. Specific Species Differences Among Milk Mucins

Electrophoresis of milk fat globule membranes from the rhesus monkey, human, cow, mouse, guinea-pig, dog, cat, horse, and goat clearly demonstrates the existence of MUC 1 in all these species (Spicer *et al.*, 1991). Mouse and guniea pig milks are reported to yield only one band, suggesting the existence of only one allele. However, it has to be considered that these results were generated by studying only a small number of animals. Milk from 115 Holstein cows, on the other hand, contained five alleles of MUC 1. The molecular weight of bovine MUC 1 ranges from 165 to 195 kD (Huott *et al.*, 1995).

Tandem repeats of the protein backbone and oligosaccharide analysis show marked interspecies variation. While murine and human MUC 1 are only 34% homologous in the tandem repeat region, they are 87% homologous in the transmembrane and cytoplasmic regions, which parts appear to be highly conserved among mammalian species (Spicer *et al.*, 1991). This suggests an important role of these regions in the regulation of milk fat globule size (Patton *et al.*, 1995). Conservation of the phosphorylation sites in the cytoplasmic tail of MUC 1 has led to the assumption that the mucin may be involved in signal transduction.

Species specific differences are best illustrated for human and bovine MUC 1 (PAS 1). The carbohydrate composition of the MUC 1 of these two species is summarized in Table 1.

In keeping with its carbohydrate composition, Concanavalin A agglutinin (*Canavalia ensiformis*) can bind only to bovine MUC 1 glycoproteins, while PNA and WGA can bind to both human and bovine MUC 1 (Patton *et al.*, 1989).

Table 1. Carbohydrate composition (mol %) of human (Shimizu and Yamauchi, 1982) and bovine (Snow *et al.*, 1977) MUC 1. Reproduced from Patton *et al.* (1995)

	Human*	Bovine*
Fucose	8.9	5.8
Galactose	4.9	15.9
Mannose	n.d.*	1.1
N-acetylglucosamine	19.8	14.0
N-acetylgalactosamine	15.0	22.3
Sialic acid	11.4	30.5

*mol% of total carbohydrate, n. d. not detected, **MUC 1 exhibits five N-glycosylation sites in the extracellular C-terminal portion outside of the tandem repeat domain. N-glycosylation of these sites has been demonstrated to occur during the first 3–4 min of MUC 1 processing in the endoplasmatic reticulum. N-linked glycans are known to contain at least 3 mannose molecules per chain (Hanisch *et al.*, 1989).

Guinea pig MUC 1 contains mannose and binds PNA. The total carbohydrate content of guinea pig mucin is 30–50%.

As was previously described, human milk fat globule membranes express numerous filaments on their surface. The same seems to be true for equine, though not for bovine and rat milks (Figure 4) (Buchheim *et al.*, 1988). However, using special conditions of preparation, evidence recently was obtained that milk fat globules of both cow and goat contain short mucin filaments (Peterson *et al.*, 1998).

Until recently, Component A/MUC X had been found only in primate and equine milks, and not in bovine or rodent milks (Shimizu *et al.*, 1986; Patton *et al.*, 1989, 1995). Recently such a mucin was found in mouse uterine epithelial cells (Pimental *et al.*, 1996).

Not much is known about species specific differences in MUC 1 quantity. Cow's milk has been reported to contain about 40 mg/l, markedly less than human milk (Patton *et al.*, 1995).

3. Protective Functions of Milk Mucins

Besides their well known general role as physical barriers at cell surfaces, mucins may play a crucial role as anti-infectious and immunoregulatory agents even though inactivation of the MUC 1 gene in the mouse produced no apparent phenotype (Spicer *et al.*, 1995). Many microorganisms express surface molecules called adhesins or lectins, which specifically recognize carbohydrate residues on the target cells (Beachy, 1981). This is a prerequisite for their adhesion to epithelial or endothelial cells, the first step towards a colonization or systemic infection. The numerous structural variants among the carbohydrate residues of the mucins produce receptor-like structures for different bacteria and viruses (Figure 5). This interaction is enhanced by the multiple repeats of certain peptide motifs (tandem repeats) and by numerous thin filaments on the milk fat globule membranes which significantly enlarge the effective surface. Due to its increased glycolysation compared to MUC 1, Component A/MUC X, may enhance this phenomenon.

Adhesion of S-fimbriae-bearing *E. coli* to human buccal epithelial cells can be significantly inhibited by membrane bound and skim milk mucins (Schroten *et al.*, 1992a, 1993). S-fimbriae are common among strains of *E. coli* implicated as a cause of sepsis and meningitis in human neonates. They bind to epithelia via sialyl(α-2-3)-galactoside structures. Even 1% (v/v) skim milk could significantly inhibit bacterial adhesion. Mucins isolated from the feces of breast-fed infants were able to inhibit bacterial adhesion to a greater extent than mucins from formula fed infants (Schroten *et al.*, 1992b). This is in keeping with the findings of Patton (1994) who demonstrated that major fragments of MUC 1 survive intestinal passage. A mucin-degrading microflora is established significantly later

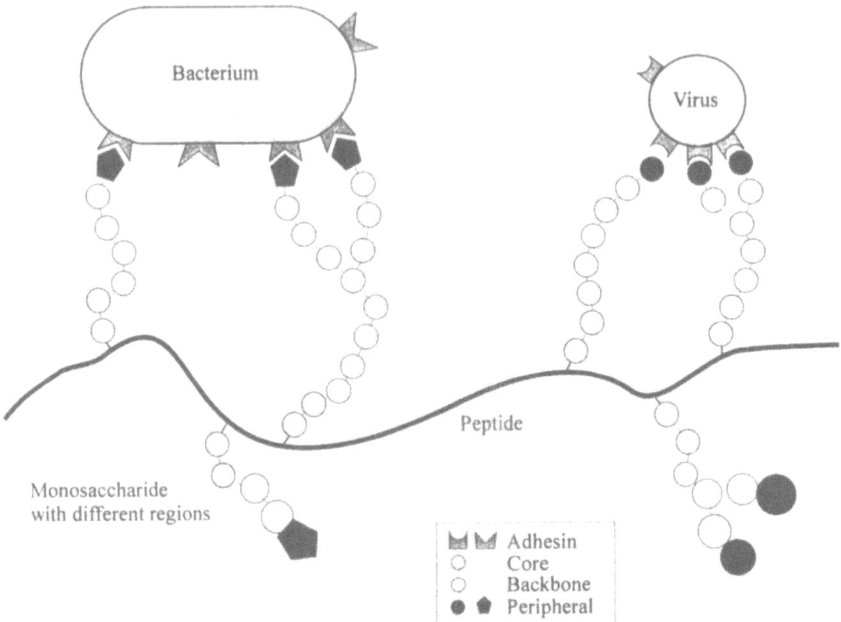

Figure 5. Interaction of a mucin molecule with bacterial and viral receptor structures.

in breast-fed than in formula-fed infants (Midtvedt *et al.*, 1994). It remains unclear whether ingested milk mucins serve nutritional purposes, but their anti-adhesive effects may well be of biologic relevance (Schroten, 1998).

The above mechanism may also be of relevance for other bacteria. Anderson *et al.* (1986) reported an inhibitory effect of a high-molecular fraction of human milk on the adhesion of *Haemophilus influenzae* to pharyngeal and buccal epithelial cells. This effect was independent of immunoglobulin. It may be speculated that this effect may be mediated by milk mucins. The same mechanism could block the adhesion of mastitis-causing bacteria to their target cells.

Yolken *et al.* (1992) showed that rotavirus binds to a 46kDa glycoprotein complexed with milk mucins. This glycoprotein, which is commonly referred to as *lactadherin*, recently has been identified as the relevant structure for protection against rotavirus (Newburg *et al.*, 1998).

GlcNAc and porcine gastric mucin (which contains GlcNAc) enhance the growth of *B. bifidum var. Pennsylvaricus* and *B. bifidum* (Petschow and Talbott, 1991). GlcNAc is also a component of milk mucins. It is therefore possible that milk mucins promote the growth of the bifidus flora, which has a major role in the establishment of an acidic pH in the feces of breast fed children. Further, an acidic pH inhibits the growth of putatively pathogenic agents in stools. Mucins also have been reported to enhance the activity of lysozyme (Inoue and Nagayama, 1970). Lysozyme, which is contained in human milk at concentrations as high as 200 mg/dl, lyses lysozyme-sensitive (Gram-positive) bacteria through hydrolysis of β-1,4-bonds between N-acetylmuraminic acid and GlcNAc in the bacterial cell wall mucopolysaccharides.

Other protective actions are attributed to milk mucins beyond their anti-infectious effects. These include a potential interaction with lymphocytes/phagocytes found in milk and also infant cells, e.g., gut lymphocytes. Mucin molecules expressed on the surface of mammary cancer cells bind to cytotoxic T-cells and (probably via regularly repeated antigenic epitopes) activate them in a major histocompatibility unrestricted fashion (Barnd *et al.*, 1989). Similarly, human B-cells react with PEM epitopes (Rughetti *et al.*, 1993). Whether these interactions between mucins and lymphocytes could bear any protective functions for the infant remains to be elucidated. At present it remains also unclear whether phagocytes are activated, which would then enhance the uptake and elimination of mucin-bound microorganisms (Patton *et al.*, 1995).

Hayes *et al.* (1990) have reported that human milk mucin interferes with the adhesion of eosinophils to immunglobulin coated targets and thus protects the epithelium against the cytotoxic effects of activated inflammatory cells. Glycosylation-dependent cell adhesion molecule 1 (GlyCAM1) mucin is expressed on endothelial surfaces, where it acts as a receptor for leukocytic L-selectin. However, experiments with GlyCAM1 mucin from milk have demonstrated that this mucin, in contrast to the one derived from the epithelium, is unable to interact with L-selectin, presumably because it is not sulfated (Dowbenko *et al.*, 1993). It remains unknown to what extent the blood group antigens expressed on milk mucins—for instance sialyl-Lewis X (Dion *et al.*, 1990)—can bind selectins and thus influence the inflammation cascade.

Future research on milk mucins will be directed toward further biochemical characterization of these highly interesting molecules and investigation of their role in the prevention of infectious diseases.

ACKNOWLEDGMENT. The author wishes to thank F.-G. Hanisch for the critical reading of the manunscript and his helpful comments.

References

Anderson, B., Porras, O., Hanson, L., Lagergard, T., and Svanborg-Eden, C. 1986. Inhibition of attachment of Streptococcus pneumoniae and Haemophilus influenzae by human milk and receptor oligosaccharides. *J. Infect. Dis.* **153**:232.

Barnd, D., Lan, M., Metzgar, R., and Finn, O. 1989. Specific, major histocompatibility complex-unrestricted recognition of tumor-associated mucins by human cytotoxic T-cells. *Proc. Natl. Acad. Sci.* **86**:7159.

Beachy, E.H. 1981. Bacterial adherence.: Adhesin—receptor interactions mediating the attachment of bacteria to mucosal surfaces. *J. Infect. Dis.* **143**:325.

Buchheim, W., Welsch, U., and Patton, S. 1988. Electron microscopy and carbohydrate histochemistry of the human milk fat globule membrane, in: *Biology of Human Milk* (L. Hanson, ed.), pp. 27, Raven Press, New York.

Burchell, J., Wang, D., and Taylor-Papadimitriou, J. 1984. Detection of the tumor-associated antigens recognized by the monoclonal antibodies HMFG-1 and -2 in serum from patients with breast cancer. *Int. J. Cancer* **34**:763.

Carlson, S.E. 1985. N-acetylneuraminic acid concentration in human milk oligosaccharides and glycoproteins during lactation. *Am. J. Clin. Nutr.* **41**:720.

Dion, A., Williams, C., Herlyn, M., and Major, P. 1990. Human milk fat globule membrane glycoproteins express blood-group-related determinants primarily on mucin-like epithelial membrane antigens and gp70. *Biochem. Internat.* **22**:295.

Dowbenko, D., Kikuta, A., Fennie, C., Gillett, N., and Lasky, L. 1993. Glycosylation–dependent cell adhesion molecule 1 (GlyCAM 1) mucin is expressed by lactating mammary gland epithelial cells and is presented in milk. *J. Clin. Invest.* **92**:952.

Feizi, T., Goori, H.C., Childs, R.A., Picard, J.K., Uemura, K., Loomes, L.M., Thorpe, S.J., and Hounsell, E.F. 1984. Mucin-type glycoproteins. *Biochem. Soc. Trans.* **12**:591.

Fischer, J., Klein, P.-J., Farrar, G.H., Hanisch, F.-G., and Uhlenbruck, G. 1984. Isolation and chemical and immunochemical characterization of the peanut-lectin-binding glycoprotein from human milk-fat-globule membranes. *Biochem. J.* **224**:581.

Gendler, S.J., Burchell, J.M., Duhig, T., Lamport, D., White, R., Parker, M., and Taylor-Papadimitriou, J. 1987. Cloning of partial cDNA encoding differentiation and tumor-associated mucin glycoproteins expressed by human mammary epithelium. *Proc. Natl. Acad. Sci.* **84**:6060.

Gendler, S.J., Lancaster, C., Taylor-Papadimitriou, J., Duhig, T., Peat, N., and Burchell, J. 1990. Molecular cloning and expression of human tumor-associated polymorphic epithelial mucin. *J. Biochem.* **265**:15286.

Hanisch, F.-G., Uhlenbruck, G., Peter-Katalinic, J., Egge, H., Dabrowski, J., and Dabrowski U. 1989. Structures of neutral O-linked poly-lactosaminoglycans on human skim milk mucins. *J. Biol. Chem.* **264**: 872.

Hanisch, F.-G., Peter-Katalinic, J., Egge, H., Dabrowski, U., and Dabrowski G. 1991. Structures of acidic O-linked polylactosaminoglycans on human skim milk mucins. *Glycoconjugate J.* **7**:525.

Hanisch, F.-G. 1993. Mucins in human milk: Structural and functional aspects, in: *New Perspectives in Infant Nutrition* (B. Renner, G. Sawatzki eds.), pp. 32, Thieme Medical Publishers Inc., New York.

Hanisch, F.-G., Uhlenbruck, G., Dienst, C., Stottrop, M., and Hippauf, E. 1985. Ca125 and Ca 19-9: Two cancer associated sialylsaccharide antigens on a mucus glycoprotein from human milk. *Eur. J. Biochem* **149**:323.

Hayes, D., Silberstein, D., Rodrique, S., and Kufe, D. 1990. DF3 antigen, a human epithelial cell mucin, inhibits adhesion of eosinophils to antibody-coated targets. *J. Immunol.* **145**:962.

Hilkens, J., Buijs, F., Hilgers, J., Hageman, P., Calafat, J., and Sonnenberg, A. 1984. monoclonal antibodies against human milk-fat globule membranes detecting differentiation antigens of the mammary gland and its tumors. *Int. J. Cancer* **34**:197.

Hilkens, J. and Buijs, F. 1988. Biosynthesis of MAM-6, an epithelial sialomucin. *J. Biol. Chem.* **263**:4215.

Huott, M., Josephson, R., Hens, J., Rogers, G., and Patton, S. 1995. Polymorphic forms of the epithelial mucin, PAS-I (MUC1), in milk of Holstein cows (bos taurus). *Comp. Biochem. Physiol.* **4**:559.

Inoue, K. and Nagayama, T. 1970. Effects of mucin upon growth of low birth-weight infants and their intestinal microflora. *Acta Paed. Jap.* **12**:15.

Kramer, M.F., Geuze, J.J., and Strous, G. 1985. Site of synthesis, intracellular transport and secretion of glycoprotein in exocrine cells. *Ciba Found Symp.* **57**:25.

Kufe, D., Inghirami, G., Abe, M., Hayes, D., Justi-Wheeler, H., and Schlom, J. 1984. Differential reactivity of a novel monoclonal antibody (DF3) with human malignant versus benign breast tumors. *Hybridoma*, **3**:223.

Lancaster, C.A., Gendler, S., Taylor-Papadimitriou, J., Duhig, T., Peat, N., Burchell, J., Pemperton, L., Lalani, E., and Wilson, D. 1990. Molecular cloning and expression of human tumor-associated polymorphic epithelial mucin. *J. Biol. Chem.* **265**:15286.

Midtvedt, A.-C., Carlstedt-Duke, B., and Midtvedt, T. 1994. Establishment of a mucin-degrading intestinal microflora during the first two years of human life. *J. Ped. Gastr. Nutr.* **18**:321.

Müller, S., Goletz, S., Packer, N., Gooley, A., Lawson, A.M., and Hanisch, F.-G. 1997. Localisation of O-glycosylation sites on glycopeptide fragments from lactation—associated MUC 1. *J. Biol. Chem.* **272**:24780.

Müller. S., Alving, K., Peter-Katalinic, J. Zachara, N., Gooley, A., and Hanisch, F.-G. 1999. High density O-glycosylation on tandem repeat peptide from secretory MUC1 of T 47D breast cancer cells. *J. Biol. Chem.* **274**:18165.

Moyer-Mileur, L. and Chan, G.M. 1989. Milk membranes—Origin, content, changes during lactation and nutritional importance, in: *Textbook of Gastroenterology and Nutrition in Infancy* (E. Lebenthal, ed.), pp. 151, Raven Press, New York.

Newburg, D., Peterson, J., Ruiz-Palacios, G., Matson, D., Morrow, A., Shults, J., Lourdes Guerrero, M. de, Chaturvedi, P., Newburg, S.O., Scallan, C.D., Taylor, M.R., Ceriani, R.L., and Pickering, L.K. 1998. Role of human-milk lactadherin in protection against symptomatic rotavirus infection. *Lancet* **351**:1160.

Neutra, M.R. and Forstner, J.F. 1987. Gastrointestinal mucus: synthesis, secretion and function, in: *Physiology of the Gastrointestinal Tract* (L.R. Johnson, ed.), pp. 975, Raven Press, New York.

Parry, G., Beck, J.C., Moss, L., Bartley, J., and Ojakian, G.K. 1990. Determination of apical membrane polarity in mammary epithelial cell cultures: the role of cell-cell, cell-substratum and membrane-cytoskeleton interactions. *Exp. Cell. Res.* **188**:302.

Patton, S. and Keenan, T.W. 1975. The milk fat globule membrane. *Biochem. Biophys. Acta* **415**:273.

Patton, S. and Huston, G.E. 1987. Differences between individuals in high-Mr glycoproteins from human mammary epithelia. *FEB* **216**:151.

Patton, S., Huston, G., Jenness, R., and Vaucher, Y. 1989. Differences between individuals in high-molecular weight glycoproteins from mammary epithelia of several species. *Biochem. Biophys. Acta* **980**:333.

Patton, S. 1994. Detection of large fragments of the human milk mucin MUC1 in feces of breast-fed infants. *J. Ped. Gastr. Nutr.* **18**:225.

Patton, S., Gendler, S.J., and Spicer, A.P. 1995. The epithelial mucin, MUC1, of milk, mammary gland and other tissues. *Biochem. Biophys. Acta* **1241**:407.

Peterson, J.A., Patton, S., and Hamosh, M. 1998. Glycoproteins of human milk fat globule in the protection of the breast–fed infant against infection. *Biol. Neonate* **74**:143.

Petschow, B. and Talbott, R.D. 1991. Response of Bifidobacterium species to growth promoters in human and cow milk. *Ped. Res.* **29**:208.

Pimental, R.A., Julian, J., Gendler, S.J., and Carson, D.D. 1996. Synthesis and intracellular trafficking of MUC 1 and mucins by polarized mouse uterine epithelial cells. *J. Biol. Chem.* **271**:28128.

Roth, J. 1984. Cytochemical localization of terminal N-acetyl-D-galactosamine residues in cellular compartiments of intestinal goblet cells: implications for the topology of O-glycosylation. *J. Cell Biol.* **98**:399.

Rughetti, A., Turchi, V., Ghetti, C., Scambia, G., Panici, P., Roncucci, G., Mancuso, S., Frati, L., and Nuti, M. 1993. Human B-cell immune response to the polymorphic epithelial mucin. *Cancer Res.* **53**:2457.

Schroten, H., Hanisch, F.-G., Plogmann, R., Hacker, J., Uhlenbruck, G., Nobis-Bosch, R., and Wahn, V. 1992a. Inhibition of adhesion of S-fimbriated E. coli to buccal epithelial cells by human milk fat globule membrane components: A novel aspect of the protective function of mucins in the nonimmunoglobuline fraction. *Infect. Immun.* **60**:2893.

Schroten, H., Lethen, A., Hanisch, F.-G., Plogmann, R., Hacker, J., Nobis-Bosch, R. and Wahn, V. 1992b. Inhibition of adhesion of S-fimbriated E. coli to epithelial cells by meconium, stool of breast-fed and formula-fed newborns—mucins are the major inhibitory component. *J. Ped. Gastr. Nutr.* **15**:150.

Schroten, H., Plogmann, R., Hanisch, F.-G., Hacker, J., Nobis-Bosch, R., and Wahn, V. 1993. Inhibition of adhesion of S-fimbriated E. coli to buccal epithelial cells by human skim milk is predominantly mediated by mucins and depends on the period of lactation. *Acta Paediatr.* **82**:6.

Schroten, H. 1998. The benefits of human milk fat globule against infection. *Nutrition* **14**:52.

Shimizu, M. and Yamauchi, K. 1982. Isolation and characterization of mucin-like glycoprotein in human milk fat globule membrane. *J. Biochem.* **91**:515.

Shimizu, M., Yamauchi, K., Miyauchi, Y., Sakurai, T., Tokugawa, K., and McIlhinney, R.A J. 1986. High-Mr glycoprotein profiles in human milk serum and fat-globule membrane. *Biochem. J.* **233**:725.

Smorodinsky, N., Weiss, M., Hartmann, M.L., Baruch, A., Harness, E., Yaakobovitz, M., Keydar, I., and Wreschner, D.H. 1996. Detection of a secreted MUC 1/SEC protein by MUC 1 isoform specific monoclonal antibodies. *Biochem. Biophys. Res. Commun.* **228**:115.

Snow, L.D., Colton, D.G., and Carraway, K.L. 1977. Purification and properties of the major sialoglycoprotein of the milk fat globule membrane. *Arch. Biochem. Biophys.* **179**:690.

Spicer, A.P., Parry, G., Patton, S., and Gendler, S. 1991. Molecular cloning and analysis of the mouse homologue of the tumor-associated mucin, MUC 1, reveals conservation of potential O-glycosylation sites, transmembrane, and cytoplasmatic domains and a loss of minisatellite-like polymorphism. *J. Biol. Chem.* **266**:15099.

Spicer, A.P., Rowse, G.J., Lidner, T.K., and Gendler, S.J. 1995. Delayed mammary tumor progression in MUC 1 null mice. *J. Biol. Chem.* **270**:30093.

Swallow, D.M., Gendler, S., Griffiths, B., Kearney, A., Povey, S., Sheer, D., Palmer, R.W., and Papadimitriou, J. 1987. The hypervariable gene locus PUM, which codes for the tumor associated epithelial mucins, is located on chromosome 1, within the region 1q21–24. *Ann. Hum. Genet.* **51**:289.

Welsch, U., Buchheim, W., Schumacher, U., Schinko, I., and Patton, S. 1988. Structural, histochemical and biochemical observations on horse milk-fat-globule membranes and casein micelles. *Histochemistry* **88**:357.

Yolken, R., Peterson, J., Vonderfecht, S., Fouts, E., Midthun, K., and Newburg, D. 1992. Human milk mucin inhibits rotavirus replication and prevents experimental gastroenteritis. *J. Clin. Invest.* **90**:1984.

Chapter 12

Antimicrobial Actions of Lactoferrin

Roberta Chierici

1. Introduction

Human milk provides infants with nutrients, hormones, enzymes and growth factors. These substances contribute to the child's health and offer protection against infectious diseases. Passively transferred components such as IgA antibodies, lysozyme and lactoferrin exert protective roles against a wide variety of microorganisms (Lönnerdal and Atkinson, 1995).

Lactoferrin (Lf) is an iron-binding glycoprotein, formerly called red milk protein, first isolated from bovine and human milk (Groves, 1960; Johansson, 1960). The protein is found in many other secretions such as tears, saliva, pancreatic juice and the secondary granules of neutrophils (Masson *et al.*, 1966, 1969). It is a single-chain metal-binding glycoprotein with a molecular weight of approximately 80,000, closely related in structure to transferrin (Tf). Like transferrin, it is a metal transporter and, as such, exists in both the iron-saturated (holo-Lf) and iron-depleted (apo-Lf) forms. Although Lf occasionally is called "lactotransferrin" the two proteins are different. The affinity constant of LF for iron is ~300 times stronger than that of Tf. Even in the presence of an iron chelator such as citrate, Lf can retain iron down to pH 3 or less, whereas Tf loses its bound iron at pH 5 (Sanchez *et al.*, 1992; Iyer and Lönnerdal,1993; Levay and Viljoen, 1995; Lönnerdal and Atkinson, 1995).

The amino acid sequence of human lactoferrin (hLf) has been studied by biochemical methods (Metz-Boutigue *et al.*, 1984) and, following the cloning and sequencing of its gene, by translating the nucleotide sequence (Rey *et al.*, 1990). Following crystallographic structure analysis at 2.8 A resolution the amino acid

---

Roberta Chierici • Department of Clinical and Experimental Medicine, Section of Neonatology, Arcispedale "S. Anna", Ferrara-Italy

sequence of 703 residues proposed by Metz-Boutigue *et al.* (1984) was revised in favor of a single polypeptide chain of 691 amino acid residues (Anderson *et al.*, 1989; Baker *et al.*, 1994). The tertiary structure of the protein has been elucidated by X-ray crystallography (Anderson *et al.*, 1989).

The molecule is folded into two globular lobes (N- and C-terminal), each one containing an iron-binding site with the capacity to bind reversibly one ferric ion. The iron atoms are coordinated to four ligands: one histidine, one aspartate and two tyrosines. The capacity of Lf to bind iron depends on the presence of bicarbonate ion, although other anions are also implicated, and is inversely related to citrate concentration. The two lobes are connected by a three-turn α helix.

Human milk Lf contains several N-glycosylation sites (Asn^{138} in the N-lobe and Asn^{479} and Asn^{624} in the C-lobe) (Spik *et al.*, 1988; van Berkel *et al.*, 1995, 1996). The carbohydrate side chains of hLf contain N-acetylneuraminic acid (sialic acid), fucose and galactose. Spik *et al.* (1988) have shown that glycosylation of Lf varies among animal species, sites of production and even among molecules produced within a single anatomical site. The influence of glycosylation on hLf functions has not been established. However, it has been suggested that it may protect the molecule against proteolysis in the gastrointestinal tract (van Berkel *et al.*, 1995).

To date, the amino acid sequences for human, bovine, mouse and porcine lactoferrins have been described (Baker *et al.*, 1994). The primary structure of bovine Lf (bLf) has been clarified by Pierce *et al.* (1991). Much of the sequence is similar to human (69%) and murine (64%) lactotransferrin. Goodman and Schanbacher (1991) determined the complete sequence of bLf mRNA isolated from mammary glands. This study revealed a complete mRNA sequence with only minor differences from those obtained by Pierce *et al.* (1991) and Mead and Tweedie (1990). The results of these studies clearly indicate that the bLf protein exhibits a striking similarity and structural homology to human Lf. In addition, human and bovine Lf share antigenic determinants and are likely to be more homologous in tertiary structure than previously suggested (Magnuson *et al.*, 1990). As for human Lf, the polypeptide chain is folded into two globular lobes, the N-and C-lobe, representing the N-terminal and C-terminal halves of the molecule (Haridas *et al.*, 1994). Each lobe is further subdivided into two domains with the specific iron binding sites in the intradomain cleft of each lobe.

Bovine Lf is characterized by glycans with α-1,3 galactose residues in the terminal non-reducing position and a glycan of the oligomannosidic type (Spik *et al.*, 1988). The precise role of these glycan chains has not been established; their removal does not seem to produce a major effect on Lf function and properties (Iyer and Lönnerdal, 1993).

2. Biological Activities of Lactoferrin

Although the structure of Lf is known, its biological function, mechanism of action and its prophylactic and therapeutic uses remain subjects of intense investigation. A variety of biological actions have been ascribed to Lf, including direct antibacterial activity, influences on iron absorption, immune responses and cellular proliferation, and growth promotion for *Bifidobacterium*. These actions are reviewed in detail elsewhere (Sanchez *et al.*, 1992; Iyer and Lönnerdal, 1993; Levay and Viljoen, 1995; Brock, 1997).

In vitro antimicrobial activities of the protein include bacteriostatic activity, probably mediated through sequestration of iron required for microbial growth, bactericidal activity, probably mediated through direct binding of Lf to bacterial cell wall components, and synergy with other antimicrobial milk proteins such as lysozyme and immunoglobulin. Lf can also exhibit fungicidal and perhaps anti-viral activity, and has also been shown to promote the growth of enterocytes as well as *Bifidobacterium* species (Brock, 1997).

A number of studies suggest that Lf may also have a role in immune and inflammatory responses. Some of these activities may depend on the ability of the protein to bind a wide variety of cells such as macrophages (Broxmeyer *et al.*, 1980), platelets (Leveugle *et al.*, 1993) and activated T and B lymphocytes (Machnicki *et al.*, 1993; Zimecki *et al.*, 1996). However, further studies are needed to clearly establish whether Lf has an immunoregulatory role.

A role for Lf as a transcriptional modulator also has been proposed because of the ability of this protein to bind with specificity and high affinity to oligonucleotide consensus sequences of DNA. Although this binding action can promote the transcription of a reporter gene in a human cell line, its biological significance remains unclear (Fleet, 1995).

3. Lactoferrin in Milk

Lactoferrin is present in the milk of most species. Human milk is particularly rich in Lf, which represents about 20% of total milk proteins (Table 1).

The mean concentration in human colostrum is particularly high (5–6 mg/ml). The level falls to about 2 mg/ml by 4 weeks, but a concentration of more than 1 mg/ml is maintained throughout lactation (Goldman *et al.*, 1982). In a recent study the concentration and ratio of Lf relative to total protein were found to be significantly higher in colostrum (5.9 g/l, 29%) than in transitional (2.9 g/l, 22%) and mature milk (2.9 g/l, 22%) in 190 samples of human milk (Cuilliere *et al.*, 1997). Synthesis of Lf in the mammary gland appears to depend on general maternal nutritional status rather than simply on maternal iron status (Houghton

Table 1. Concentration of nitrogen and proteins in human milk*

g/Liter	Time post-partum (months)			
	0–0.5	0.5–1.5	1.5–3.5	3.5–6.5
Total nitrogen	3.05 ± 0.59	1.93 ± 0.24	1.61 ± 0.21	1.48 ± 0.17
NPN	0.53 ± 0.09	0.46 ± 0.03	0.41 ± 0.04	0.38 ± 0.07
True protein**	15.80 ± 4.20	9.20 ± 1.80	7.50 ± 1.60	6.90 ± 1.20
α-Lactalbumin	3.62 ± 0.59	3.26 ± 0.47	2.78 ± 0.49	2.68 ± 0.59
Lactoferrin	3.53 ± 0.54	1.94 ± 0.38	1.65 ± 0.29	1.39 ± 0.26
Serum albumin	0.39 ± 0.06	0.41 ± 0.07	0.39 ± 0.04	0.38 ± 0.04
SigA	2.00 ± 2.50	1.00 ± 0.30	—	0.50 ± 0.10
IgM	0.12 ± 0.03	0.20	—	0.2
IgG	0.34 ± 0.01	0.05 ± 0.03	—	0.03

* From Lönnerdal and Atkinson (1995). Data are means ± SD.
** True protein = (total nitrogen-NPN) × 6.25.

et al., 1985). Nevertheless, Reddy *et al.* (1977) found no differences, even in colostral Lf level, between well-nourished and malnourished Indian mothers. Likewise, the marginal nutritional status of some Zairian mothers had no direct influence on the content of lactoferrin in their milk (Hennart *et al.*, 1991). Moreover, the level of Lf in the milk of these women during early lactation was similar to that of a group of European (Belgian) mothers.

Approximately 25–40% of the total iron in human milk is bound to lactoferrin. Human milk contains 3.6–12.5 μmol of iron/l. Of this, only 60–70% is in the whey fraction, the remainder being distributed between fat globule membrane proteins and casein. Thus, the degree of iron saturation of lactoferrin is particularly low, no more than 3–6% of its capacity, and this may be advantageous to its functions (Lönnerdal, 1984; Iyer and Lönnerdal, 1993).

Lactoferrin levels in bovine milk are much lower than those in human milk (Sanchez *et al.*, 1988). Significant amounts may be present only in colostrum (>1 mg/ml on the first postpartum day). Colostral Lf varies with breed and parity and decreases as time postpartum increases. In dairy breeds, multiparous cows produce milk with an Lf content two to three times higher than that of primiparous cows (Tsuji *et al.*, 1990). Milk from cows, guinea pigs, mice, goats and sows contains approximately the same amount of lactoferrin and transferrin (Renner *et al.*, 1989). Lf is also present in the milk of many other mammalian species with the exceptions of the dog, rabbit and rat (Masson and Heremans, 1971).

4. Antibacterial Effect of Lactoferrin

A large number of studies has indicated that Lf plays an important role in host defense. In particular, it exhibits antimicrobial activities (including bacteriostatic and, in some cases, bactericidal effects) against a wide range of microorganisms (Sanchez *et al.*, 1992; Brock, 1997). The bacteriostatic properties of Lf have been described by Bullen *et al.* (1972). These authors found that human milk and purified hLf, in combination with specific *Escherichia coli* antibody, strongly inhibited the growth of *E. coli*. This effect was abolished by saturating the protein with iron. These results were supported by experiments with suckling guinea-pigs fed *E. coli* and hematin, in which the suppressive effect of milk on *E. coli* colonization in the intestine was lost after addition of iron (hematin hydrochloride). The antimicrobial action of Lf has been attributed mainly to the ability of the protein to chelate iron, inducing an iron deficient environment that limits the growth of bacteria. A bacteriostatic effect of Lf, reversible by iron saturation, also has been demonstrated in relation to a variety of other Gram-negative bacteria (Oram and Reiter, 1968; Griffiths and Humphreys, 1977; Stephens *et al.*, 1980).

Evidence from a number of studies indicates that the antimicrobial mechanism of Lf is more complex than simple iron deprivation. Arnold *et al.* (1977) reported that Lf in the iron-free state shows direct bactericidal activity against *Streptococcus mutans, Vibrio cholerae* and other microorganisms (Arnold *et al.*, 1980). The list of susceptible organisms includes both Gram-positive and Gram-negative bacteria, rods and cocci, facultative anaerobes and aerotolerant anaerobes and yeast. The bacterial killing property of Lf requires adsorption of the protein to the target cell surface and is inhibited by iron saturation, but is not restored by exposure to an iron-poor environment alone. Culture conditions such as pH, temperature, osmolarity and metabolic phase may influence the bactericidal action of Lf (Arnold *et al.*, 1981). Dalmastri *et al.* (1988) noted an enhanced antibacterial activity toward a variety of Gram-positive and Gram-negative microorganisms when the protein was permitted to bind to the bacterial surface. Visca *et al.* (1990) investigated the bacteriostatic activity of Lf in K-12 strains of *E. coli*. In this microorganism, Lf binds to the microbial surface in a manner that does not involve iron-regulated outer membrane proteins. Its antibacterial activity is markedly reduced when the protein is separated from the bacteria by a dialysis membrane (Visca *et al.*, 1990). An important factor in the antibacterial effect of Lf is its interaction with several components of the Gram-negative bacterial outer membrane such as carbohydrates (Arnold *et al.*, 1981; Giugliano *et al.*, 1995) and lipopolysaccharides (LPS) (Ellison *et al.*, 1988).

In recent years, several groups of researchers have demonstrated that Lf has the ability to bind LPS (Ellison and Giehl, 1991; Miyazawa *et al.*, 1991; Cohen *et al.*, 1992). In studies on strains of *E. coli*, *S. typhimurium* and *V. cholerae*,

Ellison (1994) found that the interaction of human Lf with Gram-negative bacteria induces permeability alterations, with the subsequent release of LPS molecules and an increase in bacterial susceptibility to antibiotics and lysozyme. This effect is dependent on direct contact with the bacterial cell but the exact mechanism of action is unclear. In particular, there is evidence that the protein damages the membrane through its chelating activity or through a mechanism that disturbs mineral balance with regard to cations such as Ca++, Mg++ and Fe++. The latter hypothesis has been supported by observations linking the antibacterial activity of Lf to the highly cationic nature of the protein. Its cationic N-terminal domain, distinct from the chelating region of the molecule, appears to share a mechanism of action common to a large family of other cationic neutrophil-associated peptides (Ellison, 1994).

Taken together, these observations indicate that two factors are important to the growth inhibiting action of Lf: iron chelation and a direct interaction with the microbial surface. Any model of Lf interaction with the bacterial outer membrane must be in agreement with observations that it binds to the outer membrane pore-forming proteins (*porins*) of various Gram-negative organisms (Ellison, 1994; Naidu and Arnold, 1997). Specific Lf-binding receptors have been described in the human pathogens *Neisseria meningitidis*, *Neisseria gonorrhoeae*, *A. hydrophila*, *Haemophilus influenzae*, *Treponema pallidum* and *Tricomonas vaginalis* (Peterson and Alderete, 1984; Lee and Schryvers, 1988; Schryvers and Morris, 1988; Ascencio *et al.*, 1992). In these bacterial species the cell surface receptors demonstrate a greater specificity for human than for bovine Lf. Studies with iodinated Lf have shown that Lf binds to strains of *E. coli*, *Aeromonas hydrophila* and *Shigella flexneri* (Kishore *et al.*, 1991; Naidu *et al.*, 1991a; Tigyi *et al.*, 1992). Furthermore, Naidu *et al.* (1991b) have reported the interaction of bovine Lf with *S. aureus* and the coagulase-negative staphylococci associated with bovine mastitis. These authors also revealed the presence of a specific human Lf-binding protein in *S. aureus*, and established that there is a specific receptor for hLf in S. aureus strain MAS-89 (Naidu *et al.*, 1991b, 1992).

A critical issue that has only recently been identified concerns the recognition that selected bacterial pathogens have the capacity to use Lf and other iron-binding proteins as iron sources. Thus, the possession of specialized iron transport systems, perhaps including Lf receptors, may be very important to the survival of these organisms in the iron-restricted environment imposed by the host (Crosa, 1989; Guerinot, 1994). Some bacterial species have developed a sophisticated iron assimilation system represented by the production of *siderophores*, low molecular weight iron chelators with extremely high affinity for iron (Crosa, 1989). The transport of iron into the cell cytosol is mediated by a specific membrane receptor and transport systems that recognize the iron-siderophore complexes. However, the production of siderophores is not the only mechanism through which bacteria can utilize iron. Certain pathogenic bacteria, such as the families

Neisseriaceae and *Pasteurellaceae*, which do not produce siderophores, are able to use different host iron-binding proteins such as Tf, hemoglobin, and in the case of the *Neisseriacee*, Lf as a source of iron (Gray-Owen and Schryvers, 1996). These microorganisms possess outer membrane protein receptors that can recognize the Lf or Tf complex with iron. Tf receptors have been identified in a number of important human and veterinary pathogens from the families *Neisseriaceae* and *Pasteurellaceae*. In most species the Tf receptor has been shown to consist of two proteins, Tf-binding protein 1 (Tbp1) and Tf-binding protein 2 (Tbp2). In contrast, receptors for Lf have been demonstrated only in the *Neisseriaceae* family. These receptors are protein- and species-specific, and the *N. meningitidis* receptor binds human Lf but not bovine Lf or human Tf (Bonnah *et al.*, 1997; Brock, 1997). Recent studies on bacterial Lf receptors in *Neisseriaceae* suggest that, in contrast to previous conclusions, the composition of the Lf receptor is similar to that of Tf receptor and includes two proteins Lbp1 (Tbp1 homolog) and Lbp2 (Tbp 2 homolog) (Bonnah *et al.*, 1997).

To evaluate the mechanisms for scavenging iron for growth, the ability of various veterinary pathogens to sequester iron by direct binding of host Lf, Tf or other Fe-binding compounds to specific cell surface receptors has been investigated. Isolates of *Pasteurella multocida* (serotype B : 2,5 strains), an invasive bacterium associated with hemorrhagic septicaemia in domestic and feral ruminants, exhibit specific receptors only for bovine Tf. No binding of Lf was detected (Veken *et al.*, 1996). *T. vaginalis*, which lives in an Lf rich environment, so far is the only protozoan known to utilize Lf as a source of iron. Lactoferrin binds to *T. vaginalis* by specific receptors that increase in number and affinity under iron-restricted conditions (Lehker and Alderete, 1992). *Tritrichomonas foetus*, an agent responsible for a venereal disease of cattle, is able to utilize both Lf and Tf to cover its iron requirements; however, the mechanism of iron acquisition from these proteins appears to be different. Binding and internalization of Lf suggest the involvement of specific receptor-mediated endocytosis in the acquisition of Lf-bound iron, whereas the retrieval of iron from Tf seems to depend on extracellular release of the metal from this ligand (Tachezy *et al.*, 1996).

As suggested by Ellison (1994), the effect of Lf (and Tf) on organisms that are either commensals or opportunistic pathogens can be distinguished from that against highly virulent pathogens. This difference is illustrated in the example of *Pasteurella hemolytica*. This Gram-negative bacterium is pathogenic for cattle and sheep, but it is generally considered a commensal organism in humans. Ogunnariwo and Schryvers (1990) have shown that this bacterium has a Tf receptor with high affinity for bovine Tf but low affinity for human, pig or avian Tf. Thus, this organism is a "pathogen" for a species from which it can obtain iron by means of the host's iron-binding proteins, but is a "commensal" in species wherein it cannot utilize these iron sources. Further studies are needed to determine what role and contribution iron-binding proteins can make to host defense, given the possibility that micro-organisms can use the same proteins as growth factors and virulence factors.

5. Antimicrobial Peptides of Lactoferrin

Several studies have suggested that iron-free Lf exerts bactericidal activity by a mechanism distinct from iron chelation (Tomita *et al.*, 1991). To determine whether the entire protein is required for this bactericidal effect, some authors have examined the antimicrobial properties of hydrolysates prepared by enzymatic digestion of Lf (Tomita *et al.*, 1991; Bellamy *et al.*, 1992). Bovine Lf was treated with various commercial proteases. Hydrolysates prepared by cleavage of the protein with pig pepsin, cod pepsin, and acid proteases from *Penicillum duponti* showed strong activity against *E. coli* O111, whereas hydrolysates produced by trypsin, papain or other neutral proteases showed little or no activity. The peptides generated by porcine pepsin cleavage of Lf showed broad spectrum antibacterial activity, inhibiting the growth of a number of Gram-negative and Gram-positive species (Tomita *et al.*, 1991). Moreover, unlike the native protein, the peptides retained their activity in the presence of added iron. Their antibacterial potency was at least eightfold greater than that of undigested Lf regardless of bacterial species (Tomita *et al.*, 1991).

These findings indicate that active peptides, more potent than the intact Lf molecule, are produced by enzymes such as gastric pepsin (Tomita *et al.*, 1994). The pepsin hydrolysate of bovine Lf was fractionated by reverse-phase HPLC and antibacterial activity was detected in only one peak (Tomita *et al.*, 1991, 1994). The investigators named this active peptide "lactoferricin B". The antimicrobial properties of hydrolysates of human Lf were also investigated. A single active peptide, generated by pepsin cleavage, was isolated and named "lactoferricin H" (Bellamy *et al.*, 1992; Tomita *et al.*, 1994). The authors have identified the structural region responsible for the microbicidal properties of human and bovine Lf. This domain contains a potent antimicrobial sequence that consists mainly of a loop of 18 amino acid residues formed by a disulphide bond between cysteine residues 20 and 37 in human Lf and residues 19 and 36 in bovine Lf, and is located near the N-terminus of the Lf molecule (Figure 1). Both bovine (LfcinB) and human (LfcinH) lactoferricins had a stronger antimicrobial activity against *E. coli*, distinct from an iron-chelation property than the unmodified protein. Lactoferricin B, in particular, had antimicrobial properties against a wide variety of microbial species, including Gram-negative and Gram-positive bacteria and fungi. This peptide also increases bacterial susceptibility to lysozyme and causes release of LPS and rapid loss of viability in many strains of *E. coli*, *Klebsiella*, *Bacteroides*, *Staphylococcus*, *Streptococcus*, *Corynebacterium* and *Listeria* (Tomita *et al.*, 1994). The antimicrobial activities of bLf, its pepsin hydrolysate

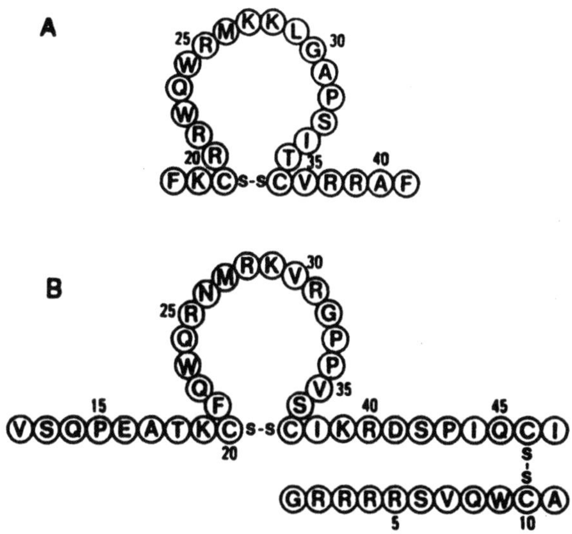

Figure 1. Primary structure of lactoferricin B (A) and lactoferricin H (B). Single-letter code is used to indicate the amino acid sequence of each peptide. Numbers indicate sequence positions in bovine and human lactoferrins. From Tomita *et al.* (1994).

(bLfH) and LfcinB also have been studied against clinical isolates of enterohemorrhagic *E. coli* O157:H7. Transmission electron microscopy suggested that Lfcin B acts on the bacterial surface to influence the level of verotoxin released by the bacterium into culture supernatant fluids (Shin *et al.*, 1998).

Further to these studies, Saito *et al.* (1991) showed that bLf lost its antigenicity and iron-binding capacity when hydrolysed at high temperature and pH 2–3. The resulting mixture of peptides, however, exhibited stronger antibacterial activity than native Lf.

6. Antimicrobial Activity of Lactoferrin in the Gut

Several epidemiological studies have indicated that breast-feeding protects infants from acute and chronic disease, including gastro-intestinal infections (American Academy of Pediatrics, 1997). One of the most striking differences between breast- and bottle fed-infants concerns the fecal pH and bacterial microflora. The feces of breast-fed infants contain predominantly a *Bifidus* flora, whereas those of bottle-fed infants are characterized by the presence of enterobacteria (Yoshioka *et al.*, 1983; Benno *et al.*, 1984; Balmer and Wharton, 1989). It has been suggested that Lf, an iron-binding protein with potent antibacterial activity, may play a role in protecting newborns and infants against gastrointestinal infections by preventing colonization of entero-pathogenic organisms (Sanchez *et al.*, 1992). The mechanism by which Lf exerts this influence on the fecal flora is difficult to understand. According to a prevalent hypothesis, the iron-scavenging action of Lf deprives bacteria of this essential nutrient in the gastrointestinal tract, where iron is scarce, and thereby prevents the growth of iron-requiring bacteria. However, it is possible that the protein has a more direct bactericidal effect on at least some intestinal bacteria.

In addition to the study of Bullen *et al.* (1972) which suggested that Lf has a protective effect in newborn guinea-pigs infected with *E. coli* (oral dose approximately 1×10^6 organisms), evidence that Lf exerts such an effect *in vivo* also has been observed in mice infected with a lethal intravenous dose of *E. coli* (2×10^8 organisms) (Zagulski *et al.*, 1989). In contrast, Czirok *et al.* (1990) reported that neither intraperitoneal, intramuscular, nor oral Lf protected mice from intraperitoneal infection with *E. coli*. However, a more recent study, using germ-free mice, indicated that bovine Lf favored a gut flora rich in bifidobacteria (Hentges *et al.*, 1992). Moreover, Teraguchi *et al.* (1997) demonstrated that orally administered bovine Lf can exert a bacteriostatic effect in the gut of milk-fed animals. When mice were fed milk with bovine Lf added the proliferation of *Enterobacteriaceae* was suppressed, as was the proliferation of orally inoculated *Clostridium ramosum* C1 strain. A similar effect was observed after oral inoculation with other species of *Clostridium*. Pepsin-hydrolyzed bLf exerted the same effect on intestinal bacteria as undigested bLf.

The bifidogenic effect of Lf was also tested in some human clinical studies using lactoferrin-supplemented infant formulae. The addition of bLf and Lf plus iron to an infant formula had little effect upon the fecal microflora and, at least during the short study period, did not change the pattern of the fecal flora to resemble a breast-fed flora (Balmer *et al.*, 1989). However, different results were reported in a study performed by Roberts *et al.* (1992). The development of infant fecal flora was followed for the first three months of life in breast-fed babies and in infants receiving an adapted formula or the same formula supplemented with 10 or 100 mg/100 ml of bLf (20% iron-saturated). Fecal samples were collected at the hospital and the flora were compared to samples from infants that were breast-fed for 7, 30, or 90 days (Roberts *et al.*, 1992). Among formula-fed infants, those fed the Lf 100 formula for 90 days exhibited a bifidus flora comparable to that of breast-fed infants. In the Lf 100 formula-fed infants, suppression of facultative microorganisms with the development of a bifidus flora occurred only at three months of life. The lateness of this effect could be due to the fact that bLf was added in relatively small amounts. Alternatively, the degree of iron saturation of the bLf was higher than that of Lf in human milk (20 vs 3–5%) and the addition of iron may have reduced the antimicrobial activity of Lf. In addition, bLf might not be an appropriate substitute for the human protein. Nevertheless, in a study designed to compare the growth promoting activity of components of human and cow's milk *in vitro*, bLf was confirmed to be a potent growth promoter of different strains of bifidobacteria (Petschow and Talbot, 1991).

To perform such functions in the gut, Lf must survive the intestinal tract. The fingerprints of native human and bovine lactotransferrins hydrolysed *in vitro* by infant gastric or duodenal secretions showed that protein was extensively digested (Spik *et al.*, 1982; Brines and Brock, 1983). The partly digested proteins maintained their ability to bind iron and hence may supplement the bacteriostatic effects of intact endogenous lactotransferrin in the intestinal tract. In this study iron-Lf survived better than the apo forms. Bovine Lf was degraded less extensively than the human form in the feces of newborn infants fed cow's milk supplemented with human or bovine Lf. Moreover, both intact Lf and its fragments have been detected in the urine of premature infants fed human milk (Hutchens *et al.*, 1991) and in the stools and urine of very low birth weight infants fed fortified human milk (Goldman *et al.*, 1990). It is unclear whether the Lf fragments were derived exclusively from ingested Lf or partially from endogenous Lf produced in the gastrointestinal tract.

Drescher *et al.* (1999) studied the prececal digestibility of Lf and casein in suckling and adult miniature pigs by using ^{15}N-labeled proteins and collecting the chyme. The animals received sow's milk supplemented with porcine Lf, bovine Lf or bovine casein. The digestibility of Lf, both porcine (84.4 ± 3.2%) and bovine (82.3 ± 4.8%), was significantly lower than that of casein (97.6 ± 0.5%) in the distal small intestine of suckling piglets. No difference was found in the digestibility of Lf of bovine and porcine origin. In adult miniature pigs there was no difference between the digestibilities of Lf and casein. Low Lf digestibility at the suckling stage of life may be important to the biological function of Lf in the intestine and, in newborn mammals may represent a defense mechanism against bacterial infections.

The use of bovine Lf in infant nutrition must be viewed with caution because the exact role of Lf in the control of the intestinal microfloral population, as well as any other possible beneficial effect on human health, has not been conclusively demonstrated. Further studies are required to determine whether gastric digestion of Lf *in vivo* generates peptides with significant antimicrobial capabilities.

7. Other Antibacterial Effects of Lactoferrin

Lf may act synergistically with secretory IgA (Stephens *et al.*, 1980) or with other antimicrobial proteins such as lysozyme (Ellison and Giehl, 1991) that are also present in human milk. Ellison and Giehl (1991) found that Lf and lysozyme together could be bactericidal for strains of *Vibrio cholerae*,

S. typhimurium and *E. coli*, whereas each protein alone was merely bacteriostatic. A recent study indicates that an iron-free IgA-Lf complex isolated from human milk exhibits a marked bactericidal activity against both *Streptococcus mutans* and *E. coli*, whereas an undialyzed IgA-Lf complex exhibited an inconsistent bactericidal effect (Akin *et al.*, 1994). In addition, it was shown that the isolated IgA-Lf complex existed, as such, in human milk. These findings confirm the presence in human milk of an IgA-Lf complex, stabilized by covalent bonds, that can provide protection against Gram-negative and Gram-positive bacteria.

Lf may also contribute to the bactericidal activity of polymorphonuclear leucocytes (PMN). This protein is known to modulate phagocyte functions such as the adhesiveness of neutrophils and phagocytosis by macrophages. Ghar *et al.* (1991) reported that, after exposure to highly purified hLf, PMN exhibited a modest enhancement in chemotactic activity and produced superoxide at a faster rate. In addition, exposure of human PMN to a high concentration of purified Lf (500 μg/ml) was followed by an increase of their random motility. Lf can bind to the surface of PMN and reduce their surface charge; the Lf-related increase in PMN motility has been ascribed to this phenomenon (Boxer *et al.*, 1982). Other PMN functions such as aggregation, degranulation and intracellular killing were not influenced by either iron-poor or iron-saturated Lf (Ghar *et al.*, 1991), although some authors have reported that Lf promotes endothelial adherence of PMN as well as aggregation of these cells (Oseas *et al.*, 1981).

Experimental evidence has accumulated for a connection between Lf and radical-based oxidative antimicrobial mechanisms. Ghar *et al.* (1991) reported that pre-incubation of PMN with hLf for 30 minutes increased the capacity of phagocytes to produce the superoxide anion. The effect was dose-dependent up to a concentration of 500 μg Lf/mL and was undetectable at Lf levels below 100 μg/mL. Early studies by Ambruso and Johnston (1981) had suggested that Lf can promote the production of hydroxyl radicals (HO·) by human neutrophils. More recent results complement this proposition by suggesting that release of Lf by activated neutrophils provides protection against indiscriminate production of HO· at sites of inflammation, through an iron-binding mechanism (Britigan *et al.*, 1994, 1997). Such protection may extend to the mononuclear phagocyte. This cell does not produce Lf, but has surface receptors specific for it. Mononuclear phagocytes permitted to bind apo-Lf exhibited a reduction in oxidative injury in the presence of iron chelates (Britigan *et al.*, 1994, 1997).

Although Lf is capable of inhibiting the Haber-Weiss mediated formation of HO·, under *in vivo* conditions the protein would likely encounter a variety of factors such as proteases or LPS that may interfere with this activity. For example, Britigan *et al.* (1997) demonstrated that cleavage of Tf and Lf, via proteases likely to be present at sites of inflammation, generates new iron chelates that are able to promote HO· formation. Thus, under conditions of high protease activity, Lf could contribute to oxidant mediated tissue injury by promoting HO· production.

These data indicate that the interaction of Lf with human phagocytes may have both anti- and pro-oxidant consequences.

8. Lactoferrin and Helicobacter Pylori

Helicobacter pylori (HP) is one of the most common agents involved in the pathogenesis of various gastro-duodenal disorders such as chronic gastritis, peptic ulcer, gastric carcinoma and gastric mucosa associated lymphoid tissue lymphoma (Parsonnet *et al.*, 1991; Blaser, 1992; Hussel *et al.*, 1993). Conflicting results are reported regarding the relationship between Lf and HP infection. Some studies indicate that hLf supports full growth of HP in media lacking other iron sources (Husson *et al.*, 1993). The ability of this organism to use Lf has been ascribed to a 70 kDa lactoferrin-binding protein identified by affinity chromatography among its outer membrane proteins (Dhaenens *et al.*, 1997). The binding action of this protein appeared to be highly specific for hLf. In contrast, bLf has been reported to inhibit the growth of HP *in vitro* and to exert an antimicrobial effect under neutral and mildly acidic conditions. However, the protein was not bactericidal and the combination of Lf and lysozyme did not further inhibit bacterial growth over that seen with Lf alone (Dial *et al.*, 1995, 1996). In addition, Nakao *et al.* (1997a) have shown that Lf is present in gastric juice and that its concentration correlates positively with HP infection. Lf concentration was high in the biopsy specimens of patients with HP-related gastritis and its level correlated with the degree of gastric inflammation. Thus, this protein may be a good marker for HP associated gastritis. According to Nakao *et al.* (1997b), although correlation does not mean causation, these studies suggest that Lf might indeed play a crucial role in the pathophysiology of HP-associated gastritis.

9. Antimicrobial Activity Against Fungi

Lf is known to exert antimicrobial activity against fungi such as *Candida albicans*, an opportunistic fungus that can colonize the mucosa and frequently causes serious infection in the immunocompromised host. The antifungal mechanism of Lf is not understood and little is known about the interaction of Lf with this organism. Human Lf in its iron-free state exhibits an antifungal effect and kills *C. albicans* in a time- and dose-dependent way. The lethal effect appears to be stronger at pH 7 than at pH 5 (Soukka *et al.*, 1992). It has also been reported that Lfcin B exhibits a potent fungicidal activity, and that the antimicrobial effect of the peptide results from its direct interaction with the cell surface (Bellamy *et al.*, 1993). This mechanism apparently corresponds to that involved in the lactoferricin killing of bacteria and is suppressed by Ca^{++} and Mg^{++}.

The fungistatic effects of bLf and its peptides in combination with several antifungal agents were investigated by Wakabayshy *et al.* (1996). The concentration of Lf, its pepsin hydrolysate (Lfhyd) and the antimicrobial peptide LfcinB required to inhibit the growth of *Candida albicans* decreased in the presence of clotrimazole (CTZ). When Lf was used alone, 200 µg/mL was needed to fully inhibit the growth of *Candida*, whereas in the presence of 3.1 or 12.5 mg CTZ/mL, the protein caused complete growth inhibition at 100 µg/ml and 50 µg/ml, respectively. In addition, the IC80 (80% growth inhibitory concentration) of Lfhyd and LfcinB decreased to one-half to one-quarter of the concentration required when used alone. In parallel with these results, the MIC of ketoconazole, another antifungal agent, decreased to one-sixteenth in the presence of 100 µg Lf//mL. These results indicate that Lf acts cooperatively with azole antifungal agents, but the mechanism of action has not been elucidated.

Okutomi *et al.* (1997) recently demonstrated that the inhibitory action of neutrophils on the growth of *C. albicans* was augmented in the presence of a physiologic concentration of Lf (30 µg/ml of bLf or 1 µg/ml of hLf). The mechanism(s) involved remain(s) to be clarified.

10. Antiviral Effects of Lactoferrin

Although the antiviral effect of milk proteins has been recognized for many years, Lf has only recently been shown to have antiviral activity. Specifically, Lf has been found to inhibit some enveloped viruses such as human immunodeficiency virus (HIV), human cytomegalovirus (HCMV) (Harmsen *et al.*, 1995; Swart *et al.*, 1996) and herpes simplex virus (HSV-1) (Marchetti *et al.*, 1996).

Harmsen *et al.* (1995) studied the anti-HIV-1 and anti-HCMV effects of a series of native and modified plasma and milk proteins tested *in vitro* in MT4 and fibroblast cells. Native lactoferrins inhibited both HCMV and HIV infections and bLf had a stronger antiviral effect than its human counterpart. When negatively charged groups were added to the milk protein by succinylation or acylation, a stronger antiviral effect against HIV types 1 and 2 was observed (Harmsen *et al.*, 1995; Swart *et al.*, 1996) but antiviral capacity against HCMV infection was decreased. It is likely that Lf exerts its effect at the level of virus adsorption or receptor-mediated binding to the target cell membrane. Heparan sulphate is a constituent of most mammalian cell plasma membranes (and of the extracellular matrix) and acts as a receptor for attachment of virions to target cells. This macromolecule also has been recently identified as a binding site for human and bovine Lf (Mann *et al.*, 1994; Wu *et al.*, 1995). Thus, Lf may interfere with the attachment of viruses to heparan sulfate receptors and, in this way, the protein may interfere with early events in viral infection. However, a direct interaction of the protein with the virus is also possible (Harmsen *et al.*, 1995).

The combined effect of Lf on HIV-1 and HCMV, which has not been previously reported, may be very important because immunosuppressed patients are often superinfected with CMV and therefore Lf could have a therapeutic role in these subjects. Moreover, it has been hypothesized that Lf may play a role in HCMV inhibition in breast milk because the decrease in Lf levels during lactation coincides with an increase in the fraction of infants infected with HCMV during the first year of life (Harmsen *et al.*, 1995). Marchetti *et al.* (1996) investigated the effect of hLf and bLf on different steps of HSV-1 multiplication in Vero cells. Both hLf and bLf were found to be potent inhibitors of HSV-1 infection and to exert their activity through inhibited attachment of virions to the cells. This action was independent of any iron withholding property, since no difference was found between lactoferrins in apo and iron-saturated form. The mechanism of the antiviral effect probably involves the capability of lactoferrins to interfere with cell receptors for HSV-1, represented by heparan sulphate and the LDL-receptor.

With regard to HCV infection, skim milk has been demonstrated to reduce the binding to target cells by two HCV envelope glycoproteins (E1 and E2) that have an important role in infection (Yi *et al.*, 1997). Human and bovine Lf both specifically bind E1 and E2 *in vitro*, as indicated by Western blots showing that E1 and E2 probes bind to bLf but not to bovine serum albumin and bind to hLf and bLF with similar affinity (Yi *et al.*, 1997). A direct interaction between E2 and Lf was also proved in intact cells, since hLf and E2 can form complexes not only within cells but also in cell culture supernatants when both are expressed by Hep G2 cells. In this study the N-terminal loop of Lf, the region important for its antibacterial activity, exerted only a minor influence on binding to HCV E2, but affected the secretion or stability of Lf in Hep G2 cells. Thus, the results of this study indicate that the N-terminal loop of hLf has little effect on E2 binding ability, although it is necessary for either secretion or stabilization of lactoferrin (Yi *et al.*, 1997).

The antiviral activity of milk proteins recently has been investigated in relation to several other viruses including rotavirus, respiratory syncytial virus, Semliki Forest virus and bovine herpes virus 2 (Grover *et al.*, 1997; Superti *et al.*, 1997). Grover *et al.* (1997) demonstrated that hLf has antiviral activity *in vitro* against RSV at a concentration similar to, and much lower than, that present in mature human milk. Antiviral activity was not changed when the Lf solution was pasteurized at 62.5°C for 30 min. Consequently, the authors suggested that Lf could be one of the unidentified antiviral factors in human milk whose activity is not affected by pasteurization treatment. Finally, Superti *et al.* (1997) investigated the capacity of some bovine milk proteins to neutralize rotavirus hemagglutination and infectivity in the enterocyte-like cell line HT-29. The proteins investigated were α-lactalbumin, β-lactalbumin, and apo- and Fe^{3+} Lf. Their antiviral action was compared with the activity of mucin, a milk glycoprotein used as a positive control for its known inhibitory activity on rotavirus infection. β-lactalbumin and both apo- and Fe^{3+} Lf inhibited the replication of rotavirus in a dose-dependent manner, apo-Lf being the more active form, displaying a selectivity index that was 27.5, 45, and 160–fold higher than that of Fe^{3+} Lf, mucin and β-lactalbumin, respectively. These results suggest that Lf may prevent rotavirus infection mainly through binding to viral particles. Thus, Lf could have an important role in the prophylaxis and treatment of virus infection in children as well as in adults.

11. Conclusions

Many studies provide evidence that Lf can contribute to antimicrobial host defense. Nevertheless, much more information is needed before we fully understand the biological role and the exact mechanisms of action of this multifunctional milk glycoprotein.

The nutritional and therapeutic applications of Lf have not been completely exploited owing to lack of an efficient and cost-effective method for the production of sufficient quantities of the purified protein. The introduction of recombinant hLf variants (Ward *et al.*, 1997) may prove helpful in this regard. New studies and clinical trials are required to explore the possible applications of Lf in humans and animals. Before advocating the use of this protein in infant nutrition, it must be clearly determined whether bovine and recombinant Lf are equivalent to human Lf.

References

American Academy of Pediatrics. 1997. Breastfeeding and the use of human milk. *Pediatrics* **100**:1035.

Akin, D.T., Lu, M.Q., Lu, S.J., Kendall, S., Rundegren, J., and Arnold, R.R. 1994. Bactericidal activity of different forms of lactoferrin, in: *Lactoferrin: Structure and Function*. Advances in Experimental Medicine and Biology, vol. 357 (T.W. Hutchens, S.V. Rumball, and B. Lönnerdal, eds.), pp. 61–70, Plenum Press, New York, London.

Ambruso, D.R. and Johnston, R.B.J. 1981. Lactoferrin enhances hydroxyl radical production by human neutrophils, neutrophil particulate fractions and an enzymatic generating system. *J. Clin. Invest.* **67**:352.

Anderson, B.F., Baker, H.M., Norris, G.E., Rice, D.W., and Baker, E.N. 1989. Structure of human lactoferrin: crystallographic structure analysis and refinement at 2.8 A resolution. *J. Mol. Biol.* **209**:711.

Arnold, R.R., Cole, M.F., and McGhee, J.R. 1977. A bactericidal effect for human lactoferrin. *Science* **197**:263.

Arnold, R.R., Brewer, M., and Gauthier, J.J. 1980. Bactericidal activity of human lactoferrin: sensitivity of a variety of microorganisms. *Infect. Immun.* **28**:893.

Arnold, R.R., Russel, J.E., Champion, J., and Gauthier, J.J. 1981. Bactericidal activity of human lactoferrin: influence of physical conditions and metabolic state of the target microorganism. *Infect. Immun.* **32**:655.

Arnold, R.R., Russel, J., Champion, W.J., Brewer, M., and Gauthier, J.J. 1981. Bactericidal activity of human lactoferrin: differentiation from the stasis due to iron deprivation. *Infect. Immu.* **35**:792.

Ascencio, F., Ljungh, A., and Wadstrom, T. 1992. Characterization of lactoferrin binding by Aeromonas Hydrophila. *Appl. Environ. Microbiol.* **58**:42.

Baker, E.N., Anderson, B.F., Baker, H.M., Day, C.D., Haridas, M., Norris, G.E., Rumball, S.V., Smith, C.A., and Thomas, D.H. 1994. Three-dimensional structure of lactoferrin in various functional states, in: *Lactoferrin: Structure and Function*. Advances in Experimental Medicine and Biology, vol. 357 (T.W. Hutchens, S.V. Rumball, and B. Lönnerdal, eds.), pp. 1–12, Plenum Press, New York and London.

Balmer, S.E. and Wharton, B.A. 1989. Diet and faecal flora in the newborn: breast milk and infant formula. *Arch. Dis. Child.* **64**:1672.

Balmer, S.E., Scott, P.H., and Wharton, B.A. 1989. Diet and faecal flora in the newborn: lactoferrin. *Arch. Dis. Child.* **64**:1685.

Bellamy, W., Takase, M., Yamauchi, K., Wakabayashi, H., Kawase, K., and Tomita, M. 1992. Identification of the bactericidal domain of lactoferrin. *Biochim. Biophys. Acta* **1121**:130.

Bellamy, W., Wakabayashi, H., Takase, M., Kawase, K., Shimamura, S., and Tomita, M. 1993. Killing of *Candida albicans* by lactoferricin B, a potent antimicrobial peptide derived from N-terminal region of bovine lactoferrin. *Med. Microbiol. Immunol.* **182**:97.

Benno, Y., Sawada, K., and Mitsuoka, T. 1984. The intestinal microflora of infants: composition of fecal flora in breast-fed and bottle-fed infants. *Microbiol. and Immunol.* **28**:975.

Blaser, M.J. 1992. Hypotheses on the pathogenesis and natural history of Helicobacter pylori-induced inflammation. *Gastroenterology* **102**:720.

Bonnah, R.A., Yu, R., and Schryvers A.B. 1997. Bacterial lactoferrin receptors in the Neisseriaceae, in: *Lactoferrin: Interaction and Biological Functions* (T.W. Hutchens and B. Lönnerdal, eds.), pp. 277–301, Humana Press, Totowa, New Jersey.

Boxer, L.A., Haak, R.A., Yang, H.-H., Wolach, J.B., Whitcomb, J.A., Butterick, C.J., and Baehner, R.L. 1982. Membrane–bound lactoferrin alters the surface properties of polymorphonuclear leukocytes. *J. Clin. Invest.* **70**:1049.

Brines, R.D. and Brock, J.H. 1983. The effect of trypsyn and chymotrypsyn on the in vitro antimicrobial and iron binding properties of lactoferrin in human milk and bovine colostrum. Unusual resistance of human apolactoferrin to proteolytic digestion. *Biochim. Biophys. Acta* **759**:229.

Britigan, B.E., Serody, J.S., and Cohen, M.S. 1994. The role of lactoferrin as an anti-inflammatory molecul, in: *Lactoferrin: Structure and Function*. Advances in Experimental Medicine and Biology, vol. 357 (T.W. Hutchens, S.V. Rumball, S.V., and B. Lönnerdal, eds.), pp. 1143–1156, Plenum Press, New York and London.

Britigan, B.E., Lewis, T.S., Olakanmi, O., McCormick, M.L., and Miller, R.A. 1997. The impact of phagocyte-lactoferrin interaction on inflammation, in: *Lactoferrin Interaction and Biological Functions* (T.W. Hutchens and B. Lönnerdal, eds.), pp. 211–232, Humana Press, Totowa, New Jersey.

Brock, J.H. 1997. Lactoferrin structure-function relationships: an overview, in: *Lactoferrin Interaction and Biological Functions* (T.W. Hutchens and B. Lönnerdal, eds.), pp. 259–275, Humana Press, Totowa, New Jersey.

Broxmeyer, H.E., DeSousa, M., Smithyman, A., Ralon, P., Hamilton, J., Kurland, J.L., and Bognacki, J. 1980. Specificity and modulation of the action of lactoferrin, a negative feedback regulator of myelopoiesis. *Blood* **55**:324.

Bullen, J.J., Rogers, H.J., and Leigh, L. 1972. Iron-binding proteins in milk and resistance to *Escherichia coli* infection in infants. *Br. Med. J.* **i**:69.

Cohen, M.S., Mao, J., Rasmussen, G.T., Serody, J.S., and Britigan, B.E. 1992. Interaction of lactoferrin and lipopolysaccharide (LPS): effect on the antioxidant property of lactoferrin and the ability of LPS to prime human neutrophils foe enhanced superoxide formation. *J. Infect. Dis.* **166**:1375.

Crosa, J.H. 1989. Genetics and molecular biology of siderophore-mediated iron transport in bacteria. *Microbiol. Rev.* **53**:517.

Cuilliere, M.L., Montagne, P., Molé, C., Béné, M.C., and Faure, G. 1997. Microparticle-enhanced nephelometric immunoassay of lactoferrin in human milk. *J. Clin. Lab. Anal.* **11**:239.

Czirok, E., Milch, H., Nemeth, K., and Gado, I. 1990. In vitro and in vivo (Ld 50) effects of human lactoferrin on bacteria. *Acta Microbiologica Hungarica* **37**:55.

Dalmastri, C., Valenti, P., Visca, P., Vittorioso, P., and Orsi, N. 1988. Enhanced antimicrobial activity of lactoferrin by binding to the bacterial surface. *Microbiologica* **11**:225.

Dhaenens, L., Szczebara, F., and Husson, M.O. 1997. Identification, characterization and, immunogenicity of the lactoferrin-binding protein from Helicobacter pylori. *Infect. Immun.* **65**:514.

Dial, E.J., Serna, J.A., and Lichtenberger, L.M. 1995. Lactoferrin inhibits the growth of Helicobacter pylori in vitro. *Gastroenterology* **108**:A82.

Dial, E.J., Hall, L.R., and Lichtenberger, L.M. 1996. Antimicrobial properties of lactoferrin which limit Helicobacter pylori (HP) growth. *Gastroenterology* **110**:A95.

Drescher, K., Roos, N., Pfeuffer, M., Seyfert, H.-M., Schrezenmeir, J., and Hagemeister, H. 1999. Recovery of ^{15}N-lactoferrin is higher than that of ^{15}N-casein in the small intestine of suckling, but not adult miniature pigs. *J. Nutr.* **129**:1026.

Ellison, R.T., Theodore, G.J., and LaForce, F.M. 1988. Damage to outer membrane of enteric gram-negative bacteria by lactoferrin and transferrin. *Infect. Immun.* **56**:2774.

Ellison, R.T. III and Giehl, T.J. 1991. Killing of Gram-negative bacteria by lactoferrin and lysozyme. *J. Clin. Invest.* **88**:1080.

Ellison, R.T. III. 1994. The effect of lactoferrin on gram-negative bacteria, in: *Lactoferrin: Structure and Function*. Advances in Experimental Medicine and Biology, vol. 357 (T.W. Hutchens, S.V. Rumball, and B. Lönnerdal, eds.), pp. 71–90, Plenum Press, New York, London.

Fleet, J.C. 1995. A new role for lactoferrin: DNA binding and transcription activation. *Nutr. Rev.* **53**:226.

Ghar, M., Speer, C.P., Damerau, B., and Sawatzki, G. 1991. Influence of lactoferrin on the function of human polymorphonuclear leucocytes and monocytes. *J. Leukoc. Biol.* **49**:427.

Giugliano, L.G., Ribeiro, S.T.G., Vainstein, M.H., and Hulhoa, G.J. 1995. Free secretory component and lactoferrin of human milk inhibit the adhesion of enterotoxigenic Escherichia coli. *J. Med. Microbiol.* **42**:3.

Goldman, A.S., Garza, C., Johnson, C.A., Nichols, B.L., and Goldblum, R.M. 1982. Immunologic factors in human milk during the first year of lactation. *J. Pediatr.* **100**:563.

Goldman, A.S., Garza, C., Schanler, R.J., and Golblum, R.M. 1990. Molecular forms of Lactoferrin in stool and urine from infants fed human milk. *Pediatr. Res.* **27**:252.

Goodman, R.E. and Schanbacher, F.L. 1991. Bovine lactoferrin mRNA: sequence, analysis and expression in the mammary gland. *Biochem. Biophys. Res. Commun.* **180**:75.

Gray-Owen, S.D. and Schryvers, A.B. 1996. Bacterial transferrin and lactoferrin receptors. *Trends in Microbiol.* **4**:185.

Griffiths, E. and Humphreys, J. 1977. Bacteriostatic effect of human milk and bovine colostrum on *Escherichia coli*: importance of bicarbonate. *Infect. Immun.* **15**:396.

Grover, M., Giouzeppos, O., Schnagl, R.D., and May, J.T. 1997. Effect of human milk prostaglandins and lactoferrin on the respiratory syncytial virus and rotavirus. *Acta Paediatr.* **86**:315.

Groves, M.L. 1960. The isolation of a red protein from milk. *J. Am. Chem. Soc.* **82**:3345.

Guerinot, M.L. 1994. Microbial iron transport. *Ann. Rev. Microbiol.* **48**:743.

Haridas, M., Anderson, B.F., Baker, H.M., Norris, G.E., and Baker, E.N. 1994. X-ray structural analysis of bovine lactoferrin at 2.5 A resolution, in: *Lactoferrin: Structure and Function*, Advances in Experimental Medicine and Biology, vol. 357 (T.W. Hutchens, S.V. Rumball, and B. Lönnerdal, eds.), pp. 235–238, Plenum Press, New York and London.

Harmsen, M.C., Swart, P.J., deBethune, M.P., Pauwels, R., De Clercq, E., The, T.H., and Meijer, D.K. 1995. Antiviral effects of plasma and milk proteins: lactoferrin shows potent activity against both human immunodeficiency virus and human cytomegalovirus replication in vitro. *J. Infect. Dis.* **172**:380.

Hennart, P.F., Brasseur, D.J., Delogne-Desnoeck, B., Dramaix, M.M., and Robyn, C.E. 1991. Lysozyme, lactoferrin, and secretory immunoglobulin A content in breast milk: influence of duration of lactation, nutrition status, prolactin status, and parity of mother. *Am. J. Clin. Nutr.* **53**:32.

Hentges, D.J., Marsh, W.W., Petschow, B.W., Thal, V.R., and Carter, M.K. 1992. Influence of infant diets on the ecology of the intestinal tract of human flora-associated mice. *J. Pediatr. Gastroenterol. Nutr.* **14**:146.

Houghton, M.R., Gracey, M., Burke, V., Bottrel, C., and Spargo, R.M. 1985. Breast milk lactoferrin levels in relation to maternal nutritional status. *J. Pediatr. Gastroenterol. Nutr.* **4**:230.

Hussel, T., Isaac son, P.F., Crabtree, J.C., and Spencer, J. 1993. The response of cells from low-grade B-cell gastric lymphomas of mucosa associated lymphoid tissue to Helicobacter pylori. *Lancet* **342**:571.

Husson, M.O., Legrand, D., Spik, G., and Leclerc, H. 1993. Iron acquisition by Helicobacter pylori: importance of human lactoferrin. *Infect. Immun.* **61**:2694.

Hutchens, T.W., Henry, J.F., and Yip, T.-T. 1991. Structurally intact (78-kDa) forms of maternal lactoferrin purified from urine of preterm infants fed human milk: identification of a trypsin-like proteolytic cleavage event *in vivo* that does not result in fragment dissociation. *Proc. Natl. Acad. Sci. USA* **88**:2994.

Iyer, S. and Lönnerdal, B. 1993. Lactoferrin, lactoferrin receptors and iron metabolism. *Eur. J. Clin. Nutr.* **47**:232.

Johansson, B. 1960. Isolation of an iron-containing red protein from human milk. *Acta Chem. Scand.* **14**:510.

Kishore, A.R., Erdei, J., Naidu, S.S., Falsen, E., Forsgren, A., and Naidu, A.S. 1991. Specific binding of lactoferrin to *Aeromonas hydrophila*. *FEMS Microbiol. Lett.* **83**:115.

Lee, B.C. and Schryvers, A.B. 1988. Specificity of the lactoferrin and transferrin receptors in *Neisseria gonorhoeae*. *Mol. Microbiol.* **2**:827.

Lehker, M.W. and Alderete, J.F. 1992. Iron regulates growth of *Trichomonas vaginalis* and the expression of immunogenic trichomonad proteins. *Mol. Microbiol.* **6**:123.

Levay, P.F. and Viljoen, M. 1995. Lactoferrin: a general review. *Haematologica* **80**:252.

Leveugle, B., Mazurier, J., Legrand, D., Mazurier, C., Montreuil, J., and Spik, G. 1993. Lactotransferrin binding to its platelet receptor inhibits platelet aggregation. *Eur. J. Biochem.* **213**:1205.

Lönnerdal, B. 1984. Iron and breast milk, in: *Iron Nutrition in Infancy and Childhood.* Nestlé Nutrition Workshop Series, vol. 4 (A. Stekel, ed.), pp. 95–117, Raven Press, New York.

Lönnerdal, B. and Atkinson, S. 1995. Nitrogenous components of milk. Human milk proteins, in: *Handbook of Milk Composition* (R.G. Jensen, ed.), pp. 351–368, Academic Press, New York, London, Tokyo.

Machnicki, M., Zimecky, M., and Zagulski, T. 1993. Lactoferrin regulates the release of tumour necrosis factor alpha and interleukin 6 *in vivo*. *Int. J. Exp. Pathol.* **74**:433.

Magnuson, J.S., Henry, J.F., Yip, T.-T., and Hutchens, T.W. 1990. Structural homology of human, bovine and porcine milk lactoferrins: evidence for shared antigenic determinants. *Pediatr. Res.* **28**:76.

Mann, D.M., Romm, E., and Migliorini, M. 1994. Delineation of the glycosaminoglycan-binding site in the human inflammatory response protein lactoferrin. *J. Biol. Chem.* **269**:23661.

Marchetti, M., Longhi, K., Conte, M.P., Pisani, S., Valenti, P., and Seganti, L. 1996. Lactoferrin inhibits herpes simplex virus type 1 absorption to Vero cells. *Antiviral Res.* **29**:221.

Masson, P.L., Heremans, J.F., and Dive, C. 1966. An iron-binding protein common to many external secretions. *Clin. Chim. Acta* **14**:735.

Masson, P.L., Heremans, J.F., and Schonne, E. 1969. Lactoferrin and iron-binding protein in neutrophilic leukocytes. *J. Exp. Med.* **130**:643.

Masson, P.L. and Heremans, J.F. 1971. Lactoferrin in milk from different species. *Comp. Biochem. Physiol.* **39B**:119.

Mead, P.E. and Tweedie J.W. 1990. cDna and bovine lactoferrin. *Nucleic Acids Res.* **18**:7167.

Metz-Boutigue, M.H., Jolles, J., Mazurier, J., Schoentgen F., Legrand, D., Spik, G., Montreuil, J., and Jolles, P. 1984. Human lactotransferrin: amino acid sequence and structural comparison with other transferrins. *Eur. J. Biochem.* **145**:659.

Miyazawa, K., Mantel, C., Lu, L., Morrison, D.C., and Broxmeyer, H.E. 1991. Lactoferrin-lipopolysaccharide interactions. Effect on lactoferrin binding to monocyte/macrophage differentiated HL-60 cells. *J. Immunol.* **146**:723.

Naidu, A.S., Erdei, J., Czirok, E., Kalfas, S., Gado, I., Thoren, A., Forsgren, A., and Naidu, A.S. 1991a. Specific binding of lactoferrin to Escherichia coli isolated from human intestinal infections. *APMIS* **99**:1142.

Naidu, A.S., Andersson, M., Miedzobrodzki, J., Forsgren, A., and Watts, J.L. 1991b. Bovine lactoferrin receptors in Staphylococcus aureus isolated from bovine mastitis. *J. Dairy Sci.* **74**:1218.

Naidu, A.S., Andersson, M., and Forsgren, A. 1992. Identification of a human lactoferrin-binding protein in Staphylococcus aureus. *J. Med. Microbiol.* **36**:177.

Naidu, A.S. and Arnold, R.R. 1997. Influence of lactoferrin on host-microbe interactions, in: *Lactoferrin Interaction and Biological Functions* (T.W. Hutchens and B. Lönnerdal, eds.), pp. 259–275, Humana Press, Totowa, New Jersey.

Nakao, K., Imoto, I., Gabazza, E.C., Yamauchi, K., Yamazaki, Y., and Taguchi, Y. 1997a. Gastric juice levels of lactoferrin and Helicobacter pylori infection. *Scand. J. Gastroenterol.* **32**:530.

Nakao, K., Imoto, I., Ikemura, N., Shibata, T., Takaji, S., and Taguchi, Y. 1997b. Relation of lactoferrin levels in gastric mucosa with Helicobacter pylori infection and with the degree of gastric inflammation. *Amer. J. Gastroenterol.* **92**:1005.

Ogunnariwo, J.A. and Schryvers, A.B. 1990. Iron acquisition in *Pasteurella haemolytica:* expression and identification of a bovine-specific transferrin receptor. *Infect Immun.* **58**:2091.

Okutomi, T., Abe, S., Tansho, S., Wakabayashi, H., Kawase, K., and Yamaguchi, H. 1997. Augmented inhibition of growth of *Candida albicans* by neutrophils in the presence of lactoferrin. *FEMS Immunol. Med. Microbiol.* **18**:105.

Oram, J.D. and Reiter, B. 1968. Inhibition of bacteria by lactoferrin and other iron-chelating agents. *Biochim. Biophys. Acta* **170**:351.

Oseas, R., Yang, H.H., Baehner, R.L., and Boxer, L.A. 1981. Lactoferrin: a promoter of polymorphonuclear leukocyte adhesiveness. *Blood* **57**:939.

Parsonnet, J., Freedman, G.D., Vandersten, D.P., Chang, V.R., Vogelman, J.H., and Orentreich, N. 1991. Helicobacter pylori infection and risk of gastric carcinoma. *N. Engl. J. Med.* **325**:1127.

Peterson, K.M. and Alderete, J.F. 1984. Iron uptake and increased intracellular enzyme activity follow host lactoferrin binding by Trichomonas vaginalis receptors. *J. Exp. Med.* **160**:398.

Petschow, B.W. and Talbot, R.D. 1991. Response of bifidobacterium species to growth promoters in human and cow milk. *Pediatr. Res.* **29**:208.

Pierce, A., Colavizza, D., Benaissa, M., Maes, P., Tartar, A., Montreuil, J., and Spik, G. 1991. Molecular cloning and sequence analysis of bovine lactotransferrin. *Eur. J. Biochem.* **196**:177.

Reddy, V., Bhaskaram, C., Raghuramulu, N., and Jagadeesan, V. 1977. Anti-microbial factors in human milk. *Acta Paediatr. Scand.* **66**:229.

Renner, E., Schaafsma, G., and Scott, K.J. 1989. Micronutrients in milk, in: *Micronutrients in Milk and Milk-based Food Products* (E. Renner, ed.), pp. 1–70, Elsevier Applied Science, London, New York.

Rey, M.W., Woloshuk, S.L., deBoer, H.A., and Pieper, F.R. 1990. Complete nucleotide sequence of human mammary gland lactoferrin. *Nucleic Acid Res.* **18**:5288.

Roberts, A.K., Chierici, R., Sawatzki, G., Volpato, S., and Vigi, V. 1992. Supplementation of an adapted formula with bovine lactoferrin: 1. Effect on the infant faecal flora. *Acta Paediatr.* **81**:119.

Saito, H., Miyakawa, H., Tamura, Y., Shimamura, S., and Tomita, M. 1991. Potent bactericidal activity of bovine lactoferrin hydrolysate produced by heat treatment at acidic pH. *J. Dairy Sci.* **74**:3724.

Sanchez, L., Aranda, P., Perez, M.D., and Calvo, M. 1988. Concentration of lactoferrin and transferrin throughout lactation in cow's colostrum and milk. *Biol. Chem. Hoppe Seyler.* **369**:1005.

Sanchez, L., Calvo, M., and Brock, J.H. 1992. Biological role of lactoferrin. *Arch. Dis. Child.* **67**:657.

Schryvers, A.B. and Morris, L.J. 1988. Identification and characterization of the human lactoferrin-binding from Neisseria meningitidis. *Infect. Immun.* **56**:1144.

Shin, K., Yamauchi K., Teraguchi, S., Hayasawa H., Tomita, M., Otsuka, Y., and Yamazaki, S. 1998. Antibacterial activity of bovine lactoferrin and its peptides against enterohaemorragic *Escherichia coli* 0157:H7. *Lett. Appl. Microbiol.* **26**:407.

Soukka, T., Tenovuo, J., and Lenander-Lumikari, M. 1992. Fungicidal effect of human lactoferrin against *Candida albicans. FEMS Microbiol. Lett.* **69**:223.

Spik, G., Brunet, B., Mazurier-Dehaine, C., Fontaine, G., and Montreuil, G. 1982. Characterization and properties of the human and bovine lactotransferrins extracted from the faeces of newborn infants. *Acta Paediatr. Scand.* **71**:979.

Spik, G., Coddville, B., and Montreuil, J. 1988. Comparative study of the primary structures of sero-lacto and ovotransferrin glycans from different species. *Biochimie* **70**:1459.

Stephens, S., Dolby, J.M., Montreuil, J., and Spik, G. 1980. Differences in inhibition of the growth of commensal and enteropathogenic strains of *Escherichia coli* by lactotranferrin and secretory immunoglobulin A isolated from human milk. *Immunology* **41**:597.

Superti, F., Ammendolia, M.G., Valenti, P., and Seganti, L. 1997. Antirotaviral activity of milk proteins: lactoferrin prevents rotavirus infection in the enterocyte-like cell line HT-29. *Med. Microbiol. Immunol.* **186**:83.

Swart, P.J., Kuipers, M.E., Smit, C., Pauwels, R., deBethune, M.P., and de Clercq, E. 1996. Antiviral effect of milk proteins: acylation results in polyanionic compounds with potent activity against human immunodeficiency virus types 1 and 2 in vitro. *AIDS Res. Hum. Retroviruses* **12**:769.

Tachezy, J., Kulda, J., Bahnikova, I., Suchan, I., Razga, J., and Schrevel, J. 1996. *Tritrichomonas foetus*: iron acquisition from lactoferrin and transferrin. *Experimental Parasitology* **83**:216.

Teraguchi, S., Shin, K., Ozawua, K., Nakamura, S., Fukuwatari, Y., Shimamura, S., and Tomita, M. 1997. Bacteriostatic effects of orally administered bovine lactoferrin on intestinal bacteria in the gut of mice fed bovine milk, in: *Lactoferrin Interaction and Biological Functions* (T.W. Hutchens, and B. Lönnerdal, eds.), pp. 303–312, Humana Press, Totowa, New Jersey.

Tigyi, Z., Kishore, A.R., Maeland, J.A., Forsgren, A., and Naidu, A.S. 1992. Lactoferrin-binding proteins in *Shigella flexneri*. *Infect. Immun.* **60**:2619.

Tomita, M., Bellamy, W., Takase, M., Yamauchi, K., Wakabayashi, H., and Kawase, K. 1991. Potent antibacterial peptides generated by pepsin digestion of bovine lactoferrin. *J. Dairy Sci.* **74**:4137.

Tomita, M., Takase, M., Wakabayashi, H., and Bellamy, W. 1994. Antimicrobial peptides of lactoferrin, in: *Lactoferrin: Structure and Function*, Advances in Experimental Medicine and Biology, vol. 357 (T.W. Hutchens, S.V. Rumball, and B. Lönnerdal, eds.), pp. 209–218, Plenum Press, New York and London.

Tsuji, S., Hirata, H., Mukai, F., and Ohtagaki, S. 1990. Comparison of lactoferrin content in colostrum between different cattle breeds. *J. Dairy Sci.* **73**:125.

van Berkel, H.C., Geerts, M.E.G., van Veen, H.A., Kooiman, P.M., Pieper, F.R., De Boer H.A., and Nuijens J.H. 1995. Glycosylated and unglycosylated human lactoferrin both bind iron and show identical affinities towards human lysozyme and bacterial lipopolysaccharide, but differ in their susceptibilities towards tryptic proteolysis. *Biochem. J.* **312**:107.

van Berkel, H.C., van Veen, H.A., Geerts, M.E.G., De Boer H.A., and Nuijens J.H. 1996. Heterogeneity in utilization on N-glycosylation sites Asn 624 and Asn 138 in human lactoferrin: a study with glycosylation-site mutans. *Biochem. J.* **319**:177.

Veken, J.W., Shah, N.H., Klaasen, P., Oudega, B., and de Graaf, F.K. 1996. Binding of host iron-binding proteins and expression of iron-regulated membrane proteins by different serotypes of *Pasteurella multocida* causing haemorrhagic septicaemia. *Microbial. Pathogenesis* **21**:59.

Visca, P., Dalmastri, C., Verzili, D., Antonini, G., Chiancone, E., and Valenti, P. 1990. Interaction of lactoferrin with *Escherichia coli* cells and correlation with antibacterial activity. *Med. Microbiol. Immunol.* **179**:323.

Wakabayashi, H., Abe, S., Okutomi, T., Tansho, S., Kawase, K., and Yamaguchi, H. 1996. Cooperative anti-Candida effects of lactoferrin or its peptides in combination with azole antifungal agents. *Microbiol. Immunol.* **40**:821.

Ward, P.P., Piddington, C.S., Cunningham, G., Zhou, X., Wyatt, R.D., and Conneely, O. 1997. Expression and functional analysis of recombinant human lactoferrin, in: *Lactoferrin:*

Interactions and Biological Functions (T.W. Hutchens and B. Lönnerdal, eds.), pp. 155–173, Humana Press, Totowa, New Jersey.

Wu, H.F., Monroe, D.M., and Church, F.C. 1995. Characterization of the glycosaminoglycan-binding region of lactoferrin. *Arch. Biochem. Biophys.* **317**:85.

Yi, M., Kaneko, S., Yu, D.Y., and Murakami, S. 1997. Hepatitis C virus envelope proteins bind lactoferrin. *J. Virol.* **71**:5997.

Yoshioka, H., Iseki, K., and Fujita, K. 1983. Development and differences of intestinal flora in the neonatal period in breast-fed and bottle-fed infants. *Pediatrics* **72**:317.

Zagulski, T., Lipinski, P., Zagulska, A., Broniek, S., and Jarzabek, Z. 1989. Lactoferrin can protect mice against a lethal dose of *Escherichia coli* in experimental infection in vivo. *Br. J. Exp. Pathol.* **70**:697.

Zimecki, M., Mazurier, J., Spik, G., and Kapp, J.A. 1966. Lactoferrin inhibits proliferative response and cytokine production of TH1 but not TH2 cell lines. *Arch. Immunol. Ther. Exp.* **44**:51.

Chapter 13

The Antimicrobial Function of Milk Lipids

Charles E. Isaacs

1. Introduction

Milk is not only a source of nutrients for the newborn but also a source of immunoglobulins and non-immunoglobulin secretory products, similar to those of other parts of the secretory immune system, which play an important role in protecting mucosal surfaces from infection. Infants who are breast-fed have been found to have a lowered incidence of gastrointestinal infections than infants fed formula or cow's milk (Larsen and Homer, 1978; Cunningham, 1979; Myers *et al.*, 1984). The incidence of any infection in very low birthweight infants is significantly lower in infants fed human milk than in infants fed formula (Hylander *et al.*, 1998). While early studies of milk attributed this protective action to immunoglobulins, primarily secretory IgA (sIgA), other studies suggest that nonspecific protective factors play an important role in the initial defense against infection (Welsh and May, 1979).

Nonspecific secretory protective factors in milk are part of an innate defense system that phylogenetically precedes the adaptive immune system, e.g., antibodies, which is found solely in vertebrates (Fearon, 1997). These nonspecific protective factors include lipids, lactoferrin, lactoperoxidase, lysozyme and receptor oligosaccharides (Welsh and May, 1979; Kabara, 1980; Ogra and Losonsky, 1984; Goldman *et al.*, 1985; Newburg *et al.*, 1992; Newburg, 1997). Such nonspecific factors have been shown to inactivate or reduce the infectivity of viruses, bacteria, protozoa and other microorganisms (Welsh and May, 1979;

Charles E. Isaacs • Department of Developmental Biochemistry, Institute for Basic Research, 1050 Forest Hill Road, Staten Island, NY 10314

Ogra and Losonsky, 1984; Goldman *et al.*, 1985). Non-immunoglobulin secretory products can provide a first line of protection at mucosal surfaces when mother and infant lack sufficient concentrations of antibodies to protect against a particular pathogen. Microbial inactivation measured using individual purified antimicrobial factors may underestimate the effectiveness of these factors *in vivo* because antimicrobial factors in milk interact. For example, sIgA, peroxidase and lactoferrin produce greater antimicrobial activity than the sum of their activities measured individually (Adinolfi and Glynn, 1979; Reiter, 1981; Moldoveanu *et al.*, 1983; Watanabe *et al.*, 1984).

Nonspecific protective factors such as lipids can prevent or inhibit the establishment, multiplication and spread of pathogenic microorganisms in the host (Mandel and Ellison, 1985). Milk lipids inactivate enveloped viruses (Isaacs *et al.*, 1986), protozoa (Hernell *et al.*, 1986) and bacteria (Isaacs *et al.*, 1990). Enveloped viruses inactivated by human milk lipids include herpes simplex virus (HSV) (Isaacs *et al.*, 1986), respiratory syncytial virus (Laegreid *et al.*, 1986), measles virus (Isaacs *et al.*, 1986), vesicular stomatitis virus (VSV) (Isaacs *et al.*, 1986), visna virus (Isaacs *et al.*, 1986), mouse mammary tumor virus (Sarkar, 1973), dengue virus types 1–4 (Falkler *et al.*, 1975), cytomegalovirus (CMV) (Welsh *et al.*, 1979) and Semliki forest virus (Welsh *et al.*, 1978, 1979). Milk lipids also inactivate both Gram positive bacteria (*Staphylococcus epidermidis*) and Gram negative bacteria (*Escherichia coli*) (Isaacs et al., 1990) as well as the protozoal pathogen Giardia lamblia (Hernell *et al.*, 1986; Reiner *et al.*, 1986; Rohrer *et al.*, 1986). Milk lipids therefore have the potential to inactivate pathogens present in the gastrointestinal tract. They may be especially important for malnourished or low birthweight infants, since even in an adequately nourished normal birthweight infant T lymphocytes have lower levels of immune interferon production, T cell mediated cytotoxicity and lymphokine production relative to the T cell activities of adolescents and adults (Bryson *et al.*, 1980; Winter *et al.*, 1983).

2. Source of Lipid-Dependent Antimicrobial Activity

Milk lipids are not unique in possessing antimicrobial activity. Human epidermis-derived skin lipids, especially free fatty acids, have been found to inactivate *Staphylococcus aureus* (Aly *et al.*, 1972; Kearney *et al.*, 1984; Miller *et al.*, 1988; Bibel *et al.*, 1989). Extraction of skin with acetone permitted added bacteria to persist on the skin of volunteers and re-application of the extract inactivated the bacteria. Similarly, lung surfactant from humans, dogs, rats and guinea pigs contains free fatty acids that are bactericidal for pneumococci (Coonrod and Yoneda, 1983; Coonrod, 1986, 1987). Inhaled pneumococci were killed extracellularly in rats (Coonrod, 1987) and the antibacterial activity was found to reside in the long-chain polyunsaturated free fatty acids purified from lung surfactant providing further support for the potential antimicrobial role of surfactant free fatty acids. Since surfactant-like particles, which are biochemically similar to lung surfactant, have been isolated from adult rat intestine (Eliakim *et al.*, 1989), both endogenous and milk-derived free fatty acids may have an antimicrobial function in the intestine of the suckling neonate.

Free fatty acids also have been found to have antimicrobial activity in nine species of brown algae (Rosell and Srivastava, 1987). Both Gram positive and Gram negative bacteria were inactivated by algal lipids. In another study on Caribbean marine algae, over 70% of the lipid extracts from approximately 100 algae tested had antibacterial activity (Ballantine *et al.*, 1987). Fatty acids, therefore, not only have a protective role at the mucosal surfaces of vertebrates but also appear to have antibacterial activity in phylogenetically primitive eucaryotes.

Welsh *et al.* (1978, 1979) suggested that milk lipid-dependent antimicrobial activity resulted from the hydrolysis of triglycerides catalysed by endogenous milk lipases. Studies on the stomach contents of suckling rabbits established that their "stomach oil" had strong antibacterial and antifungal activity (Canas-Rodriguez and Smith, 1966; Smith, 1966). The most active antimicrobial factors isolated from rabbit "stomach oil" were free n-decanoic acid and n-octanoic acid. Antimicrobial activity was not found in rabbit milk, suggesting that this activity originated in the stomach.

3. Lipid-Dependent Antimicrobial Activity in Human Milk

Fat in milk is present mainly in the form of globules, which can be transported in an aqueous medium (Jensen *et al.*, 1990; Jensen, 1996). The fat globule has a triglyceride core that comprises 98% of total lipid in milk (Hamosh, 1991a,b). Following milk expression the lipids in human milk are not immediately antimicrobial (Isaacs *et al.*, 1986). Moreover, if milk is frozen at –80°C (Isaacs *et al.*, 1986) the lipid fraction does not became active against microorganisms. However, sIgA-depleted human milk stored at 4°C for intervals varying between 2 and 7 days inactivated enveloped RNA and DNA viruses (measles virus, HSV and VSV) but not nonenveloped viruses (polio virus and vaccinia virus) (Table 1). Further, human milk stored at 4°C consistently produced significant lipid-dependent antiviral activity within 5 days, whereas antiviral activity in milk stored at –20°C took as long as 50 days to appear. The antiviral activity that appeared following milk storage was localized in the lipid fraction following centrifugation and was not lost following treatment with antibodies directed against secretory IgA. Human milk lipids diluted five-fold following storage at 4°C still reduced the concentration of infectious enveloped viruses by at least

Table 1. Human Milk Inactivates Enveloped Viruses following Storage at 4°C[a]

Source of milk	Herpes simplex virus[b]	Measles virus[b]	Poliovirus[c]
Freshly expressed	0	0	0
Stored at −80°C	0	0	0
Stored at 4°C	99.9%[3]	99.9%	0

[a] Percentage reduction in viral titer; [b] enveloped virus; [c] nonenveloped virus.

1,000-fold. Thus, milk storage studies using temperatures ranging from −80°C to 23°C indicated that the lipid fraction of human milk became antimicrobial more rapidly as the temperature was raised suggesting that an enzymatic activity was involved in the process. Published studies (Isaacs *et al.*, 1986) showed that when the lipid and aqueous fractions of human milk were separated prior to storage, antiviral activity was not produced whereas the lipid fraction was antiviral following combined storage.

Neither triglycerides nor diglycerides inactivate microbes. However, human milk contains two lipases capable of converting milk triglycerides to free fatty acids and monoglycerides that are antimicrobial (Hernell and Blackberg, 1985; Freed *et al.*, 1987). The primary lipolytic activity in human milk is bile salt-stimulated lipase (BSSL) and there are also traces of lipoprotein lipase (LPL) (Hernell and Blackberg, 1985; Freed *et al.*, 1987). BSSL lipolytic activity in most human milk samples is several hundred-fold greater than that of LPL (Isaacs *et al.*, 1986). BSSL has been shown to resist acid inactivation in the stomach and to function in the milk-fed infant's intestinal lumen, whereas LPL is inactivated in the stomach and has not been found to function in the gastrointestinal tract. It has been suggested that LPL leaks from damaged mammary gland cells during milk secretion (Hamosh and Scow, 1971).

All of the human milk samples had sufficient BSSL activity to release fatty acids from human milk triglycerides in the presence of bile salts, but the samples that developed antimicrobial activity following storage at 4°C were those with relatively high levels of LPL (Table 2) (Isaacs, *et al.*, 1986). If purified LPL is added to milk samples with naturally low levels of LPL, antimicrobial activity appears at the same level as in samples with endogenous LPL activity. That BSSL is not required for this process is also indicated by studies on fresh unpasteurized cow's milk, which contains LPL but not BSSL and which becomes antimicrobial during storage (Isaacs *et al.*, 1986).

To determine whether milk-derived lipid-dependent antimicrobial activity is present *in vivo*, the stomach contents from infants fed unpasteurized human milk via a nasogastric tube were examined for antiviral activity (Isaacs *et al.*,

Table 2. Evidence That the Appearance of Antiviral Activity in Stored Human Milk Requires Lipoprotein Lipase Activity

Sample	Bile salt-stimulated lipase (units)	Lipoprotein lipase (units)	Decrease in viral titer
1	1.55	159	99.97%
2	9.51	20	0
2[a]	9.51	220	99.9%

[1] Purified lipoprotein lipase (200 units) was added to the sample prior to storage at 4°C for 9 days (Isaacs *et al.*, 1986).

1986, 1990). The milk fed to the infants lacked lipid-dependent antiviral activity. At one hour after feeding all samples of stomach contents had antiviral activity, which was localized in the lipid fraction as it was in stored milk (Isaacs *et al.*, 1986). Lipid-dependent antiviral activity could be detected in stomach contents for as long as 3 hours after feeding. The release of antimicrobial lipids in the infant's gastrointestinal tract is likely due to the combined actions of lingual, gastric, pancreatic and milk lipases (Hamosh, 1991b).

The appearance of lipid-dependent antimicrobial activity in stored human milk is probably the result of milk handling procedures that disrupt the milk fat globule membrane and make milk triglycerides available to milk lipase. Fatty acid release in the stomach and proximal small intestine is much more rapid than in stored milk in which antimicrobial activity does not appear for at least two days at 4°C. This is undoubtedly due to the greater activity of lipases at 37°C than at 4°C and to the efficient penetration of the milk fat globule by lingual and gastric lipases, initiating the hydrolysis of the core triglyceride (Hamosh, 1991b).

There have been reports that BSSL itself has antimicrobial activity (Gillin and Reiner, 1983) and is directly able to inactivate the parasitic intestinal protozoan *Giardia lamblia*. However, subsequent studies have shown that protozoal inactivation is due to the release of antimicrobial fatty acids following stimulation of BSSL by bile salts, and that even when bile salts are added to purified BSSL in the presence of *G. lamblia* no protozoal killing is found (Hernell *et al.*, 1986). Therefore, the lipid-dependent antimicrobial activity in human milk appears to be due to the hydrolysis of milk triglycerides by lipase activity.

4. Antimicrobial Fatty Acids and Monoglycerides Derived from Human Milk

The antimicrobial activity produced by hydrolysis of milk triglycerides can be duplicated using purified fatty acids and monoglycerides (Thormar *et al.*,

1987). The short-chain fatty acids present in milk (butyric, caproic and caprylic) and the long-chain acids (palmitic and stearic), have little or no antiviral effect even at exceedingly high concentrations (Thormar *et al.*, 1987). The medium-chain saturated and long-chain unsaturated fatty acids are all antiviral but at different concentrations. Table 3 shows the lowest concentration of fatty acids causing a ≥10,000-fold reduction in the concentration of enveloped viruses. Incubation of the nonenveloped poliovirus with capric, lauric, myristic, palmitoleic, oleic, linoleic and arachidonic acids did not significantly reduce viral concentration.

These findings suggest that the viral envelope of enveloped viruses such as HSV and human immunodeficiency virus (HIV) is the target for lipid-dependent viral inactivation (Isaacs and Thormar, 1990). Electron microscopic studies confirmed this indication and showed that the envelope of VSV is completely destroyed by linoleic acid (Thormar *et al.*, 1987). Other purified lipids that are also products of the hydrolysis of milk triglycerides by lipases, e.g., 1-monoglycerides, also effectively inactivate enveloped viruses (Thormar *et al.*, 1987). Studies with purified monoglycerides indicate that they may have an antimicrobial role in the infant's gastrointestinal tract. In most instances the monoglyceride is antiviral at a concentration 5 to 10 times lower than the corre-

Table 3. Minimum Fatty Acid Concentration Required to Effectively Reduce Enveloped Virus Titers[a]

Fatty acid	Concentration (mM)
Butyric (4:0)[b]	113[c]
Caproic (6:0)	86[c]
Caprylic (8:0)	69[c]
Capric (10:0)	22
Lauric (12:0)	10
Myristic (14:0)	16
Palmitic (16:0)	78[c]
Palmitoleic (16:1)	15
Stearic (18:0)	70[c]
Oleic (18:1)	7
Linoleic (18:2)	3.5
Linolenic (18:3)	3.6
Arachidonic (20:4)	1.6

[a]This table is based on data from Thormar *et al.* (1987); [b]carbon atoms: double bonds; [c]highest concentration tested for fatty acids with little or no antiviral activity.

sponding fatty acid (Thormar *et al.*, 1987). Monoglycerides also have strong antibacterial activity (Table 4). Gram positive bacteria are inactivated equally well by monoglycerides with medium-chain saturated and long-chain unsaturated fatty acids, whereas the Gram-negative bacteria *E. coli* and *Salmonella enteritidis* are unaffected by monoglycerides containing long-chain unsaturated fatty acids and monolaurin. However, *Haemophilus influenzae*, a Gram-negative bacterium that causes otitis media in infants, is inactivated by monolaurin. Hence there are varying sensitivities to monoglyceride inactivation among Gram negative bacteria. Differences in susceptibility to inactivation may be due to compositional variability in bacterial cell walls and membranes. Scanning electron micrographs of *S. epidermidis* treated with monolaurin showed that the bacterial membrane was completely destroyed. Studies with enveloped viruses, bacteria and other microorganisms suggest that the mechanism of lipid-dependent antimicrobial activity involves a generalized membrane destabilizing effect in susceptible microorganisms. While the exact mechanism by which antimicrobial fatty acids and their active derivatives destabilize membranes has not been established, lipids which are membrane active have been shown by electron microscopy (Noseda *et al.*, 1989) to produce ruffling of plasma membranes as well as holes and ruptures in the membrane.

Dietary manipulation can alter the fatty acid composition of human and rat milk (Ross *et al.*, 1985; Muskiet *et al.*, 1987). In both species, long-chain unsaturated fatty acids could be replaced by medium-chain saturated fatty acids and vice versa. However, the antimicrobial activity of fatty acids and monoglycerides is additive and therefore it is the total concentration of antimicrobial lipids in milk that is critical for inactivating pathogens (Isaacs and Thormar, 1990). Since lipid-dependent antimicrobial activity resides in both medium-chain saturated and

Table 4. Inactivation of Gram Positive and Gram Negative Bacteria by Monoglycerides

Monoglyceride	E. coli[a]	S. enteritidis[a]	H. influenzae[a]	S. epidermidis[b]	Group B streptococcus[b]	
Monocaprylin	I[c]	—	[d]	I	I	—
Monocaprin	I	—		I	I	I
Monolaurin	N[e]	N	I	I	I	
Monoolein	N	N	—	I	—	
Monoeicosenoin	N	N	—	I	—	

[a] Gram negative; [b] gram positive; [c] inactivated; [d] not done; [e] no effect.
The data in this table is based on work previously published (Isaacs *et al.*, 1995).

long-chain unsaturated fatty acids (Kabara, 1978), changes in the balance between these two groups of fatty acids in milk should not affect its antimicrobial activity. Data in the literature (Kabara, 1978) and our own studies (Thormar *et al.*, 1987) indicate that the most active chain length for medium-chain satuated fatty acids is 12 carbons and the most active unsaturated fatty acids are 16 and 18 carbons. It is not possible to compare the antiviral and antibacterial activities of these two groups of fatty acids directly because some pathogens are sensitive to medium-chain and short-chain fatty acids, e.g. Gram negative bacteria, but resistant to long-chain unsaturated fatty acids (Shibasaki and Kato, 1978) whereas Gram positive bacteria are quite sensitive to inactivation by long-chain unsaturated fatty acids (Kabara, 1978).

5. Potential Therapeutic Role of Milk Lipids

Despite the presence of multiple protective factors in milk, HIV and other enveloped viruses have been recovered from human milk (Table 5) and are known to be transmitted to suckling infants (Ruff, 1994). Many of these viruses, such as HIV and human T-cell lymphotrophic virus type-1 (HTLV-1), are retroviruses that can be transmitted inside leucocytes and epithelial cells in milk (Southern and Southern, 1998). Cytomegalovirus (CMV), which is not a retrovirus, is also found both in milk cells and in extracellular milk compartments (Asanuma *et al.*, 1996; Numazaki, 1997). Both cellular and extracellular CMV in milk have been implicated in viral transmission to the suckling neonate. The mechanisms by

Table 5. Enveloped Viruses Transmitted through Milk

Human immunodeficiency virus Type 1 (HIV)[a]
Human T-lymphotrophic virus Type 1 (HTLV-1)[b]
Human T-lymphotrophic virus Type 2 (HTLV-2)[c]
Cytomegalovirus (CMV)[d]
Murine leukemia virus[e]
Feline immunodeficiency virus[f]
Maedi/Visna virus[g]
Caprine arthritis-encephalitis virus[h]
Mouse mammary tumor virus[i]

[a]Kreiss, 1997; [b]Southern and Southern, 1998; [c]Lal *et al.*, 1993 [d]Numazaki, 1997; [e]Sharpe *et al.*, 1989; [f]Sellon *et al.*, 1994; [g]Dawson, 1987; [h]Adams *et al.*, 1983; [i]Hainaut *et al.*, 1985; [a–d]Ruff, 1994.

which extracellular virus particles escape inactivation by protective factors in milk have not been established. In this connection, however, viruses may already have been transmitted to the neonate by the time antimicrobial lipids have been released from milk by lipolytic activity in the stomach (Roubinian and Blair, 1980). If lipid-dependent antimicrobial activity were induced in milk at the time of ingestion it might be possible to reduce the transmission of enveloped viruses in milk.

Lipolytic activity in the infant's gastrointestinal tract may hydrolyze triglycerides of formula as well as of breast milk to yield antimicrobial products (Isaacs *et al.*, 1992). Purified lipases generate antiviral and antibacterial activity when added to infant formulas containing varying proportions of triglycerides of medium-chain saturated and long-chain unsaturated fatty acids (Isaacs *et al.*, 1992). When fatty acids and monoglycerides are added directly to human milk, bovine milk or infant formula they can inactivate HSV, respiratory syncytial virus, *H. influenzae* and Group B streptococcus (Isaacs *et al.*, 1995). In the latter study medium-chain saturated fatty acids exhibited stronger antiviral and antibacterial activity than long-chain unsaturated fatty acids and monoglycerides (Isaacs *et al.*, 1995). These results indicate that the presence of antimicrobial lipids, regardless of their source, may reduce the concentration of enveloped viruses in milk. When breast feeding is unavoidable in populations with high rates of vertical HIV transmission, it may be possible to collect milk and add purified antimicrobial lipids to reduce the concentration of pathogens prior to consumption. While this could increase the risk of breast milk contamination in the non-industrialized world, the risk of HIV transmission is so great through breast milk, as much as one-third to one-half of vertically transmitted HIV (Datta *et al.*, 1994; Bertolli *et al.*, 1996; Kreiss, 1997), that adding antimicrobial lipids to milk may provide a net benefit in these populations.

Another potential therapeutic use of antimicrobial lipids similar to those found in milk may be as part of a topical vaginal microbicide designed to reduce vertical transmission of pathogens during birth. It is estimated that 40–80% of the children who acquire HIV-1 from their mothers become infected during delivery (Mostad *et al.*, 1997) by exposure of the skin or mucous membranes to virus (Baba *et al.*, 1994). Approximately 80% of all cases of neonatal HSV infection are transmitted from mother to infant by the same means (Whitley, 1994). This results from direct contact of the fetus with infected maternal genital secretions.

Most efforts to develop antimicrobials have been focused on producing compounds which act systemically following oral administration during infection. Microbicidal agents such as the protective factors in human milk function by direct inactivation of pathogens. It is possible that preparations of antimicrobial milk lipids could be developed which would provide protection against infection during breast or bottle feeding as well as during birth. This protection would apply to vertically transmitted viruses and to bacteria present in the birth canal. For example, *Chlamydia trachomatis* has one of the highest incidences of infection of any sexually transmitted disease. Infants have a 33 to 50% risk of contracting the infection from an infected mother during vaginal delivery (Eng and Butler, 1997). Of those infants infected by *C. trachomatis*, approximately 50% develop conjunctivitis and 20% develop pneumonia (Alexander *et al.*, 1982; Beem and Saxon, 1982). The lipids present in human milk have been shown to inactivate *C. trachomatis* and prevent infection of susceptible cells (Elbagir *et al.*, 1990). Antimicrobial lipids based on those found in milk are stable, inactivate susceptible pathogens and are not toxic at mucosal surfaces.

6. Summary

Milk lipids serve not only as nutrients but as antimicrobial agents that constitute a defense system against microbial infections that occur at mucosal surfaces. The lipid fraction of milk develops antimicrobial activity in the gastrointestinal tract of suckling neonates as a result of lipolytic activity which converts milk triglycerides to antimicrobial fatty acids and monoglycerides. Antimicrobial milk lipids may be particularly important in protecting infants with an inadequate secretory immune response from infection. The lipid-dependent antimicrobial activity of milk is due to medium-chain saturated and long-chain unsaturated fatty acids and their respective monoglycerides released by lipases in the gastrointestinal tract. The antimicrobial activity of fatty acids and monoglycerides is additive and consequently it is their combined concentration that determines the lipid-dependent antimicrobial activity of milk. Microbial inactivation occurs rapidly by membrane destabilization. The antimicrobial activity of milk lipids can be duplicated using purified fatty acids and monoglycerides. It should be possible, therefore, to supplement banked human milk to provide lipid-dependent antimicrobial activity from the moment of ingestion (Schanler *et al.*, 1986). This could reduce the risk of viral transmission from mother to infant through milk. Milk lipids also could be adapted for use at mucosal surfaces other than those in the gastrointestinal tract to reduce vertical transmission of pathogens during birth.

ACKNOWLEDGMENTS. This work was supported by grants from the National Institutes of Health (HL41179 and AI39061) and by the Office of Mental Retardation in Developmental Disabilities of New York State. Thanks are given to Anna Parese for help in preparation of this manuscript.

References

Adams, D.S., Klevjer-Anderson, P., Carlson, J.L., McGuire, T.C., and Gorham, J.R. 1983. Transmission and control of caprine arthritis-encephalitis virus. *Am. J. Vet. Res.* **44**:1670–1675.

Adinolfi, M. and Glynn, A. 1979. The interaction of antibacterial factors in breast milk. *Develop. Med. Child. Neuorol.* **21**:808–819.

Alexander, E.R., Harrison, H.R., Lewis, M., Sim, D.A., and Podgore, J.K. 1982. Strategies for prevention of infant chlamydial disease, in: *Chlamydial Infections*, (P.-A. Mardh, K.K. Holmes, J.D. Oriel, P. Piot, and J. Schachter, eds.), pp. 225–228, Elsevier Biomedical Press, Amsterdam.

Aly, R., Maibach, H.I., Shinefield, H.R., and Strauss, W.G. 1972. *Survival of pathogenic microorganisms on human skin. J. Invest. Dermatol.* **58**:205–210.

Asanuma, H., Numazaki, K., Nagata, N., Hotsubo, T., Horino, K., and Chiba, S. 1996. Role of milk whey in the transmission of human cytomegalovirus infection by breast milk. *Microbiol. Immunol.* **40**(3):201–204.

Baba, T.W., Koch, K.J., Mittler, E.S., Greene, M., Wyand, M., Penninck, D., and Ruprecht, R.M. 1994. Mucosal infection of neonatal rhesus monkeys with cell-free SIV. *Aids Res. Human Retroviruses* **10**:351–357.

Ballantine, D.L., Gerwick, W.H., Velez, S.M., Alexander, E., and Guevara, P. 1987. Antibiotic activity of lipid-soluble extracts from Caribbean marine algae. 1987. Antibiotic activity of lipid-soluble extracts from Caribbean marine algae, in: *Hydrobiologia*, (M.A. Ragan and C.J. Bird, eds.), Twelfth International Seaweed Symposium, pp. 463–469, Dr. W. Junk Publishers, Dordrecht, Netherlands.

Beem, M.O. and Saxon, E.M. 1982. Chlamydia trachomatis infection in infants, in: *Chlamydial Infections*, (P.-A. Mardh, K.K. Holmes, J.D. Oriel, P. Piot., and J. Schachter, eds.), pp. 199–212, Elsevier Biomedical Press, Amsterdam.

Bertolli, J., St. Louis, M.E., Simonds, R.J., Nieburg, P., Kamenga, M., Brown, C., Tarande, M., Quinn, T., and Ou, C.-Y. 1996. Estimating the timing of mother-to-child transmission of human immunodeficiency virus in a breast-feeding population in Kinshasa, Zaire. *J. Infect. Dis.* **174**:722–726.

Bibel, D.J., Miller, S.J., Brown, B.E., Pandey, B.B., Elias, P.M., Shinefield, H.R., and Aly, R. 1989. Antimicrobial activity of stratum corneum lipids from normal and essential fatty acid-deficient mice. *J. Invest. Dermatol.* **92**:632–638.

Bryson, Y.J., Winter, H.S., Gard, S.E., Fischer, T.J., and Stiehm, E.R. 1980. Deficiency of immune interferon production by leukocytes of normal newborns. *Cell Immunol.* **55**:191–200.

Canas-Rodriguez, A. and Smith, H.W. 1966. The identification of the antimicrobial factors of the stomach contents of suckling rabbits. *Biochem. J.* **100**:79–82.

Coonrod, J.D. 1986. The role of extracellular bactericidal factors in pulmonary host defense. *Sem. Respir. Infect.* **1**:118–129.

Coonrod, J.D. 1987. Role of surfactant free fatty acids in antimicrobial defenses. *Eur. J. Respir. Dis.* **71**:209–214.

Coonrod, J.D. and Yoneda, K. 1983. Detection and partial characterization of antibacterial factor(s) in alveolar lining material of rats. *J. Clin. Invest.* **71**:129–141.

Cunningham, A.S. 1979. Morbidity in breast-fed and artificially fed infants. II. *J. Pediatr.* **59**:685–689.

Datta, P., Embree, J.E., Kreiss, J.K., Ndinya-Achola, J.O., Braddick, M., Temmerman, M., Nagelkerke, N.J.D., Maitha, G., Holmes, K.K., Piot, P., Pamba, H.O., and Plummer, F.A. 1994. Mother-to-child transmission of human immunodeficiency virus type 1: Report from the Nairobi study. *J. Infect. Dis.* **170**:1134–1140.

Dawson, M. 1987. Pathogenesis of maedi-visna. *Vet. Rec.* **120**:451–454.

Elbagir, A., Petterson, M., Lindahl, M., Genc, M., Froman, G., and Mardh, P.-A. 1990. Influence of whole human milk, and fractions thereof, on inclusion-formation of Chlamydia trachomatis in McCoy cells. *APMIS* **98**:609–614.

Eliakim, R., DeSchryver-Kecskemetis, K., Nogee, L., Stenson, W.F., and Alpers, D.H. 1989. Isolation and characterization of a small intestinal surfactant-like particle containing alkaline phosphatase and other digestive enzymes. *J. Biol. Chem.* **264**:20614–20619.

Eng, T.R. and Butler, W.T. 1997. In: *Confronting Sexually Transmitted Diseases, The Hidden Epidemic*, pp. 1–432, National Academy Press, Washington, DC.

Falkler, W.A., Jr., Diwan, A.R., and Halstead, S.B. 1975. A lipid inhibitor of Denque virus in human colostrum and milk; with a note on the absence of anti-Dengue secretory antibody. *Arch. Virol.* **47**:3–10.

Fearon, D.T. 1997. Seeking wisdom in innate immunity. *Nature* **388**:323–324.

Freed, L.M., York, C.M., Hamosh, P., Mehta, N.R., and Hamosh, M. 1987. Bile salt-stimulated lipase of human milk: Characteristics of the enzyme in the milk of mothers of premature and full-term infants. *J. Pediatr. Gastroenterol. Nutr.* **6**:598–604.

Gillin, F.D. and Reiner, D.S. 1983. Human milk kills parasitic intestinal protozoa. *Science* **221**:1290–1292.

Goldman, A.S., Ham Pong, A.J., and Goldblum, R.M. 1985. Host defenses: Development and maternal contributions, in: *Advances in Pediatrics*, (L.A. Barnes, ed.), pp. 71–100, Year Book Publication, Chicago.

Hainaut, P., Vaira, D., Francois, C., Calberg-Bacq, C.-M., and Osterrieth, P.M. 1985. Natural infection of Swiss mice with mouse mammary tumor virus (MMTV): Viral expression in milk and transmission of infection. *Arch. Virol.* **83**:195–206.

Hamosh, M. 1991a. Lipid metabolism, in: *Neonatal Nutrition and Metabolism*, (W.H. Hay, Jr., ed.), pp. 122–142, Mosby Year Dock, St. Louis.

Hamosh, M. 1991b. Free fatty acids and monoglycerides: Anti-infective agents produced during the digestion of milk fat by the newborn, in: *Immunology of Milk and the Neonate*, (J. Mestecky *et al.*, eds.), pp. 151–158, Plenum Press, New York.

Hamosh, M. and Scow, R.O. 1971. Lipoprotein lipase activity in guinea pig and rat milk. *Biochim. Biophys. Acta* **231**(2):283–289.

Hernell, O. and Blackberg, L. 1985. Lipolysis in human milk: Causes and consequences, in: *Composition and Physiological Properties of Human Milk*, (J. Schaud, ed.), pp. 165–177, Elsevier Science Publishers, B.V. (Biomedical Division).

Hernell, O., Ward, H., Blackberg, L., and Pereira, M.E.A. 1986. Killing of Giardia lamblia by human milk lipases: An effect mediated by lipolysis of milk lipids. *J. Infect. Dis.* **153**:715–720.

Hylander, M.A., Strobino, D.M., and Dhanireddy, R. 1998. Human milk feeding and infection among very low birth weight infants. *Pediatrics* **102**(3):E38.

Isaacs, C.E. and Thormar, H. 1990. Human milk lipids inactivate enveloped viruses, in: *Breastfeeding, Nutrition, Infection and Infant Growth in Developed and Emerging Countries*, (S.A. Atkinson, L.A. Hanson, and R.K. Chandra, eds.), pp. 161–174, ARTS Biomedical Publishers and Distributors, St. John's, Newfoundland, Canada.

Isaacs, C.E., Kashyap, S., Heird, W.C., and Thormar, H. 1990. Antiviral and antibacterial lipids in human milk and infant formula feeds. *Arch. Dis. Child.* **65**:861–864.

Isaacs, C.E., Litov, R.E., and Thormar, H. 1995. Antimicrobial activity of lipids added to human milk, infant formula, and bovine milk. *J. Nutr. Biochem.* **6**:362–366.

Isaacs, C.E., Litov, R.E., Marie, P., and Thormar, H. 1992. Addition of lipases to infant formulas produces antiviral and antibacterial activity. *J. Nutr. Biochem.* **3**:304–308.

Isaacs, C.E., Thormar, H., and Pessolano, T. 1986. Membrane-disruptive effect of human milk: Inactivation of enveloped viruses. *J. Infect. Dis.* **154**:966–971.

Jensen, R.G. 1996. The lipids in human milk, in: *Prog. Lipid Res.*, Elsevier Science Ltd., Great Britain, pp. 53–92.

Jensen, R.G., Ferris, A.M., Lami-Keefe, C.J., and Henderson, R.A. 1989. Lipids of bovine and human milks: A comparison. *J. Dairy Sci.* **73**:223–240.

Kabara, J.J. 1978. Fatty acids and derivatives as antimicrobial agents. A review, in: *The Pharmacological Effect of Lipids*, (J.J. Kabara, ed.), pp. 1–14, The American Oil Chemists Society, St. Louis.

Kabara, J.J. 1980. Lipids as host-resistance factors of human milk. *Nutr. Reviews* **38**:65–73.

Kearney, J.N., Ingham, E., Cunliffe, W.J., and Holland, K.T. 1984. Correlations between human skin bacteria and skin lipids. *Br. J. Dermatol.* **110**:593–599.

Kreiss, J. 1997. Breastfeeding and vertical transmission of HIV-1. *Acta Paediatr. Suppl.* **421**:113–117.

Laegreid, A., Kolstootnaess, A.-B., Orstavik, I., and Carlsen, K.H. 1986. Neutralizing activity in human milk fractions against respiratory syncytial virus. *Acta Paediatr. Scand.* **75**:696–701.

Lal, R.B., Gongora-Biachi, R.A., Pardi, D., Switzer, W.M., Goldman, I., and Lal, A.A. 1993. Evidence for mother-to-child transmission of human T lymphotropic virus type II. *J. Infect. Dis.* **168**:586–591.

Larsen, S.A., Jr. and Homer, D.R. 1978. Relation of breast versus bottle feeding to hospitalization for gastroenteritis in a middle-class U.S. population. *J. Pediatr.* **92**:417–418.

Mandel, I.D. and Ellison, S.A. 1985. The biological significance of the nonimmunoglobulin defense factors, in: *The Lactoperoxidase System Chemistry and Biological Significance*, (M. Pruitt and J.O. Tenovuo, eds.), pp. 1–14, NY, Marcel Dekker, Inc.

Miller, S.J., Aly, R., Shinefeld, H.R., and Elias, P.M. 1988. In vitro and in vivo antistaphylococcal activity of human stratum corneum lipids. *Arch. Dermatol.* **124**:209–215.

Moldoveanu, Z., Tenovuo, J., Pruitt, K.M., Mansson-Rahemtulla, B., and Mestecki, J. 1983. Antibacterial properties of milk: IgA-peroxidase-lactoferrin interactions. *Ann. NY Acad. Sci.* **409**:848–850.

Mostad, S.B., Overbaugh, J., DeVange, D.M., Welch, M.J., Chohan, B., Mandaliya, K., Byange, P., Martin, H.L., Jr., Ndinya-Achola, J., Bwayo, J.J., and Kreiss, J.K. 1997. Hormonal contraception, vitamin A deficiency, and other risk factors for shedding of HIV-1 infected cells from the cervix and vagina. *Lancet* **350**:922–927.

Muskiet, F.A.J., Hutter, N.H., Martini, I.A., Jonxis, J.H.P., Offringa, P.J., and Boersma, E.R. 1987. Comparison of the fatty acid composition of human milk from mothers in Tanzania, Curacao and Surinam. *Human Nutr: Clin. Nutr.* **41C**:149–159.

Myers, M.G., Fomon, S.J., Koontz, F.P., McGuinness, G.A., Lachenbruch, P.A., and Hollingshead, R. 1984. Respiratory and gastrointestinal illnesses in breast- and formula-fed infants. *Am. J. Dis. Child.* **138**:629–632.

Newburg, D.S. 1997. Do the binding properties of oligosaccharides in milk protect human infants from gastrointestinal bacteria? *J. Nutr.* **127**(5 Suppl):980S–984S.

Newburg, D.S., Viscidi, R.P., Ruff, A., and Yolken, R.H. 1992. A human milk factor inhibits binding of human immunodeficiency virus to the CD4 receptor. *Pediatr. Res.* **31**(1): 22–28.

Noseda, A., White, J.G., Godwin, P.L., Jerome, W.G., and Modest, E.J. 1989. Membrane damage in leukemic cells induced by ether and ester lipids: An electron microscopic study. *Exper. Molec. Pathol.* **50**:69–83.

Numazaki, K. 1997. Human cytomegalovirus infection of breast milk. *FEMS Immun. Med. Microbiol.* **18**:91–98.

Ogra, P.L. and Losonsky, G.A. 1984. Defense factors in products of lactation, in: *Nutritional and Immunological Interactions*, (P.L. Ogra, ed.), pp. 67–87, Grune and Stratton, Orlando, FL.

Reiner, D.S., Wang, C.-S., and Gillin, F.D. 1986. Human milk kills Giardia lamblia by generating toxic lipolytic products. *J. Infect. Dis.* **154**:825–832.

Reiter, B. 1981. The contribution of milk to resistance to intestinal infection in the newborn, in: *Immunological Aspects of Infection in the Fetus and Newborn*, (H.P. Lambert and C.B.S. Wood, eds.), Academic Press, London: New York.

Rohrer, L., Winterhalter, K.H., Eckert, J., and Kohler, P. 1986. Killing of Giardia lamblia by human milk is mediated by unsaturated fatty acids. Antimicrob. *Agents Chemother.* **30**:254–257.

Rosell, K.-G. and Srivastava, L.M. 1987. Fatty acids as antimicrobial substances in brown algae, in: *Hydrobiologia*, (M.A. Ragan and C.J. Bird, eds.), pp. 471–475, Twelfth International Seaweed Symposium, Dr. W. Jung Publishers, Dordrecht, Netherlands.

Ross, A.C., Davila, M.E., and Cleary, M.P. 1985. Fatty acids and retinyl esters of rat milk: Effects of diet and duration of lactation. *J. Nutr.* **115**:1488–1497.

Roubinian, J.R. and Blair, P.B. 1980. Short gastric veins as the major portal of entry for milk-borne murine mammary tumor virus. *JNCI* **65**:795–800.

Ruff, A.J. 1994. Breastmilk, breastfeeding, and transmission of viruses to the neonate. *Seminars in Perinatology* **18**:510–516.

Sarkar, N.H., Charney, J., Dion, A.S., and Moore, D.H. 1973. Effect of human milk on the mouse mammary tumor virus. *Cancer Res.* **33**:626–629.

Schanler, R.J., Goldblum, R.M., Garza, C., and Goldman, A.S. 1986. Enhanced fecal excretion of selected immune factors in very low birth weight infants fed fortified human milk. *Pediatr. Res.* **20**:71–715.

Sellon, R.K., Jordan, H.L., Kennedy-Stoskopf, K., Tompkins, M.B., and Tompkins, W.A.F. 1994. Feline immunodeficiency virus can be experimentally transmitted via milk during acute maternal infection. *J. Virol.* **68**:3380–3385.

Sharpe, A.H., Hunter, J.J., Ruprecht, R., and Jaenisch, R. 1989. Maternal transmission of retroviral disease and strategies for preventing infection of the neonate. *J. Virol.* **63**:1049–1053.

Shibasaki, I. and Kato, N. 1978. Combined effects on antibacterial activity of fatty acids and their esters against gram-negative bacteria, in: *The Pharmacological Effect of Lipids*, (J.J. Kabara, ed.), pp. 15–24, The American Oil Chemists Society, St. Louis.

Smith, H.W. 1966. The antimicrobial activity of the stomach contents of suckling rabbits. *J. Pathol. Bacteriol.* **91**:1–9.

Southern, S.O. and Southern, P.J. 1998. Persistent HTLV-1 infection of breast luminal epithelial cells: A role in HTLV transmission? *Virol.* **241**:200–214.

Thormar, H., Isaacs, C.E., Brown, H.R., Barshatzky, M.R., and Pessolano, T. 1987. Inactivation of enveloped viruses and killing of cells by fatty acids and monoglycerides. *Antimicrob. Agents Chemother.* **31**:27–31.

Watanabe, T., Nagura, H., Watanabe, K., and Brown, W.R. 1984. The binding of human milk lactoferrin to immunoglobulin A. *FEBS Letts.* **168**:203–207.

Welsh, J.K. and May, J.T. 1979. Anti-infective properties of breast milk. *J. Pediatr.* **94**:1–9.

Welsh, J.K., Arsenakis, M., Coelen, R.J., and May, J.T. 1979. Effect of antiviral lipids, heat, and freezing on the activity of viruses in human milk. *J. Infect. Dis.* **140**:322–328.

Welsh, J.K., Skurrie, I.J., and May, J.T. 1978. Use of Semliki forest virus to identify lipid-mediated antiviral activity and anti-alphavirus immunoglobulin A in human milk. *Infect. Imm.* **19**:395–401.

Whitley, R.J. 1994. Herpes simplex virus infections of women and their offspring: Implications for a developed society. *Proc. Natl. Acad. Sci.* **91**:2441–2447.

Winter, H.S., Gard, S.E., Fischer, T.J., Bryson, Y.J., and Stiehm, E.R. 1983. Deficient lymphokine production of newborn lymphocytes. *Pediatr. Res.* **17**:573–578.

Chapter 14

The Antimicrobial and Immunomodulating Actions of Milk Leukocytes

Gerd Riedel-Caspari

1. Introduction

The ability of highly developed organisms to defend themselves against invading microorganisms depends on recognizing and destroying dangerous substances such as microbes and their toxins. Recognition and destruction are tasks of the immune system, primarily of the leukocytes. Subpopulations of leukocytes variously release antimicrobial substances such as immunoglobulins, lysozyme, complement factors and reactive oxygen metabolites into the cellular environment upon exposure to an antigenic stimulus, and they also ingest foreign materials or secrete cytokines which regulate the immune response. During lactation a mother can pass immunogenic substances to her offspring through her mammary secretions. These compounds in colostrum and milk are an important, if not decisive, part of the adoptive transfer of immunity from mother to offspring in many species. Most scientific investigations into the adoptive transfer of immunity through milk have focused on immunoglobulins rather than on leukocytes. However, adoptive transfer of the cells in milk also appears to occur. It is important to explain how immunocompetent maternal leukocytes can modulate immune responses in the newborn without recognizing and destroying the neonate's leukocytes and tissues as nonself elements that bear different, paternally-derived major histocompatibility complex (MHC) antigens. Conversely, why don't the immunocompetent leukocytes of the newborn attack and eliminate milk-derived cells that

Gerd Riedel-Caspari • PlantaVet, Postfach 1339, Finkenweg 13, D-88339 Bad Waldsee, Germany bear a maternal MHC? An important purpose of this chapter is to address these intriguing questions to the extent that current data permit. Another important question is whether mammary leukocytes have any protective role or whether the milk is simply a convenient route of excretion for these cells. A full understanding of the anti-microbial action of mammary gland leukocytes for all mammalian species must await more exhaustive study.

2. Cytology

Leukocytes, the main cellular component of mammary secretions, include mononuclear phagocytes (macrophages), polymorphonuclear neutrophils (PMN) and lymphocytes (Table 1). These cell types normally make up more than 90% of the somatic cells in colostrum and milk. The remaining fraction may contain a varying number of epithelial cells. Leukocytes and epithelial cells in milk are termed body, or somatic, cells to differentiate them from contaminant microbial cells. Milk somatic cell counts (MSCC) range from several hundred to several million cells per ml, depending on the species, the stage of lactation and, above all, the level of inflammation. The highest MSCC from non-infected glands of women (Ogra and Ogra, 1978; Crago *et al.*, 1979) and cows (Meriläinen *et al.*, 1979; Jensen and Eberhart, 1981) were observed immediately after delivery. Bovine colostrum contained concentrations of viable leukocytes similar to cell counts in the peripheral blood, i.e., from about 0.3 to 3.0×10^6 per ml (Riedel-Caspari, 1993).

The first role postulated for mammary leukocytes was to protect the gland itself against infection. This view was substantiated by Schalm *et al.* (1976), who turned a chronic *Staphylococcus aureus* mastitis into a gangrenous disease by treating the infected cows with an anti-bovine leukocyte serum. A second protective function was proposed by Newby and Bourne (1977) from the results of infusion experiments and by Sheldrake *et al.* (1985a), who observed an increase of immunoglobulin-positive cells after specific antigens were injected into the mammary gland. These authors concluded that mammary B lymphocytes may play a role in the local synthesis of immunoglobulins, thereby contributing to the adoptive transfer of immunity to the newborn. Other cell types also may play a role. The neonate ingests large numbers of macrophages, PMNs and T lymphocytes with its first meals. However, the significance of these cells in the adoptive transfer of immunity is still unclear.

Misinterpretations of cytological findings have been one reason for underestimating the protective potential of milk cells. In particular, based on light microscopic examination the predominant cell type was thought to be sloughed-off epithelial cells. This was due mainly to underestimation of macrophage numbers because, in milk, these cells differ morphologically from

Table 1. Differential cell counts of mammary secretions*

Species	Stage of lactation (time post partum)**	PMN	Macro-phages	Lympho-cytes	Epithelial cells	Authors
Human	colostrum (d1–3)	5	77	11	n.d. ***	Diaz-Jouanen and Williams, 1974
Human	prepartum	21	66	11	n.d.	Ho *et al.*, 1979
	colostrum	42	48	7	n.d.	
	milk (d8)	20	72	5	n.d.	
Human	colostrum (d1–4)	54	37	7	0	Crago *et al.*, 1979
Porcine	prepartum	72	1	26	n.d.	Evans *et al.*, 1982
	colostrum (d3)	55	15	23	6	
	milk (d10)	39	15	14	31	
Porcine	colostrum (d1–2)	61	7	11	20	Schollenberger *et al.*, 1986a
	milk (d3–31)	47	10	12	31	
Porcine	prepartum	93	3	2	41	Lee *et al.*, 1983
	colostrum (d1)	62	21	8	7	
	milk (d2–21)	37	20	12	41	
Porcine	colostrum (d1)	70	2	n.d.	1	Williams, 1993
Porcine	colostrum	50	n.d.	n.d.	n.d.	Persson *et al.*, 1996
	milk (d22)	43	n.d.	n.d.	n.d.	
Bovine	colostrum (d1–2)	9	79	12	n.d.	Meriläinen *et al.*, 1979
	milk (d6)	12	74	.14	n.d.	
Bovine	dry period	3	89	7	1	Lee *et al.*, 1980
	colostrum (d1,4)	62	35	4	0	
	milk	3	80	16	2	
Bovine	dry period	22	48	30	n.d.	Jensen and Eberhart, 1981
	colostrum (d2)	30	42	28	n.d.	
	milk (d3–6)	35	35	30	n.d.	
Bovine	colostrum	38	40	23	n.d.	Duhamel *et al.*, 1987
Bovine	colostrum fm (d2)	23	70	7	n.d.	Östensson, 1993
	colostrum rm (d2)	39	51	11	n.d.	
	milk fm (m3)	17	79	4	n.d.	
	milk rm (m3)	32	45	23	n.d.	
Ovine	involution	40	51	9	0	Lee and Outteridge, 1981
	colostrum (d1)	66	24	34	<1	
	milk	2	84	13	2	

* Means expressed as percent of identified leukocytes and epithelial cells; data partially converted to enhance inter-study comparisons.

** Approximate period during which the mammary secretions were collected, expressed in days (d) or months (m) post partum; foremilk (fm) or residual milk (rm).

*** Not determined.

the mononuclear phagocytes found in the blood (monocytes) and most other organs. The reason for this morphologic dissimilarity has been revealed by electron microscopy to be a result of uptake by milk macrophages of cell fragments, fat, casein micelles and lipid droplets from the glandular environment. It is now generally accepted that most of the somatic cells in mammary secretions are viable leukocytes and that milk macrophages form part of the mononuclear phagocyte system (MPS), formerly called the reticulo-endothelial system (Lee *et al.*, 1980; Mielke and Koblenz, 1980a). Nevertheless, epithelial cells also may be present in appreciable numbers, e.g., in mammary secretions of women (Ho *et al.*, 1979) and sows (Lee *et al.*, 1983; Schollenberger *et al.*, 1986a).

The MSCC and the milk differential cell counts (MDCC) are influenced by species, breed, variations between individuals, period of lactation or gravidity, milk fraction (foremilk or residual milk) and, above all, the presence of pathogenic or non-pathogenic microorganisms in the gland (Smith and Schultz, 1977; Ogra *et al.*, 1978; Crago *et al.*, 1979; Lee *et al.*, 1983; Harmon and Heald, 1982; Mielke and Koblenz, 1981; Östensson 1993). Even a weak inflammatory stimulus, such as the instillation of phosphate-buffered saline into the gland, can provoke an alteration in the number and activity of mammary leukocytes (Heyermann and Senft, 1985). Another expeimental procedure used to produce a sterile intramammary inflammation has been to insert plastic devices into the gland cisterns of the bovine udder. Raising the MSCC by this procedure provided protection against experimental challenge with *S. aureus* (Schultze and Paape, 1984) and *S. uberis* (Paape *et al.*, 1988), but failed to reduce natural infections of the udder in field conditions (Corlett *et al.*, 1984). In other studies, clinical symptoms of acute mastitis were produced by infusion of endotoxins from Gram-negative bacteria into the bovine udder. The symptoms were comprised of the general signs of acute inflammation, including release of chemotactic factors by macrophages and massive migration of neutrophils from the blood to sub-epithelial sites. Within a few hours the neutrophils penetrated the basal lamina, migrated through the epithelium into the milk and dominated the MDCC (Nickerson and Pankey, 1984; Guidry *et al.*, 1983; Mattila and Frost, 1989).

The most significant physiological changes in MSCC and MDCC occur at parturition when leukocytes accumulate in the mammary gland. This is possibly due to hormonal changes. With the onset of lactation, leukocytes leave perivascular areas of the connective tissue and settle at sites that permit easy passage into alveolar areas and the milk. In most of the species investigated, the absolute somatic cells counts declined from parturition until the end of lactation, at which time the counts rose dramatically (Seelig, 1980; Campbell *et al.*, 1950).

2.1. Macrophages

The differential cell counts show significantly fewer lymphocytes and many more mononuclear phagocytes in mammary secretions than in the peripheral blood. The mere presence of macrophages in milk and colostrum is not surprising because the mammary gland, like many other organs, harbors a large number of mononuclear phagocytes. However, the extent to which these cells dominate the somatic cell count of secretions from healthy glands is surprising. In contrast to the macrophages of most other tissues, milk mononuclear phagocytes exhibit a pronounced vacuolization. Electron microscopic analysis reveals that the vacuoles contain lipid droplets, casein micelles, debris and degraded neutrophils (Smith and Goldman, 1968; Mielke and Koblenz, 1980b; Lee *et al.* 1983). Vacuolization of milk macrophages is intensified during the colostral period, ebbs during lactation and increases again at the beginning of the dry period (Lee *et al.*, 1969, 1980). It has been postulated that macrophages (as well as PMN) contribute to the reabsorption of milk lipids during the involution of the mammary gland and that these phagocytes serve as "street cleaners" in the gland during lactation. Nevertheless, it remains unclear why their numbers rise at the beginning of lactation and what the importance is of their large scale phagocytosis of non-pathogenic particles, which could also be removed mechanically with the flow of milk. In fact, the ingested material may compromise the ability of the milk macrophage to ingest and destroy microorganisms. Thus, at first glance, it seems doubtful that these garbage-laden macrophages would be of benefit either to the gland or to the suckling neonate.

In vitro studies on the macrophages of mammary secretions have revealed a wide range of activities similar to those of other members of the MPS. These activities include an enhanced random migration by mammary macrophages exposed to a soluble factor in milk (Mushtaha *et al.*, 1989), the presence of Fc and C3b surface receptors and at least some capacity for ingestion and destruction of opsonized and non-opsonized microorganisms (Lascelles *et al.*, 1969; Mohr *et al.*, 1970; Smith and Goldman, 1968; Jensen and Eberhart, 1975; Robinson *et al.*, 1978; Pitt, 1979). In addition, milk macrophages can present antigen to lymphocytes (Smith and Goldman, 1970; Smith *et al.*, 1971; Crago *et al.*, 1979; Mielke and Koblenz, 1981) and produce cytokines such as MIF and LIF (Mohr *et al.*, 1970) as well as antibacterial substances including lysozyme, lactoferrin and the complement factors C3 and C4 (Murillo and Goldman, 1970; Pitt, 1979). It may be assumed that mononuclear phagocytes are an important, if not the major, source of these anti-microbial factors in colostrum and milk. Perhaps these cells depend on the stimulus of exposure to particles such as debris or micelles to generate and release antimicrobial products into the milk.

Another feature of mammary phagocytes renders them active contributors to the adoptive transfer of immunity to the neonate (Pittard *et al.*, 1977). Milk phagocytes, both macrophages and neutrophils, take up lactoferrin and immunoglobulins from the surrounding fluid (Moro *et al.*, 1983; Laven *et al.*, 1981). As IgA dominates the immunoglobulins in the milk of most species, this immunoglobulin class usually predominates in the vacuoles of mammary phagocytes. Moreover, these cells avidly retain ingested immunoglobulin (Pitt, 1979; Weaver *et al.*, 1981). Even after days in culture, the intracellular IgA concentration of human colostral macrophages remained five-fold higher than that in the culture medium. In contrast, addition of ingestible particles to the cultures, e.g., Candida, *E. coli* or latex beads, induced a significant release of IgA into the culture medium within 15 to 30 minutes. Colostral immunoglobulins are essential for all mammalian species with an epitheliochorial placenta, which precludes the intrauterine transfer of antibodies to the fetus. Even in species such as humans, clinical evidence clearly demonstrates that under suboptimal hygienic conditions colostrum, and particularly its secretory IgA, prevent enteropathogenic microorganisms from adhering to the intestinal mucosa and provoking diarrhea. Hence the attachment of immunoglobulin-laden leukocytes to the mucosa of the intestinal tract, followed by the release of the ingested antibody, is an important adjunct in support of the neonate's intestinal defense.

2.2. Neutrophils

Mammary phagocytic cells also include varying numbers of PMN. In contrast to the long-lived, antigen-presenting and immuno-regulatory macrophages, neutrophils are short-lived and have only one purpose, which is to invade sites of acute inflammation and perform phagocytic microbicide, after which the phagocyte dies. PMN may approach 100% of the somatic cell count in the secretions of severely inflamed glands (Schalm,1977). In nearly all species investigated, PMN comprise a higher proportion of the somatic cell count in colostrum than in mature milk. This may reflect an inflammation-like condition in the gland at parturition or simply an easy passage through the vascular wall at this stage. As a short-lived and non-dividing leukocyte population, PMN enter the mammary gland directly from the blood and do not undergo morphological changes in the manner of the mononuclear phagocyte. From electron microscopic studies it is apparent that neutrophils are also able to concentrate globules and debris in cytoplasmic vacuoles, thereby contributing to mammary involution, although PMN seem to be less effective in phagocytosis than macrophages (Lee *et al.*, 1969, 1983; Ho *et al.*, 1979). However, in contrast to macrophages, PMN are able to release large quantities of oxygen radicals into mammary secretions. Thus the PMN may be the main source of reactive oxygen metabolites for the extracellular lactoperoxidase system, one of the most important humoral defense mechanisms against bacteria, at least in the bovine udder (Hallén Sandgren *et al.*, 1991).

Neutrophils from milk have been subjected to a variety of functional tests and comparisons have been made with their counterparts in the blood. It is generally agreed that the phagocytic activity of milk neutrophils is low in comparison with that of blood neutrophils. Moreover, milk PMN contain fewer primary granules and less glycogen than blood neutrophils, ingest fewer bacteria than blood PMN *in vitro* and kill ingested organisms less effectively (Smith and Goldman, 1968; Paape *et al.*, 1975). The reduced microbicidal activity of milk PMN in these studies can be attributed, at least in part, to a reduced oxidative burst.

The uptake of milk lipids, casein and antibody-antigen complexes is considered to be the reason for the low phagocytic activity of milk PMN. These cells appear to become "exhausted" by the massive uptake of particulate material. This view is supported by *in vitro* findings with blood neutrophils, which show similarly depressed phagocytic activity when they are cultivated in a medium that contains milk (Russel and Reiter, 1975; Russel *et al.*, 1976; Ho and Lawton, 1978; Csorba *et al.*, 1979; Pickering *et al.*, 1980; Heyermann and Senft, 1985; Schollenberger *et al.*, 1986c; Targowski and Niemialtowski, 1986). Moreover, two additional experiments support the hypothesis that the milk medium, rather than the advanced age of milk PMN, is responsible for the low phagocytic activity of these cells. Firstly, neutrophils from residual bovine milk, i.e., cells that spent less time in the gland, ingested more opsonized yeast particles *in vitro* and showed higher chemiluminescence than stripping milk neutrophils which had been present in the gland for at least half a day (Hallén Sandgren *et al.*, 1991). Secondly, PMN harvested directly from the bovine teat cistern by a surgical method that precluded contact with milk ingested more particles and released more myeloperoxidase (determined by luminol-dependent chemiluminescence) than blood neutrophils from the same cow. Therefore, it appears that neutrophils entering the mammary gland constitute a cell population that is highly active, with a permanently activated NADPH oxidase as well as a continuous release of myeloperoxidase (Hallén Sandgren *et al.*, 1992). As in the case of the macrophage, some functions of the PMN in the mammary gland appear to become down-regulated in a process that has the main purpose of enriching mammary secretions with soluble antimicrobial substances. Such substances may offer more immediate antimicrobial protection than phagocytic processes both to the mammary gland and to the suckling neonate.

2.3. Lymphocytes

Lymphocytes represent the phylogenetically modern part of the immune system. Only a few primed lymphocytes are required to modulate immune responses decisively through clonal expansion, antibody formation or lymphokines. Thus maternal lymphocytes may be a means of transferring immunologic experience from mother to offspring. It is therefore worthwhile to examine mammary gland lymphocyte reactions.

Several lymphocyte subpopulations, particularly T and B lymphocytes, have been identified in mammary secretions. However, a rigorous interpretation of these results requires critical assessment of methodology. For example, surface immunoglobulins do not conclusively identify B lymphocytes because T lymphocytes, colostral macrophages, PMN and even non-cellular elements can stain positively for immunoglobulins.

Between 50% and 90% of lymphocytes from mammary secretions of different species are T cells, as determined by spontaneous rosette formation with heterologous erythrocytes (Diaz-Jouanen and Williams, 1974; Ogra *et al.*, 1978; Schore *et al.*, 1981), reaction with Helix pomatia lectin (Concha *et al.*, 1978a,b, 1980) or reaction with monoclonal antibodies (Duhamel *et al.*, 1987; Jain *et al.*, 1991). Generally, only about 4% of milk lymphocytes have been identified as B cells using the method of erythrocyte-antibody-complement (EAC) rosettes (Bush and Beer, 1979; Schore *et al.*, 1981; Salmon and Delouis, 1982; Williams, 1993). An exception is the sheep, in which 30% of milk lymphocytes are reported to be B cells on the basis of EAC rosetting (Lee and Outteridge, 1981). Using B cell-specific monoclonal antibodies, 3% to 30% of milk lymphocytes stained positive in human (Jain *et al.*, 1991) and bovine (Duhamel, 1987) colostra. Despite the postulated local synthesis of immunoglobulins in the mammary gland, B lymphocytes constitute a minority of milk somatic cells and plasma cells have not been detected in milk (Crago *et al.*, 1979) except for one report on sheep milk (Lee and Outteridge, 1981). The absence of plasma cells in milk may be due to a low requirement for these highly productive cells in the synthesis of immunoglobulins or to their location at sites from which they are not readily mobilized.

Monoclonal antibodies directed to differentiation antigens of the main immunoregulatory lymphocyte, the T cell, reveal several distinct subsets in mammary secretions, including OKT4(CD4), OKT8(CD8) and γ/δ positive cells (Crago *et al.*, 1979; Richie *et al.*, 1982; Moro *et al.*, 1985; Bertotto *et al.*, 1990). The numbers of OKT4 and OKT8 positive T cells were more nearly equal in colostrum than in the peripheral blood of humans (Richie *et al.*, 1982), an observation that remains unexplained. Bertotto *et al.* (1990) demonstrated that non-sensitized "naive" T lymphocytes, identified by monoclonal antibodies against a series of differentiation antigens, were less frequent in human breast milk than in blood. Thus, antigen-primed memory T lymphocytes appear to prevail in human milk, and from these results it may be concluded that memory and effector T cells preferentially pass into the gland.

In vitro, mammary lymphocytes exhibit a proliferative response to a variety of stimuli. A high proportion of human milk cells exhibited transformation when they were stimulated with a polyclonal phytomitogen (PHA), but a small proportion also responded to specific antigens such as diphtheria antigen, tetanus toxoid or tuberculin (Smith and Goldman, 1968). The antigen-specific blastogenic response of milk lymphocytes to tuberculosis antigens (purified protein derivative) *in vitro* was shown to correlate with that of peripheral lymphocytes and with the skin reactivity of the donor to the same antigen (Mohr *et al.*, 1970). These early findings that sensitized T cells are simultaneously present in blood and milk have been confirmed in a series of publications on mammary secretions in humans (Diaz-Jouanen and Williams, 1974; Parmely *et al.*, 1976; Ogra and Ogra, 1978; Meggs and Beer, 1979) and other species (Smith and Schultz, 1977; Concha *et al.*, 1978a,b,1980; Schore *et al.*, 1981; Salmon, 1987). It was generally noted that milk lymphocytes were less responsive in terms of cellular proliferation than blood lymphocytes. However, the converse also has been reported, e.g., in response to the K1-antigen of *E. coli* (Parmely *et al.*, 1976), the cytomegalovirus (Meggs and Beer, 1979) and endotoxin (Nonnecke and Kehrli, 1985). An enhanced responsiveness to "gut-specific" antigens on the part of milk lymphocytes, together with some clinical observations (Allardyce *et al.*, 1974; Goldblum *et al.*, 1975) and lymphocyte homing experiments in mice (Roux *et al.*, 1977; Lamm *et al.*, 1978; Weisz-Carrington *et al.*, 1978; McDermott and Bienenstock, 1979), gave rise to speculations that certain clones of gut-derived lymphocytes might have preferential access to the mammary gland and should, therefore, be regarded as part of the common mucosal immune system (Bienenstock *et al.*, 1978; Parmely and Beer, 1977). This hypothesis remains controversial.

The results of a series of experiments on rats, swine, cows and sheep have indicated that sensitized lymphocytes from gut-associated lymphoid tissue (GALT) may colonize the mammary gland pre-partum, but without any clearcut preferential access comparable to that of other peripheral blood lymphocytes (Manning and Parmely, 1980; Sheldrake *et al.* 1985a; Salmon 1987; Harp and Moon, 1987; Harp *et al.*, 1988). Two studies in lactating and non-lactating ewes (Sheldrake and Husband, 1985; Sheldrake *et al.*, 1985b) confirmed that, without an intramammary antigenic stimulus, only a marginal migration of intestinally primed GALT B lymphocytes to the mammary gland can be expected. Furthemore, following immunization of ewes an increase in specific antibody-containing cells of intestinal origin was apparent in the respiratory tract but not in the mammary gland (Scicchitano *et al.*, 1984), indicating that the lymphoid tissue of the lung forms part of the common mucosal immune system whereas that of the mammary gland does not. It is probable that the mammary gland provides offspring with both GALT-specific and systemic lymphocytes without any preference regarding the origin of the lymphocytes.

Very small numbers of cytotoxic T cells and natural killer (NK) cells have also been reported in mammary secretions, the latter on the basis of the surface marker HNK-1 (Leu 7/CD56) (Moro *et al.*, 1985). Functional tests of antibody-independent cytotoxicity or the mixed leukocyte reaction have demonstrated that, in general, the activity of the responsible leukocyte populations from milk was lower than that of the corresponding blood leukocytes (Kohl *et al.*, 1978a,b, 1980; Parmely *et al.*, 1976; Parmely and Williams, 1979).

3. Adoptive Transfer of Cellular Immunity Through the Colostrum

Pitt *et al.* (1977) demonstrated that, in a *Klebsiella pneumoniae* infection model, colostral leukocytes provided effective protection for newborn rats against necrotizing enterocolitis. When isolated colostral leukocytes (albeit not in defined numbers) were added to milk formula or to milk subjected to freezing and thawing, the infected neonatal rats survived the experimental infection, as did control rat pups given complete fresh rat milk. In contrast, frozen/thawed rat milk or the formula alone resulted in 90% and 100% mortality, respectively. *In vitro* experiments confirmed the antibacterial potential of leukocytes in the milk of rats as well as the inadequacy of milk formula and frozen milk in this regard.

Beneficial properties of colostral leukocytes were also apparent in a trial involving twenty colostrum-fed calves (Riedel-Caspari, 1993). The animals were experimentally infected with an hemolytic "attaching and effacing" strain of *E. coli* three hours post partum. Immediately after infection, as well as on three consecutive feedings, the calves received either cell-depleted pooled colostrum or the same pooled colostrum supplemented with the colostral cells from their own dams. During the first week of life the calves that had received the cell-supplemented colostrum excreted significantly fewer infectious bacteria in their feces than the calves fed the cell-depleted colostrum. Also, complete elimination of the infectious organisms was accelerated by brief supplementation with colostral leukocytes immediately after birth. It is noteworthy that cells from two glands that developed clinical mastitis seemed to be the most effective in promoting elimination of the *E. coli* infection.

Unfortunately, the impressive results with experimental *Klebsiella* and *E. coli* infections could not be confirmed in colostrum-deprived calves which were infected with Rota-Corona virus and pathogenic *E. coli* under natural conditions (Riedel-Caspari and Schmidt, 1991a; Riedel-Caspari *et al.*, 1991). In these studies, calves deprived of colostrum were fed twice daily either milk substitute or substitute supplemented with the complete cell harvest from 2 liters of their own dam's colostra. Two of seven calves that received colostral leukocytes survived the first four weeks of life, and three of six non-cell-supplemented calves survived. In addition, one of seven supplemented calves, but three of five non-supplemented calves, developed a significant interferon-α concentration in the blood serum on day 3 of life, the presumptive peak of viral infection. Moreover, the bactericidal activity of whole blood against *E. coli* was low and the beginning of immunoglobulin synthesis (as judged by the rise of blood immunoglobulin concentration during the first four weeks of life) was delayed in the supplemented animals compared to the non-supplemented animals. Contrary to the colostrum-deprived animals, none of thirty-three calves in two additional experimental groups kept under the same conditions and fed either cell-supplemented or cell-depleted colostrum died or even developed significant diarrhea. Clearly colostral immunoglobulin is of vital significance for an agammaglobulinemic neonate such as the calf and cannot be replaced by leukocytes. Moreover, the modestly reduced survival rate and the impaired immune competence of cell-supplemented calves given milk substitute indicate that isolated leukocytes could even have detrimental effects if they are administered to an animal without concomitant adoptive transfer of immunoglobulins.

3.1. Materno-Fetal Transfer of Systemic Immune Functions

Apart from the possibility that cells in milk may support local defense in the neonatal intestine, the more exciting question is whether the maternal cells in colostrum are able to cross the intestinal wall and modulate systemic immune responses of the neonate. Early studies suggested that T cell functions could be transferred from mother to offspring via the milk. Type 4 delayed hypersensitivity reactions, such as the tuberculin (PPD) skin reaction, are known to depend on sensitized T lymphocytes and adoptive transfer is possible only by passage of sensitized cells. Mohr (1973) and Schlesinger and Covelli (1977) compared the PPD skin sensitivity of twenty breast-fed and fifteen non-breast-fed children of mothers who were PPD-positive but clinically free of tuberculosis. More than half of the breast-fed children exhibited anti-PPD skin sensitivity, whereas only one of the non-breast-fed infants (and none of more than eighty breast-fed children of PPD-negative mothers) did so. However, the conclusion that cell-mediated immunity to PPD was transferred to the infants through colostral T lymphocytes is questionable because reinfusion of fluorescently labeled maternal leukocytes showed that granulocytes and lymphocytes crossed the placental barrier and could be detected in the cord blood of more than 50% of infants at delivery (Desai and Creger 1963; Field and Caspary, 1971). Sensitized T cells could therefore have crossed the easily penetrable hemochorial human placenta, although epidemiological evidence for this hypothesis was weak. However, support for the postulated colostral transfer of sensitized T lymphocytes has come from studies on other species. Duhamel (1986) showed that PPD sensitivity was passed to calves consuming colostrum from PPD-positive cows. This is noteworthy because the impermeable bovine epitheliochorial placenta excludes not only leukocyte traffic but even the transfer of immunoglobulins.

Colostral transfer of T helper cell functions has been demonstrated in the nude mouse (Hale *et al.*, 1976). B lymphocytes of homozygous nude mice (nu/nu) are unable to produce antibodies against antigens that need the mediation of T lymphocytes, e.g., heterologous erythrocytes. In contrast, heterozygous mice (nu/+) can generate T-dependent antibody responses. When homozygous (nu/nu) offspring were suckled on heterozygous (nu/+) foster mothers, they developed the ability to form antibodies against sheep erythrocytes. The results of this experiment, as well as from studies on the transfer or suppression of graft-versus-host reactions in rodents (Beer *et al.*, 1972; Beer and Billingham, 1973; Head *et al.*, 1977; Uphoff, 1977), are consistent with the concept of the passage of antigen-reactive cells from mother to offspring via colostrum and milk. Opposite findings (Silvers and Poole, 1975) may be due to unidentified variables in the experiments.

In addition to the foregoing, Head and Beer (1979) demonstrated transfer of tumor resistance via colostrum to susceptible newborn mice. The different susceptibility of two strains of mice, C57BL/6 and A, to a Leydig cell tumor is explained by strain-specific differences in the activities of cytotoxic T cells, NK cells and macrophages in the elimination of neoplastic cells. When fostered on dams of the resistant strain, about half of newborn mice of the susceptible C58BL/6 strain resisted experimental tumor induction following inoculation with tumor cells. This finding is in contrast to the 100% mortality observed when newborns of the susceptible strain were suckled on dams of the same strain. Cross-fostering to the resistant strain clearly provided the adopted offspring with leukocytes that modulated tumor resistance. However, it cannot be determined from this experiment whether the enhanced resistance observed resulted from the transfer of maternal cells to the neonate or from the absorption of milk cytokines by the gut. Milk leukocytes release cytokines such as LIF and MIF (Mohr *et al.*, 1970), interferon (Emodi and Just, 1974; Lawton *et al.*, 1979) and tumor necrosis factor-α (Mushtaha *et al.*, 1989) in vitro. It is possible that these molecules can modulate immune responses of the newborn by way of absorption through the gut, although clear evidence for this is lacking.

3.2. Passage of Milk Cells Across the Intestinal Wall

Adoptive transfer of immune functions by way of the milk would be substantially explained if viable maternal leukocytes were demonstrated in the neonate. Demonstrating passage of milk leukocytes across the intestinal wall, however, has been a challenge because of the small numbers of maternal cells that locate in the tissues of the newborn. In three studies, leukocytes from various sources were injected directly into intestinal segments after laparotomy and were detected in the intestinal wall and lymphoid tissue close to the sites of application within a short time after administration. Radio-labeled lymph node cells from Fisher (Fi) × Dark Agouti (DA) F1 hybrids or from purebred Fisher rat donors were detected by autoradiography in the epithelium and the lamina propria of adult Fi × DA hybrid recipients 24 to 48 hours after infusion (Seelig and Billingham, 1981). Sheldrake and Husband (1985) used cells from mesenteric lymph nodes of syngeneic donors and confirmed such translocation of leukocytes into the intestine of the neonatal rat. These authors also infused radio-labeled allogeneic blood leukocytes from ewes into different gut segments of recipient lambs. Within two hours the labeled cells were detected in the lacteals and the mesenterial lymph nodes of the recipients. Finally, allogeneic colostral cells with a fluorescein isothyocyanate (FITC) label were injected into intestinal loops of calves and were subsequently detected using fluorescent antibodies applied to sections of the epithelium and lamina propria of the recipients (Liebler-Tenorio and Riedel-Caspari, unpublished data).

In apparent contrast to the foregoing studies, Miller (1981) failed to detect orally administered colostral leukocytes in a morphological study of the intestinal mucosa of suckling mice. However, additional studies have shown that orally administered leukocytes can, indeed, locate in the recipient intestine. Tuboly *et al.* (1988) fed technetium-labelled porcine colostral cells to piglets and detected these cells by autoradiography in intestinal sections of the recipients. Likewise, Tuboly *et al.* (1995) demonstrated intestinal uptake of colostral cells by lambs, whether the cells were administered orally or intraduodenally. Williams (1993) demonstrated that FITC-labeled colostral leukocytes were present in the peripheral blood of colostrum-deprived newborn piglets from two hours to twenty-four hours after oral administration. Colostral cells were also detected in the spleen, liver, lungs, and lymph nodes. In contrast, leukocytes prepared from the peripheral blood of the dams did not penetrate the intestinal barrier. In addition, use of intestinal explant cultures from colostrum-deprived neonatal piglets indicated intercellular passage of colostral leukocytes, but not of peripheral blood mononuclear cells, between epithelial cells of the duodenal and jejunal mucosa. At present, this finding appears inconsistent with the results of studies on the passage of leukocytes from the mesenteric lymph nodes of rats and the peripheral blood of lambs (Sheldrake and Husband, 1985).

Apart from the anatomic origin of maternal leukocytes, the genetic relationship between the donor and recipient may play a role in the passage of maternal cells across the intestinal wall. Studying outbred piglets, Tuboly *et al.* (1988) observed that colostral leukocytes only from a recipient piglet's own dam passed through the intestinal wall; cells from other dams remained trapped in the intestinal mucosa. A similar observation was reported by Kmetz *et al.* (1970), who administered virus-infected bovine lymphocytes to newborn rabbits and detected these xenogeneic cells in the intestinal wall, but not in the mesenteric lymph nodes, of the recipients. From the present literature it is hard to determine how close the genetic relationship must be to permit the passage of colostral leukocytes into or through the neonatal intestine. Passage between two closely related inbred mouse strains, as reported by Hale *et al.* (1976), is not surprising. However, transfer of leukocytes from purebred foster mothers of Dark Agouti and Fisher rats to the reciprocal offspring seems to occur at an early stage after birth, despite the fact that the strains are highly histoincompatible and develop runting disease and a graft-versus-host reaction when given reciprocal skin allografts (Beer *et al.*, 1975; Head *et al.*, 1977). Moreover, prolonged survival of skin grafts from the reciprocal strain was observed when the recipient rat pups were suckled on foster mothers of the reciprocal strain immediately after birth (Beer *et al.*, 1975). The latter experiments, although not confirmed by Silvers and Poole (1975), provide evidence that maternal immunocompetent cells not only gain access to the immune system of the neonate but influence immune functions despite MHC incompatibility. This phenomenon currently can be explained only in terms of tolerance induction in the neonate.

Maternal cells may persist in offspring for a considerable period of time. For example, various authors have described leukocyte chimerism on the part of male children bearing cells with a 46 XX karyotype in their blood. In one case, such chimerism was found to persist in a male infant up to 39 weeks after birth (Loke, 1978). Such apparent tolerance may also explain the classic observation of Peer (1958) that maternal skin grafts given to human infants survive for several months, whereas paternal grafts are rejected within days. Tolerance depends on the stage of immunologic maturation of the recipient and can still be induced after birth. This is critical for the neonates of all species with a placental barrier that is impermeable to maternal cells.

4. Transfer of Pathogenic Microorganisms Via Milk Leukocytes

Colostral leukocytes have occasionally been implicated as vehicles for pathogenic microorganisms, although the evidence for this mechanism is slight. Mitsuda *et al.* (1989) investigated the hypothesis that sporadic vaccination failure against hepatitis B virus (HBV) in children of HBV-positive mothers results from the transfer of virus-laden colostrum and colostral leukocytes. HBV-specific DNA was detected by polymerase chain reaction in eight samples of whey and colostral cells from 10 antigen-positive mothers. Only one infant developed HBV carrier status as indicated by detection of cord blood HBV DNA. These results are consistent with the report of Beasley *et al.* (1975) that the transfer of virus-containing colostrum and colostral leukocytes is a minor factor in infection of newborn infants with HBV. Kmetz *et al.* (1970) likewise failed to demonstrate the transmission of bovine leukemia virus when cultured bovine leukocytes were fed to newborn rabbits. However, the reason for the negative outcome of this experiment might have been the fact that the xenogeneic leukocytes were unable to cross the intestinal mucosa of the recipients.

Transmission of intracellular blood parasites such as *Cowdria ruminantium* which transmits heartwater disease to ruminants in East Africa, generally depends on the cooperation of blood-sucking arthropod vectors that transfer the parasitized erthrocytes from a carrier to another suscetible organism. However, in light of new findings, colostral cells may provide a mode of vertical transmission of this parasite. When isolated colostral cells from cows living in a heartwater-endemic area were injected intravenously into five susceptible goats, three of the recipient animals developed clinical signs of the disease (Deem *et al.*, 1996).

5. Colostral Leukocytes and the Neonate—Friends or Foes?

It is obvious that colostral leukocytes may support antimicrobial defense and immune responses of the offspring, including functions such as the synthesis of T cell-dependent antibodies and clonal expansion of lymphocytes. Subpopulations of mammary leukocytes produce soluble factors, including lysozyme, complement components, reactive oxygen metabolites and immunoglobulins, which impart anti-microbial resistance locally and systemically. Even resistance to certain tumors may be transferred via colostrum. On the other hand, colostral leukocytes have been shown to contribute to the transmission of graft-versus-host reactions, and apparently they can impair antimicrobial resistance if not integrated appropriately into the immune system of the neonate. Finally, it remains possible that intracellular pathogens can be transferred from mother to offspring by colostral leukocytes. Thus, whether colostral and milk leukocytes act as friends or foes to the suckling offspring depends on a variety of factors which have not yet been fully identified. The relationship between the donor and recipient, however, seems to play a decisive role.

References

Allardyce, R.A., Shearman, D.J.C., McClelland, D.B.L., Marwick, K., Simpson, A.J., and Laidlaw, R.B. 1974. Appearance of specific colostrum antibodies alter clinical infection with Salmonella typhimurium. *Br. Med. J.* **3**:3079.

Beasley, R.P., Stevens, C.E., Shiao, I.S., and Meng, H.C. 1975. Evidence against breast-feeding as a mechanism of vertical transmission of hepatitis B. *Lancet* **2**:740.

Beer, A.E., Billingham, R.E., and Yang, S.L. 1972. Maternally induced transplantation immunity, tolerance and runt disease in rats. *J. Exp. Med.* **135**:808.

Beer, A.E. and Billingham, R.E. 1973. Maternally acquired runt disease. *Science* **179**:240.

Beer, A.E., Stott, J.R., and Billingham, R.E. 1975. Histoimcompatibility and maternal immunological status as determinants of fetoplacental weight and litter size in rodents. *J. Exp. Med.* **142**:180.

Bertotto, A., Gerli, R., Fabietti, G., Crupi, S., Arcangeli, C., Scolise, F., and Vaccaro, R. 1990. Human breast milk T lymphocytes display the phenotype and functional characteristics of memory T cells. *Eur. J. Immunol.* **20**:1877.

Bienenstock, J., McDermott, M., Befus, D., and O'Neill, M. 1978. A common mucosal immunologic system involving the bronchus, breast and bowel. *Adv. Exp. Med. Biol.* **107**:53.

Bush, J.F. and Beer, A.E. 1979. Analysis of complement receptors on B-lymphocytes in human milk. *Am. J. Obstet. Gynecol* **133**:708.

Campbell, B., Porter, R.M., and Petersen, W.M. 1950. Plasmacytosis of the bovine udder during colostrum secretion and experimental cessation of milking. *Nature* **166**:913.

Concha, C., Holmberg, O., and Morein, B. 1978a. Proportion of B- and T-lymphocytes in normal bovine milk. *J. Dairy Res.* **45**:287.

Concha, C., Holmberg, O., and Morein, B. 1978b. Characterization of bovine mammary lymphocytes at different periods of lactation. *Adv. Exp. Med. Biol.* **137**:806.

Concha, C., Holmberg, O., and Morein, B. 1980. Characterization and response to mitogens of mammary lymphocytes from the bovine dry period secretion. *J. Dairy Res.* **47**:305.

Corlett, N.J., Peters, R.R., Paape, M.J., and Schultze, W.D. 1984. Effect of intramary device on new infection rate, milk yield, and somatic cell counts in Maryland dairy herds. *J. Dairy Sci.* **67**:2571.

Crago, S.S., Prince, S.J., Pretlow, G.T., McGhee, J.R., and Mestecky, J. 1979. Human colostral cells. I. Separation and characterization. *Clin. Exp. Immunol.* **38**:585.

Csorba, S., Nagy, B., Varga, S., Marodi, L., and Jezerniczky, J. 1979. Leukocyte function in colostrum. *Mschr. Kinderheilk.* **127**:557.

Deem, S.L., Norval, R.A., Donachie, P.L., and Mahan, S.M. 1996. Demonstration of vertical transmission of Cowdria ruminantum, the causative agent of heartwater, from cows to their calves. *Vet. Parasitol.* **61**:119.

Desai, R.G. and Creger, W.P. 1963. Maternal-fetal passage of leucocytes and platelets in man. *Blood* **21**:665.

Diaz-Jouanen, E. and Williams, R.C., Jr. 1974. T and B lymphocytes in human colostrum. *Clin. Immunol. Immunopathol.* **3**:248.

Duhamel, G.E. 1986. Characterization of bovine mammary lymphocytes and their effects on neonatal calf immunity. *Diss. Abstr. Intern. B* **47**:1477.

Duhamel, D.E., Bernocco, D., Davis, W.C., and Osburn, B.I. 1987. Distribution of T- and B-lymphocytes in mammary dry secretions, colostrum and blood of adult dairy cattle. *Vet. Immunol. Immunopathol.* **14**:101.

Evans, P.A., Newby, T.J., Stokes, C.R., and Bourne, F.J. 1982. A study of cells in the mammary secretions of sows. *Vet. Immunol. Immunopathol.* **3**:515.

Emödi, G. and Just, M. 1974. Interferon production by lymphocytes in human milk. *Scand. J. Immunol.* **3**:157.

Field, E.J. and Caspary, E.A. 1971. Is maternal lymphocyte sensitization passed to the child? *Lancet* **2**:337.

Goldblum, R.M., Ahlstedt, S., Carlsson, B., Hanson, L.A., Jodal, U., Lidin-Janson, G., and Sohl-Akerlund, A. 1975. Antibody-forming cells in human colostrum alter oral immunization. *Nature* **257**:297.

Guidry, A.J., Ost, M., Mather, I.H., Shainline, W.E., and Weinland, B.T. 1983. Sequential response of milk leukocytes, albumin, immunoglobulins, monovalent ions, citrate, and lactose in cows given infusion of Escherichia coli endotoxin into the mammary gland. *Am. J. Vet. Res.* **44**:2262.

Hale, M.L., Hanna, E.E., and Hansen, C.T. 1976. Nude mice from homozygous nude parents show smalter PFC responses to sheep erythrocytes than nude mice from heterozygous mothers. *Nature* **260**:44.

Hallén Sandgren, C., Nordling, K., and Björk, I. 1991. Isolation and phagocytic properties of neutrophils and other phagocytes from nonmastitic bovine milk. *J. Dairy Sci.* **74**:2965.

Hallén Sandgren, C., Larsson, I., and Persson, K. 1992. Bovine neutrophils recruited by endotoxin to a teat cistern continuously produce oxygen radicals and show increased phagocytosis and extracellular chemiluminescence. *Inflammation* **16**:117.

Harmon, R.J. and Heald, C.W. 1982. Migration of polymorphonuclear leukocytes into the bovine mammary gland during experimentally induced Staphylococcus aureus mastitis. *Am. J. Vet. Res.* **43**:992.

Harp, J.A. and Moon, H.W. 1987. Lymphocyte localization in lymph nodes of pubescent, prepartum, and postpatum sheep. *Vet. Immunol. Immunopathol.* **15**:297.

Harp, J.A., Runnels, P.L., and Pesch, B.A. 1988. Patterns of localization by mammary and mesenteric lymph node lymphocytes. *Vet. Immunol. Immunopathol.* **20**:31.

Head, J.R., Beer, E.A., and Billingham, R.E. 1977. Significance of the cellular component of the maternal immunologic endowment in milk. *Transplant. Proc.* **9**:1465.

Head, J.R. and Beer, E.A. 1979. In vivo and in vitro assessment of the immunologic role of leukocytic cells in milk, in: *Immunology of Breast Milk* (P.L. Ogra and D. Dayton, eds.), pp. 207–225, Raven Press, New York.

Heyermann, H. and Senft, B. 1985. Der Einfluss autologer Milch- und Blutseren auf die Phagozytoseaktivität neutrophiler Granulozyten des Blutes und der Milch von Kühen. *Zbl. Vet. Med. B* **32**:488.

Ho, F.C. and Lawton, J.W. 1978. Human colostral cells: phagocytosis and kiling of E. coli and C. albicans. *J. Pediatr.* **93**:910.

Ho, F.C.S., Wong, R.L.C., and Lawton, J.W.M. 1979. Human colostral and breast milk cells. *Acta Paediatr. Scand.* **68**:389.

Jain, N., Mathur, N.B., Sharma, V.K., and Dwarkadas, A.M. 1991. Cellular composition indicating lymphocyte subsets in preterm and full term human colostrum and milk. *Acta Paediatr. Scand.* **80**:395.

Jensen, D.L. and Eberhart, R.J. 1975. Macrophages in bovine milk. *Am J. Vet. Res.* **36**:619.

Jensen, D.L. and Eberhart, R.J. 1981. Total and differential cell counts in secretion of the non lactating bovine mammary gland. *Am. J. Vet. Res.* **42**:743.

Kmetz, M., Dunne, H.W., and Schultz, R.D. 1970. Leukocytes as carriers in the transmission of bovine leukaemia: invasion of the digestive tract of the newborn by ingested, cultured, leukocytes. *Am. J. Vet. Res.* **31**:637.

Kohl, S., Maloy, M.M., Pickering, L.K., Morriss, F., Adcock, E.W., and Walters, D.L. 1978a. Human colostral cytotoxcicity. I. Antibodydependent cellular cytotoxicity against Herpes simplex viralinfected cells mediated by colostral cells. *J. Clin. Lab. Immunol.* **1**:221.

Kohl, S., Shaban, S.S., Starr, S.E., Wood, P.A., and Nahmias, A.J. 1978b. Human neonatal and maternal monocyte-macrophage and lymphocyte mediated antibody-dependent cytotoxicity to cells infected with Herpes simplex. *J. Pediatr.* **93**:2060.

Kohl, S., Pickering, L.K., Cleary, T.G., Steinmetz, K.D., and Loo, L.S. 1980. Human colostral cytotoxicity. II. Relative defects in colostral leukocyte cytotoxicity and inhibition of peripheral blood leukocyte cytotoxicity by colostrum. *J. Infect. Dis.* **142**:884.

Lamm, M.E., Weisz-Carrington, P., Roux, M.E., McWilliams, M., and Phillips-Quagliata, J.M. 1978. Development of the IgA system in the mammary gland. *Adv. Exp. Med. Biol.* **107**:35.

Lascelles, A.K., Gruner, B.W., and Coombs, R.R.A. 1969. Some properties of human colostral cells. *Austr. J. Exp. Biol. Med. Sci.* **47**:349.

Laven, G.T., Crago, S.S., Kutteh, H.W., and Mestecky, J. 1981. Hemolytic plaque formation by cellular and noncellular elements of human colostrum. *J. Immunol.* **127**:1967.

Lawton, J.W.M., Shortridge, K.F., Wong, R.L.C., and Ng, M.H. 1979. Interferon synthesis by human colostral leukocytes. *Arch. Dis. Child.* **54**:127.

Lee, C.S., McDowell, G.H., and Lascelles, A.K. 1969. The importance of macrophages in the removal of fat from the involution mammary gland. *Res. Vet. Sci.* **10**:34.

Lee, C.S., Wooding, P., and Kemp, P. 1980. Identification, properties, and differential counts of cell populations using electron microscopy of dry cows secretions, colostrum and milk from normal cows. *J. Dairy Res.* **47**:39.

Lee, C.S. and Outteridge, P.M. 1981. Leukocytes of sheep colostrum, milk and involution secretion, with particular reference to ultrastructure and lymphocyte sub-populations. *J. Dairy Res.* **48**:225.

Lee, C.S., McCauly, I., and Hartmann, P.E. 1983. Light and electron microscopy of cells in pig colostrum, milk and involution secretion. *Acta Anat.* (*Basel*) **116**:126.

Loke, Y.W. 1978. *Immunology and Immunopathology of the Human Foetal-Maternal Interaction.* pp. 88–89. Elsevier/North-Holland Biomedical Press, Amsterdan, New York, Oxford.

Manning, L.S. and Parmely, M.J. 1980. Cellular determinants of mammary cell-mediated immunity in the rat. l. The migration of radioisotopically labeled T lymphocytes. *J. Immunol.* **125**:2508.

Mattila, T. and Frost, A.J. 1989. Induction by endotoxin of the inflammatory response in the lactating and dry bovine mammary gland. *Res. Vet. Sci.* **46**:238.

McDermott, M.R. and Bienenstock, J. 1979. Evidence for a common mucosal immunologic system. l. Migration of B immunoblasts into intestinal, respiratory, and genital tissues. *J. Immunol.* **122**:1892.

Meggs, P.D. and Beer, A.E. 1979. In vitro stimulation of human colostral lymphocytes by cytomegalovirus. *Am. J. Obstet. Gynecol.* **133**:703.

Meriläinen, V., Mayra, A., Korhonen, H., Antila, M., and Uusi-Rauva, A. 1979. Cells in bovine colostrum and properties of lymphocyte population. *Meijeriteellinen Aikakausirja* **37**:45.

Mielke, E. and Koblenz, C. 1980a. Zur Einteilung und Differenzierung der Milchzellen eutergesunder und euterkranker Kühe. Mh. *Vet. med.* **35**:367.

Mielke, E. and Koblenz, C. 1980b. Struktur und Funktion der Makrophagen der Kuhmilch. Mh. *Vet. med.* **35**:376.

Mielke, E. and Koblenz, C. 1981. Herkunft und Verhalten der Makrophagen der Milch eutergesunder und euterkranker Kühe. *Arch. Exp. Vet. med.* **35**:1.

Miller, S.C. 1981. Failure to demonstrate morphologically the presence of colostral or milk cells in the wall of the gastrointestinal tract of the suckling neonatal mouse. *J. Reprod. Imunol.* **3**:187.

Mitsuda, T., Yokota, S., Mori, T., Ibe, M., Ookawa, N., Shimizu, H., Aikara, Y., Yoshida, N., Kosuge, K., and Matsuayma, S. 1989. Demonstration of mother-to-infant transmission of hepatitis B virus by means of polymerase chain reaction. *Lancet* **2 (8668)**:886.

Mohr, J.A., Leu, R., and Mabry, W. 1970. Colostral leukocytes. *J. Surg. Oncol.* **2**:163.

Mohr, J.A. 1973. The possible induction and/or acquisition of cellular hypersensitivity associated with ingestion of colostrum. *J. Pediatr.* **82**:1062.

Moro, I., Crago, S.S., and Mestecky, J. 1983. Localization of IgA and IgM in human colostral elements using electron microscopy. *J. Clin. Immunol.* **3**:382.

Moro, I., Abo, T., Crago, S.S., Komiyama, U., and Mestecky, J. 1985. Natural killer cells in human colostrum. *Cell. Immunol.* **93**:467.

Murillo, G.J. and Goldman, A.S. 1970. The cells of human colostrum. II. Synthesis of IgA and β1c. *Pediatr. Res.* **4**:71.

Mushtaha, A.A., Schmalstieg, F.C., Hughes, T.K., Rajaraman, S., Rudloff, H.E., and Goldman, A.S. 1989. Chemokinetic agents for monocytes in human milk: possible role of tumor necrosis factor-alpha. *Pediatr. Res.* **25**:629.

Newby, T.J. and Bourne, J. 1977. The nature of the local immune system of the bovine mammary gland. *J. Immunol.* **118**:461.

Nickerson, S.C. and Pankey, J.W. 1984. Neutrophil migration through teat end tissues of bovine mammary quarters experimentally challenged with Staphylococcus aureus. *J. Dairy Sci.* **67**:826.

Nonnecke, B.J. and Kehrli, M.E., Jr. 1985. Isolation of mononuclear cells from bovine milk by continuous-flow and density gradient centrifugation: response of cells to mitogens. *Am. J. Vet. Res.* **46**:1259.

Östensson, K. 1993. Trafficing of leukocytes and immunoglobulin isotypes in the bovine udder. Dissertation University of Agricultural Sciences, Uppsala, Sweden.

Ogra, S.S. and Ogra, P.L. 1978. Immunologie aspects of human colostrum and milk. II. Characteristics of lymphocyte reactivity and distribution of E-rosette forming cells at different times alter the onset of lactation. *J. Pediatr.* **29**:550.

Ogra, S.S., Weintraub, D., and Ogra, P.L. 1978. Immunologie aspects of human colostrum and milk: Interaction with the intestinal immunity of the neonate. *Adv. Exp. Med. Biol.* **107**:957.

Paape, M.J., Guidry, A.J., Kirk, T.S., and Dold, D.J. 1975. Measurement of phagocytosis of P-labeled Staphylococcus aureus by bovine leukocytes: lysostaphin digestion and inhibitory effect of cream. *Am. J. Vet. Res.* **36**:1737.

Paape, M.J., Schultze, W.D., Corlett, N.J., and Weinland, B.T. 1988. Effect of abraded intrammary device on outcome in lactating cows after challenge exposure with Streptocossus uberis. *Am. J. Vet. Res.* **49**:790.

Parmely, M.J., Beer, A.J., and Billingham, R.E. 1976. In vitro studies on the T-lymphocyte population of human milk. *J. Exp. Med.* **144**:358.

Parmely, M.J. and Beer, A.E. 1977. Colostral cell-mediated immunity and the concept of a common secretory immune system. *J. Dairy Sci.* **60**:655.

Parmely, M.J. and Williams, S.B. 1979. Selective expression of immunocompetence in human colostrum: preliminary evidence for the control of cytotoxic T-lymphocytes including those specific for paternal alloantigens, in: *Immunology of Breast Milk* (P.L. Ogra and D.H. Dayton, eds.), pp. 173–183, Raven Press, New York.

Peer, L.A. 1958. Behaviour of skin grafts exchanged between parents and offsprings. *Ann. N.Y. Acad. Sci.* **73**:584.

Persson, A., PedersenMorner, A., and Kuhl, W. 1996. A long-term study of the health status and performance of sows on different feed allowances during late pregnanc. II. The Total cell content and its percentage of polymorphonuclear leucocytes in pathogen-free colostrum and milk collected from clinically healthy sows. *Acta Vet. Scand.* **37**:279.

Pickering, L.K., Cleary, T.G., Kohl, S., and Geiz, S. 1980. Polymorphonuclear leukocytes of human colostrum. l. Oxidative metabolism and kinetics of killing of radiolabeled Staphylococcus aureus. *J. Infect. Dis.* **142**:685.

Pitt, J., Barlow, B., and Heird, W.C. 1977. Protection against experimental necrotizing enterocolitis by maternal milk. I. Route of milk leukocytes. *Pediatr. Res.* **11**:906.

Pitt, J. 1979. The milk mononuclear phagocyte. *Pediatr.* **64**:745.

Pittard, W.B., Polmar, S.H., and Fanaroff, A. 1977. The breast milk macrophage: A potential vehicle for immunoglobulin transport. *Pediatr. Res.* **11**:492.

Richie, E.R., Steinmetz, K.D., Meistrich, M.L., Ramirez, I., and Hilliard, J.K. 1980. T-lymphocytes in colostrum and peripheral blood differ in their capacity to form thermostable E-rosettes. *J. Immunol.* **125**:2344.

Riedel-Caspari, G. 1993. The influence of colostral leukocytes on the course of an experimental Escherichia coli infection and serum antibodies in neonatal calves. *Vet. Immunol. Immunopathol.* **35**:275.

Riedel-Caspari, G. and Schmidt, F.-W. 1991a. The influence of colostral leukocytes on the immune system of the neonatal calf. II. Effects on passive and active immunization. *Dtsch. tierärztl. Wschr.* **98**:190.

Riedel-Caspari, G., Schmidt, F.W., and Marquardt, J. 1991. The influence of colostral leukocytes on the immune system of the neonatal calf. IV. Effects on bactericidity, complement and interferon. *Dtsch. tierärztl. Wschr.* **98**:395.

Robinson, J.M., Harvey, B.A., and Soothil, J.F. 1978. Phagocytosis and killing of bacteria and yeast by human milk cells alter opsonization in aqueous phase milk. *Br. Med. J.* **1**:1443.

Roux, M.E., McWilliams, M., Phillips-Quagliata, J.M., Weisz-Carrington, P., and Lamm, M.E. 1977. Origin of IgA secreting plasma cells in the mammary gland. *J. Exp. Med.* **146**:1311.

Russel, M.W. and Reiter, D. 1975. Phagocytic defínciency of bovine milk leukocytes: An effect of casein. *J. Reticuloendothel. Soc.* **18**:1.

Russell, M.W., Brooker, B.E., and Reiter, B. 1976. Inhibition of the bactericidal activity of bovine polymorphonuclear leukocytes and related systems by casein. *Res. Vet. Sci.* **20**:30.

Salmon, H. and Delouis, C. 1982. Cinetique de sous-populations lymphocytaires et des pplasmocytes dans la mammelle de truie primipaire en relation avec la gestation et la lactation. *Ann. Rech. Vet.* **13**:42.

Salmon, H. 1987. The intestinal and mammary immune system in pigs. *Vet. Immunol. Immunopathol.* **17**:367.

Schalm, O.W. 1977. Pathologic changes in the milk and the udder of cows with mastitis. *J. Am. Vet. Med. Assoc.* **170**:1137.

Schalm, O.W., Lasmanis, J., and Jain, N.C. 1976. Conversion of chronic staphylococcal mastitis to acute gangrenous mastitis after neutropenia in blood and bone marrow produced by an equine anti-bovine leukocyte serum. *Am. J. Vet. Res.* **37**:885.

Schlesinger, J.J. and Covelli, H.D. 1977. Evidence for the transmission of lymphocyte responses to tuberculin by breast-feeding. *Lancet* **2**:529.

Schollenberger, A., Degorski, A., Frymus, T., and Schollenberger, A. 1986a. Cells of sow mammary secretions. I. Morphology and differential counts during lactation. *J. Vet. Med. Assoc.* **33**:31.

Schollenberger, A., Degorski, A., Frymus, T., and Schollenberger, A. 1986c. Cells of sow mammary secretions. III. Some properties of phagocytic cells. *J. Vet. Med. Assoc.* **33**:353.

Schore, C.E., Osburn, B.I., Jasper, D.E., and Tyler, D.E. 1981. B- and T-lymphocytes in the bovine mammary gland: Rosette formation and mitogen response. *Vet. Imunol. Imunopathol.* **2**:561.

Schultze, W.D. and Pappe, M.J. 1984. Effect on outcome of intramammary challenge with Staphylococcus aureus of somatic cell concentration and presence of an intramammary device. *Am. J. Vet. Res.* **45**:420.

Scicchitano, R., Husband, A.J., and Clancy, R.L. 1984. Contribution of intraperitoneal immunization to the local immmune response in the respiratory tract of sheep. *Immunol.* **53**:375.

Seelig, L.L. 1980. Dynamics of leukocytes in rat mammary epithelium during pregnancy and lactation. *Biol. Reprod.* **22**:1211.

Seelig, L.L. and Billingham, R.E. 1981. Capacity of "transplanted" lymphocytes to transverse the intestinal epithelium of adult rats. *Transpl.* **32**:308.

Sheldrake, R.F. and Husband, A.J. 1985. Intestinal uptake of intact maternal lymphocytes by neonatal rats and lambs. *Res. Vet. Sci.* **39**:10.

Sheldrake, R.F., Husband, A.J., and Watson, D.L. 1985a. Specific antibody-containing cells in the mammary gland of non-lactating sheep following intraperitoneal and intramammary immunization. *Res. Vet. Sci.* **39**:312.

Sheldrake, R.F., Husband, A.J., Watson, D.L., and Cripps, A.W. 1985b. The effect of intraperitoneal and intramammary immunization of sheep on the number of antibody-containing cells in the mammary gland and the antibody titers in blood serum and mammary secretions. *Immunol.* **56**:605.

Silvers, W.K. and Poole, T.W. 1975. The influence of foster nursing on the survival and immunologic competence of mice and rats. *J. Immunol.* **115**:1117.

Smith, C.W. and Goldman, A.S. 1968. The cells of human colostrum. I. In vitro studies of morphology and functions. *Pediatr. Res.* **2**:103.

Smith, C.W. and Goldman, A.S. 1970. Interactions of leukocytes and macrophages from human colostrum: characteristics of the interacting lymphocyte. *J. Reticuloendothel. Soc.* **8**:91.

Smith, C.W., Goldman, A.S., and Yates, R.D. 1971. Interaction of lymphocytes and macrophages from human colostrum. Exp. Cell Res. **69**:409.

Smith, J.W. and Schultz, R.D. 1977. Mitogen and antigen-responsive milk lymphocytes. *Cell. Immunol.* **29**:165.

Targowski, S.P. and Niemialtowski, M. 1986. Inhibition of lacteal leukocyte phagocytosis by colostrum, non-lactating secretion, and mastitic milk. *Am. J. Vet. Res.* **47**:1940.

Tuboly, S., Bernath, S., Glavits, R., and Medveczky, I. 1988. Intestinal absorption of colostral lymphoid cells in new-born piglets. *Vet. Immunol. Immunopathol.* **20**:75.

Tuboly, S., Bernath, S., Glavits, R., Kovacs, A., and Megyeri, Z. 1995. Intestinal absorption of colostral lymphocytes in new-born lambs and their role in the development of the immune status. *Acta Vet. Hung.* **43**:105.

Uphoff, D.E. 1977. Runt disease in suckling hybridmice: Substrain differences and litter seriation. *Transplant. Proc.* **9**:1455.

Weaver, E.A., Goldblum, R.M., Davis, C.P., and Goldman, A.S. 1981. Enhanced immunoglobulin A release from human colostral cells during phagocytosis. *Infec. Immun.* **34**:498.

Williams, P.P. 1993. Immunomodulating effects of intestinal absorbed maternal colostral leukocytes by neonatal pigs. *Can. J. Vet. Res.* **57**:1.

Weisz-Carrington, P., Roux, M.E., Phillips-Quagliata, J.M., and Lamm, M.E. 1978. Hormonal induction of the secretory immune system in the mammary gland. *Proc. Nat. Acad. Sci.* **75**:2928.

Part IV

Activity of Milk in Promoting Development of Immunocompetence in the Suckling Neonate

Chapter 15

Maturation of Immunocompetence in Breast-Fed vs. Formula-Fed Infants

U. I. Dahlgren, L. Å. Hanson, and E. Telemo

1. Introduction

The most important task for the developing immune system is to learn how to react adequately to the environment. This is achieved by an active recognition and response to all antigens that are encountered. These include dietary antigens, commensal and pathogenic micro-organisms, non-replicating environmental antigens and autoantigens. To maintain homeostasis the immune system needs to discriminate between harmful and innocuous antigens.

The main interface between the individual and his or her surroundings is the intestinal epithelium. In congruence with this the main part of the immune system is localized in the connective tissue underlying the intestinal epithelium which is the largest mucosal surface and the one most highly exposed to antigenic materials. The maternal milk, together with the bacteria normally colonizing the gut of the newborn, constitute the first major contacts of the gut associated lymphoid tissue. The content and composition of the milk have been carefully designed by evolution to promote proper maturation and tuning of the immune system. Replacing natural milk with any other formulation may increase the risk for inflammatory events, primarily in the gastrointestinal tract, caused either by invading microorganisms or by noxious substances. Such interference can disturb the normal immunological development of the neonate. Major components of maternal milk that can influence the immune system of the child and aid the development of immunological homeostasis are immunoglobulins, other functional

U. I. Dahlgren, L. Å. Hanson, and E. Telemo • Göteborg University, Department of Clinical Immunology, Guldhedsgatan 10, S-413 46 Göteborg, Sweden

proteins, various protective components including oligosaccharides, growth factors, cytokines, fatty acids and living cells like T and B lymphocytes.

The protective capacity of milk against intestinal bacteria is primarily mediated by secretory IgA (sIgA) antibodies, which are produced by plasma cells in the mammary glands. These plasma cells, in turn, arise from the Peyer's patches, where the B-lymphocytes have been selected by the antigens taken up from the mother's intestine (Goldblum *et al.*, 1975; Goldblum, 1990). Thus, most of the sIgA antibodies are directed against antigens displayed by intestinal bacteria and have been shown to protect against infection in experimental systems (Winner *et al.*, 1991). Apart from directly binding to bacteria via the antigen specific part of the antibody, sIgA can agglutinate bacteria nonspecifically through oligosaccharide side chains on the α-chain of the immunoglobulin molecule. These oligosaccharides bind to fimbrial lectins, e.g., of *E. coli* and other bacteria (Wold *et al.*, 1990). This agglutination may aid in the control of colonization and facilitate the elimination of certain bacteria. The rich and diverse content of oligosaccharide in human milk, not found in cow's milk, may also contribute in this control (McVeagh and Miller, 1997).

The specific antibodies in the immunoglobulin fraction of mother's milk are important for protection against bacterial infections and especially diarrheal diseases, which are a major cause of morbidity and mortality in children in developing countries. The rate of mortality from diarrhea in non-breast-fed infants is 4 to 25 times higher than in exclusively breast-fed babies (Feachem and Koblinsky, 1984; Victora *et al.*, 1987). However, breast-feeding may not be consistently protective against all types of diarrhea. There are studies suggesting a higher risk of rotavirus diarrhea in breast-fed children after the cessation of breast-feeding than in non-breast-fed children (Mitra and Rabbani, 1995). Other studies have indicated that the incidence of rotavirus infection is similar in breast- and bottle-fed infants but that the course of the disease is milder in breast-fed infants (Duffy *et al.*, 1986a,b). This was also found in a recent study on rotavirus infection in Nicaraguan children (Espinoza *et al.*, 1997). Long term breast-feeding was associated with asymptomatic rotavirus infections in these children. Another recent study implies that protection against rotavirus by human milk is associated with the glycoprotein lactadherin (Newburg *et al.*, 1998). An elevated risk of cytomegalovirus and human immunodeficiency virus transmission due to breast-feeding also has been reported (Minamishima *et al.*, 1994; Dunn *et al.*, 1998).

Breast-feeding seems to prevent deaths from lower respiratory tract infections and necrotizing enterocolitis and to lower the incidence of urinary tract infections and otitis media (Saarinen, 1982; Màßrild *et al.*, 1989, 1990; Lucas and Cole, 1990; Pisacane *et al.*, 1992). In a Brazilian study it was found that bottle feeding led to 3.6 times higher risk of death from respiratory infections than complete breast-feeding (Victora *et al.*, 1987).

2. Indirect Effects of Breast-Feeding on the Development of the Intestinal Microflora and the Immune System of the Offspring

Microorganisms in the intestine of the newborn have a great impact on the ontogeny of the immune system (Karlsson *et al.*, 1999). Breast-feeding indirectly influences development by controlling the species and number of microorganisms that the mucosal surfaces are exposed to and thereby the severity of infections. In support of this we have shown that rats colonized by a bacterial strain at birth become tolerant to antigens associated with that organism. Newborn rats colonized at birth with a transgenic *Escherichia coli* strain producing ovalbumin do not produce any anti-ovalbumin antibodies when immunized with ovalbumin at adult age (Karlsson *et al.*, 1999). Adult animals, on the other hand, do not become tolerant; instead they are primed by colonization. The reported beneficial effect of breast-feeding on Crohn's disease (Bergstrand and Hellers, 1983) and ulcerative colitis (Koletzko *et al.*, 1991) suggests that breast milk is beneficial in the development of tolerance to the normal flora. Human breast milk has also proven to be anti-inflammatory in a rat model of colitis where infant formula was ineffective (Grazioso *et al.*, 1997).

The intestinal microflora is different in breast-fed and bottle-fed infants. In particular, bottle-fed children have more haemolysin- and diarrhea-producing *E. coli* than breast-fed infants (Giugliano *et al.*, 1993). However, in the breast-fed baby, other strains of *E. coli* may be favored by way of the ability of strains of this organism to bind nonspecifically to oligosaccharide side chains of IgA. For example, individuals with IgA deficiency have low numbers of *E. coli* with type 1 fimbriae exhibiting oligosaccharide-specific adhesins (Friman *et al.*, 1996). This type of mechanism may promote colonization of the intestine of the breast-fed infant by some strains of *E. coli* (Slavikova *et al.*, 1995). Only minor effects on other bacteria have been found, with higher incidence of *Staphylococcus epidermidis* in breast-fed babies but equal frequency of *Bifidobacterium* and *Bacteroides* species (Lundequist *et al.*, 1985).

3. Direct Effects of Breast-Feeding on the Development of the Immune System

Besides the direct protection against pathogens provided by the antibodies in the milk, it has been shown that anti-idiotypic antibodies in milk, and in the circulation of the mother, can prime the offspring's immune system against the antigen that the idiotype is directed to. A likely explanation of the priming effect of anti-idiotypic antibodies is that they resemble the original antigen. An antibody to a polysaccharide antigen thus can give rise to an anti-antipolysaccharide antibody that, although a protein, mimics the polysaccharide antigen. Monoclonal anti-K13 and anti-anti-K13 polysaccharide antibodies given by the peritoneal route to young mice, or provided via the milk, primed the young animals for a protective response to an *E. coli* K13 bacterial vaccine (Stein and Söderström, 1984). Oral administration of an anti-idiotype also mediated protection to cholera toxin in a rat model (Lucas *et al.*, 1991). Likewise, mouse pups nursed by mothers immunized with antibodies against anti-F-glycoprotein of respiratory syncytial virus exhibited anti-F-glycoprotein antibodies in serum and antibody producing cells in the spleen (Okamoto *et al.*, 1989). In this way the newborn receives "educational" antibodies via the placenta early in development as well as via the milk. Those antibodies represent the immunological experience of the mother and guide the immune system of the offspring to mount an appropriate response to the antigenic repertoire in her surroundings.

The possible effect of anti-idiotypic antibodies or other immunomodulating factors in milk may go far beyond the period of breast-feeding. It was recently demonstrated that breast-fed children between two and three years of age were better protected against *Haemophilus influenzae* type b infections than non-breast-fed children (Silfverdal *et al.*, 1997). The protective effect lasted for at least 10 years (Silfverdal *et al.*, 1999). This is congruent with the results of Howie *et al.* (1990), who found that breast-feeding for 13 weeks was still protective against gastroenteritis at age 2 years. Both recent and earlier studies have reported similar long term effects of breast-feeding against respiratory tract infections, otitis media and wheezing bronchitis (Saarinen, 1982; Burr *et al.*, 1993; Wilson *et al.*, 1998).

Milk contains high numbers of various leukocytes such as macrophages, neutrophils, T-lymphocytes and B-lymphocytes (Wold and Hanson, 1994). Several studies have shown that these white blood cells are taken up by the offspring (Weiler *et al.*, 1983; Jain *et al.*, 1989). The transient tuberculin reaction seen in breast-fed infants is indirect evidence that the cells passing from the maternal milk are functionally active (Mohr, 1972; Schlesinger and Covelli, 1977). However, this finding was questioned in a later study (Keller *et al.*, 1987). Tuboly *et al.* (1995) found that milk lymphocytes were taken up by newborn lambs and were present in mesenteric lymph nodes. Such cells could, presumably through idiotypic/anti-idiotypic interactions, transfer priming to vaccinations previously given to the ewe. Another indication that the cells in milk shape the developing immune system is that breast-feeding enhances the take and survival of kidney grafts from mother to offspring and between siblings differing in one haplotype (Campbell *et al.*, 1984; Kois *et al.*, 1984). Foster nursing experiments in rats have confirmed that breast-feeding leads to a decreased reactivity to cells with transplantation molecules identical to those in foster dams (Deroche *et al.*, 1993). Furthermore, breast-fed infants have fewer T cell precursors reactive with maternal transplantation molecules than non-breast-fed infants, suggesting tolerance induction against the maternal transplantation molecules (Zhang *et al.*, 1991). In addition, there is a significant pool of MHC class II molecules in the milk fat globule fraction of human milk (Wiman *et al.*, 1979; Newman *et al.*, 1980). Another striking finding is that the thymus of breast-fed infants at 4 months of age is significantly larger than in bottle-fed infants, indicating that breast milk also has a generalized immuno-stimulatory effect (Hasselbalch *et al.*, 1997).

Whether breast-feeding has beneficial effects on the response to vaccines has been discussed for some time. In favour of this view, studies have shown that breast-fed infants respond better to *Haemophilus influenzae* type b conjugate vaccine and BCG vaccination than bottle-fed infants (Pabst *et al.*, 1989; Pabst and Spady, 1990). Serum antibodies as well as SIgA and SIgM antibody responses in saliva and stools after vaccination with oral polio virus, parenteral tetanus and diphtheria toxoid were also higher in breast-fed than in non-breast-fed infants (Hahn-Zoric *et al.*, 1990). However in most other studies there was no difference between breast- and bottle-fed infants in the ability to respond to parenteral vaccines (Zoppi *et al.*, 1983; Stephens *et al.*, 1984; Scheifele *et al.*, 1992; Karron *et al.*, 1995). Further, the response to per-oral live virus vaccines seems to be better in bottle- than in breast-fed infants (Pichichero, 1990; WHO, 1995; Rennels, 1996). However, this may be due to giving live viral vaccine too close to a virus neutralizing breast meal as found a long time ago by Sabin (1978). It may also be due to differences in the immune response of the mother, which may be transferred to the offspring via idiotypic/anti-idiotypic antibodies via the milk, the placenta or both (Stein and Söderström, 1984; Okamoto *et al.*, 1989; Hahn-Zoric *et al.*, 1990). Such effects may even be transferred to a second generation (Lundin *et al.*, 1999).

Among other factors in milk that may influence the developing immune system are inflammatory and anti-inflammatory cytokines (Wold and Hanson, 1994), lactoferrin and the protein hormone anti-secretory factor (Lönnroth *et al.*, 1988). Lactoferrin and the anti-secretory factor both have anti-inflammatory activity (Mattsby-Baltzer *et al.*, 1996; Johansson *et al.*, 1997a,b). Cytokines in the milk, among which one can find IL-1, TNF-α, IFN-γ, IL-6, TGF-β and IL-10, survive the passage through the gastrointestinal tract and may be taken up by the infant and affect the immune system (Söder, 1987; Rudloff *et al.*, 1992; Bocci *et al.*, 1993; Rollwagen and Baqar, 1996). Milk cells also can be induced to produce cytokines such as IL-2, IL-3, IL-4, IL-10, IFN-γ, TGF-β and TNF-α (Skansen-Saphir *et al.*, 1993; Letterio *et al.*, 1994). The cytokines in milk not only influence the immune system but are also important for the normal development of the newborn. Thus, neonatal mice with a disrupted TGF-β gene developed cardiac abnormalities and died if they were nursed by a mother that was unable to produce TGF-β but not if nursed by a mother that was able to produce this cytokine (Letterio *et al.*, 1994). This study suggests that milk cytokines can be absorbed by the offspring in biologically active form. It should also be noted that TGF-β, which is abundant in breast milk, has profound anti-inflammatory activity (Wahl, 1992; Clark and Coker, 1998) and is an important switch factor for IgA (Edmiston and Lebman, 1997; Shi and Stavnezer, 1998).

In general, it seems that the immune system of bottle-fed children has a higher baseline activity than breast-fed infants, since spontaneous T cell proliferation and IFN-γ production are increased after bottle-feeding (Pabst *et al.*, 1997). On the other hand, breast-fed infants respond with more IFN-γ than bottle-fed infants do to infections with respiratory syncytial virus (Chiba *et al.*, 1987). Thus it seems as if breast-fed children have a lower basal activity of the immune system, but that they respond better to specific antigenic stimulation.

4. Breast-Feeding and Allergic Disease

The possible protective effect of breast-feeding on the risk of developing atopic disease has been debated for many years. Kramer (1988) claimed that most studies concerning the influence of breast-feeding on allergy development were deficient in both biological and methodological aspects. Some studies have indicated that exclusively breast-fed high-risk children have a reduced frequency of wheezing, atopic dermatitis, vomiting/diarrhea and infantile colic (Kajosaari and Saarinen, 1983; Halken *et al.*, 1992; Saarinen and Kajosaari, 1995). Other investigations have failed to show a protective effect of breast-feeding against the development or severity of atopic eczema (Shohet *et al.*, 1985). The fatty acid composition of the milk seems to play a role in atopic eczema (Businco *et al.*, 1993; Duchen *et al.*, 1998; Yu *et al.*, 1998).

In one unconfirmed study, breast-feeding preceded by early feeding of cow's milk formula reduced the incidence of allergy symptoms before 18 months of age in infants with a family history of allergy (Lindfors and Enocksson, 1988). Other studies have suggested that the risk of developing allergy to cow's milk is inversely related to the level of antibodies to cow's milk protein in the mother's milk (Machtinger and Moss, 1986; Casimir *et al.*, 1989). This was not confirmed by Fälth-Magnusson (1989). In a recent study, newborn babies were supplemented with cow's milk during the first three days of life and compared with babies given placebo substitute during the same time (de Jong *et al.*, 1998). There was no difference in development of atopy either at one or two years of age. Hence this study did not confirm the findings by Høst *et al.* (1988) that there was an increased risk of atopic disease after formula feeding during the first 3 days of life or the finding of Lindfors and Enocksson (1988) that formula had a protective effect before breast-feeding. Further, children with a family history of atopic disease ran the same risk of subsequent atopic disease whether they were fed formula or breast milk alone in the maternity ward (Gustafsson *et al.*, 1992). In this study cow's milk formula did not increase the risk of developing atopic disease.

To reduce suspected allergenic activity, hydrolyzed artificial formulas are given to atopy-prone newborns. A study by Oldaeus *et al.* (1997) indicated an allergy preventive effect of an extensively hydrolyzed formula, but not of a partially hydrolyzed formula, during the first 18 months of life in high risk infants. Vaarala *et al.* (1995) found that infants given non-hydrolyzed cow's milk based formula, but not those given hydrolyzed formula, developed antibodies and T cell responses to cow's milk protein. Interestingly, it was found that exposure to cow's milk proteins after the age of 9 months resulted in depressed cellular and humoral responsiveness to these proteins. Extensively hydrolyzed milk formulas do not stimulate β-casein specific T cell clones while partially hydrolyzed formulas do (Szepfalusi *et al.*, 1998). Residual immunological reactivity to partially hydrolyzed milk proteins has been shown in rats fed partially hydrolyzed β-lactoglobulin. The partially hydrolyzed material induced tolerance whereas the extensively hydrolyzed material did not (Fritsché *et al.*, 1997).

Reduced exposure of infants to allergenic foods caused by food allergen avoidance on the part of the reactive mother has been reported to decrease food sensitization and allergy primarily during the first year of life (Zeiger *et al.*, 1989). Consumption of cow's milk by mothers and of cow's milk-based formula by infants elevates the risk of allergies, while consumption of hypoallergenic formula by pregnant and lactating women and by their infants was helpful in preventing allergy development in the infants (Fukushima *et al.*, 1997). Antigen avoidance solely by the mother in late pregnancy has been found not to influence the development of atopic disease in the infant (Fälth-Magnusson and Kjellman, 1992). However, there was a significant reduction in allergic manifestations when the avoidance was extended into the lactation period (Lovegrove *et al.*, 1996) or applied solely during the first 3 months of breast-feeding (Hattevig *et al.*, 1990). This was apparent even at a 4-year follow up (Sigurs *et al.*, 1992). Interestingly, the presence of cow's milk protein in human breast milk has been shown to induce hypersensitivity reactions in solely breast-fed infants (Jakobsson and Lindberg, 1978). Whether this manifestation requires prior sensitization *in utero* has not been investigated (Piastra *et al.*, 1994; Jones *et al.*, 1996).

Long-term effects of breast-feeding on other allergens have been suggested by a study on children followed for 17 years. Initially they had significantly fewer food allergies and later significantly fewer inhalant allergies if they had been breast-fed for more than one month. No other factor was associated with their lower prevalence of allergic disease (Saarinen and Kajosaari, 1995).

Careful, large epidemiological studies of celiac disease in early childhood have shown that breast-feeding seems to protect against celiac disease (Fälth-Magnusson *et al.*, 1996). An earlier, smaller study has shown that children breast-fed for 3 months or more have a marked delay in the onset of the disease and a longer latency time from gluten introduction to appearance of disease, when compared with bottle-fed children (Greco *et al.*, 1985).

5. Breast-Feeding and Autoimmune Diseases

Some studies have suggested that breast-feeding may protect against type 1 diabetes (Borch-Johnsen *et al.*, 1984; Mayer *et al.*, 1988; Virtanen *et al.*, 1991, 1992; Dosch *et al.*, 1993). Other studies have not revealed a beneficial effect (Fort *et al.*, 1986). It has been proposed that differences seen between breast-fed and non-breast-fed infants are due to biases in methodology (Norris and Scott, 1996). Other studies have suggested that breast-feeding may decrease the risk of developing multiple sclerosis and rheumatoid arthritis but not autoimmune thyroiditis (Fort *et al.*, 1990; Pisacane *et al.*, 1994; Brun *et al.*, 1995). Obviously these possible relationships must be investigated further.

6. Conclusions

The immune system of the newborn expands rapidly after birth, primarily stimulated by the microbial flora colonizing the mucous membranes of the gut. Breast-feeding influences the composition of the intestinal flora of the suckling infant, albeit to a lesser degree than is often suggested. Nevertheless, in this way breast-feeding influences the nature of developmental stimuli imposed on the immune system through the colonizing bacteria. The high concentration in the milk of SIgA antibodies against the microflora present in the mother's milieu help to protect the infant against the mucosal flora. Numerous other defense mechanisms in the milk also provide passive protection to the infant.

In addition to passive protection, there is now good evidence of a prolonged enhancement of protection after the termination of breast-feeding. This may be due to the effects of anti-antibodies, lymphocytes and certain cytokines from the milk. Breast-feeding generally has not been proven to affect the risk of developing immunological diseases such as allergy and autoimmune conditions, although some preliminary evidence exists, and breast-feeding seems to protect against development of celiac disease in early life.

References

Bergstrand, O. and Hellers, G. 1983. Breast-feeding during infancy in patients who later develop Crohn's disease. *Scand. J. Gastroenterol.* **18**:903.

Bocci, V., von Bremen, K., Corradeschi, F., Franchi, F., Luzzi, E., and Paulesu, L. 1993. Presence of interferon-gamma and interleukin-6 in colostrum of normal women. *Lymphokine Cytokine Res.* **12**:21.

Borch-Johnsen, K., Joner, G., Mandrup-Poulsen, T., Christy, M., Zachau-Christiansen, B., Kastrup, K., and Nerup, J. 1984. Relation between breast-feeding and incidence rates of insulin-dependent diabetes mellitus. A hypothesis. *Lancet* **2**:1083.

Brun, J.G., Nilssen, S., and Kvåle, G. 1995. Breast feeding, other reproductive factors and rheumatoid arthritis. A prospective study. *Br. J. Rheumatol.* **34**:542.

Burr, M.L., Limb, E.S., Maguire, M.J., Amarah, L., Eldridge, B.A., Layzell, J.C., and Merrett, T.G. 1993. Infant feeding, wheezing, and allergy: a prospective study. *Arch. Dis. Child.* **68**:724.

Businco, L., Ioppi, M., Morse, N.L., Nisini, R., and Wright, S. 1993. Breast milk from mothers of children with newly developed atopic eczema has low levels of long chain polyunsaturated fatty acids. *J. Allergy Clin. Immunol.* **91**:1134.

Campbell, D.A., Jr., Lorber, M.I., Sweeton, J.C., Turcotte, J.G., Niederhuber, J.E., and Beer, A.E. 1984. Breast feeding and maternal-donor renal allografts. Possibly the original donor-specific transfusion. *Transplantation* **37**:340.

Casimir, G.J., Duchateau, J., Cuvelier, P., and Vis, H.L. 1989. Maternal immune status against beta-lactoglobulin and cow milk allergy in the infant. *Ann. Allergy* **63**:517.

Chiba, Y., Minagawa, T., Mito, K., Nakane, A., Suga, K., Honjo, T., and Nakao, T. 1987. Effect of breast feeding on responses of systemic interferon and virus-specific lymphocyte transformation in infants with respiratory syncytial virus infection. *J. Med. Virol.* **21**:7.

Clark, D.A. and Coker, R. 1998. Transforming growth factor-beta (TGF-beta). *Int. J. Biochem. Cell Biol.* **30**:293.

de Jong, M.H., Scharp-vand der Linden, V.T., Aalberse, R.C., Oosting, J., Tijssen, J.G., and de Groot, C.J. 1998. Randomised controlled trial of brief neonatal exposure to cows' milk on the development of atopy. *Arch. Dis. Child.* **79**:126.

Deroche, A., Nepomnaschy, I., Torello, S., Goldman, A., and Piazzon, I. 1993. Regulation of parental alloreactivity by reciprocal F1 hybrids. The role of lactation. *J. Reprod. Immunol.* **23**:235.

Dosch, H.M., Martin, J.M., Robinson, B.H., Åkerblom, H.K., and Karjalainen, J. 1993. An immunological basis for disproportionate diabetes risks in children with a type I diabetic mother or father. *Diabetes Care* **16**:949.

Duchen, K., Yu, G., and Bjorksten, B. 1998. Atopic sensitization during the first year of life in relation to long chain polyunsaturated fatty acid levels in human milk. *Pediatr. Res.* **44**:478.

Duffy, L.C., Byers, T.E., Riepenhoff-Talty, M., La Scolea, L.J., Zielezny, M., and Ogra, P.L. 1986a. The effects of infant feeding on rotavirus-induced gastroenteritis: a prospective study. *Am. J. Public Health* **76**:259.

Duffy, L.C., Riepenhoff-Talty, M., Byers, T.E., La Scolea, L.J., Zielezny, M.A., Dryja, D.M., and Ogra, P.L. 1986b. Modulation of rotavirus enteritis during breast-feeding. Implications on alterations in the intestinal bacterial flora. *Am. J. Dis. Child.* **140**:1164.

Dunn, D.T., Tess, B.H., Rodrigues, L.C., and Ades, A.E. 1998. Mother-to-child transmission of HIV: implications of variation in maternal infectivity. *Aids* **12**:2211.

Edmiston, J.S. and Lebman, D.A. 1997. A transforming growth factor-beta regulable RNA-binding protein interacts specifically with germline Ig alpha transcripts. *Int. Immunol.* **9**:427.

Espinoza, F., Paniagua, M., Hallander, H., Svensson, L., and Strannegård, Ö. 1997. Rotavirus infections in young Nicaraguan children. *Pediatr. Infect. Dis. J.* **16**:564.

Feachem, R.G. and Koblinsky, M.A. 1984. Interventions for the control of diarrhoeal diseases among young children: promotion of breast-feeding. *Bull. WHO.* **62**:271.

Fort, P., Lanes, R., Dahlem, S., Recker, B., Weyman-Daum, M., Pugliese, M., and Lifshitz, F. 1986. Breast feeding and insulin-dependent diabetes mellitus in children. *J. Am. Coll. Nutr.* **5**:439.

Fort, P., Moses, N., Fasano, M., Goldberg, T., and Lifshitz, F. 1990. Breast and soy-formula feedings in early infancy and the prevalence of autoimmune thyroid disease in children. *J. Am. Coll. Nutr.* **9**:164.

Friman, V., Adlerberth, I., Connell, H., Svanborg, C., Hanson, L.Å., and Wold, A.E. 1996. Decreased expression of mannose-specific adhesins by *Escherichia coli* in the colonic microflora of immunoglobulin A-deficient individuals. *Infect. Immun.* **64**:2794.

Fritsché, R., Pahud, J.J., Pecquet, S., and Pfeifer, A. 1997. Induction of systemic immunologic tolerance to beta-lactoglobulin by oral administration of a whey protein hydrolysate. *J. Allergy Clin. Immunol.* **100**:266.

Fukushima, Y., Iwamoto, K., Takeuchi-Nakashima, A., Akamatsu, N., Fujino-Numata, N., Yoshikoshi, M., Onda, T., and Kitagawa, M. 1997. Preventive effect of whey hydrolysate formulas for mothers and infants against allergy development in infants for the first 2 years. *J. Nutr. Sci. Vitaminol.* (*Tokyo*) **43**:397.

Fälth-Magnusson, K. 1989. Breast milk antibodies to foods in relation to maternal diet, maternal atopy and the development of atopic disease in the baby. *Int. Arch. Allergy Appl. Immunol.* **90**:297.

Fälth-Magnusson, K., Franzen, L., Jansson, G., Laurin, P., and Stenhammar, L. 1996. Infant feeding history shows distinct differences between Swedish celiac and reference children. *Pediatr. Allergy Immunol.* **7**:1.

Fälth-Magnusson, K. and Kjellman, N.I. 1992. Allergy prevention by maternal elimination diet during late pregnancy—a 5-year follow-up of a randomized study. *J. Allergy Clin. Immunol.* **89**:709.

Giugliano, L.G., Meyer, C.J., Arantes, L.C., Ribeiro, S.T., and Giugliano, R. 1993. Mannose-resistant haemagglutination (MRHA) and haemolysin (Hly) production of strains of Escherichia coli isolated from children with diarrhoea: effect of breast feeding. *J. Trop. Pediatr.* **39**:183.

Goldblum, R.M. 1990. The role of IgA in local immune protection. *J. Clin. Immunol.* **10**:64.

Goldblum, R.M., Ahlstedt, S., Carlsson, B., Hanson, L.A., Jodal, U., Lidin-Janson, G., and Sohl-Akerlund, A. 1975. Antibody-forming cells in human colostrum after oral immunisation. *Nature* **257**:797.

Grazioso, C.F., Werner, A.L., Alling, D.W., Bishop, P.R., and Buescher, E.S. 1997. Antiinflammatory effects of human milk on chemically induced colitis in rats. *Pediatr. Res.* **42**:639.

Greco, L., Mayer, M., Grimaldi, M., Follo, D., De Ritis, G., and Auricchio, S. 1985. The effect of early feeding on the onset of symptoms in celiac disease. *J. Pediatr. Gastroenterol. Nutr.* **4**:52.

Gustafsson, D., Löwhagen, T., and Andersson, K. 1992. Risk of developing atopic disease after early feeding with cows' milk based formula. *Arch. Dis. Child.* **67**:1008.

Hahn-Zoric, M., Fulconis, F., Minoli, I., Moro, G., Carlsson, B., Böttiger, M., Räihä, N., and Hanson, L.Å. 1990. Antibody responses to parenteral and oral vaccines are impaired by conventional and low protein formulas as compared to breast-feeding. *Acta Paediatr. Scand.* **79**:137.

Halken, S., Høst, A., Hansen, L.G., and Osterballe, O. 1992. Effect of an allergy prevention programme on incidence of atopic symptoms in infancy. A prospective study of 159 "high-risk" infants. *Allergy* **47**:545.

Hasselbalch, H., Jeppesen, D.L., Ersboll, A.K., Engelmann, M.D., and Nielsen, M.B. 1997. Thymus size evaluated by sonography. A longitudinal study on infants during the first year of life. *Acta Radiol.* **38**:222.

Hattevig, G., Kjellman, B., Sigurs, N., Grodzinsky, E., Hed, J., and Björksten, B. 1990. The effect of maternal avoidance of eggs, cow's milk, and fish during lactation on the development of IgE, IgG, and IgA antibodies in infants. *J. Allergy Clin. Immunol.* **85**:108.

Howie, P.W., Forsyth, J.S., Ogston, S.A., Clark, A., and du Florey, C. 1990. Protective effect of breast feeding against infection. *Brit. Med. J.* **300**:11.

Høst, A., Husby, S., and Østerballe, O. 1988. A prospective study of cow's milk allergy in exclusively breast-fed infants. Incidence, pathogenetic role of early inadvertent exposure to cow's milk formula, and characterization of bovine milk protein in human milk. *Acta Paediatr. Scand.* **77**:663.

Jakobsson, I. and Lindberg, T. 1978. Cow's milk proteins cause infantile colic in breast-fed infants. *Lancet* **2**:437.

Johansson, E., Jennische, E., Lange, S., and Lönnroth, I. 1997a. Antisecretory factor suppresses intestinal inflammation and hypersecretion. *Gut* **41**:642.

Johansson, E., Lange, S., and Lönnroth, I. 1997b. Identification of an active site in the antisecretory factor protein. *Biochim. Biophys. Acta* **1362**:177.

Jones, A.C., Miles, E.A., Warner, J.O., Colwell, B.M., Bryant, T.N., and Warner, J.A. 1996. Fetal peripheral blood mononuclear cell proliferative responses to mitogenic and allergenic stimuli during gestation. *Pediatr. Allergy Immunol.* **7**:109.

Kajosaari, M. and Saarinen, U.M. 1983. Prophylaxis of atopic disease by six months' total solid food elimination. Evaluation of 135 exclusively breast-fed infants of atopic families. *Acta Paediatr. Scand.* **72**:411.

Karlsson, M., Hanson, L.Å., Kahu, H., Telemo, E., and Dahlgren, U.I. 1999. Neonatal colonization of rats induces immunological tolerance to bacterial antigens. *Eur. J. Immunol.* **29**:109.

Karron, R.A., Steinhoff, M.C., Subbarao, E.K., Wilson, M.H., Macleod, K., Clements, M.L., Fries, L.F., and Murphy, B.R. 1995. Safety and immunogenicity of a cold-adapted influenza A (H1N1) reassortant virus vaccine administered to infants less than six months of age. *Pediatr. Infect. Dis. J.* **14**:10.

Keller, M.A., Rodgriguez, A.L., Alvarez, S., Wheeler, N.C., and Reisinger, D. 1987. Transfer of tuberculin immunity from mother to infant. *Pediatr. Res.* **22**:277.

Kois, W.E., Campbell, D.A., Jr., Lorber, M.I., Sweeton, J.C., and Dafoe, D.C. 1984. Influence of breast feeding on subsequent reactivity to a related renal allograft. *J. Surg. Res.* **37**:89.

Koletzko, S., Griffiths, A., Corey, M., Smith, C., and Sherman, P. 1991. Infant feeding practices and ulcerative colitis in childhood. *Br. Med. J.* **302**:1580.

Kramer, M.S. 1988. Does breast feeding help protect against atopic disease? Biology, methodology, and a golden jubilee of controversy. *J. Pediatr.* **112**:181.

Letterio, J.J., Geiser, A.G., Kulkarni, A.B., Roche, N.S., Sporn, M.B., and Roberts, A.B. 1994. Maternal rescue of transforming growth factor-beta 1 null mice. *Science* **264**:1936.

Lindfors, A. and Enocksson, E. 1988. Development of atopic disease after early administration of cow milk formula. *Allergy* **43**:11.

Lovegrove, J.A., Morgan, J.B., and Hamptom, S.M. 1996. Dietary factors influencing levels of food antibodies and antigens in breast milk. *Acta Paediatr.* **85**:778.

Lucas, A. and Cole, T.J. 1990. Breast milk and neonatal necrotising enterocolitis. *Lancet* **336**:1519.

Lucas, G.P., Cambiaso, C.L., and Vaerman, J.P. 1991. Protection of rat intestine against cholera toxin challenge by monoclonal anti-idiotypic antibody immunization via enteral and parenteral route. *Infect.Immun.* Oct.:3651.

Lundequist, B., Nord, C.E., and Winberg, J. 1985. The composition of the faecal microflora in breastfed and bottle fed infants from birth to eight weeks. *Acta Paediatr. Scand.* **74**:45.

Lundin, B.S., Dahlman-Höglund, A., Pettersson, I., Dahlgren, U.I., Hanson, L.Å., and Telemo, E. 1999. Antibodies given orally in the neonatal period can affect the immune response for two generations: evidence for active maternal influence on the newborn's immune system. *Scand. J. Immunol.* In press.

Lönnroth, I., Martinsson, K., and Lange, S. 1988. Evidence of protection against diarrhoea in suckling piglets by a hormone-like protein in the sow's milk. *Zentralbl. Veterinarmed.* [*B*]. **35**:628.

Machtinger, S. and Moss, R. 1986. Cow's milk allergy in breast-fed infants: the role of allergen and maternal secretory IgA antibody. *J. Allergy Clin. Immunol.* **77**:341.

Mattsby-Baltzer, I., Roseanu, A., Motas, C., Elverfors, J., Engberg, I., and Hanson, L.Å. 1996. Lactoferrin or a fragment thereof inhibits the endotoxin-induced interleukin-6 response in human monocytic cells. *Pediatr. Res.* **40**:257.

Mayer, E.J., Hamman, R.F., Gay, E.C., Lezotte, D.C., Savitz, D.A., and Klingensmith, G.J. 1988. Reduced risk of IDDM among breast-fed children. The Colorado IDDM Registry. *Diabetes* **37**:1625.

McVeagh, P. and Miller, J.B. 1997. Human milk oligosaccharides: only the breast. *J. Paediatr. Child. Health* **33**:281.

Minamishima, I., Ueda, K., Minematsu, T., Minamishima, Y., Umemoto, M., Take, H., and Kuraya, K. 1994. Role of breast milk in acquisition of cytomegalovirus infection. *Microbiol, Immunol.* **38**:549.

Mitra, A.K. and Rabbani, F. 1995. The importance of breast feeding in minimizing mortality and morbidity from diarrhoeal diseases: the Bangladesh perspective. *J. Diarrhoeal Dis. Res.* **13**:1.

Mohr, J.A. 1972. Lymphocyte sensitisation passed to the child from the mother. *Lancet* **1**:688.

Mà6rild, S., Jodal, U., and Hanson, L.Å. 1990. Breast feeding and urinary-tract infection. *Lancet* **336**:942.

Mà6rild, S., Jodal, U., and Mangelus, L. 1989. Medical histories of children with acute pyelonephritis compared with controls. *Pediatr. Infect. Dis. J.* **8**:511.

Newburg, D.S., Peterson, J.A., Ruiz-Palacios, G.M., Matson, D.O., Morrow, A.L., Shults, J., Guerrero, M.L., Chaturvedi, P., Newburg, S.O., Scallan, C.D., Taylor, M.R., Ceriani, R.L., and Pickering, L.K. 1998. Role of human-milk lactadherin in protection against symptomatic rotavirus infection. *Lancet* **351**:1160.

Newman, R.A., Ormerod, M.G., and Greaves, M.F. 1980. The presence of HLA-DR antigens on lactating human breast epithelium and milk fat globule membranes. *Clin. Exp. Immunol.* **41**:478.

Norris, J.M. and Scott, F.W. 1996. A meta-analysis of infant diet and insulin-dependent diabetes mellitus: do biases play a role? *Epidemiology* **7**:87.

Okamoto, Y., Tsutsumi, H., Kumar, N.S., and Ogra, P.L. 1989. Effect of breast feeding on the development of anti-idiotype antibody response to F glycoprotein of respiratory syncytial virus in infant mice after post-partum maternal immunization. *J. Immunol.* **142**:2507.

Oldaeus, G., Anjou, K., Björkstén, B., Moran, J.R., and Kjellman, N.I. 1997. Extensively and partially hydrolysed infant formulas for allergy prophylaxis. *Arch. Dis. Child.* **77**:4.

Pabst, H., Grace, M., Godel, J., Cho, H., and Spady, D. 1989. Effect on breast-feeding on antibody response to BCG vaccination. *Lancet* **I**:295.

Pabst, H.F. and Spady, D.W. 1990. Effect of breast-feeding on antibody response to conjugate vaccine. *Lancet* **336**:269.

Pabst, H.F., Spady, D.W., Pilarski, L.M., Carson, M.M., Beeler, J.A., and Krezolek, M.P. 1997. Differential modulation of the immune response by breast- or formula-feeding of infants. *Acta Paediatr.* **86**:1291.

Piastra, M., Stabile, A., Fioravanti, G., Castagnola, M., Pani, G., and Ria, F. 1994. Cord blood mononuclear cell responsiveness to beta-lactoglobulin: T-cell activity in 'atopy-prone' and 'non-atopy-prone' newborns. *Int. Arch. Allergy Immunol.* **104**:358.

Pichichero, M.E. 1990. Effect of breast-feeding on oral rhesus rotavirus vaccine seroconversion: a metaanalysis. *J. Infect. Dis.* **162**:753.

Pisacane, A., Graziano, L., Mazzarella, G., Scarpellino, B., and Zona, G. 1992. Breast-feeding and urinary tract infection. *J. Pediatr.* **120**:87.

Pisacane, A., Impagliazzo, N., Russo, M., Valiani, R., Mandarini, A., Florio, C., and Vivo, P. 1994. Breast feeding and multiple sclerosis. *Br. Med. J.* **308**:1411.

Rennels, M.B. 1996. Influence of breast-feeding and oral poliovirus vaccine on the immunogenicity and efficacy of rotavirus vaccines. *J. Infect. Dis.* **174** Suppl 1:S107.

Rollwagen, F.M. and Baqar, S. 1996. Oral cytokine administration. *Immunol. Today* **17**:548.

Rudloff, H.E., Schmalstieg, F.C., Jr., Mushtaha, A.A., Palkowetz, K.H., Liu, S.K., and Goldman, A.S. 1992. Tumor necrosis factor-alpha in human milk. *Pediatr. Res.* **31**:29.

Saarinen, U.M. 1982. Prolonged breast feeding as prophylaxis for recurrent otitis media. *Acta Paediatr. Scand.* **71**:567.

Saarinen, U.M. and Kajosaari, M. 1995. Breastfeeding as prophylaxis against atopic disease: prospective follow-up study until 17 years old. *Lancet* **346**:1065.

Sabin, A.B. 1978. Poliomyelitis vaccination. Evaluation and direction in continuing application. *Am. J. Clin. Pathol.* **70**:136.

Scheifele, D., Björnson, G.J., Guasparini, R., Friesen, B., and Meekison, W. 1992. Breastfeeding and antibody responses to routine vaccination in infants. *Lancet* **340**:1406.

Schlesinger, J.J. and Covelli, H.D. 1977. Evidence for transmission of lymphocyte responses to tuberculin by breast-feeding. *Lancet* **2**:529.

Shi, M.J. and Stavnezer, J. 1998. CBF alpha3 (AML2) is induced by TGF-beta1 to bind and activate the mouse germline Ig alpha promoter. *J. Immunol.* **161**:6751.

Shohet, L., Shahar, E., and Davidson, S. 1985. Breast feeding as prophylaxis for atopic eczema: a controlled study of 368 cases. *Acta Paediatr. Hung.* **26**:35.

Sigurs, N., Hattevig, G., and Kjellman, B. 1992. Maternal avoidance of eggs, cow's milk, and fish during lactation: effect on allergic manifestations, skin-prick tests, and specific IgE antibodies in children at age 4 years. *Pediatrics* **89**:735.

Silfverdal, S.A., Bodin, L., Hugosson, S., Garpenholt, O., Werner, B., Esbjörner, E., Lindquist, B., and Olcén, P. 1997. Protective effect of breastfeeding on invasive *Haemophilus influenzae* infection: a case-control study in Swedish preschool children. *Int. J. Epidemiol.* **26**:443.

Silfverdal, S.A., Bodin, L., and Olcén, P. 1999. Protective effect of breastfeeding: an ecologic study of Haemophilus influenzae meningitis and breastfeeding in a Swedish population. *Int. J. Epidemiol.* **28**:152.

Skansen-Saphir, U., Lindfors, A., and Andersson, U. 1993. Cytokine production in mononuclear cells of human milk studied at the single-cell level. *Pediatr. Res.* **34**:213.

Slavikova, M., Lodinova-Zadnikova, R., Adlerberth, I., Hanson, L.A., Svanborg, C., and Wold, A.E. 1995. Increased mannose-specific adherence and colonizing ability of Escherichia coli 083 in breast-fed infants. *Adv. Exp. Med. Biol.* **371A**:421.

Stein, K.E. and Söderström, T. 1984. Neonatal administration of idiotype or antiidiotype primes for protection against *Escherichia coli* K13 infection in mice. *J. Exp. Med.* **160**:1001.

Stephens, S., Kennedy, C.R., Lakhani, P.K., and Brenner, M.K. 1984. In-vivo immune responses of breast- and bottle-fed infants to tetanus toxoid antigen and to normal gut flora. *Acta Paediatr. Scand.* **73**:426.

Szepfalusi, Z., Nentwich, I., Jost, E., Gerstmayr, M., Ebner, C., Frischer, T., and Urbanek, R. 1998. Cord blood mononuclear cells and milk-specific T-cell clones are tools to evaluate the residual immunogenicity of hydrolyzed milk formulas. *J. Allergy Clin. Immunol.* **101**:514.

Söder, O. 1987. Isolation of interleukin-1 from human milk. *Int. Arch. Allergy Appl. Immunol.* **83**:19.

Tuboly, S., Bernath, S., Glavits, R., Kovacs, A., and Megyeri, Z. 1995. Intestinal absorption of colostral lymphocytes in newborn lambs and their role in the development of immune status. *Acta Vet. Hung.* **43**:105.

Vaarala, O., Saukkonen, T., Savilahti, E., Klemola, T., and Åkerblom, H.K. 1995. Development of immune response to cow's milk proteins in infants receiving cow's milk or hydrolyzed formula. *J. Allergy Clin. Immunol.* **96**:917.

Victora, C.G., Smith, P.G., Vaughan, J.P., Nobre, L.C., Lombardi, C., Teixeira, A.M., Fuchs, S.M., Moreira, L.B., Gigante, L.P., and Barros, F.C. 1987. Evidence for protection by breast-feeding against infant deaths from infectious diseases in Brazil. *Lancet* **2**:319.

Virtanen, S.M., Rasanen, L., Aro, A., Lindström, J., Sippola, H., Lounamaa, R., Toivanen, L., Tuomilehto, J., and Åkerblom, H.K. 1991. Infant feeding in Finnish children less than 7 yr of age with newly diagnosed IDDM. *Diabetes Care* **14**:415.

Virtanen, S.M., Rasanen, L., Aro, A., Ylonen, K., Lounamaa, R., Tuomilehto, J., and Åkerblom, H.K. 1992. Feeding in infancy and the risk of type 1 diabetes mellitus in Finnish children. *Diabet. Med.* **9**:815.

Wahl, S.M. 1992. Transforming growth factor beta (TGF-beta) in inflammation: a cause and a cure. *J. Clin. Immunol.* **12**:61.

Weiler, I.J., Hickler, W., and Sprenger, R. 1983. Demonstration that milk cells invade the suckling neonatal mouse. *Am. J. Reprod. Immunol.* **4**:95.

WHO Collaborative. 1995. Factors affecting the immunogenicity of oral poliovirus vaccine: a prospective evaluation in Brazil and the Gambia. World Health Organization Collaborative Study Group on Oral Poliovirus Vaccine. *J. Infect. Dis.* **171**:1097.

Wilson, A.C., Forsyth, J.S., Greene, S.A., Irvine, L., Hau, C., and Howie, P.W. 1998. Relation of infant diet to childhood health: seven year follow up of cohort of children in the Dundee infant feeding study. *Br. Med. J.* **316**:21.

Wiman, K., Curman, B., Trägårdh, L., and Peterson, P.A. 1979. Demonstration of HLA-DR-like antigens on milk fat globule membranes. *Eur. J. Immunol.* **9**:190.

Winner, L.D., Mack, J., Weltzin, R., Mekalanos, J.J., Kraehenbuhl, J.P., and Neutra, M.R. 1991. New model for analysis of mucosal immunity: intestinal secretion of specific monoclonal immunoglobulin A from hybridoma tumors protects against Vibrio cholerae infection. *Infection & Immunity* **59**:977.

Wold, A.E. and Hanson, L.Å. 1994. Defence factors in human milk. *Curr. Opin. Gastroenterol.* **10**:652.

Wold, A.E., Mestecky, J., Tomana, M., Kobata, A., Ohbayashi, H., Endo, T., and Svanborg-Edén, C. 1990. Secretory immunoglobulin A carries oligosaccharide receptors for *Escherichia coli* type 1 fimbrial lectin. *Infect. Immun.* **58**:3073.

Yu, G., Duchen, K., and Björkstén, B. 1998. Fatty acid composition in colostrum and mature milk from non-atopic and atopic mothers during the first 6 months of lactation. *Acta Paediatr.* **87**:729.

Zeiger, R.S., Heller, S., Mellon, M.H., Forsythe, A.B., O'Connor, R.D., Hamburger, R.N., and Schatz, M. 1989. Effect of combined maternal and infant food-allergen avoidance on development of atopy in early infancy: a randomized study. *J. Allergy Clin. Immunol.* **84**:72.

Zhang, L., van Bree, S., van Rood, J.J., and Claas, F.H. 1991. Influence of breast feeding on the cytotoxic T cell allorepertoire in man. *Transplantation* **52**:914.

Zoppi, G., Gasparini, R., Mantovanelli, F., Gobio-Casali, L., Astolfi, R., and Crovari, P. 1983. Diet and antibody response to vaccinations in healthy infants. *Lancet* **2**:11.

Chapter 16

Milk Components with Immunomodulatory Potential

Suzanne M. Filteau

1. Introduction

Most discussions of the immunological and infant health benefits of breast-feeding have focused on the passive protection of the infant, either by provision of various factors in milk or by avoidance of contaminated water and food. However, recent evidence suggests that the infant is not simply a passive recipient but also actively adapts its own immune system under the influence of breast milk. Immunological differences between breast- and formula-fed infants for which passive immune protection seems an insufficient explanation include increased antibody responses to some vaccinations (Hahn-Zoric *et al.*, 1990; Pabst and Spady, 1990; Pickering *et al.*, 1998), transfer of specific responsiveness to tuberculin (Mohr, 1973; Schlesinger and Covelli, 1977; Pabst *et al.*, 1989), increased immunoglobulin A (IgA) production (Prentice, 1987; Goldblum *et al.*, 1989), altered development of lymphoid organs, lymphocyte subpopulations and cytokine production (Carver *et al.*, 1991a; Hasselbalch *et al.*, 1996; Pabst *et al.*, 1997), survival of maternal, but not paternal, allografts (Campbell *et al.*, 1984), and decreased prevalence and severity of allergy (Juto *et al.*, 1982; Saarinen and Kajosaari, 1995). Although the decreased susceptibility of breast-fed infants to infectious, particularly diarrheal, diseases can be largely explained on the basis of passive protection by milk factors, active protection may also be involved. These mechanisms are not always separable; for example, the

Suzanne M. Filteau • Centre for International Child Health, Institute of Child Health, 30 Guilford Street, London WC1N 1EH, UK

passive immunity provided by the improved gut integrity of breast-fed infants (Weaver *et al.*, 1987) will reduce the amount of dietary antigen reaching their systemic immune system. Systemic antigenic stimulation is likely responsible for the increased proliferative responses to several mitogens or antigens seen in formula-fed infants after the early neonatal period (Juto *et al.*, 1982; Pabst *et al.*, 1997).

The present chapter will focus on components of milk which may trigger these adaptive changes. Immunomodulatory components in milk function in two major ways: to promote immune reactivity in order that the infant develops protection against infection, and to down regulate inappropriate reactivity that might lead to the development of allergy. Components such as growth factors which promote immune maturation through effects on gut development or on gut microflora are beyond the scope of this chapter.

Milk contains numerous cells and soluble factors which may influence maturation of the infant's immune system. To understand which components actually do affect infant immune development, it is necessary to consider several questions:

1. Is the factor present in colostrum or milk in biologically active quantities?
2. Does the factor remain functionally intact after passing through the stomach?
3. Are there cells responsive to the factor in the gut or does the factor enter the infant's circulation?
4. What other immune components in milk at the same time augment or inhibit the activity of the factor in question?
5. What conditions influence the amounts of the factor in milk, and do these have any relationship with infant health?

For most candidate immunomodulatory components of milk, the answers to these questions are limited or unavailable. The available evidence is divided almost equally between studies in humans and in dairy cows, with occasional information on laboratory animals. Although there are many similarities between lactation in humans and in cows, there are important differences, for example, in the concentrations of protein and bulk minerals (Lawrence, 1994). Therefore, it cannot simply be assumed that immunological components, their functions and their importance will be identical in different species.

2. Colostral and Milk Leukocytes

Macrophages, neutrophils and lymphocytes in milk and colostrum are often considered candidate immunomodulators in infants. It is clear from many *in vitro* studies that colostral and milk leukocytes are functional although some of their activities, for example, proliferation following stimulation and cytokine secretion, are not completely equivalent to these activities in blood leukocytes. However, it is less clear that such complex entities as living cells can retain these functions after passage through the infant stomach and intestine, even though in early infancy stomach pH is relatively high (Euler *et al.*, 1979) and activity of digestive enzymes is lower than in older children (Lebenthal and Lee, 1980). As determined by their ability to exclude trypan blue, human milk leukocytes were able to survive for more than an hour *in vitro* wide ranges of aqueous phase osmolality and protein content, pH ranges from 3–7, and temperature ranges from 30–40°C (Paxson and Cress, 1979). However, human milk macrophages have exhibited greatly reduced viability, also estimated by trypan blue exclusion, when subjected to pH less than 6 for 30 minutes (Blau *et al.*, 1983). *In vivo* evidence that lymphocytes are still functional after passage through the neonatal stomach is the ability of blood or milk cells from immunised rats, when given orally to rat pups, to transmit immunity to the intestinal worm *Trichinella spiralis* (Kumar *et al.*, 1989).

Gut maturation and closure vary considerably among species. Consequently, interpretation of the evidence for milk cell transfer across the gut epithelium must take the species into account. There is both histological and functional evidence for such cell transfer in sheep. Fluorescently-labelled sheep colostral or blood leukocytes fed to newborn or 3-day-old lambs could be detected several hours later in the lamb's blood in numbers largely proportional to the dose given, lymphocytes being preferentially transferred compared with monocytes, macrophages or neutrophils (Schnorr and Pearson, 1984). Radiolabelled colostral leukocytes from their own mother, but not from another ewe, could be detected in the gut epithelium and mesenteric lymph nodes after oral dosing of newborn lambs, and there was evidence that such cells could transfer immunity to tetanus from ewe to lamb (Tuboly *et al.*, 1995).

For ethical reasons similar studies have not been conducted in humans, so evidence for milk cell transfer to infants is limited. A study on the *in vivo* distribution of human milk cells given orally to baboons (Jain *et al.*, 1989) is widely quoted as evidence. Baboons delivered by caesarean section at 140 days, rather than at the normal gestational period of 184 days, were given a dose of radiolabelled human milk cells by nasogastric tube within 12 hours of delivery and no other food until death at 72 hours of age. Key results were that considerable radioactivity persisted in the gut lumen until death (that is, 60 hours after isotope dosing) and that liver and spleen took up more isotope than was found in blood or bone marrow. This outcome suggested an uptake of intact labelled cells, since previous work showed that bone marrow was a major destination of free isotope (Jain *et al.*, 1989). In addition to experimental constraints such as unphysiological food deprivation and, importantly, the use of milk leukocytes from a different species, the main problem with extrapolating this work to normal human infants is the use of extremely premature baboons. It is well recognised that premature human infants differ from term infants in the maturity and function of both the gut (Roberton *et al.*, 1982) and the immune system (Ballow *et al.*, 1986).

Studies of transplantation between different strains of rats (Beer and Billingham, 1975) provide less direct evidence for transfer to the pups of milk cells which contain intact major histocompatibility (MHC) antigens, although they are not necessarily viable functional cells. Cross-fostering of different combinations of rat strains, followed by careful studies of skin transplantation, led to the conclusion that offspring can be made tolerant to otherwise foreign MHC antigens if exposed to them in their early diet (Beer and Billingham, 1975). Likewise, previously breast-fed humans had fewer cytotoxic T cell precursors reactive to maternal lymphocytes than did individuals who had not been breast-fed, whereas no such difference was apparent in relation to paternal lymphocytes (Zhang *et al.*, 1991). This could provide an explanation for the increased survival of maternal, but not paternal, kidney transplants in individuals previously breast-fed (Campbell *et al.*, 1984). Transplants between siblings who had both been breast-fed also exhibited improved survival, possibly because of down-regulation of reactivity to the maternal antigens (Kois *et al.*, 1984).

Phagocytes—macrophages and neutrophils—are the most numerous cells in bovine (Taylor *et al.*, 1997) and human milk (Lawrence, 1994). Both cell types differ in numerous ways from their peripheral blood counterparts and, in both, these differences can be attributed to greater activation on the part of the milk cells. Human milk neutrophils exhibit surface markers, for example low levels of L-selectin and high levels of CD11b (Keeney *et al.*, 1993), which are characteristic of activated neutrophils. Blood neutrophil surface marker expression could be rendered similar to that of milk neutrophils by stimulation with formylmethionyl-leucyl-phenylalanine or by incubation with human milk (Keeney *et al.*, 1993). Neutrophils that have already been activated appear hypofunctional in many standard *in vitro* assays. Human colostral neutrophils, compared with blood neutrophils, exhibited lower unstimulated motion, chemotaxis, phagocytosis, and reduction of nitroblue tetrazolium (Ozkaragoz *et al.*, 1988; Buescher and McIlheran, 1993). Exposure of blood neutrophils to the aqueous fraction of colostrum resulted in similar decreases in function (Buescher and McIlheran, 1993). This decrease is typical of what the authors described as 'used' neutrophils, that is, those that have exited the blood to carry out their functions in tissues. It is interesting that similarly depressed activities were found in oral exudate neutrophils (Buescher and McIlheran, 1993). The authors concluded that, due either to characteristics within milk neutrophils themselves or to other factors in the colostral environment, neutrophils in milk are inactive even before passage through the infant stomach. The main role of these cells may be anti-inflammatory in the infant gut (Ozkaragoz *et al.*, 1988; Buescher and McIlheran, 1993) or possibly antimicrobial within the mammary gland.

In contrast to milk neutrophils, unstimulated milk macrophages were more mobile in collagen gel than were blood monocytes (Ozkaragoz *et al.*, 1988). Colostral macrophages could kill *Escherchia coli* if serum or the aqueous fraction of colostrum was present to opsonise the bacteria (Honorio-Franca *et al.*, 1997). Incubation of cow's milk macrophages with *Staphyloccus aureus* increased their synthesis of a factor chemotactic for bovine neutrophils (Politis *et al.*, 1991a). The authors concluded that the active agent was not interleukin-1 (IL-1) since the macrophages synthesized little of this cytokine. They did not investigate more likely neutrophil chemotactic factors such as IL-8. There is evidence suggesting that milk macrophages can influence development of specific immune responses. Milk macrophages from women whose infants developed cow's milk allergy exhibited lower expression of the surface antigen HLA-DR involved in antigen presentation (Jarvinen *et al.*, 1999). The authors suggested that poor antigen presentation might inhibit normal development of oral tolerance to dietary antigens, including traces of cow's milk protein found in some human milk.

The majority of lymphocytes in both human (Lawrence, 1994) and cow's milk (Taylor *et al.*, 1997) are T cells. There is evidence for a T cell population in milk that is distinct from that in blood, the milk population being more like mucosal lymphocytes such as gut intraepithelial lymphocytes (Cummins and Thompson, 1997). Selective homing of mucosal lymphocytes may preferentially transfer immunity to gut pathogens to the mammary gland. Compared with blood, more human milk T cells bear the γ/δ form of the T cell receptor (Bertotto *et al.*, 1990a; Eglinton *et al.*, 1994; Lindstrand *et al.*, 1997). In humans, the ratio of CD4/CD8 positive T cells in milk is lower than in blood (Wirt *et al.*, 1992). However, the CD4/CD8 ratio increases dramatically during mastitis in cows (Taylor *et al.*, 1997). A large proportion of milk T cells express a memory phenotype (Bertotto *et al.*, 1990b; Eglinton *et al.*, 1994) and more milk than blood T cells bear surface markers indicating activation (Wirt *et al.*, 1992; Eglinton *et al.*, 1994). Milk T cell proliferative responses measured *in vitro* reflect this activated memory phenotype in that milk T cells respond poorly to polyclonal mitogens but as vigorously as blood T cells to antibodies against CD2 or CD3 (Bertotto *et al.*, 1990b). Memory T cells presumably help prime infant immunity to locally important antigens such as *Mycobacterium tuberculosis* (Bertotto *et al.*, 1993). This could be a key mechanism whereby milk lymphocytes can actively promote infant immune development in ways to improve resistance to local infections.

3. Immunomodulatory Activities of the Aqueous Fraction of Milk or Colostrum

Considerable research has addressed immunodulatory activities of the whole aqueous fraction of milk and colostrum. When systemic (blood or spleen) leukocytes have been incubated with these fractions, both immunoenhancing and inhibitory activities have been found, generally with enhancement at low milk or colostrum content in the medium and inhibition at higher concentration. For example, the aqueous fractions of both colostrum and milk increased proliferation and immunoglobulin (probably IgG) secretion by mouse spleen cells treated *in vitro* with a low dose of *E. coli* lipopolysaccharide (LPS); with colostrum there was a linear dose-response up to a plateau at 2.5% in the medium (Juto, 1985). These results are similar to the linear dose-response increase in peripheral blood cell production of IL-1 and IL-3 with up to 5% colostral fraction in the medium and the stimulation of IL-6 production with up to 2.5% colostrum but inhibition at higher concentrations (Bessler *et al.*, 1996). Colostrum at 10% in the medium inhibited IL-2 production by human blood cells (Hooton *et al.*, 1991). Adding a crude protein fraction from human colostrum enhanced peripheral blood lymphocyte mitogen-induced proliferation at low fraction concentrations but inhibited it at high concentrations, whereas mature milk contained only inhibitory activity (Mincheva-Nilsson *et al.*, 1990). Human colostrum and milk at high concentrations (usually 50%) are anti-inflammatory, generally decreasing systemic neutrophil functions (Murphey and Buescher, 1993; Buescher and McIlheran, 1993; Grazioso and Buescher, 1996). However, as discussed above, neutrophils in milk itself appear to be already non-functional at the point of secretion. It is unclear what milk concentrations are relevant as far as influencing infant neutrophil functions in the gut or periphery. Milk concentrations experienced by gut neutrophils may approach the 50% associated with consistent inhibition of neutrophil functions, but milk would presumably have been greatly diluted before having a chance to influence systemic neutrophils. Similar considerations may also apply to the other concentration-dependent activities of the aqueous fractions of milk and colostrum.

Some attempts have been made to identify the active factors in the aqueous fraction of colostrum or milk. Protein fractionation indicated that lactoferrin may have been responsible for enhancing the proliferation of blood lymphocytes whereas the inhibitory activity appeared to be due to an unidentified glycoprotein that bound concanavalin A (Mincheva-Nilsson *et al.*, 1990). A human milk component that increased LPS responsiveness of mouse spleen cells was destroyed by heating at 56°C and it did not pass through dialysis tubing, suggesting that it was a heat labile protein (Juto, 1985). These properties apply to lactoferrin, complement components, and many cytokines. The aqueous anti-inflammatory factor in human colostrum resided mainly in the fraction with molecular weight <10 KDa (Murphey and Buescher, 1993) and thus could not have been known milk anti-inflammatory agents such as lactoferrin (MW = 77 KDa; Lee *et al.*, 1993) or transforming growth factor-β (TGFβ; MW = 25 KDa; Grande, 1997). Similarly, the factor in human colostrum that inhibited IL-2 production was shown not to be TGFβ (Hooton *et al.*, 1991).

In an interesting study, the whey fraction of cow's milk inhibited mitogen-induced proliferation of bovine peripheral blood mononuclear cells to varying degrees which were related to the cow's udder health (Barta *et al.*, 1990). Inhibitory activity was lowest in whey from healthy udder quarters, highest in those with clinical mastitis, and intermediate in subclinical mastitis, diagnosed on the basis of cell count and lymphocyte blastogenesis (Barta *et al.*, 1990). These results suggest that the inhibitory activity was due to one or more of the cytokines that increase in concentration in milk during mastitis (see below). The importance of these mastitic changes for the health and development of suckling offspring has not been investigated.

In summary, studies of the aqueous fractions of milk and colostrum clearly show that they contain factors with immunomodulatory potential which can be seen under particular, usually *in vitro*, experimental conditions. A dependence on concentration suggests the presence of inhibitors whose activities can be diluted out, allowing enhancing activities present to predominate at low concentrations of milk or colostrum. The identities of the factors involved are unclear, but lactoferrin and various cytokines are likely candidates.

3.1. Cytokines

In the last decade, research on immunoregulation in general has featured the discovery and characterisation of an ever increasing number of cytokines. Thus, it is not surprising that researchers interested in infant immune development induced by breast-feeding have begun to investigate cytokines in milk. Table 1 gives a list of the cytokines that have been detected in human or bovine milk.

A major constraint for quantitation of all soluble immune factors in milk is the variability of many of the assays used. This variability is greater than for similar assays in plasma and is most severe for components present in low amounts such as cytokines (Goldman *et al.*, 1985; Filteau *et al.*, 1999a, 1999b). The milk fat globules likely contribute to this increased variability but removing them may render quantitation inaccurate since some cytokines, for example IL-10 (Garofalo *et al.*, 1995) and TGFβ (Filteau *et al.*, 1999b), or other aqueous components (Goldman *et al.*, 1985; Schroten *et al.*, 1999) occur partly in the lipid layer.

Table 1. Cytokines in milk or colostrum

Factor	Sources		Probable functions
	leukocytes	epithelium	
IL-1β	Y	N	inflammation, mammary protection
IL-1RA	Y	?	anti-inflammatory
IL-6	Y	Y	inflammation, increase IgA, mammary protection
IL-8	Y	Y	chemotactic for neutrophils, mammary protection
IL-10	N	?	anti-inflammatory, increase IgG,A,M
TGFβ	Y	N	anti-inflammatory, tolerance, increase IgA, tissue repair
TNFα	Y/N	Y	mammary protection
IFNγ	Y/N	N	inflammation
M-CSF	Y	Y	leukocyte development
GM-CSF	Y	?	mammary protection

Y = yes; N = no; ? = no information.

Most of the data on cytokines have appeared in single reports by only a few investigators. Reports contain disagreements, for example about the presence (Garofalo *et al.*, 1995; Sone *et al.*, 1997) or absence (Srivastava *et al.*, 1996) of IL-1, IL-6, IL-10 and tumour necrosis factor-α (TNFα) in milk. It is not clear whether these disagreements result from different women being studied or from different assays used. The identity of the cytokines in milk and their concentrations vary considerably depending on whether protein or mRNA production by milk cells was measured, whether the former was assessed using bioassays or immunoassays, and on the specificity of the antibodies or responding cells used (Whiteside, 1994). Hence it is impossible to cite 'normal' concentrations in Table 1. In addition, inter-individual differences are likely; for example, IL-8 concentrations are very high in milk from a sizeable minority of women but are low in milk of most women (Srivastava *et al.*, 1996; Filteau *et al.*, 1999a, 1999b).

Although cytokines are generally found in very small (ng/L–μg/L) quantities in milk, these concentrations are sufficient for biological activity. To date, however, virtually all research on milk cytokine activities has been conducted *in vitro* using milk that has never been subjected to the infant's stomach acid or intestinal enzymes. Nevertheless, there is evidence that at least some cytokines given orally, for example, IL-2 and IL-6, retain their activity and exert physiological effects at the gut mucosa or systemically (Rollwagen and Baqar, 1996). Furthermore, milk contains many factors that could synergise with or inhibit cytokine activities, for example, adhesion molecules and soluble receptors for cytokines (Buescher and Malinowska, 1996). There may be a similar complex regulatory soup produced in the infant. Thus, although cytokines are among the most likely milk candidates for modulators of infant immune function, there is very little evidence that they actually do so *in vivo*.

3.1.1. Interleukin-1β. Interleukin-1β (IL-1β) is an important mediator of inflammation found in human milk (Sone *et al.*, 1997). This cytokine is not produced by cultured mammary epithelial cells (Basolo *et al.*, 1993; Palkowetz *et al.*, 1994) but can be made by leukocytes from human colostrum (Munoz *et al.*, 1990) and mature milk (Srivastava *et al.*, 1996). IL-1β production increases in human (Munoz *et al.*, 1990; Skansen-Saphir *et al.*, 1993) and bovine (Politis *et al.*, 1991b) milk macrophages in response to stimulation *in vitro* with LPS or heat-killed *S. epidermidis*. In both healthy cows and those with naturally occurring mastitis, most IL-1β produced *in vitro* remained in the milk macrophage (Politis *et al.*, 1991b). Thus, macrophages may sequester and protect IL-1β during passage through the stomach, enabling it to be released in active form when the cells are broken down during digestion.

3.1.2. Interleukin-1 Receptor Antagonist. Interleukin-1 receptor antagonist (IL-1RA) is a natural competitor of IL-1β. It is produced by human milk macrophages *in vitro* in response to LPS (Skansen-Saphir *et al.*, 1993) and is found in colostrum and milk in amounts (1–10 ng/L) that appear more than sufficient to inhibit the inflammatory actions of the amounts of IL-1β present (Srivastava *et al.*, 1996; Buescher and Malinowska, 1996). This example emphasises the complexity of cytokine interactions and the difficulties involved in relating the presence and concentration of a cytokine in milk to its activity and influence in the infant. In a model of acetic acid colitis in rats, feeding human milk compared to rat chow or infant formula, decreased colonic inflammation, assessed histologically and from elevations in colonic myeloperoxidase activity (Grazioso *et al.*, 1997). Interestingly, addition of recombinant IL-1RA to the formula resulted in anti-inflammatory activity similar to that in human milk (Grazioso *et al.*, 1997). However, the concentration used, 100 μg/L, was several orders of magnitude greater than the levels found in human milk, so the importance of IL-1RA in anti-inflammatory activity within the gut of the breast-fed infant remains speculative.

3.1.3. Interleukin-6. Interleukin-6 is an inflammatory mediator in human colostrum and milk (Saito *et al.*, 1991; Sone *et al.*, 1997). This cytokine can be produced by cultured mammary epithelial cells (Basolo *et al.*, 1993; Palkowetz *et al.*, 1994) and by rat (Na and Seelig, 1994) and bovine (Taylor *et al.*, 1997) milk leukocytes. Milk leukocyte expression of IL-6 mRNA increased during natural bovine mastitis (Taylor *et al.*, 1997). IL-6 activity in the infant is probably affected by the presence of a soluble IL-6 receptor in colostrum and milk (Buescher and Malinowska, 1996).

In addition to its inflammatory activity, IL-6 may function to promote IgA production. Antibody to IL-6 decreased IgA production by colostral mononuclear cells stimulated *in vitro* with *S. aureus* (Saito *et al.*, 1991). However, IgA occurs in milk as sIgA and must be synthesized before binding to secretory component, produced in the mammary epithelium, and secretion. It is unknown whether IL-6 is involved in stimulating IgA production by lymphocytes within the gland or whether it exerts this function in the infant gut.

3.1.4. Interleukin-8. Interleukin-8 is a major chemotactic factor for neutrophils and an important mediator of epithelial inflammation, for example in the gut (Fiocchi *et al.*, 1994). It can be produced by human milk leukocytes, spontaneously (Skansen-Saphir *et al.*, 1993; Srivastava *et al.*, 1996) and in response to LPS (Skansen-Saphir *et al.*, 1993) and by cultured mammary epithelial cells (Basolo *et al.*, 1993; Palkowetz *et al.*, 1994). In most samples of transitional and mature milk of Bangladeshi women IL-8 concentration was low (2–30 ng/L); however, 20% of samples taken around 2 weeks postpartum and 10% of those taken at 3 months had IL-8 concentrations between 100 ng/L and 1000 ng/L (Filteau *et al.*, 1999a). IL-8 concentrations correlated with the concentration ratio of sodium to potassium in milk, even after controlling for the number of days postpartum that the sample was collected. The sodium/potassium ratio was used as a measure of subclinical breast inflammation which induces paracellular pathways through the mammary epithelium permitting plasma sodium to leak into milk (Prentice *et al.*, 1985). Thus, IL-8 may mediate some of the inflammatory processes of mastitis. As for other cytokines believed to mediate mastitis (see below) it is unknown whether IL-8 retains its inflammatory activity, with potentially adverse consequences, once it reaches the infant gut (Goldman *et al.*, 1986).

3.1.5. Interleukin-10. Interleukin-10 has anti-inflammatory activity which is mediated through its ability to down-regulate production of inflammatory cytokines such as interferon-γ (IFNγ), TNFα, and IL-8 (Chernoff *et al.*, 1995). Genetically engineered mice lacking IL-10 develop colitis (Schreiber, 1997), suggesting that the anti-inflammatory functions of this cytokine are particularly important in the gut. IL-10 can increase IgM, IgG and IgA production from appropriately stimulated B cells (Defrance *et al.*, 1992). IL-10 is found in considerable amounts in colostrum and has been determined to be active in decreasing the proliferation of human blood lymphocytes (Garofalo *et al.*, 1995). Whereas most researchers looking for cytokines in milk assess only the aqueous fraction, Garofalo *et al.* (1995) also detected IL-10 in the lipid fraction and found that amounts measured by immunoassay in both fractions were increased after treatment with sodium taurocholate. These results suggest that IL-10 may be protected by compartmentalisation during passage through the infant stomach and be released in the intestine. The source of IL-10 in milk is unclear since milk leukocytes do not produce it (Skansen-Saphir *et al.*, 1993; Garofalo *et al.*, 1995). A mammary epithelial cell line constitutively expressed mRNA for IL-10 but produced little IL-10 protein (Garofalo *et al.*, 1995).

3.1.6. Tumour Necrosis Factor-α. Tumour necrosis factor-α (TNFα) is an early trigger of inflammatory responses (Callard and Gearing, 1994). It has been reported in human colostrum in highly variable quantities (Munoz *et al.*, 1990; Rudloff *et al.*, 1991; Sone *et al.*, 1997). Milk contains high concentrations of soluble TNFα receptors (Buescher and Malinowska, 1996) which may contribute to the variability in measured TNFα. TNFα is produced by human (Skansen-Saphir *et al.*, 1993; Sone *et al.*, 1997) and rat milk leukocytes (Na and Seelig, 1994). There is evidence for (Basolo *et al.*, 1993) and against (Palkowetz *et al.*, 1994) its production by cultured human mammary epithelial cells. TNFα may be involved in the response to mastitis in cows (Shuster *et al.*, 1995) and sheep (Waller *et al.*, 1997), but its importance in infant immune development has not been investigated.

3.1.7. Transforming Growth Factor-β. Transforming growth factor-β is perhaps the cytokine for which there are the most data available regarding a role in infant immune development. It is produced by human milk leukocytes (Saito *et al.*, 1993; Srivastava *et al.*, 1996) but not by cultured mammary epithelial cells (Palkowetz *et al.*, 1994). TGFβ, predominantly TGFβ2, is found in bovine (Pakkanen, 1998) and porcine colostrum (Xu *et al.*, 1999) and in human and rat milk in high concentration throughout lactation (Saito *et al.*, 1993; Srivastava *et al.*, 1996; Penttila *et al.*, 1998). It is activated, rather than destroyed, at pH 2–3.5 (Saito *et al.*, 1993; Srivastava *et al.*, 1996; Grande, 1997; Xu *et al.*, 1999), similar to the pH of infant stomach (Euler *et al.*, 1979). Perhaps this is the reason that human milk given to young adult mice and then sucked from their stomachs contained more TGFβ bioactivity than the original dose (Ishizaka *et al.*, 1994). However, we have found that treatment of human milk with sodium taurocholate increases the amount detectable immunologically (Filteau *et al.*, 1999b), suggesting that, as for IL-10 (Garofalo *et al.*, 1995), it is protected by compartmentalisation during passage through the stomach and thus it is unclear how much TGFβ is exposed to stomach acid. TGFβ1 fed to TGFβ1 null mice could be detected in several organs, indicating that it was able to pass in immunologically active form into the regions where it could modulate systemic immune functions (Letterio *et al.*, 1994). Similar evidence that TGFβ can be absorbed from the gut with its functions intact comes from a study in which a *Salmonella* construct expressing TGFβ was fed to mice and shown to reduce inflammation and alter cytokine expression in their carrageenan-injected footpads (Ianaro *et al.*, 1995).

TGFβ appears to act as an immunomodulator both locally in the gut and systemically (Ianaro *et al.*, 1995; Neurath and Buschenfelde, 1996; Grande,

1997). It is involved in induction of tolerance to food antigens (Weiner, 1997) and may provide a mechanism for reduced risk of allergy in breast-fed infants. In gut epithelial cell cultures, TGFβ played a key role in promoting tissue repair following injury (Ciacci *et al.*, 1993).

Two oral doses of human milk given to young adult mice together with oral sheep red blood cells (SBRC) decreased anti-SRBC antibody-producing cells in spleen and peripheral blood (Ishizaka *et al.*, 1994). When the SRBC were given by the intraperitoneal, intravenous, or intrabronchial routes, oral human milk increased splenic and blood plaque forming cells. The active factor was identified as TGFβ by passage of milk through an anti-TGFβ affinity column and comparing the activities in the effluent and eluate. Dose-response curves for effects of both human milk and purified TGFβ were seen. There was no effect of increasing the number of days of dosing with human milk to simulate regular milk consumption. These results support a dual role for TGFβ in down-regulating infant gut inflammation and allergic responses to oral antigens, as well as increasing systemic immune responses to antigens that enter through other routes.

TGFβ synergises with IL-2 or IL-5 to promote *in vitro* IgA synthesis and decrease IgG and IgM synthesis by LPS-stimulated mouse spleen cells (Coffman *et al.*, 1989; Kim and Kagnoff, 1990). Together with IL-10 and stimulation by cross-linking CD40 using anti-CD40 antibodies, TGFβ induced tonsullar B cells to differentiate into IgA-secreting cells (Defrance *et al.*, 1992). The relation between these *in vitro* results showing that TGFβ decreases IgG production and those showing increased splenic (presumably) IgG production after oral TGFβ administration (Ishizaka *et al.*, 1994) is difficult to determine. This lack of fit between *in vitro* and *in vivo* results emphasises the importance of *in vivo* work for a true understanding of how milk factors influence infant immune development.

3.1.8. Other Cytokines. Other cytokines that have been found in human milk in potentially active amounts include IFNγ (Basolo *et al.*, 1993; Eglinton *et al.*, 1994; Srivastava *et al.*, 1996), macrophage colony stimulating factor (M-CSF; Hara et al., 1995; Srivastava *et al.*, 1996) and granulocyte-macrophage colony stimulating factor (GM-CSF; Gasparoni et al., 1996). Although the functions of these cytokines have been examined in other systems (Callard and Gearing, 1994), their importance for infant immune development remains untested.

3.2. Other Soluble Aqueous Factors

3.2.1. Lactoferrin. Lactoferrin is found in abundance in human milk, as much as 1 g/L, but there are large species differences (Lawrence, 1994). Milk lactoferrin is found in both the aqueous fraction and within neutrophil granules.

Lactoferrin can pass intact through the infant gut, as evidenced by its presence in the feces of breast-fed infants (Prentice *et al.*, 1987). Moreover, this protein can be transported intact from the apical to the basolateral side of intestinal epithelial cells *in vitro* (Mikogami *et al.*, 1994). Many cell types, including human blood lymphocytes (Mazurier *et al.*, 1989) and intestinal brush border cells from young adult mice (Hu *et al.*, 1988), bear receptors for lactoferrin.

Lactoferrin has well-recognised immunomodulatory properties (Brock, 1995). Its immunoregulatory activities apply across species as indicated by the ability of human lactoferrin to induce maturation of mouse B cells (Zimecki *et al.*, 1995) and the ability of both human and bovine lactoferrin to decrease IL-6 production by LPS-stimulated human monocytes (Mattsby-Baltzer *et al.*, 1996). Other work on inhibition of IL-1, IL-2 and TNFα production by lactoferrin *in vitro* revealed effects only below 10^{-10}mol/L (Crouch *et al.*, 1992) and thus is likely of little relevance considering the large amounts of lactoferrin in milk. Lactoferrin was identified as the probable colostral protein that enhanced proliferation of human blood lymphocytes in response to T cell mitogens (Mincheva-Nilsson *et al.*, 1990). Further work is required to determine if and how milk lactoferrin regulates immune function in infants.

3.2.2. Prolactin. Prolatin, a hormone that possesses immunomodulatory activity (Reber, 1993), is found in considerable quantities and different forms in human colostrum and milk (Ellis and Picciano, 1995). There is evidence for prolactin receptors in the gastrointestinal tract and transformed gut cell lines from several mammalian species (Nagano *et al.*, 1995). Prolactin can be activated by the intestinal enzyme alkaline phosphatase (Ellis and Picciano, 1995) and can be absorbed intact from the intestine of suckling rats (Gonnella *et al.*, 1989). In lactating rats injected with bromocriptine from days 2–5 postpartum in doses that decrease milk prolactin with neglible effects on milk volume, low prolactin concentration in milk was associated with alterations in proliferative responses of the pups' spleen and thymus cells to mitogens (Grove *et al.*, 1991). The effects were highly age-dependent, the pups of bromocriptine-treated dams showing greater proliferation at 10 days of age, less at 15 days, and no difference at 21 days, the age of weaning and intestinal closure in the rat (Grove *et al.*, 1991). There were also transient effects on thymocyte surface antigen expression, pups of bromocriptine-treated dams having significantly lower numbers of CD4+ and CD5+ and higher numbers of CD2+ cells in the thymus at 10 days of age only (Grove *et al.*, 1991). Thus, although prolactin appears capable of modulating infant immune function, further work is required to clarify whether such modest and transient effects are physiologically important.

Although the concept of interactions between the endocrine and immune systems is well established and thus, in theory, numerous hormones in milk could promote infant immune development, to date there is little evidence for immunomodulatory activity on the part of any milk hormones other than prolactin.

3.2.3. Nucleotides. Nucleotides, as well as nucleosides and nucleic acids, are abundant constituents of human milk but not of cow's milk (Thorell *et al.*, 1996). The evidence for the importance of nucleotides in immune function in human infants is strong. Both local effects of nucleotides on infant intestinal growth and repair (Uauy *et al.*, 1990) and systemic immunostimulation may contribute to the beneficial effects of nucleotides.

Nucleotide supplementation of cow's milk formula decreased the incidence of diarrhea among American infants (Pickering *et al.*, 1998) and among infants from low socioeconomic Chilean families (Brunser *et al.*, 1994). There were no effects on incidence of respiratory or other infections among the Chilean infants. There was no evidence for decreased intestinal permeability in premature infants given a nucleotide-supplemented formula (Martinez-Augustin *et al.*, 1997). Hence, it is probable that reduced diarrhea resulted from mechanisms other than a direct effect on gut integrity.

Systemic immunoregulatory actions of dietary nucleotides include upregulation of mouse splenic IFNγ and down-regulation of IL-4 production (Nagafuchi *et al.*, 1997). Human infants given nucleotide-supplemented formulas had higher blood natural killer cell activity (Carver *et al.*, 1991b), an increased serum concentration of IgG specific to β-lactalbumin (Martinez-Augustin *et al.*, 1997) and increased antibody titres to diphtheria toxoid and hemophilus influenzae B vaccines but not to tetanus and oral polio vaccines (Pickering *et al.*, 1998). The variable effects on antibody responses are interesting because they indicate, in the case of oral polio vaccine, that the increased response of breast-fed compared with formula-fed infants in developed countries (Hahn-Zoric *et al.*, 1990; Pickering *et al.*, 1998) is not due to nucleotides in human milk. Insensitivity to nucleotide supplementation on the part of the response to tetanus toxoid is not surprising in view of the robust anti-tetanus response that is seen even in the face of generalised undernutrition (Dietz *et al.*, 1997).

3.2.4. Anti-Idiotypic Antibodies. Anti-idiotypic antibodies are naturally occurring antibodies with specificity against other autologous specific antibodies. Immunization with anti-idiotypes has been used experimentally to manipulate immune responses, for example, autoimmunity (Shoenfield *et al.*, 1994). Although human infants obtain most of their maternal blood antibodies prenatally, there is evidence that some dietary proteins can be absorbed postnatally (Jakobsson *et al.*, 1986). Human colostrum can contain IgA specific for antibodies to polio virus (Hahn-Zoric *et al.*, 1993). Such anti-idiotypic antibodies may be absorbed so as to prime breast-fed infants for polio responses. This could explain the higher antibody responses to certain vaccines in breast-fed compared with formula-fed infants (Hahn-Zoric *et al.*, 1990; 1993; Pabst and Spady, 1990; Pickering *et al.*, 1998). Neonatal mice primed with anti-idiotypic antibody given either by injection or by suckling milk from their primed mothers were protected against *E. coli* infection in young adulthood (Stein and Soderstrom, 1984).

4. What Conditions Influence the Concentrations of Immune Factors in Milk and Do They Relate to Infant Health?

The answer to this query may shed some light on the question of whether the potentially immunomodulatory factors discussed above actually perform in this role *in vivo*. Unfortunately, conditions affecting the concentrations of most immune factors in milk have been little studied.

The most striking changes in the concentrations of all such factors in milk are temporal ones dependent on the time postpartum. Most milk components are at their highest concentrations in colostrum, decline rapidly during the first few days postpartum and then undergo slow decreases throughout the rest of lactation (Lawrence, 1994). The high concentrations in colostrum occur at the time in the life of infants when they are most vulnerable to infection, when there is greatest potential for uptake through the gut epithelium and when the volume the infant can consume is at its minimum. A striking exception to this pattern is lysozyme which increases in concentration with time of lactation (Hennart *et al.*, 1991; Lawrence, 1994). It is difficult to account teleologically for this distinction.

A second condition that affects the concentration of most immune factors in milk is the level of ambient infection. Women from low-income countries where water supplies are limited and unclean, hygiene is relatively poor and infections are prevalent, have much higher concentrations of many immune factors in their milk than do women from developed countries (Lönnerdal *et al.*, 1976; Hennart *et al.*, 1991). However, most of the factors found to be high in the milk of women from contaminated environments are those which appear to provide mainly passive protection to infants. There is little information about environmental effects on cytokines and other components of milk believed to play a more active role in promoting infant immune development. An exception is lactoferrin which is reported in higher concentrations in Gambian than in British milk (Prentice *et al.*, 1983), and in Ethiopian than in Swedish milk (Lönnerdal *et al.*, 1976).

A third condition that induces major changes in the concentrations of immune factors is mastitis. Immune responses during mastitis have been extensively studied by dairy researchers because of the large economic losses that can result from this condition. It has been well documented that milk leukocyte numbers increase during this inflammation (Shuster *et al.*, 1993; 1995; Waller *et al.*, 1997; Lawrence, 1994). In cows, mastitis is associated with udder edema and increased concentrations of serum albumin in the milk (Shuster *et al.*, 1993; 1995). In women it is associated with increased milk sodium levels (Prentice *et al.*, 1985). These changes are the result of local mammary epithelial inflammation and leakage of plasma constituents from the capillaries. Both leukocyte recruitment and the opening of tight junctions appear to be mediated by increased production of inflammatory and chemotactic cytokines. Studies in cows (Shuster *et al.*, 1993; 1995; 1997; Barber and Yang, 1998), sheep (Waller *et al.*, 1997) and humans (Filteau *et al.*, 1999a; 1999b) indicate that the cytokines responsible for these changes are likely to include IL-1, IL-6, IL-8, TNFα, and GM-CSF. Cytokine production and leukocyte recruitment appear to be part of an active response by the mammary epithelium that presumably occurs to protect the breast or udder. The cytokines in milk seem to be generated within the mammary gland since they are generally in much lower concentrations in the plasma (Shuster *et al.*, 1993; 1995). Leukocytes which migrate into the mammary gland may themselves produce immune factors, the levels of which then rise in milk (Prentice *et al.*, 1985). Thus, many changes in the concentrations of milk components arise from a response primarily designed for maternal protection. This conclusion implies that milk immune factors should not be presumed, *ipso facto*, to have a role in infant protection. It is even possible that cytokines or cells produced as part of the response to mastitis may, by stimulating inflammation in the infant gut, compromise infant gut integrity and health. This hypothesis appears not to have been tested. However, there is little evidence of pathology in infants whose mothers have mastitis and milk contains soluble receptors or antagonists for inflammatory cytokines that may inhibit damage to the infant's gut (Buescher and Malinowska, 1996). International policy recommends that women continue to breast-feed during episodes of mastitis (Akre, 1989) but this policy is directed largely toward improving maternal rather than infant health.

5. Conclusions

With respect to the question whether milk or colostral factors are capable of promoting immune development in breast-fed infants, it is apparent that there are many components, both soluble and cellular, with such potential. The ability of these factors to survive infant stomach acid and digestive enzymes has not been well documented but it appears that nucleotides, lactoferrin, prolactin, anti-idiotypic antibodies, and at least some cytokines may survive functionally intact.

Leukocytes seem less likely to survive in a functional state but it is possible that key surface molecules can remain antigenically intact and thereby modulate infant responses to maternal MHC antigens as seen in transplantation studies. Considering the wide array of potentially responsive cells, both epithelial and leukocytic, in the infant gut, it is likely that most potential immunomodulators which survive that far could influence infant immune development at the level of the gut. However, fewer factors seem to be absorbed from the gut to influence infant immune functions in the periphery. Reasonably good evidence exists only for lactoferrin, prolactin, nucleotides, anti-idiotypes and TGFβ. The question whether other factors in milk or in the infant gut may augment or inhibit functions of the putative immunomodulators has been scarcely addressed. Many researchers have acknowledged the possible presence of such factors, but most continue to study the action of potential immunomodulators in isolation. This subject needs to be addressed in order to understand the relative importance of the many candidate immunomodulators found in milk. Finally, the clear importance of many cytokines and other immunomodulators for the protection of the mammary gland, rather than the infant, needs to be kept in mind. For example, it is likely that the most important functions of the milk neutrophils, IL-1, IL-8 and TNFα lie in protecting the mother from infection.

What clinical and epidemiological differences in immune function observed between breast- and formula-fed infants require active immunostimulation and cannot be explained simply by passive protection provided by breast milk? Increased IgA production could be due to the action of several cytokines, including IL-6, IL-10 and TGFβ. The increased response to certain specific antigens could be part of such generalised cytokine- or nucleotide-induced increases in immunoglobulin synthesis but could also result from specific priming by anti-idiotypes. Differences in lymphocyte subpopulations and in cytokine production between breast- and formula-fed infants could be due, in part, to factors such as cytokines and lactoferrin in milk but also to the increased intestinal absorption of food antigens by formula-fed infants (Weaver *et al.*, 1987; Jacobsson *et al.*, 1986). The transfer of responsiveness to tuberculin and of tolerance to maternal transplantation antigens are likely to require transfer of antigenically intact cell surface proteins from milk leukocytes. Finally, the decreased prevalence and severity of allergy in breast-fed infants may result from promotion of oral tolerance by TGFβ in milk.

The study of breast milk immunomodulators and their promotion of infant immune development is a new field and many gaps in our knowledge remain. Future studies need to combine the virtues of earlier work using whole milk fractions, such as the aqueous or cellular components, with later work on concentrations and activities of specific cytokines or other factors. It is important to look not only at the components of milk but also at their inhibitors and their net ability to affect infant immune functions *in vivo*. Finally, it will be important to consider whether the major changes in milk inflammatory cytokines and leukocytes that occur during mastitis have any effect on infant immune system function and development.

ACKNOWLEDGMENTS. I am grateful to J. Willumsen, Institute of Child Health, K. Bodnam-Smith, London School of Hygiene and Tropical Medicine, and L.A. Hanson, University of Goteberg, for helpful comments on drafts of this manuscript.

References

Akre, J., ed. 1989. Infant feeding: the physiological basis. *Bull. WHO* **67, supp.**:44.

Ballow, M., Cates, K.L., Rowe, J.C., Goetz, C., and Desbonnet, C. 1986. Development of the immune system in very low birth weight (less than 1500 g) premature infants: concentrations of plasma immunoglobulins and patterns of infections. *Pediatr. Res.* **20**:899.

Barber, M.R. and Yang, T.J. 1998. Chemotactic activities in nonmastitic and mastitic mammary secretions: presence of interleukin-8 in mastitic but not nonmastitic secretions. *Clin. Diag. Lab. Immunol.* **5**:82.

Barta, O., Barta, V.D., Crisman, M.V., and Akers, R.M. 1990. Lymphocyte blastogenesis inhibition by milk whey as an indicator of mastitis. *J. Dairy Sci.* **73**:2112.

Basolo, F., Conaldi, P.G., Fiore, L., Calvo, S., and Toniolo, A. 1993. Normal breast epithelial cells produce interleukins 6 and 8 together with tumor-necrosis factor: defective IL6 expression in mammary carcinoma. *Int. J. Cancer* **55**:926.

Beer, A.E. and Billingham, R.E. 1975. Immunologic benefits and hazards of milk in maternal-perinatal relationship. *Ann. Int. Med.* 83:865.

Bertotto, A., Castellucci, G., Fabietti, G., Scalise, F., and Vaccaro, R. 1990a. Lymphocytes bearing the T cell receptor γ/δ in human breast milk. *Arch. Dis. Child.* **65**:1274.

Bertotto, A., Gerli, R., Fabietti, G., Crupi, S., Arcangeli, C., Scalise, F., and Renato. 1990b. Human breast milk T lymphocytes display the phenotype and functional characteristics of memory T cells. *Eur. J. Immunol.* **20**:1877–1880.

Bertotto, A., Gerli, R., Castellucci, G., Crupi, S., Scalise, F., Spinozzi, F., Fabietti, G., Forenza, N., and Vaccaro, R. 1993. Mycobacteria-reactive γ/δ T cells are present in human colostrum from tuberculin-positive, but not tuberculin-negative nursing mothers. *Am. J. Reprod. Immunol.* **29**:131.

Bessler, H., Straussberg, R., Hart, J., Notti, I., and Sirota, L. 1996. Human colostrum stimulates cytokine production. *Biol. Neonate* **69**:376.

Blau, H., Passwell, J.H., Levanon, M., Davidson, J., Kohen, F., and Ramot, B. 1983. Studies on human milk macrophages: effect of activation on phagocytosis and secretion of prostaglandin E2 and lysozyme. *Pediatr. Res.* **17**:241.

Brock, J. 1995. Lactoferrin: a multifunctional immunoregulatory protein? *Immunol. Today* **16**:417.

Brunser, O., Espinoza, J., Araya, M., Cruchet, S., and Gil, A. 1994. Effect of dietary nucleotide supplementation on diarrhoeal disease in infants. *Acta Paediatr.* **83**:188.

Buescher, E.S. and Malinowska, I. 1996. Soluble receptors and cytokine antagonists in human milk. *Pediatr. Res.* **40**:839.

Buescher, E.S. and McIlheran, S.M. 1993. Polymorphonuclear leukocytes and human colostrum: effects of *in vivo* and *in vitro* exposure. *J. Pediatr. Gastroenterol. Nutr.* **17**:424.

Callard, R.E. and Gearing, A.J.H. 1994. *The Cytokine Factsbook*, Academic Press, London.

Campbell, D.A., Lorber, M.I., Sweeton, J.C., Turcotte, J.G., Niederhuber, J.E., and Beer, A.E. 1984. Breast feeding and maternal-donor renal allografts. *Transplant.* **37**:340.

Carver, J.D., Pimental, B., Wiener, D.A., Lowell, N.E., and Barness, L.A. 1991a. Infant feeding effects on flow cytometric analysis of blood. *J. Clin. Lab. Anal.* **5**:54.

Carver, J.D., Pimental, B., Cox, W.I., and Barness, L.A. 1991b. Dietary nucleotide effects upon immune function in infants. *Pediatrics* **88**:359.

Chernoff, A.E., Granowitz, E.V., Shapiro, L., Vannier, E., Lonnemann, G., Angel, J.B., Kennedy, J.S., Rabson, A.R., Wolff, S.M., and Dinarello, C.A. 1995. A randomized, controlled trial of IL-10 in humans: inhibition of inflammatory cytokine production and immune responses. *J. Immunol.* **154**:5492.

Ciacci, C., Lind, S.E., and Podolsky, D.K. 1993. Transforming growth factor-B regulation of migration in wounded rat intestinal epithelial monolayers. *Gastroenterology* **105**:93.

Coffman, R.L., Lebman, D.A., and Shrader, B. 1989. Transforming growth factor-B specifically enhances IgA production by lipopolysaccharide-stimulated murine B lymphocytes. *J. Exp. Med.* **170**:1039.

Crouch, S.P.M., Slater, K.J., and Fletcher, J. 1992. Regulation of cytokine release from mononuclear cells by the iron-binding protein lactoferrin. *Blood* **80**:235.

Cummins, A.G. and Thompson, F.M. 1997. Postnatal changes in mucosal immune response: a physiological perspective of breast feeding and weaning. *Immunol. Cell Biol.* **75**:419.

Defrance, T., Vanbervliet, B., Briere, F., Durand, I., Rousset, F., and Banchereau, J. 1992. Interleukin 10 and transforming growth factor-B cooperate to induce anti-CD40-activated naive human B cells to secrete immunoglobulin A. *J. Exp. Med.* **175**:671.

Dietz, V., Galazka, A., van Loon, F., and Cochi, S. 1997. Factors affecting the immunogenicity and potency of tetanus toxoid: implications for the elimination of neonatal and non-neonatal tetanus as public health problems. *Bull. WHO* **75**:81.

Eglinton, B.A., Roberton, D.M., and Cummins, A.G. 1994. Phenotype of T cells, their soluble receptor levels, and cytokine profile of human breast milk. *Immunol. Cell Biol.* **72**:306.

Ellis, L.A. and Picciano, M.F. 1995. Bioactive and immunoreactive prolactin variants in human milk. *Endocrinology* **136**:2711.

Euler, A.R., Byrne, W.J., Meis, P.J., Leake, R.D., and Ament, M.E. 1979. Basal and pentagastrin-stimulated acid secretion in newborn human infants. *Pediatr. Res.* **13**:36.

Filteau, S.M., Rice, A.L., Ball, J.J., Chakraborty, J., Stoltzfus, R., de Francisco, A., and Willumsen, J.F. 1999a. Breast milk immune factors in Bangladeshi women supplemented postpartum with retinol or beta-carotene. *Am. J. Clin. Nutr.* **69**:953.

Filteau, S.M., Lietz, G., Mulokozi, G., Bilotta, S., Henry, C.J.K., and Tomkins, A.M. 1999b. Milk cyokines and subclinical breast inflammation in Tanzanian women: effects of dietary red palm oil or sunflower oil supplementation. *Immunology* **97**:595.

Fiocchi, C., Binion, D.G., and Katz, J.A. 1994. Cytokine production in the human gastrointestinal tract during inflammation. *Curr. Opin. Gastroenterol.* **2**:639.

Garofalo, R., Chheda, S., Mei, F., Palkowetz, K.H., Rudloff, H.E., Schmalstieg, F.C., Rassin, D.K., and Goldman, A.S. 1995. Interleukin-10 in human milk. *Pediatr. Res.* **37**:444.

Gasparoni, A., Chirico, G., de Amici, M., Ravagni-Probizer, M.F., Ciardelli, L., Marchesi, M.E., Rondini, G. 1996. Granulocyte-macrophage colony-stimulating factor in human milk. *Eur. J. Pediatr.* **155**:69.

Goldman, A.S., Carlsson, B., Pickering, L., Keller, M. and Paape, M.J. 1985. Immunochemical assays, in *Human Lactation: Milk Components and Methodologies*, (R.G. Jensen and M.C. Neville, eds.), pp. 47, Plenum, New York and London.

Goldman, A.S., Thorpe, L.W., Goldblum, R.M., and Hanson, L.A. 1986. Anti-inflammatory properties of human milk. *Acta Paediatr. Scand.* **75**:689.

Goldblum, R.M., Schanler R.J., Garza, C., and Goldman, A.S. 1989. Human milk feeding enhances the urinary excretion of immunologic factors in low birth weight infants. *Pediatr. Res.* **25**:184.

Gonnella, P.A., Harmatz, P., and Walker, W.A. 1989. Prolactin is transported across the epithelium of the jejunum and ileum of the suckling rat. *J. Cell. Physiol.* **140**:138.

Grande, J.P. 1997. Role of transforming growth factor-B in tissue injury and repair. *Proc. Soc. Exp. Biol. Med.* **214**:27.

Grazioso, C.F. and Buescher, E.S. 1996. Inhibition of neutrophil function by human milk. *Cell. Immunol.* **168**:125.

Grazioso, C.F., Werner, A.L., Alling, D.W., Bishop, P.R., and Buescher, E.S. 1997. Antiinflammatory effects of human milk on chemically induced colitis in rats. *Pediatr. Res.* **42**:639.

Grove, D.S., Bour, B., Kacsoh, B., and Mastro, A.M. 1991. Effect of neonatal milk-prolactin deprivation on the ontogeny of the immune system of the rat. *Endocrine Reg.* **25**:111.

Hahn-Zoric, M., Fulconis, F., Minoli, I., Moro, G., Carlsson, B., Bottiger, M., Raiha, N., and Hanson, L.A. 1990. Antibody responses to parenteral and oral vaccines are impaired by conventional and low protein formulas as compared to breast-feeding. *Acta Paediatr. Scand.* **79**:1137.

Hahn-Zoric, M., Carlsson, B., Jeansson, S., Ekre, H.P., Osterhaus, A.D.M.E., Roberton, D., and Hanson, L.A. 1993. Anti-idiotypic antibodies to poliovirus antibodies in commercial immunoglobulin preparations, human serum, and milk. Pediatr. Res. **33**:475.

Hara, T., Irie, K., Saito, S., Ichijo, M., Yamada, M., Yanai, N., and Miyazaki, S. 1995. Identification of macrophage colony-stimulating factor in human milk and mammary gland epithelial cells. *Pediatr. Res.* **37**:437.

Hasselbalch, H., Jeppesen, D.L., Engelmann, M.D.M., Michaelsen, K.F., Nielsen, M.B. 1996. Decreased thymus size in formula-fed infants compared with breastfed infants. *Acta Paediatr.* **85**:1029.

Hennart, P.F., Brasseur, D.J., Delonge-Desnoeck, J.B., Dramaix, M.M., and Robyn, C.E. 1991. Lysozyme, lactoferrin, and secretory immunoglobulin A content in breast milk: influence of duration of lactation, nutrition status, prolactin status, and parity of mothers. *Am. J. Clin. Nutr.* **53**:32.

Honorio-Franca, A.C., Carvalho, M.P.S.M., Isaac, L., Trabulsi, L.R., and Carneiro-Sampaio, M.M.S. 1997. Colostral mononuclear phagocytes are able to kill enteropathogenic *Escherichia coli* opsonized with colostral IgA. *Scand. J. Immunol.* **46**:59.

Hooton, J.W.L., Pabst, H.F., Spady, D.W., and Paetkau, V. 1991. Human colostrum contains an activity that inhibits the production of IL-2. *Clin. Exp. Immunol.* **86**:520.

Hu, W.L., Mazurier, J., Sawatzki, G., Montreuil, J., and Spik, G. 1988. Lactotransferrin receptor of mouse small-intestinal brush border: characteristics of membrane-bound and Triton X-100-solubilized forms. *Biochem. J.* **249**:435.

Ianaro, A., Xu, D., O'Donnell, C.A., Di Rosa, M., and Liew, F.Y. 1995. Expression of TGF-B in attenuated *Salmonella typhimurium*: oral administration leads to the reduction of inflammation, IL-2, and IFN-γ, but enhancement of IL-10, in carrageenin-induced oedema in mice. *Immunol.* **84**:8.

Ishizaka, S., Kimoto, M., Tsujii, T., and Saito, S. 1994. Antibody production system modulated by oral administration of human milk and TGF-B. *Cell. Immunol.* **159**:77.

Jain, L., Vidyasagar, D., Xanthou, M., Ghai, V., Shimada, S., and Blend, M. 1989. *In vivo* distribution of human milk leucocytes after ingestion by newborn baboons. *Arch. Dis. Child.* **64**:930.

Jakobsson, I., Lindberg, T., Lothe, L., Axelsson, I., and Benediktsson, B. 1986. Human alpha-lactalbumin as a marker of macromolecular absorption. *Gut* **27**:1029.

Jarvinen, K.-M., Juntunen-Backman, K., and Suomalainen, H. 1999. Relation between weak HLA-DR expression on human breast milk macrophages and cow milk allergy in suckling infants. *Pediatr. Res.* **45**:76.

Juto, P. 1985. Human milk stimulates B cell function. *Arch. Dis. Child.* **60**:610.

Juto, P., Moller, C., Engberg, S., and Bjorksten, B. 1982. Influence of type of feeding on lymphocyte function and development of infantile allergy. *Clin. Allergy* **12**:409.

Keeney, S.E., Schmalstieg, F.C., Palkowetz, K.H., Rudloff, H.E., Le, B.-M., and Goldman, A.S. 1993. Activated neutrophils and neutrophil activators in human milk: increased expression of CD11b and decreased expression of L-selectin. *J. Leuko. Biol.* **54**:97.

Kim, P.-H. and Kagnoff, M.F. 1990. Transforming growth factor-B1 is a costimulator for IgA production. *J. Immunol.* **144**:3411.

Kois, W.E., Campbell, D.A., Lorber, M.I., Sweeton, J.C., and Dafoe, D.C. 1984. Influence of breast feeding on subsequent reactivity to a related renal allograft. *J. Surg. Res.* **37**:89.

Kumar, S.N., Stewart, G.L., Steven, W.M., and Seelig, L.L. 1989. Maternal to neonatal transmission of T-cell mediated immunity to *Trichinella spiralis* during lactation. *Immunol.* **68**:87.

Lawrence, R.A. 1994. *Breastfeeding: A Guide for the Medical Profession*, 4th Edition. Mosby, St. Louis.

Lebenthal, E. and Lee, P.C. 1980. Development of functional response in human exocrine pancreas. *Pediatrics* **66**:556.

Lee, G.R., Bithell, T.C., Foerster, J., Athens, J.W., Lukens, J.N. 1993. *Wintrobe's Clinical Hematology*, 9th Edition. Lea and Febiger, Philadelphia.

Letterio, J.J., Geiser, A.G., Kulkarni, A.B., Roche, N.S., Sporn, M.B., and Roberts, A.B. 1994. Maternal rescue of transforming growth factor-B1 null mice. *Science* **264**:1936.

Lindstrand, A., Smedman, L., Gunnlaugsson, G., and Troye-Blomberg, M. 1997. Selective compartmentalization of γ/δ T lymphocytes in human breastmilk. *Acta. Paediatr.* **86**:890.

Lönnerdal, B., Forsum, E., Gebre-Medhin, M., and Hambraeus, L. 1976. Breast milk composition in Ethiopian and Swedish mothers. II. Lactose, nitrogen, and protein contents. *Am. J. Clin. Nutr.* **29**:1134.

Martinez-Augustin, O., Boza, J.J., Del Pino, J.I., Lucena, J., Martinez-Valverde, A., and Gil, A. 1997. Dietary nucleotides might influence the humoral immune response against cow's milk proteins in preterm neonates. *Biol. Neonate* **71**:215.

Mattsby-Baltzer, I., Roseanu, A., Motas, C., Elverfors, J., Engberg, I., and Hanson, L.A. 1996. Lactoferrin or a fragment thereof inhibits the endotoxin-induced interleukin-6 response in human monocytic cells. *Pediatr. Res.* **40**:257.

Mazurier, J., Legrand, D., Hu, W.L., Montreuil, J., and Spik, G. 1989. Expression of human lactotransferrin receptors on phytohemagglutinin-stimulated human peripheral blood lymphocytes: isolation of the receptors by anti-ligand affinity chromatography. *Eur. J. Biochem.* **179**:481.

Mikogami, T., Heyman, M., Spik, G., and Desjeux, J.-F. 1994. Apical-to-basolateral transepithelial transport of human lactoferrin in the intestinal cell line HT-29cl.19A. *Am. J. Physiol.* **267**:G308.

Mincheva-Nilsson, L., Hammarstrom, M.-L., Juto, P., and Hammarstrom, S. 1990. Human milk contains proteins that stimulate and suppress T lymphocyte proliferation. *Clin. Exp. Immunol.* **79**:463.

Mohr, J.A. 1973. The possible induction and/or acquisition of cellular hypersensitivity associated with ingestion of colostrum. *J. Pediatr.* **82**:1062.

Munoz, C., Endres, S., van der Meer, J., Schlesinger, L., Arevalo, M., and Dinarello, C. 1990. Interleukin-1-B in human colostrum. *Res. Immunol.* **141**:505.

Murphey, D.K. and Buescher, E.S. 1993. Human colostrum has anti-inflammatory activity in a rat subcutaneous air pouch model of inflammation. *Pediatr. Res.* **34**:208.

Na, H.R. and Seelig, L.L. 1994. Effect of maternal ethanol consumption on *in vitro* tumor necrosis factor, interleukin-6 and interleukin-2 production by rat milk and blood leukocytes. *Alcoholism: Clin. Exp. Res.* **18**:398.

Nagafuchi, S., Katayanagi, T., Nakagawa, E., Takahashi, T., Yajima, T., Yonekubo, A., and Kuwata, T. 1997. Effects of dietary nucleotides on serum antibody and splenic cytokine production in mice. *Nutr. Res.* **17**:1163.

Nagano, M., Chastre, E., Choquet, A., Bara, J., Gespach, C., and Kelly, P.A. 1995. Expression of prolactin and growth hormone receptor genes and their isoforms in the gastrointestinal tract. *Am. J. Physiol.* **268**:G431.

Neurath, M.F. and Buschenfelde, K.-H.M. 1996. Protective and pathogenic roles of cytokines in inflammatory bowel diseases. *J. Invest. Med.* **44**:516.

Ozkaragoz, F., Rudloff, H.B., Rajaraman, S., Mushtaha, A.A., Schmalstieg, F.C., and Goldman, A.S. 1988. The motility of human milk macrophages in collagen gels. *Pediatr. Res.* **23**:449.

Pabst, H.F. and Spady, D.W. 1990. Effect of breast-feeding on antibody response to conjugate vaccine. *Lancet* **336**:269.

Pabst, H.F., Godel, J., Grace, M., Cho, H., and Spady, D.W. 1989. Effect of breast-feeding on immune response to BCG vaccination. Lancet **1**:295.

Pabst, H.F., Spady, D.W., Pilarski, L.M., Carson, M.M., Beeler, J.A., and Krezolek, M.P. 1997. Differential modulation of the immune response by breast- or formula-feeding of infants. *Acta Paediatr.* **86**:1291.

Pakkanen, R. 1998. Determination of transforming growth factor-β2 in bovine colostrum samples. *J. Immunoassay* **19**:23.

Palkowetz, K.H., Royer, C.L., Garofalo, R., Rudloff, H.E., Schmalstieg, F.C., and Goldman, A.S. 1994. Production of interleukin-6 and interleukin-8 by human mammary gland epithelial cells. *J. Reprod. Immunol.* **26**:57.

Paxson, C.L. and Cress, C.C. 1979. Survival of human milk leukocytes. *J. Pediatr.* **94**:61.

Penttila, I.A., van Spriel, A.B., Zhang, M.F., Xian, C.J., Steeb, C.B., Cummins, A.G., Zola, H. and Read, L.C. 1998. Transforming growth factor-β levels in maternal milk and expression in postnatal rat duodenum and ileum. *Pediatr. Res.* **44**:524.

Pickering, L.K., Granoff, D.M., Erickson, J.R., Masor, M.L., Cordle, C.T., Schaller, J.P., Winship, T.R., Paule, C.L., and Hilty, M.D. 1998. Modulation of the immune system by human milk and infant formula containing nucleotides. *Pediatrics* **101**:242.

Politis, I., Zhao, X., McBride, B.W., and Burton, J.H. 1991a. Examination of chemotactic properties of bovine mammary macrophages. *Can. J. Vet. Res.* **55**:321.

Politis, I., McBride, B.W., Burton, J.H., Zhao, X., and Turner J.D. 1991b. Secretion of interleukin-1 by bovine milk macrophages. *Am. J. Vet. Res.* **52**:858.

Prentice, A. 1987. Breast feeding increases concentrations of IgA in infants' urine. *Arch. Dis. Child.* **62**:792.

Prentice, A., Prentice, A.M., Cole, T.J., and Whitehead, R.G. 1983. Determinants of variations in breast milk protective factor concentrations of rural Gambian mothers. *Arch. Dis. Child.* **58**:518.

Prentice, A., Prentice, A.M., and Lamb, W.H. 1985. Mastitis in rural Gambian mothers and the protection of the breast by milk antimicrobial factors. *Trans. Roy. Soc. Trop. Med. Hyg.* **79**:90.

Prentice, A., Ewing, G., Roberts, S.B., Lucas, A., MacCarthy, A., Jarjou, L.M.A., and Whitehead, R.G. 1987. The nutritional role of breast-milk IgA and lactoferrin. *Acta Paediatr. Scand.* **76**:592.

Reber, P.M. 1993. Prolactin and immunomodulation. *Am. J. Med.* **95**:637.

Roberton, D.M., Paganelli, R., Dinwiddie, R., and Levinsky, R.J. 1982. Milk antigen absorption in the preterm and term neonate. *Arch. Dis. Child.* **57**:369.

Rollwagen, F.M. and Baqar, S. 1996. Oral cytokine administration. *Immunol. Today* **17**:548.

Rudloff, H.E., Schmalstieg, F.C., Mushtaha, A.A., Palkowetz, K.H., Liu, S.K., and Goldman, A.S. 1991. Tumor necrosis factor-α in human milk. *Pediatr. Res.* **31**:29.

Saarinen, U.M. and Kajosaari, M. 1995. Breastfeeding as prophylaxis against atopic disease: prospective follow-up until 17 years old. *Lancet* **346**:1065.

Saito, S., Maruyama, M., Kato, Y., Moriyama, I., and Ichijo, M. 1991. Detection of IL-6 in human milk and its involvement in IgA production. *J. Reprod. Immunol.* **20**:267.

Saito, S., Yoshida, M., Ichijo, M., Ishizaka, S., and Tsujii, T. 1993. Transforming growth factor-B in human milk. *Clin. Exp. Immunol.* **94**:220.

Schlesinger, J.J. and Covelli, H.D. 1977. Evidence for transmission of lymphocyte responses to tuberculin by breast-feeding. *Lancet* **2**:529.

Schnorr, K.L. and Pearson, L.D. 1984. Intestinal absorption of maternal leucocytes by newborn lambs. *J. Reprod. Immunol.* **6**:329.

Schreiber, S. 1997. Interleukin-10 in the intestine. *Gut* **41**:274.

Schroten, H., Bosch, M., Nobis-Bosch, R., Kohler, H., Hanisch, F.-G., and Plogmann, R. 1999. Secretory immunoglobulin A is a conponent of the human milk fat globule membrane. *Pediatr. Res.* **45**:82.

Shoenfeld, Y., Amital, H., Ferrone, S., and Kennedy, R.C. 1994. Anit-idiotypes and their application under autoimmune, neoplastic, and infectious conditions. *Int. Arch. Allergy Immunol.* **105**:211.

Shuster, D.E., Kehrli, M.E., and Stevens, M.G. 1993. Cytokine production during endotoxin-induced mastitis in lactating dairy cows. *Am. J. Vet Res.* **54**:80.

Shuster, D.E., Kehrli, M.E., and Baumrucker, C.R. 1995. Relationship of inflammatory cytokines, growth hormone, and insulin-like growth factor-1 to reduced performance during infectious disease. *Proc. Soc. Exp. Biol. Med.* **210**:140.

Shuster, D.E., Kehrli, M.E., Rainard, P., and Paape, M. 1997. Complement fragment C5a and inflammatory cytokines in neutrophil recruitment during intramammary infection with *Escherichia coli. Infect. Immun.* **65**:3286.

Skansen-Saphir, U., Lindfors, A., and Andersson, U. 1993. Cytokine production in mononuclear cells of human milk studied at the single-cell level. *Pediatr. Res.* **34**:213.

Sone, S., Tsutsumi, H., Takeuchi, R., Matsuda, K., Imai, S., Ogra, P.L., and Chiba, S. 1997. Enhanced cytokine production by milk macrophages following infection with respiratory syncytial virus. *J. Leuko. Biol.* **61**:630.

Srivastava, M.D., Srivastava, A., Brouhard, B., Saneto, R., Groh-Wargo, S., and Kubit, J. 1996. Cytokines in human milk. *Res. Commun. Molec. Pathol. Pharmacol.* **93**:263.

Stein, K.E. and Soderstrom, T. 1984. Neonatal administration of iodotype or antiidiotype primes for protection against *Escherichia coli* K13 infection in mice. *J. Exp. Med.* **160**:1001.

Taylor, B.C., Keefe, R.G., Dellinger, J.D., Nakamura, Y., Cullor, J.S., and Stott, J.L. 1997. T cell populations and cytokine expression in milk derived from normal and bacteria-infected bovine mammary glands. *Cell. Immunol.* **182**:68.

Thorell, L., Sjoberg, L.-B., and Hernell, O. 1996. Nucleotides in human milk: sources and metabolism by the newborn infant. *Pediatr. Res.* **40**:845.

Tuboly, S., Bernath, S., Glavits, R., Kovacs, A., and Megyeri, Z. 1995. Intestinal absorption of colostral lymphocytes in newborn lambs and their role in the development of immune status. *Acta Vet. Hungarica* **43**:105.

Uauy, R., Stringel, G., Thomas, R., and Quan, R. 1990. Effect of dietary nucleosides on growth and maturation of the developing gut in the rat. *J. Pediatr. Gastroenterol. Nutr.* **10**:497.

Waller, K.P., Colditz, I.G., Flapper, P., and Seow, H.-F. 1997. Leukocyte and cytokine accumulation in the ovine teat and udder during endotoxin-induced inflammation. *Vet. Res. Commun.* **21**:101.

Weaver, L.T., Laker, M.F., Nelson, R., and Lucas, A. 1987. Milk feeding and changes in intestinal permeability and morphology in the newborn. *J. Pediatr. Gastroenterol. Nutr.* **6**:351.

Weiner, H.L. 1997. Oral tolerance: immune mechanisms and treatment of autoimmune diseases. *Immunol. Today* **18**:335.

Whiteside, T.L. 1994. Cytokine measurements and interpretation of cytokine assays in human disease. *J. Clin. Immunol.* **14**:327.

Wirt, D.P., Adkins, L.T., Palkowetz, K.H., Schmalstieg, F.C., and Goldman, A.S. 1992. Activated and memory T lymphocytes in human milk. Cytometry **13**:282.

Xu, R.-J., Doan, Q.C., and Regester G.O. 1999. Detection and characterisation of transforming growth factor-beta in porcine colostrum. *Biol. Neonate* **75**:59.

Zhang, L., van Bree, S., van Rood, J.J., Claas, F.H.J. 1991. Influence of breast feeding on the cytotoxic T cell allorepertoire in man. *Transplant.* **52**:914.

Zimecki, M., Mazurier, J., Spik, G., and Kapp, J.A. 1995. Human lactoferrin induces phenotypic and functional changes in murine splenic B cells. *Immunol.* **86**:122.

Chapter 17

Dietary Whey Proteins and Immunocompetence in the Post-Weaning Stages of Life

C. W. Wong

1. Introduction

In the latter part of the twentieth century, the attention of industrialized nations became directed increasingly to "specific" health-promoting properties of nutrients and foods (Brennan *et al.*, 1986; Marshall, 1994). In this context, there is a great impetus for research on amino acids, peptides and proteins that may promote host defense against stressful or pathological conditions. Milk constitutes an important source of high quality dietary proteins that are considered to have health-promoting properties. Research and development in this area have benefitted greatly from recent improvements to protein separation technologies in the dairy and pharmaceutical industries. However, the health-promoting advantages, including immunological benefits, of consuming milk proteins during the postweaning stages of life need further exploration.

Bovine milk whey includes a unique heterogeneous group of high-quality proteins and peptides that remain soluble following precipitation of casein (Eigel *et al.*, 1984). Major whey proteins of bovine milk include β-lactoglobulin, α-lactalbumin, immunoglobulin, and serum albumin, among which β-lactoglobulin is the largest component (approximately 50% of the total protein). There are also a number of minor proteins/peptides present in whey proteins, and these include lactoferrin, lactoperoxidase, various growth factors, cytokines and hormone-like

C. W. Wong • Animal Science, University of New England, Armidale, NSW 2351, Australia

peptides. Many proteins and peptides found in whey have been reported to possess immune-promoting activities (Ogra and Ogra, 1978; Duncan and McArthur, 1981; Newby *et al.*, 1982; Juto, 1985; Stoeck *et al.*, 1989; Mincheva-Nilsson *et al.*, 1990; Barta *et al.*, 1991; Politis *et al.*, 1991; Watson *et al.*, 1992). Such reports continue to attract interest in the possibility that dietary whey proteins can ameliorate the course of various diseases, especially neoplasia and infections, by exerting an impact on immune competence (Bounous and Kongsgavn, 1985; Bounous *et al.*, 1981, 1983a,b, 1988a,b, 1989, 1991; Papenburg *et al.*, 1990; Wong and Watson, 1995). The purpose of this chapter is to assess information concerning the immunomodulatory properties of dietary bovine whey proteins during the post-weaning stages of life. The discussion is sub-divided as follows:

- Dietary whey protein concentrates and immune responses
- Possible immunomodulatory activities of individual whey proteins
- Conclusions and Outlook

Information identified as the topic of other chapters will not be elaborated here. This includes the biological properties of immunoglobulins and the anti-microbial activity of lactoferrin.

2. Dietary Whey Protein Concentrates and Immune Responses

2.1. Humoral Immune Responses

Dietary whey proteins are reported to enhance humoral immune competence in mice during the early postweaning stage of life (Bounous *et al.*, 1985, 1988b; Bounous and Kongshavn, 1982; Wong and Watson, 1995). The primary antibody response of the mouse (mainly IgM) against sheep red blood cells (SRBC), a T cell-dependent antigen, was enhanced by diets containing several levels (12%, 20%, or 28%) of bovine whey protein concentrate (WPC) compared to the response of mice fed diets containing comparable levels of casein (Bounous and Kongshavn, 1982; Bounous *et al.*, 1985). Subsequently, Bounous *et al.* (1988b) showed that the primary anti-SRBC antibody response was greater in mice fed 20% WPC than in animals fed diets containing this level of other protein types such as soybean, wheat, corn, egg albumin, beef and fish proteins. Likewise, Wong and Watson (1995) reported that, compared to soy proteins, dietary WPC (20%) could enhance the secondary antibody response (mainly IgG) to the T-cell dependent antigen, ovalbumin, in mice (Figure 1). Thus, these results appear consistent through both primary and secondary responses. Moreover, the antibody responses were assessed both as antibody-producing cell numbers

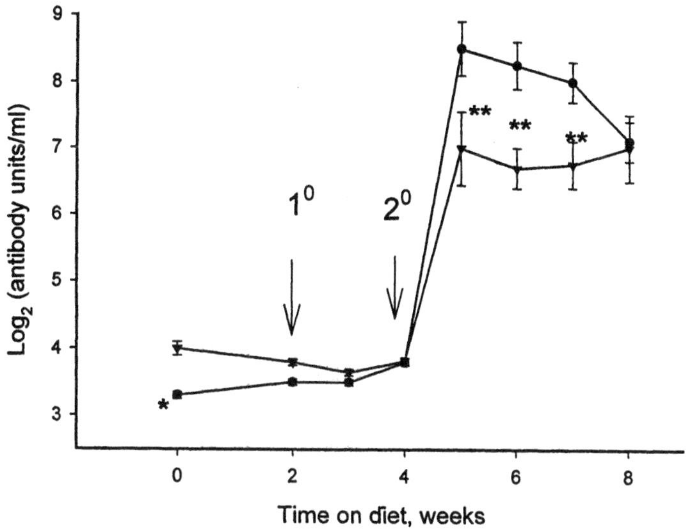

Figure 1. Serum anti-ovalbumin concentrations in mice fed diets containing 20 g/100 g diet of ● whey protein concentrate or ▼ soybean protein isolate. Values are means with SEM. Arrows show timing of primary (1^0) and secondary (2^0) immunization with ovalbumin. Levels of significant difference between groups are * $P < 0.01$, ** $P < 0.001$ (Wong and Watson, 1995).

(Bounous *et al.*, 1985, 1988b; Bounous and Kongshavn, 1982) and as serum titer (Wong and Watson, 1995), the latter providing a measure related directly to humoral defense capacity.

2.2. Cell-Mediated Immune Responses

The influence of dietary WPC on cell-mediated immunity in animals has also been studied using both *in vitro* and *in vivo* assays (Bounous *et al.*, 1983a; Bounous and Kongshavn, 1985; Wong and Watson, 1995). Wong and Watson (1995) reported that WPC could enhance cell-mediated immune responses in mice, relative to dietary soy proteins, as revealed by the delayed-type hypersensitivity (DTH) footpad response to SRBC. In the same study, WPC also enhanced the *in vitro* spleen cell response to the non-specific T cell mitogen, Concanavalin A. In contrast, Bounous and Kongshavn (1985) found no effect of WPC on cell-mediated immune competence despite use of the same footpad DTH assay. The possibility of mouse strain-related differences should not be ignored. However, it is noteworthy that the latter study involved a shorter feeding period (2 weeks) than the 5-week study period of Wong and Watson (1995). In addition, Bounous and Kongshavn (1985) used Freund's Complete Adjuvant (FCA) in eliciting experimental cell-mediated immunity, and the strong and prolonged immunostimulatory effect of FCA might have masked the impact of dietary WPC. In this connection, the DTH-induced footpad swelling (range: 2.8–3.4 mm) of mice in the study of Bounous and Kongshavn (1985) was large compared to the results (range: 0.3–1.0 mm) reported by Wong and Watson (1995). Furthermore, it is worthwhile to note that, whereas Bounous *et al.* (1983a) and Wong and Watson (1995) reported enhanced spleen cell responses to non-specific T cell mitogens in WPC-fed mice, Bounous and Kongshavn (1985) could not reproduce this finding. In these studies, WPC was compared to casein (Bounous *et al.*, 1983a; Bounous and Kongshavn, 1985) and to soybean protein (Wong and Watson, 1995), and the contrasting outcomes relative to dietary casein were reported in studies of identical design. These results suggest that not all WPC preparations are active or, at least, equally active in promoting immune competence. Bounous and Gold (1991) have indicated that the configuration of whey protein molecules may affect their biological activity. Thus, it is possible that the level of the biological activity of whey proteins may vary among batches depending on the milk source and processing methods. Such batch-related variation is not confined to WPC.

2.3. Mechanisms and Other Considerations

Bounous and Kongshavn (1985) proposed that the influence of dietary WPC on the humoral response resulted from a direct impact on the B cell. This was because these authors noted an influence on a thymus-independent antibody response whereas several indices of cell-mediated immune competence were unaffected by WPC. However, it is noteworthy that the response to the type of thymus-independent antigen used by Bounous and Kongshavn (1985) is sensitive to T cell cytokines (Kuby, 1997). Therefore, in view of reports that dietary WPC can promote cell-mediated immune competence, it is possible that all of the various reported influences of dietary WPC on immune competence are effected through T cells, antigen presenting cells and/or their cytokines.

The characteristics that are responsible for the immunomodulatory effects of dietary WPC are not clear, but an important possibility pertains to the unique amino acid composition of WPC. Thus, Bounous and Kongshavn (1982) showed that, when compared to a casein hydrolysate, a dietary WPC hydrolysate enhanced the humoral response to SRBC as well as the splenic T cell blastogenesis response in mice. The low phenylalanine concentration of the WPC hydrolysate appeared to be particularly critical to its immunomodulatory activity. Another interesting point from this study was that WPC hydrolysate had little or no enhancing effect on T- or B-cell blastogenesis unless the animals were challenged with an antigen. The response to an antigenic challenge appears to include diversion of nutrients and energy away from growth or other physiological functions so as to support immune responses (Davis, 1998). In this regard, therefore, it is possible that the amino acid content of WPC (and/or some other characteristic of this preparation) is particularly supportive of such nutrient partitioning. However, knowledge of the metabolic impact of dietary WPC is scant.

In view of the complexity of human diets, results of animal experiments based on WPC as the sole protein source must be interpreted with caution. However, both Bounous *et al.* (1988b) and Wong and Watson (1995) showed that the artificially high levels of supplementation common to many experimental designs are unnecessary for discernment of an impact of dietary WPC on the humoral response in mice.

It is probable that the immunomodulatory activity of a WPC preparation results from interactions among the individual protein components. Thus, bovine lactoferrin forms non-covalent complexes with other whey proteins including albumin, immunoglobulins and β-lactoglobulin (Oram and Reiter, 1968; Butler, 1973; Lampreave *et al.*, 1990). The immunological significance of these chemical interactions is suggested by Wong *et al.* (1996, 1997a,b) who showed that the influence of lactoferrin on lymphocytes and neutrophils was much affected by the presence of the whey protein, lactoperoxidase, or by the mixture of proteins in a WPC preparation.

3. Possible Immunomodulatory Activities of Individual Whey Proteins

Recent studies have demonstrated that the immunomodulatory activity of individual bovine whey proteins *in vitro* becomes more clear-cut as the proteins become progressively more purified (Wong *et al.*, 1996b, 1997a,b). These findings demonstrate the differences in immunological activity among individual whey proteins and lend support to the proposition that the immunomodulatory activity of dietary whey proteins is at least partly attributable to the intestinal absorption of intact whey proteins and peptides (Hutchens *et al.*, 1994; Wong *et al.*, 1996b, 1997a,b, 1998).

3.1. Lactoferrin

There is little doubt that lactoferrin is the best-studied minor whey protein. Information on the structure and functions of this protein has been reviewed in detail by Hutchens *et al.* (1994). The concentration of lactoferrin in milk varies greatly among species, and this range includes species in which the protein has, thus far, been undetectable (Masson and Heremans, 1971). In addition, the iron-binding and antimicrobial activities of lactoferrin obtained from the milk of different species differ significantly (Shimazaki *et al.*, 1994). The underlying physiological reason for such inter-species variation is unclear, but it may be of significance when considering the immunomodulatory potential of lactoferrin. In addition, the immunological activity of bovine lactoferrin appears to depend on the route whereby the protein is administered (Machnicki *et al.*, 1993; Zimecki and Machnicki, 1994; Wong *et al.*, 1996, 1997b, 1998; Adamik *et al.*, 1998; Yoo *et al.*, 1997, 1998; Zimecki *et al.*, 1998a,b, 1999). For a variety of reasons, therefore, this discussion focuses on the immunological effects of bovine lactoferrin administered as a dietary component. Lactoferrin binds to specific receptors on a variety of cells including neutrophils, lymphocytes, monocytes, macrophages and intestinal epithelial cells (Bennett and Davis, 1981; Hu *et al.*, 1988; Sato *et al.*, 1990; Hutchens *et al.*, 1994; Hagiwara *et al.*, 1995). The molecular characteristics of lactoferrin receptors have been extensively studied in several species, particularly mice, pigs and humans (Hutchens *et al.*, 1994). Although species differences exist, bovine lactoferrin can bind to lactoferrin receptors of other animal species (Hu *et al.*, 1988; Hutchens *et al.*, 1994). This provides some basis for anticipating that bovine lactoferrin could exert immunomodulatory influences when included in diets for humans or other non-bovine species. It is also interesting that neither surface nor intracellular lactoferrin receptors are expressed by resting lymphocytes, but that activation of T cells by mitogens such as phytohemagglutinin triggers the expression of these receptors (Mazurier *et al.*, 1989). The physiological significance of this phenomenon is not clear, but its implications regarding the influence of dietary lactoferrin or intestinal T cells may be important.

Debbabi *et al.* (1998) reported that daily oral administration of bovine lactoferrin to mice (0.05–1 mg/g body weight for 28 days) increased the level of immunoglobulins in intestinal mucus as well as the *in vitro* proliferative response of lymphocytes from the Peyer's patches and the spleen to antigenic stimulation. In apparent contrast, Zimecki *et al.* (1998b) observed that an oral dose of lactoferrin (10 mg/day for 5 alternate days) reduced the inflammatory response of the rat when carrageenan was ingested into the footpad. In the same study, oral lactoferrin inhibited *in vitro* production of IL-6 and TNFα by endotoxin-stimulated spleen cells. Thus oral lactoferrin is likely to exert selective effects on different components of the immune system, e.g., T cells and macrophages. This could include influences on components of innate defense. In this regard, Sato *et al.* (1996) reported an uncontrolled study of the efficacy of topical bovine lactoferrin in the treatment of intractable feline stomatitis. The results were interpreted as showing that lactoferrin promotes the phagocytic activity of the neutrophil. This interesting outcome requires confirmation in studies with a more rigorously controlled design.

Information pertaining to the immunological influences of lactoferrin given to human subjects appears limited to uncontrolled trials. Yamauchi *et al.* (1998)

monitored immune indices of ten healthy male volunteers, 31–55 years of age, before, during and after a 28-day course of oral bovine lactoferrin (2 g/day). Increased phagocytic activity was reported on the part of blood neutrophils together with an increase in the number of natural killer cells in the blood. No index of adaptive immune competence was assessed in this trial. Similarly, Zimecki *et al.* (1998a) conducted a trial using seventeen healthy volunteers (8 females and 9 males, 26–59 years of age), each of whom was given a commercial bovine lactoferrin preparation at a dosage of 40 mg, daily, for 10 days. The preparation also contained other potential immunomodulators such as selenium and several vitamins, a feature compounding the difficulties associated with the unconrolled design of the study. Influences on a number of indices of non-specific immune competence (increases in blood neutrophil and lymphocyte counts, decreases in eosinophil and monocyte counts, a decrease in serum TNF-α level, and decreases in spontaneous production of IL-6 and TNF-α by blood mononuclear cells *in vitro*) were apparent following the 10-day treatment period. In particular, effects were apparent on the monocyte and neutrophil count fourteen days after the termination of lactoferrin treatment, thus indicating a carry-over effect which should receive further research attention. Although the lymphocyte proliferative response to mitogenic stimulation was studied and found transiently decreased by the treatment, the investigation would have benefitted from inclusion of specific immune responses. Importantly, Zimecki *et al.* (1999) reported results consistent with these findings in a controlled trial involving different dose regimens of 2, 10, or 50 mg lactoferrin per day for 7 days. However, a need clearly exists for studies that permit unambiguous attribution of outcomes to lactoferrin.

3.2. Proteins Other Than Lactoferrin

Numerous whey proteins such as β-lactoglobulin, α-lactalbumin, lactoperoxidase and some growth factors can be absorbed through the gut by transcytosis and subsequently may exert immunomodulatory actions as intact molecules (Gardner, 1988; Marcon-Genty *et al.*, 1989; Caillard and Tome, 1995; Tome, 1998). Although information is scant and, in some cases entirely speculative, regarding the immunomodulatory effects of these whey proteins, some comment is worthwhile.

Alpha-lactalbumin and β-lactoglobulin are not generally considered to be immunologically active in their own right. However, Wong *et al.* (1997a) demonstrated that intact α-lactalbumin could enhance IL-1β production *in vitro* by sheep lung macrophages in response to *E. coli* endotoxin challenge. Moreover, recent studies on human and bovine α-lactalbumin have revealed that some of their conformational variants are capable of antimicrobial activities and of inducing apoptosis in thymocytes and peripheral lymphocytes (Hakansson *et al.*, 1995,

2000; Svensson *et al.*, 1999, 2000). Antibacterial activities of some bovine α-lactalbumin fragments produced by different proteolytic enzymes have also been identified (Pellegrini *et al.*, 1999). The quantity of intact α-lactalbumin that can be absorbed through the intestine is unknown. In this connection, however, it is worth noting that dietary α-lactalbumin is reported to promote lactose production in animals (Grimble and Mansaray, 1987), and the mechanism appears to involve association of the intact α-lactalbumin molecule with galactosyltransferases (Brew, 1969; O'Keefe *et al.*, 1980; Heine *et al.*, 1991; Permyakova and Berliner, 2000). These enzymes are found on cellular surfaces and are therefore positioned to influence a variety of cellular processes (Ram and Munjal, 1985).

A number of whey-based opioid peptides such as α- and β-lactorphins derived from α-lactalbumin and β-lactoglobulin, respectively, have been identified during the last decade (Teschemacher and Koch, 1991; Marshall, 1994; Teschemacher *et al.*, 1997; Belem *et al.*, 1999; Bordenave *et al.*, 1999; FitzGerald and Meisel, 1999). Neither the metabolism nor the physiological role(s) of these peptides are known. However, other opioid peptides influence immune competence (Meisel and Bockelmann, 1999), and the opioids in milk, therefore, deserve investigation as potential whey-derived immunomodulators.

4. Conclusions and Outlook

Animal studies (Bounous and Kongshavn, 1985; Wong and Watson, 1995) have demonstrated unequivocally that dietary bovine WPC is superior to several other sources of high-quality dietary protein in enhancing specific immune responses to novel antigens. Unfortunately, evidence is lacking to show that this translates to superiority in promoting immune resistance to infections in either animals or humans. The term, immunocompetence, is used to mean the capacity of an individual to develop an immune response following stimulation by antigen. It is widely considered that a high level of immunocompetence, as usually assessed, leads to better resistance against infectious and neoplastic diseases. However, the definition of immunocompetence is vague and its relationship with disease resistance is ill-defined quantitatively. There is a need to establish immunological biomarkers, including normal ranges for these indices, that can be applied in assessing the capacity of foods (or food components such as WPC) to promote real improvements in disease resistance. In addition, studies are needed in which infectious disease incidence is an outcome measure. Despite the value of animal-based investigations, recommendations for humans ultimately must derive from well-designed studies of efficacy in human subjects. Inevitably this includes a need for lengthy and quite costly trials. As only one example,

however, certainly this is feasible among the elderly (both free-living and institutionalized) who represent a group at high risk of immune deficiency and infection that is responsive to dietary intervention (Mazari and Lesourd, 1998; Lesourd, 1999).

Large-scale commercial production of high-value whey-derived preparations such as WPC and lactoferrin is now a reality. Technologies to ensure a consistent product are also being developed. Therefore, it is realistic to foresee the capability to engineer whey protein products so as to maximize their immune-boosting properties. In order to take advantage of such technological capabilities, it will be necessary to develop a much-improved understanding of the mechanism(s) whereby dietary WPC, and possibly individual whey proteins, exert influence on immune competence.

References

Adamik, B., Zimecki, M., Wlaszczyk, A., Berezowicz, P., and Kubler, A. 1998. Lactoferrin effects on the in vitro immune response in critically ill patients. *Arch. Immunol. Ther. Exp. (Warsz).* **46**:169.

Barta, O., Barta, V.D., Crisman, M.V., and Akers, R.M. 1991. Inhibition of lymphocyte blastogenesis by whey. *Am. J. Vet. Res.* **52**:247.

Belem, M.A., Gibbs, B.F., and Lee, B.H. 1999. Proposing sequences for peptides derived from whey fermentation with potential bioactive sites. *J. Dairy Sci.* **82**:486.

Bennett, R.M. and Davis, J.J. 1981. Lactoferrin binding to human peripheral blood cells: an interaction with a B-enriched population of lymphocytes and a subpopulation of adherent mononuclear cells. *J. Immunol.* **127**:1211.

Bordenave, S., Sannier, F., Ricart, G., and Piot, J.M. 1999. Continuous hydrolysis of goat whey in an ultrafiltration reactor: generation of alpha-lactorphin. *Prep. Biochem. Biotechnol.* **29**:189.

Bounous, G., Stevenson, M.M., and Kongshavn, P.A. 1981. Influence of dietary lactalbumin hydrolysate on the immune system of mice and resistance to salmonellosis. *J. Infect. Dis.* **144**:281.

Bounous, G. and Kongshavn, P.A. 1982. Influence of dietary proteins on the immune system of mice. *J. Nutr.* **112**:1747.

Bounous, G., Letourneau, L., and Kongshavn, P.A. 1983a. Influence of dietary protein type on the immune system of mice. *J. Nutr.* **113**:1415.

Bounous, G., Letourneau, L., and Kongshavn, P.A. 1983b. The effect of concentration and type of dietary protein hydrolysates on tumour growth in mice. *Clin. Invest. Med.* **6**:197.

Bounous, G. and Kongshavn, P.A. 1985. Differential effect of dietary protein type on the B-cell and T-cell immune responses in mice. *J. Nutr.* **115**:1403.

Bounous, G., Shenouda, N., Kongshavn, P.A., and Osmond, D.G. 1985. Mechanism of altered B-cell response induced by changes in dietary protein type in mice. *J. Nutr.* **115**:1409.

Bounous, G., Papenburg, R., Kongshavn, P.A., Gold, P., and Fleiszer, D. 1988a. The immunoenhancing property of dietary whey protein concentrate. *Clin. Invest. Med.* **11**:213.

Bounous, G., Kongshavn, P.A., and Gold, P. 1988b. The immunoenhancing property of dietary whey protein concentrate. *Clin. Invest. Med.* **11**:271–278.

Bounous, G., Batist, G., and Gold, P. 1989. Immunoenhancing property of dietary whey protein in mice: role of glutathione. *Clin. Invest. Med.* **12**:154.

Bounous, G. and Gold, P. 1991. The biological activity of undenatured dietary whey proteins: role of glutathione. *Clin. Invest. Med.* **14**:296.

Bounous, G., Batist, G., and Gold, P. 1991. Whey proteins in cancer prevention. *Cancer Lett.* **57**:91.

Brennan, M.F., Cerra, F., Daly, J.M., Fischer, J.E., Moldawer, L.L., Smith, R.J., Vinnars, E., Wannemacher, R., and Young, V.R. 1986. Report of a research workshop: branched-chain amino acids in stress and injury. *J. Parenter. Enteral. Nutr.* **10**:446.

Butler, J.E. 1973. The occurrence of immunoglobulin fragments, two types of lactoferrin and a lactoferrin-IgG2 complex in bovine colostral and milk whey. *Biochim. Biophys. Acta.* **295**:341.

Caillard, I. and Tome, D. 1995. Transport of beta-lactoglobulin and alpha-lactalbumin by enterocyte-like Caco-2 cells. *Reprod. Nutr. Dev.* **3**:653.

Davis, S.L. 1998. Environmental modulation of the immune system via the endocrine system. *Domest. Anim. Endocrinol.* **15**:283.

Debbabi, H., Dubarry, M., Rautureau, M., and Tome, D. 1998. Bovine lactoferrin induces both mucosal and systemic immune response in mice. *J. Dairy Res.* **65**:283.

Duncan, R.L. and McArthur, W.P. 1981. Lactoferrin-mediated modulation of mononuclear cell activities. 1. Suppression of murine in vitro primary antibody response. *Cell Immunol.* **63**:308.

Eigel, W.N., Butler, J.E., Ernstrom, C.A., Farrell, H.M., Harwalker, V.R., Jenness, R., and Whitney, R.M. 1984. Nomenclature of proteins of cow's milk, fifth revision. *J. Dairy Sci.* **67**:1599.

FitzGerald, R.J. and Meisel, H. 1999. Lactokinins: whey protein-derived ACE inhibitory peptides. *Nahrung* **43**:165.

Gardner, M.L.G. 1988. Gastrointestinal absorption of intact proteins. *Ann. Rev. Nutr.* **8**:329.

Grimble, R.F. and Mansaray, Y.K. 1987. Effects in rats of dietary protein inadequacy on lactose production, milk volume and components of lactose synthetase complex (EC 2.4.1.22). *Ann. Nutr. Metab.* **31**:179.

Hagiwara, T., Shinoda, I., Fukuwatari, Y., and Shimamura, S. 1995. Effects of lactoferrin and its peptides on proliferation of rat intestinal epithelial cell line, IEC-18, in the presence of epidermal growth factor. *Biosci. Biotechnol. Biochem.* **59**:1875.

Hu, W.L., Mazurier, J., Sawatzki, G., Montreuil, J., and Spik, G. 1988. Lactotransferrin receptor of mouse small-intestinal brush border. Binding characteristics of membrane-bound and triton X-100-solubilized forms. *Biochem. J.* **249**:435.

Hutchens, T.W., Rumball, S.V., and Lönnerdal, B. 1994. *Lactoferrin—Structure and Function.* Plenum Press, New York and London.

Juto, P. 1985. Human milk stimulates B cell function. *Arch. Dis. Childhood* **60**:610.

Kuby, J. 1997. B-cell maturation, activation, and differentiation, in: *Immunology*, 3rd edition, pp. 195–222. W.H. Freeman and Co., New York.

Lampreave, F., Pineiro, A., Brock, J.H., Castillo, H., Sanchez, L., and Calvo, M. 1990. Interaction of bovine lactoferrin with other proteins of milk whey. *Int. J. Biol. Macromol.* **12**:2.

Lesourd, B.M. 1999. Immune response during disease and recovery in the elderly. *Proc. Nutr. Soc.* **58**:85.

Machnicki, M., Zimecki, M., and Zagulski, T. 1993. Lactoferrin regulates the release of tumour necrosis factor alpha and interleukin 6 *in vivo*. *Int. J. Exp. Pathol.* **74**:433.

Marcon-Genty, D., Tome, D., Kheroua, O., Dumontier, A.-M., Heyman, M., and Desjeux, J.-F. 1989. Transport of bb-lactoglobulin across rabbit ileum *in vitro*. *Am. J. Physiol.* **256**:G943.

Marshall, W.E. 1994. Amino acids, peptides, and Proteins, in: *Functional Foods—Designer Foods, Pharmafoods, Nutraceuticals* (I Goldberg, ed.), pp. 242–260, Chapman and Hall, New York.

Masson, P.L. and Heremans, J.F. 1971. Lactoferrin in milk from different species. *Comp. Biochem. Physiol.* **39**:119.

Mazari, L. and Lesourd, B.M. 1998. Nutritional influences on immune response in healthy aged persons. *Mech. Ageing Dev.* **104**:25.

Mazurier, J., Legrand, D., Hu, W.L., Montreuil, J., and Spik, G. 1989. Expression of human lactotransferrin receptors in phytohemagglutinin-stimulated human peripheral blood lymphocytes. Isolation of the receptors by antiligand-affinity chromatography. *Eur. J. Biochem.* **179**:481.

Meisel, H. and Bockelmann, W. 1999. Bioactive peptides encrypted in milk proteins: proteolytic activation and thropho-functional properties. *Antonie Van Leeuwenhoek.* **76**:207.

Mincheva-Nilsson, L., Hammarströöm, M.L., Juto, P., and Hammarströöm, S. 1990. Human milk contains proteins that stimulate and suppress T lymphocyte proliferation. *Clin. Exp. Immunol.* **79**:463.

Newby, T.J., Stokes, C.R., and Bourne, F.J. 1982. Immunological activity of milk. *Vet Immunol. Immunopathol.* **3**:67.

Ogra, S.S. and Ogra, R.A. 1978. Immunologic aspects of human colostrum and milk. II Characteristics of lymphocyte reactivity and distribution of E-rosette forming cells at different times after the onset of lactation. *J. Pediatr.* **92**:550.

Oram, J.D. and Reiter, B. 1968. Inhibition of bacteria by lactoferrin and other iron-chelating agents. *Biochim. Biophys. Acta* **170**:351.

Papenburg, R., Bounous, G., Fleiszer, D., and Gold, P. 1990. Dietary milk proteins inhibit the development of dimethylhydrazine-induced malignancy. *Tumour Biol.* **11**:129.

Pellegrini, A., Thomas, U., Bramaz, N., Hunziker, P., and von Fellenberg, R. 1999. Isolation and identification of three bactericidal domains in the bovine α-lactalbumin molecule. *Biochim. Biophys. Acta* **1426**:439.

Politis, I., Zhao, X., McBride, B.W., and Burton, J.H. 1991. Effect of bovine skim milk and whey on monocyte function. *J. Dairy Sci.* **74**:2467.

Sato, K., Shinomoto, H., Tanimoto, M., Dosako, S., and Nakajima, I. 1990. Uptake and secretion of human lactoferrin by B lymphocytes. *Agri. Biol. Chem.* **54**:1275.

Sato, R., Inanami, O., Tanaka, Y., Takase, M., and Naito, Y. 1996. Oral administration of bovine lactoferrin for treatment of intractable stomatitis in feline immunodeficiency virus (FIV)-positive and FIV-negative cats. *Am. J. Vet. Res.* **57**:1443.

Shimazaki, K., Oota, K., Nitta, K., and Ke, Y. 1994. Comparative study of the iron-binding strengths of equine, bovine and human lactoferrins. *J. Dairy Res.* **61**:563.

Stoeck, M., Ruegg, C., Miescher, S., Carrel, S., Cox, D., Von Fliedner, V., and Alkan, S. 1989. Comparison of the immunosuppressive properties of milk growth factor and transforming growth factors bb1 and β2. *J. Immunol.* **143**:3258.

Teschemacher, H., Koch, G., and Brantl, V. 1997. Milk protein-derived opoid receptor ligands. *Biopolymers* **43**:99.

Teschemacher, H. and Koch, G. 1991. Opioids in the milk. *Endocr. Regul.* **25**:147.

Tome, D. 1998. Bioactive peptides and proteins from milk—potential health benefits. In: *Milk and Heath—Proceedings of 25th International Dairy Congress, 21–24 September*, (Ravn, A., ed.), pp. 163–180, The Danish National Committee of the IDF, Frederiks.

Watson, D.L. Francis, G.L., and Ballard, F.J. 1992. Factors in ruminant colostrum that influence cell growth and murine IgE antibody responses. *J. Dairy Res* **59**:369.

Wong, C.W. and Watson, D.L. 1995. Immunomodulatory effects of dietary whey proteins in mice. *J. Dairy Res.* **62**:359.

Wong, C.W., Regester, G.O., Francis, G.L., and Watson, D.L. 1996. Immunomodulatory activities of whey fractions in efferent prefemoral lymph of sheep. *J. Dairy Res.* **63**:257.

Wong, C.W., Seow, H.F., Husband, A.J., Regester, G.O., and Watson, D.L. 1997a. Effects of purified bovine whey factors on cellular immune functions in ruminants. *Vet. Immunol. Immunopathol.* **56**:85.

Wong, C.W., Liu, A.H, Regester, G.O., Francis, G.L., and Watson, D.L. 1997b. Influence of whey and purified whey proteins on neutrophil functions in sheep. *J. Dairy Res.* **64**:281.

Wong, C.W., Watson, D.L., Regester, G.O., and Smithers, G.W. 1998. Comparison of immunomodulatory properties of purified lactoferrin, lactoperoxidase and β-casein in cannulated sheep. *J. Dairy Res.* **65**:697.

Yamauchi, K., Wakabayashi, H., Hashimoto, S., Teraguchi, S., Hayasawa, H., and Tomita, M. 1998. Effects of orally administered bovine lactoferrin on the immune system of healthy volunteers. *Adv. Exp. Med. Biol.* **443**:261.

Yoo, Y.C., Watanabe, S., Watanabe, R., Hata, K., Shimazaki, K., and Azuma, I. 1997. Bovine lactoferrin and lactoferricin, a peptide derived from bovine lactoferrin, inhibit tumor metastasis in mice. *Jpn J. Cancer Res.* **88**:184.

Yoo, Y.C., Watanabe, S., Watanabe, R., Hata, K., Shimazaki, K., and Azuma, I. 1998. Bovine lactoferrin and Lactoferricin inhibit tumor metastasis in mice. *Adv. Exp. Med. Biol.* **443**:285.

Zimecki, M. and Machnicki, M. 1994. Lactoferrin inhibits the effector phase of the delayed type hypersensitivity to sheep erythrocytes and inflammatory reactions to M. bovis (BCG). *Arch. Immunol. Ther. Exp. (Warsz)* **42**:171.

Zimecki, M., Wlaszczyk, A., Cheneau, P., Brunel, A.S., Mazurier, J., Spik, G., and Kubler, A. 1998a. Immunoregulatory effects of a nutritional preparation containing bovine lactoferrin taken orally by healthy individuals. *Arch. Immunol. Ther. Exp. (Warsz)* **46**:231.

Zimecki, M., Miedzybrodzki, R., and Szymaniec, S. 1998b. Oral treatment of rats with bovine lactoferrin inhibits carrageenan-induced inflammation; correlation with decreased cytokine production. *Arch. Immunol. Ther. Exp. (Warsz)* **46**:361.

Zimecki, M., Spiegel, K., Wlaszczyk, A., Kubler, A., and Kruzel, M.L. 1999. Lactoferrin increases the output of neutrophil precursors and attenuates the spontaneous production of TNF-alpha and IL-6 by peripheral blood cells. *Arch. Immunol. Ther. Exp. (Warsz)* **47(2)**:113.

Part V

External Influences on Defense Components of Milk

Chapter 18

Maternal Modulation of Specific and Non-Specific Immune Components of Colostrum and Mature Milk

Peter E. Hartmann, Mark D. Cregan, and Leon R. Mitoulas

1. Introduction

The neonates of most mammalian species cannot survive on an adult diet or exposure to an array of adult pathogens. Lactation provides a transitional period for the immature newborn to adapt to the adult environment. However, there is immense variation in the metabolic maturity of mammals at birth, as well as in the rate of postnatal growth and development during the lactation period. As a result, large differences in milk composition have evolved among mammals (Oftedal and Iverson, 1995). Furthermore, large progressive changes in milk composition occur during lactation to meet the changing metabolic needs of the young (Nicholas *et al.*, 1997). For example, the Tammar Wallaby has a relatively short gestation period, during which Phase 1 of mammary development occurs. The animal weighs only 400 mg at birth and remains permanently attached to the teat for the next 130 days, while growing to 100 g (Phase 2A). The pouch young (PY) then detaches from the teat and sucks intermittently to reach a weight of 400–500 g by 200 days of age (Phase 2B). The PY then leaves the pouch intermittently until 260 days when it exits the pouch permanently, finally weaning at 300 days, at a weight of 1.6–2.5 kg (Phase 3). The increase in body weight of the young from about 0.5 g to 2000 g during lactation is associated with major changes in

Peter E. Hartmann, Mark D. Cregan and Leon R. Mitoulas • Department of Biochemistry, The University of Western Australia, Nedlands, WA 6907, Australia

milk composition that define phases 2A, 2B and 3 of lactation. Both mammary gland size and milk production increase commensurate with the rate of growth of the young.

In other mammals (e.g., human, goat, elk) whose young are metabolically more mature at birth there are only relatively minor changes in milk composition during lactation (Green, 1984). Thus, the milk of each species is uniquely adapted to facilitate optimal rates of growth and development of its neonates, which are born with functionally immature systems (digestive, hepatic, neural, renal, blood vascular, visual, skeletal and immune) and which exhibit very different rates of postnatal growth and maturation (Hartmann *et al.*, 1997).

The immune system presents consistent levels of immaturity at birth (Goldman *et al.*, 1998) because the near sterile conditions of the uterus do not provide antigenic stimulation for the development of specific (acquired) immunity. Thus, the main difference among species lies in whether passive humoral immune protection by IgG is transferred across the placenta before birth (e.g., human infant) or transferred into colostrum (e.g., bovine calf) and absorbed from the small intestine after birth (Cowie *et al.*, 1980). The mechanisms by which milk provides defense against pathological micro-organisms include the development of immunity to specific antigens (specific immunity), passive immune protection (humoral protection, IgG; surface protection, sIgA) as well as non-specific (natural/innate) immunity e.g., conferred by lactoferrin, oligosaccharides, lipids, anti-bacterial enzymes, substrate binding proteins, anti-microbial proteins, anti-inflammatory agents and immuno-modulators (Goldman and Goldblum, 1995). Colostrum and milk compensate for the immaturity of the newborn's immune system, but the overall protection gained before weaning depends not only on the factors that affect the concentration of these components in milk but also on the total amount of milk delivered to the offspring each day (milk production). While milk production determines the "dose" of all components of the specific and non-specific immune systems delivered to the suckling offspring, the effect of maternal dietary and physiological factors on the concentration of a number of these components is unknown. Diseases (e.g., mastitis) (Prosser and Hartmann, 1983) and certain medications (e.g., ergot alkaloids) (Kulski *et al.*, 1978) can affect milk composition and production, but are outside the scope of this review.

2. Milk Production

Much of the research into the control of milk production has been sponsored by the dairy industry with the aim of maximizing profits. Consequently, dairy cows and goats are milked with the objective of maintaining milk production at physiologically maximum limits for the duration of the lactation period. In contrast, milk production represents a balance between the needs of the young for optimum growth, health and development and its cost to the metabolic economy of the mother (Peaker and Wilde, 1987). Various strategies can be used to achieve this balance. For example, the female Brown Antechinus, a carnivorous marsupial, has two litters during its lifetime and destroys all female offspring in the first litter and all male offspring in the second litter. Males move to new territories after weaning and do not compete with their mother for food, whereas females remain in the mother's territory and, therefore, would compete with their mother for food after the first but not the second litter. By that means the mother's reproductive potential is maximized and the balance between males and females is maintained. On the other hand, most women can produce more than enough milk to meet their baby's needs and milk production is regulated locally within each breast to meet the varying demands of the infant's appetite (Hartmann *et al.*, 1995a). Therefore, caution must be exercised in making generalizations from one species to another as well as from dairy to suckling mammals.

2.1. Initiation of Lactation

The initiation of lactation develops in two stages (Hartmann, 1973). The mammary gland differentiates during pregnancy to develop the ability to produce unique milk proteins (lactogenesis I) and then initiates copious milk secretion at, or just after, parturition (lactogenesis II). Relatively small delays in the onset of lactogenesis II can greatly influence the amount of immunoglobulins and other host defense agents provided to the newborn during the critical first few days after birth. Indeed, after changes in milk volume intake have been taken into account, 4 g of IgA is ingested by the newborn in the first 24 hours of breast-feeding compared to 1 g/day after day 4 (Xanthou, 1998). While a number of the antimicrobial agents may be present in milk primarily to protect the lactating mammary gland (Fetherston *et al.*, 2001), many undoubtedly provide protection to the neonate over this vulnerable period (Renner *et al.*, 1989; Goldman and Goldblum, 1995).

In most species studied to date (Cowie *et al.*, 1980; Hartmann *et al.*, 1995a), lactogenesis II is controlled by the withdrawal of progesterone in the presence of permissive hormones such as prolactin/growth hormone, glucocorticoids and insulin. In women, the fall in circulating progesterone due to loss of its major site of synthesis (the placenta), triggers lactogenesis II approximately 30–40 hours after delivery. However, in the rare event that viable fragments of placenta are retained in the uterus, the initiation of lactation is delayed until the fragments are removed (Neifert *et al.*, 1981). Furthermore, in sows even slight variations in the rate of withdrawal of progesterone over the farrowing period can produce significant impairment of the onset of lactogenesis II, decreasing growth and increasing the mortality of piglets at one week of age (De Passillé *et al.*, 1993).

In western countries health professionals focus on the difficulties most women experience in the positioning and attachment of their babies to the breast (Riordan and Auerbach, 1993; Brodribb, 1997). Consequently, the possibility of subtle endocrine abnormalities suppressing lactogenesis II is often overlooked. It is known that lactogenesis II is delayed by about 24 h in women with insulin-dependent diabetes (Neubauer *et al.*, 1993; Arthur *et al.*, 1994) but the influence of subtle changes in the withdrawal of progesterone has not been investigated. The influence of exogenously administered glucocorticoids on lactogenesis II also should be considered. Betamethasone is administered to women with at-risk pregnancies to precociously mature the lungs of the fetus (Jobe *et al.*, 1998). Apart from the potential for this treatment to influence the development of the fetal immune system, glucocorticoids administered during pregnancy have been shown to override progesterone inhibition and prematurely initiate lactation in a number of mammals (Cowie *et al.*, 1980). Furthermore, Brandon *et al.* (1975) have shown that the administration of glucocorticoids to cows 6–8 weeks before the expected time of parturition completely abolished the characteristic increase in IgG transfer into colostrum. Thus, the administration of betamethasone to pregnant women could influence both the protective properties and quantity of colostrum available to the vulnerable premature baby. It is of interest that most studies in which the milk of mothers who delivered prematurely (preterm milk) was compared with that of those who delivered at term (term milk), indicated that preterm milk has a unique nutrient composition (Atkinson, 1995). Many possible explanations have been proposed for the lower milk production as well as higher concentrations of nitrogen, immunoglobulins, lipid, medium chain fatty acids (MCFA), energy, vitamins, some minerals and trace elements in preterm compared to term milk (Atkinson, 1995). However, the possibility that therapeutic betamethasone may alter mammary gland development, lactogenesis II and immunological competence should be given serious consideration.

2.2. Established Lactation

It has been generally concluded from studies on dairy cows that maternal diet plays a key role in determining milk production. However, some mammals (bears, seals and baleen whales) fast during lactation (Oftedal, 1993). For example, female blue whales feed in summer in high latitude waters, increasing in body weight by 50% or more, and then migrate to warmer lower latitude waters to calve in midwinter. The females feed little from midwinter until they wean their calves upon returning to the feeding grounds and, yet, the calf increases in body weight by about 17,000 kg during the seven months of lactation. Likewise, maternal dietary intake has little influence on milk production by women in either developed or developing countries (Prentice *et al.*, 1986; Hartmann *et al.*, 1995b; Jensen, 1995). In fact, when women breast-feed to their infants' needs (feeding on demand), it is the infant's appetite that regulates its mother's milk supply (Dewey *et al.*, 1991; Daly *et al.*, 1993a).

The maternal regulation of milk production in established lactation appears to be under three levels of control (metabolic, endocrine and autocrine) and the contribution of each level depends on the species and its environment. Early *in vivo* and *in vitro* studies showed that systemic support from hormones such as prolactin, growth hormone, oxytocin, insulin and glucocorticoids was essential for the maintenance of lactation (Lyons, 1958; Cowie *et al.*, 1980). On the other hand, recent research in women suggests that local regulation within individual breasts occurs, as the short-term (between breast-feeds) rate of milk synthesis is very sensitive to the degree of fullness of the breast (Daly *et al.*, 1993a; Cox *et al.*, 1996). Linzell and Peaker (1971) first proposed that milk production in goats could be modulated by a component(s) in the milk and they hypothesised that removal of this inhibition could promote increased milk synthesis. Current evidence suggests that the local regulation of milk synthesis is multifactorial, involving autocrine factors (Wilde *et al.*, 1995) and possibly free fatty acids (Heesom *et al.*, 1992). The action of these local inhibitors as well as hormones such as prolactin (Streuli and Edwards, 1998) and possibly oxytocin (Moore *et al.*, 1987) may be modulated by interactions between integrin, laminin, the extracellular matrix and the secretory mammary epithelial cells (lactocytes) as the alveoli fill and empty of milk. Indeed, the rate of milk synthesis appears to be inhibited when the lactocytes tend towards a squamous morphology either in distended alveoli *in vivo* or in conventional monolayer cultures of primary lactocytes (Streuli and Edwards, 1998).

The average daily milk production in women remains relatively constant from 1–6 months of lactation (Hartmann *et al.*, 1995b; Kent *et al.*, 1999). However, there is large variation (coefficient of variation 25%) in milk production even among mothers who are exclusively breast-feeding their babies at a similar stage of lactation. Therefore, there could be very large differences in the daily intakes of immunoglobulins, lymphocytes, and other anti-microbial agents depending upon the milk intake of individual babies. In Australia some aboriginal communities breast-feed their children to their sixth year of life (Thomson, 1983) but the weaning age of 3.5 years for the Kung hunter-gatherers of Botswana and Namibia (Konner and Worthman, 1980) is closer to the normal physiological length of human lactation (Dettwyler, 1995).

Currently, in Australia (NH&MRC, 1996) and the USA (AAP, 1997) as in other countries, it is recommended that infants should be exclusively breast-fed for 4–6 months and that breast-feeding should continue into the infant's second

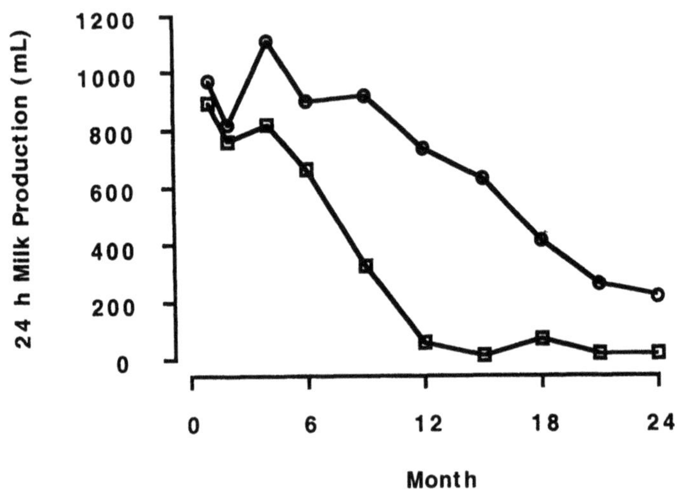

Figure 1. The 24 h milk production (mL) for Mother 6 (□) and Mother 7 (O) over 24 months of lactation.

year of life. Unfortunately, there is very limited information on milk production and composition beyond the first 6 months of lactation (Hartmann *et al.*, 1985). However, our observations suggest that there is even greater variation in milk production among mothers at this time (Kent *et al.*, 1999). Indeed, one woman produced less than 50 ml/24 h from 12 to 24 months of lactation, whereas in another woman production decreased gradually from 735 ml/24 h to 216 ml/24 h over the same period (Figure 1). Therefore, care must be exercised in the interpretation of epidemiological studies on the role of extended breast-feeding in the protection of the infant against pathological microorganisms. Quantitative rather than qualitative measures of breast-feeding are required for these studies. It is likely that important protective properties of breast milk are being overlooked because of large error variances associated with the statistical analysis of both prospective and cross-sectional epidemiological studies based upon questionnaires containing only qualitative assessments of breast-feeding.

3. Milk Composition

The milk of different mammals contains the same unique classes of proteins, carbohydrates and lipids, but the relative amounts of these compounds vary greatly among species (Oftedal and Iverson, 1995; Goldman and Goldblum,

1995). It is difficult to increase the concentration of economically desirable components in the milk of dairy cows (Welch *et al.*, 1997). The most cost-effective intervention to date has been to produce more milk of the same composition by the administration of growth hormone (Collier, 1997). However, very high milk production appears to compromise mammary defense mechanisms and there is argument that udder health may be adversely affected (Coghlan, 1999).

3.1. Initiation of Lactation

The composition of colostrum varies greatly among species, but consistently has a higher concentration of protein than mature milk due to the dominant presence of immunoglobulins. Each species appears to have a particular pattern for the classes of immunoglobulins present, but in general in species dependent on placental transfer for passive humoral protection, sIgA predominates, whereas in those dependent on colostrum IgG predominates (Cowie *et al.*, 1980). In this context, the mother's previous antigenic experience is paramount in determining both surface (sIgA) and humoral (IgG) passive protection provided to her young during the suckling period.

The offspring of species with predominant transfer of passive immunity across the mammary gland are particularly at risk of disease and death if the ingestion of colostrum is impeded. Furthermore, maternal age, parity, genotype, nutritional status, premature parturition and stage of lactation (McFadden *et al.*, 1997) influence the immunoglobulin content of the mammary secretion. Variation in the concentration of colostral immunoglobulins among breeds of dairy cows (Muller and Ellinger, 1981) as well as between dairy and beef animals (Petrie *et al.*, 1984), suggests that there is the potential to select for immunoglobulin levels in ruminant colostrum (McFadden *et al.*, 1997). Unilateral prepartum milking of dairy cows has demonstrated a high degree of local, intra-mammary gland, control of the active transfer of IgG into milk (Guy *et al.*, 1994).

Colonisation of the gastrointestinal tract of the newborn occurs within the first few days after birth. Intestinal bacteria can produce D-lactate, whereas mammalian metabolism produces only L-lactate. The concentration of D-lactate was found to be low in the blood of piglets at birth ($22 \pm 2\,\mu mol/l$), to increase significantly by 72 h and to continue to increase thereafter, reaching $423 \pm 79\,\mu mol/l$ at 120 h (Thompson *et al.*, 1997). This suggests that the anti-bacterial properties of colostrum are important not only in providing defense against pathogenic bacteria, but also in promoting the colonisation of the gut by non-pathogenic bacteria. Indeed, Adlerberth *et al.* (1991) found that non-breast-fed infants in Pakistan were rapidly colonised by changing populations of several potentially pathogenic strains of aerobic bacteria whereas exclusively breast-fed Swedish infants showed a stable, more slowly developing intestinal flora dominated mostly by a single strain of non-virulent aerobic bacteria. Therefore, it is probable that this early colonisation induces the immune tolerance to one's own intestinal bacteria that is observed later in life (Duchmann *et al.*, 1995; Florin, 1997). It is possible that intestinal flora, dietary experiences and infection history of the breast-feeding mother could influence the development of the infant's intestinal flora and thereby the risk of diarrhea and other infections.

3.2. Established Lactation

The concentration of some minor components of milk (e.g., sIgA, IgG, long chain polyunsaturated fatty acids) can vary greatly among individual mothers, from day to day and with stage of lactation (Renner *et al.*, 1989). High concentrations of immunoglobulins are present in colostrum, and although the concentration of these proteins decreases precipitously as lactation establishes, the daily output in human milk shows only a slight decrease (Kulski and Hartmann, 1981).

Petersen was the first to describe not only the production of immune milk from the local antigenic stimulation of the mammary gland (Campbell and Petersen, 1955) but also the milk ejection reflex (Ely and Petersen, 1941). Although controversy followed regarding the experimentation that led to both of these discoveries (Folley, 1956; Bilger, 1996), Petersen's overall conceptualisation of his findings has withstood the challenge of time. Indeed, it is now clear that a specific sIgA response can be elicited both physiologically and by intraductal vaccination of the mammary gland, particularly during the non-lactating (dry) period (Outteridge *et al.*, 1968). Thus, the enterobronchomammary pathway (Newman, 1995; Xanthou, 1998) provides a physiological means for the passive transfer of specific sIgA protection to the infant. If the mother develops a bronchial or gastrointestinal tract infection, the lymphoid cells from the local lymph nodes provide a sIgA response. The lymphocytes that have been specifically sensitised in the bronchial tree and gut pass into the blood and preferentially locate to the mammary gland. These cells then secrete sIgA into the milk to provide protection for the baby against respiratory and gastrointestinal tract infection. In this connection, the licking of the offspring's anal area, commonly observed in sucking mammals, presumably is designed to prime protection by way of the enteromammary axis, but human attitudes towards hygiene in western countries may limit the effectiveness of this protective mechanism. However, our subjective observations suggest that breast-feeding mothers often exhibit a more casual attitude towards their baby's feces than those who feed their babies formula.

Because of their high rates of cell division and rapid protein turnover, lymphoid tissues are extremely vulnerable to malnutrition (Chandra, 1993). It is widely accepted that protein-energy malnutrition causes atrophy of lymphoid tissue and a loss of lymphoid cells (Moore, 1998). The protective elements in the mammary gland also appear to be affected by maternal nutritional status. Saha *et al.* (1987) found that the total cell counts in colostrum from under-nourished Indian mothers were significantly lower than for well-nourished mothers. Chang (1990) reported that the concentrations of maternal IgG and cord IgG were not significantly different between well-nourished and under-nourished Chinese mothers, whereas the concentration of sIgA in the milk of the undernourished mothers during the first 7 days of lactation was only half that of the well-nourished group. In dairy cows, modification of the frequency of milk removal can temporarily alter the total cell count and permanently change the composition of the cells in milk (Stelwagen and Lacy-Hulbert, 1996). It is not known if the suckling frequency in other species, including humans, influences the total cell count of their milk.

Microminerals and vitamins can also influence both the maternal and infant immune systems. Dietary deficiencies of zinc, iron and magnesium can lead to impairment of antibody production and disrupted neutrophil function (Moore, 1998) whilst copper deficiency has been shown to cause neutropenia (Karimbakas *et al.*, 1998) and impaired resistance to infection (Prohaska and Failla, 1993) in many species. However, the concentrations of these and other minerals (with the exceptions of iodine, fluorine and selenium) in milk are not affected greatly by the nutritional status of the mother (Casey *et al.*, 1995). Likewise, vitamins B6, C and folate can affect various aspects of the immune system, e.g., lymphocyte counts and antibody production (Moore, 1998), but the concentrations of vitamin C and folate in milk appear independent of maternal dietary intake (Casey *et al.*, 1995; Emmett and Rogers, 1997), whereas the concentration of vitamin B6 in milk responds to maternal intake (Harzer and Haschke, 1989; Chang and Kirksey, 1990). In general, milk water-soluble vitamin content is maintained during the development of maternal depletion or during supplementation of mothers judged to be nutritionally adequate, and a linear relationship between milk levels and either degree of deficiency or level of supplementation is apparent only when maternal stores are depleted (Casey *et al.*, 1995). With regard to fat-soluble vitamins, vitamin A deficiency is associated with severe immuno-deficiency (Hanson *et al.*, 1996). However, the effect of maternal vitamin A intake on the concentration of vitamin A in breast milk is unclear (Casey *et al.*, 1995; Emmett and Rogers, 1997). Thus, in many cases, maternal dietary deficiency does not greatly influence the levels of microminerals and vitamins in milk, and a direct effect on the infant's immune system is thereby precluded. However, their deficiency in the maternal diet can lead to impairment of the mother's immune system that in turn could indirectly affect the infant by influencing the passage of specific immunoglobulin protection and leukocytes into milk.

Generally, it is assumed that breast milk contains proteins derived from the mother's diet and, therefore, even exclusively breast-fed infants can come into contact with trophallergens (food-derived allergens) such as cow's milk, egg, wheat and parasite proteins (Moneret-Vautrin, 1994; Koletzko *et al.*, 1998). Indeed, experience has demonstrated that breast-fed infants can have allergic reactions to certain antigens contained in the mother's diet (Minchin, 1985; Lawrence, 1994). However, food allergens are mainly proteins with molecular masses of 15–40 kDa and it is difficult to understand how antigenic amounts of these large molecules could be absorbed by the mother and then gain access to breast milk during established lactation when tight junctions between the lactocytes have closed (Kulski and Hartmann, 1981). Anti-idiotypic antibodies (anti-antibodies) against polio-virus have been detected in human milk (Hahn-Zoric *et al.*, 1993). Therefore, the possibility exists that malfunction of the entero-mammary axis could inadvertently transfer the antigenic determinant of a trophallergen in the form of an anti-antibody into breast milk and thereby stimulate an allergic reaction in the infant. The specific binding assays normally used for the analysis of proteins in milk may not discriminate between trophallergens and anti-antibodies.

Maternal influences on the concentrations of most of the non-specific immune components of milk have not been as extensively investigated as the influences on specific immune components. However, it has been shown that lysozyme and κ-casein contribute to the antibacterial properties of human milk, whilst bile salt dependant lipase hydrolyzes the triglycerides of human milk in the intestine of the newborn and releases free fatty acids with antiprotozoan, antiviral and antimicrobial properties (Hamosh, 1998). In addition, lactoferrin is an iron binding glycoprotein present in the granules of neutrophils and found in high concentrations in human milk (Iyer and Lönnerdal, 1993). Since it was first isolated in 1960, many bacteriostatic and bactericidal actions have been attributed to this protein (Renner *et al.*, 1989; Hamosh, 1998; Xanthou, 1998). However, recent research has focused on the immunomodulatory activity of peptides derived from proteolytic digests of bovine lactoferrin (Miyauchi *et al.*, 1997; Hamosh, 1998). The synthesis of milk proteins appears to be closely coupled, so that their relative concentrations in milk are maintained over a wide range of conditions (Davies *et al.*, 1983). However, the synthesis of lactoferrin is not coupled to that of milk proteins such as α-lactalbumin and casein. Indeed, the *in situ* hybridization studies of Molenaar *et al.* (1992; 1996) show that the expression of lactoferrin messenger RNA is prominent in distended alveoli filled with lipid globules, whereas that of casein and α-lactalbumin is confined to collapsed (filling) alveoli. Since these alveoli tend to be grouped into lobules it is possible that the functional unit for the local regulation of milk synthesis is the lobule rather than the whole gland. The rate of synthesis of lactoferrin appears to be related to the shape of the lactocyte, but the relationship is the inverse of that for casein and α-lactalbumin. Furthermore, the synthesis of lactoferrin was not confined to the lactocytes but also was detected in the cells lining the ducts.

Therefore, it is not surprising that the concentration of lactoferrin in the mammary secretion increases to very high levels at the cessation of breast-feeding (Hartmann and Kulski, 1978).

Carbohydrates have an important role in milk as non-specific defense agents. The disaccharide lactose is found in the milk of all mammals, its concentration ranging from 50–70 g/l in primates, and 30–50 g/l in rodents down to trace amounts in pinnipeds, cetaceans, bear and marsupials (Newburg and Neubauer, 1995). It has been speculated that one of its functions may be to encourage the colonization of the gut by favorable microflora (Newburg and Neubauer, 1995; Newburg, 1997). In species in which lactose is the major osmotic component of milk (e.g., 60–70% of the total osmotic pressure in human milk), lactose provides an osmotic pressure gradient for the passage of water into the Golgi vesicles and thereby into milk to maintain osmotic equilibrium with blood plasma. Therefore, alterations in the rate of synthesis of lactose would be more likely to change milk production rather than milk composition (Peaker, 1977). Indeed, the concentration of lactose in milk shows little response to changes in either the diet or the physiological state of the mother such as severe exercise (Lovelady *et al.*, 1990), parity (Prentice, 1985), delivery by Caesarean section (Kulski *et al.*, 1981) or lactation for more than two years post-partum (Boediman *et al.*, 1979).

Oligosaccharides constitute the other major class of carbohydrate present in milk and are thought to provide protection against infections of the respiratory and digestive tracts by preventing binding of pathogens to target cells (Peterson *et al.*, 1998). Although the concentrations of oligosaccharides vary widely among species, with the highest concentrations (up to 100 g/l) being found in marsupial milk, high concentrations (12–14 g/l) are also found in human milk (Newburg and Neubauer, 1995; McVeagh and Brand Miller, 1997). Indeed, oligosaccharides constitute the third largest solid component in human milk, being present in higher concentrations than protein. The oligosaccharide content of human milk varies four-fold among women at the same stage of lactation and also shows temporal variation (Brand Miller *et al.*, 1994). The oligosaccharides in human milk consist of over a hundred different molecular structures that bear structural homology to cell surface glycoconjugates. Viverge *et al.* (1990) found that women with different blood group types had distinctive patterns for the composition of oligosaccharides in their milk. Subsequently, Newburg and Neubauer (1995) speculated that genetic differences among women in the ability to produce milk oligosaccharides might influence their breast-fed infants' susceptibilities to particular enteric diseases. In this connection it would be of interest to determine if genetic variation in the oligosaccharide composition of breast milk is related to "nursing bottle caries" (a pattern of extensive decay in newly erupted maxillary incisors) in breast-fed infants (Brams and Maloney, 1983).

The lipid in milk is not only very complex, containing thousands of different molecules (Patton and Jensen, 1976), but also is the most variable component of the major milk constituents. The average lipid content of milk varies from 2 g/l in the black rhinoceros (Gregory *et al.*, 1965) to 600 g/l in some phocid seals (Iverson *et al.*, 1993). Furthermore, there can be large differences among individuals of the same species. For example, the lipid content of a group of 16 mothers averaged 40.7 ± 9.6 g/l from one to six months of lactation (Hartmann *et al.*, 1998). However, one mother averaged <30 g/l whereas another averaged >60 g/l over this period.

The lipid content also increases from fore-milk (before a feed) to hind-milk (at the completion of a feed) even in sows (Atwood and Hartmann, 1992), a species that produces milk that does not cream (Whittlestone, 1953). Daly *et al.* (1993a) used a Computerised Breast Measurement system to determine the fullness of the breasts of women after every breast-feed over a 24 h period. They found (Daly *et al.*, 1993b) that the difference in the lipid content of fore-milk and hind-milk was related to the degree of fullness of the breast ($r^2 = 0.69$) rather than to either the amount of milk removed from the breast or the length of the interval between breast-feeds. This finding has important implications for sampling milk from mammals suckling their young, as the average lipid content will vary according to the degree of fullness of the gland when the sample was taken.

In all species studied, 97–99% of the milk lipid is composed of fatty acids esterified into triacylglycerols (Oftedal and Iverson, 1995). Compared with adipose tissue, there is a much greater range in the chain length of the fatty acids in milk. In general, the short- and medium-chain length fatty acids (C4:0 to C14:0) are synthesised *de novo* in the mammary gland and the long-chain fatty acids (>C16:0) are derived preformed from either the diet or adipose tissue stores (Davies *et al.*, 1983). Thompson and Smith (1985) found that the overall contribution of endogenous mammary gland fatty acid synthesis to human milk fat was probably less than 20%. In support of this, Hachey *et al.* (1987) found that, for women in energy balance, medium chain length saturated fatty acids (i.e., those synthesised in the mammary gland) represented approximately 10–12% of the total fatty acids in milk. In addition, the remaining fatty acids (i.e., >C16:0) were derived from the diet (29%), tissue synthesis and adipose stores (59%). Whereas protein and lactose are secreted through the Golgi apparatus, milk lipid is secreted as globules that form at the endoplasmic reticulum of the lactocyte. These globules migrate to the apical membrane of the lactocyte where they are enveloped and then extruded into the lumen of the alveoli. As a result, lipid secretion is not tightly coupled to protein and lactose secretion. For example, dairy cows consuming high-energy low-roughage diets may produce milk with a low lipid content (Davies *et al.*, 1983).

Although intact milk lipids are not active, the products of digestion of milk lipids in the alimentary tract are regarded as important components of the non-specific immune system. Hydrolysis of milk triglycerides is promoted by esterases and lipases in milk as well as in the saliva, gastric and pancreatic secretions of the offspring (Goldman and Goldblum, 1995). Some of the non-esterified fatty acids (C8:0-C12:0 and C18:2) and their monoglycerides released by hydrolysis of the triglycerides are able to disrupt enveloped viruses (Issacs *et al.*, 1986; Thormar *et al.*, 1987) as well as to kill parasitic protozoa (Gillin *et al.*, 1985). Furthermore, glycoproteins present in the lipid globule membrane provide protection against a range of microorganisms (Peterson *et al.*, 1998).

Burr and Burr (1929) observed that rats fed a strictly fat free diet after weaning had impaired growth and developed a scaly tail. It took more than 40 years for the first clinical report of the human equivalent of scaly tail (Collins *et al.*, 1971) and subsequently the fatty acids linoleic (C18:2n6; LA) and α-linolenic (C18:3n3; ALA) were added to the list of essential nutrients for the human diet. Mammals can synthesize palmitic acid (C16:0) from acetyl-CoA, elongate it to stearic acid (C18:0) and then desaturate it to produce oleic acid (C18:1n9). However, mammals, unlike plants, cannot further desaturate oleic acid to produce either LA or ALA. Thus these essential fatty acids (EFA) must be consumed in the diet. Mammals can elongate LA to arachidonic acid (C20:4n6; AA), and ALA to eicosapentaenoic acid (C20:5n3; EPA) and docosahexaenoic acid (C22:6n3, DHA) through a series of common enzymatic reactions. Although the n6 and n3 families of long-chain polyunsaturated fatty acids (LCPUFA) compete for a common elongation pathway, they are not metabolically interconvertible. Therefore the ratio of n6 to n3 EFA in the diet may influence both the rate and the amount of synthesis of each family of LCPUFA.

Miles and Calder (1998) have recently reviewed the effects of the amount and type of fat in the diet on immune cell function and documented the suppression of lymphocyte function by the n3 family of LCPUFA in particular. Perhaps the enhanced vaccine responses and the down-regulation of other immune reactions in breast-fed compared to non-breast-fed infants, e.g., transplant rejection (Hanson *et al.*, 1996) and atopy (Saarinen and Kajosaari, 1995), is attributable partly to modulation of programmed immune responses in the infant by n3 LCPUFA in breast milk.

Compared to bovine milk (Jensen and Newburg, 1995), the lipid in human milk contains a much higher proportion of LCPUFA (Jensen *et al.*, 1995). The proportions of LA, ALA, AA, EPA and DHA in the milk of the two Australian women previously mentioned (see Figure 1) did not change with stage of lactation, even up to 24 months postpartum (Figure 2). In contrast, the 24h production of these fatty acids remained relatively constant to 24 months of lactation for one woman whereas, in the other, production decreased by almost ten-fold from 6–15 months postpartum (Figure 2). Thus the conventional presentation of the fatty acid analysis of milk in terms of a percentage of the total fatty acids can be misleading when related to potential functional outcomes for the infant.

The fatty acid content of milk is particularly sensitive to the fatty acid composition of the diet in monogastric mammals, whereas pre-gastric bio-

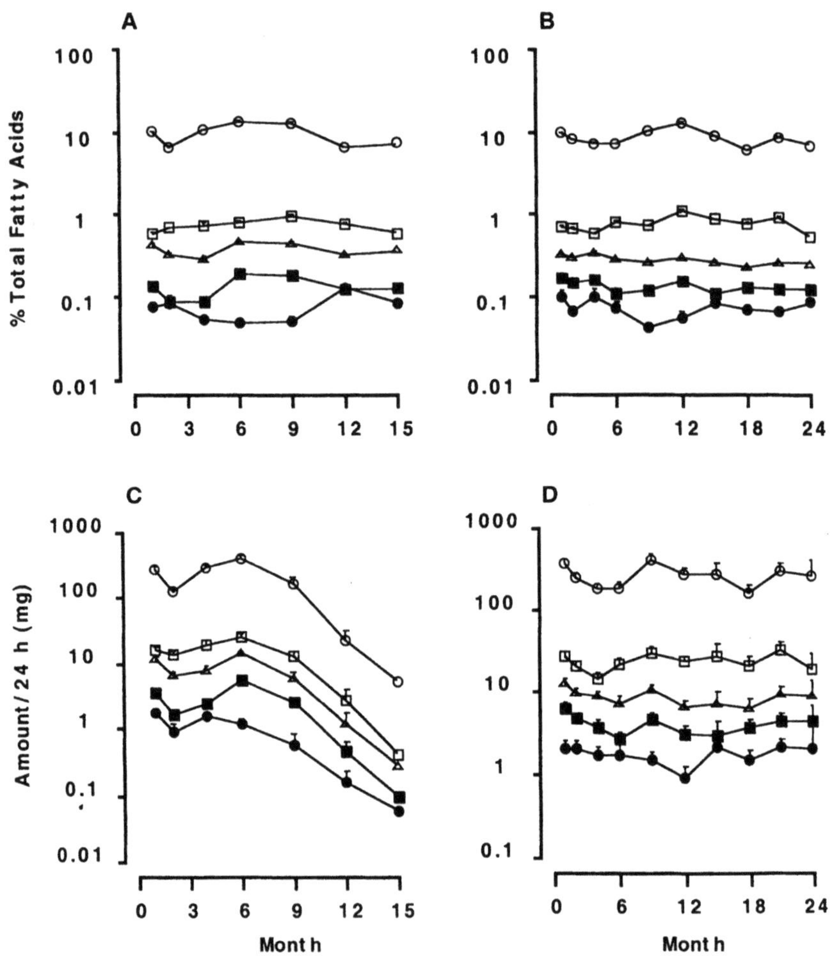

Figure 2. Changes in the percentage of total fatty acids (A and B) and amount (mg) produced per 24h (C and D) of linoleic (○; C18:2n6), α-linolenic (□; C18:3n3), arachidonic (△; C20:4n6), eicosapentaenoic (●; C20:5n3) and docosahexaenoic (■; C22:6n3) acids in milk for Mother 6 over 15 months of lactation (A and C) and Mother 7 over 24 months of lactation (B and D). Values are expressed as mean ± SEM.

hydrogenation of fatty acids by the anaerobic microbial flora of the rumen protects ruminant milk against marked variations due to the fatty acid composition of the diet. In women, the sensitivity of milk fatty acids to diet has been shown using dietary extremes. For example, mothers consuming vegan diets (low in DHA) have been shown to produce milk that is low in DHA (Sanders and Reddy, 1992; Sanders *et al.*, 1978). On the other hand, mothers consuming diets high in fish (high in DHA) produce milk that is high in DHA (Kneebone *et al.*, 1985; Innis and Kuhnlein, 1988). Furthermore, diets high in carbohydrate and low in fat have been shown to increase the proportion of MCFA (C8:0-C14:0) in human milk at the expense of the long-chain fatty acids (Insull *et al.*, 1959; Read *et al.*, 1965). These studies and others (Harris *et al.*, 1984; Makrides *et al.*, 1996) have shown how it is possible to alter specific fatty acids in milk by maternal dietary manipulation. When the oleic acid content of the diet of four lactating women was increased there was a significant decrease in the n6 to n3 ratio (25%, $P = 0.02$) in their milk within a week. Thus, increasing the content of a non-essential, n9 long-chain fatty acid in the diet can alter the n6 to n3 ratio in the milk. Therefore alteration of the fatty acid content of the diet can have both direct and indirect effects on the fatty acid composition of milk lipid, and it is difficult to predict the outcome of dietary manipulations involving different mixtures of fatty acids.

Obviously there is considerable potential for the dietary manipulation of both the quantity and composition of long-chain fatty acids in the milk of women and probably other monogastric mammals. Furthermore, there is a naturally occurring reciprocal relationship between MCFA (C10:0, C12:0 and C14:0) and long-chain fatty acids (C18:0, C18:1n9 and C18:2n6) in women (Daly *et al.*, 1993b). This implies that the anti-microbial and immunomodulatory actions of the fatty acids in breast milk may be counter balanced. Indeed, an inverse relationship is reported between lauric acid (C12:0, an anti-microbial fatty acid) and ALA (an immunomodulatory fatty acid) at one month of lactation (Figure 3). It is not known whether these relationships occur as a consequence of variations in maternal lipid metabolism or to facilitate the development of the infant's immune system. However, it is clear that in addition to maternal antigenic experience, dietary and metabolic factors can alter the lipid composition of breast milk and influence the development of the infants' immune system.

4. Considerations for the Future

Barker (1994) advanced the hypothesis of "the fetal and infant origins of adult disease—metabolic programming". There is now substantial evidence that there is an increased risk of developing a number of common adult diseases if

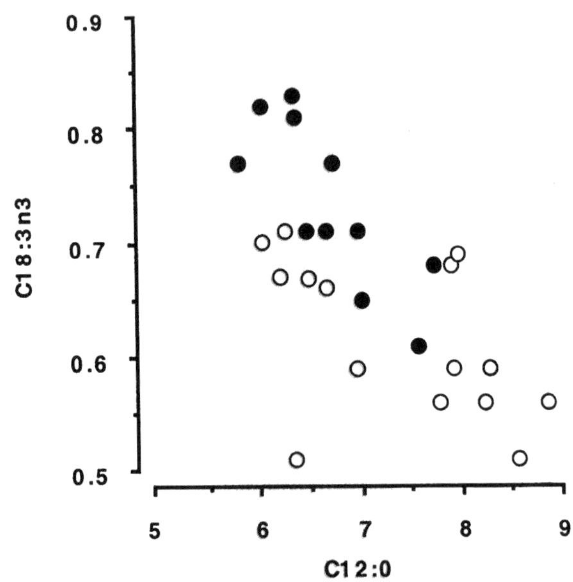

Figure 3. Correlation ($r = -0.56$, $p < 0.01$, $n = 26$) between percentages of lauric acid (C12:0) and α-linolenic acid (C18:3n3) in milk from each breast-feed over a 24 h period at 1 month of lactation for Mother 6 (○) and Mother 7 (●).

prenatal and postnatal development is impaired (Koletzko *et al.*, 1998; Moore, 1998; Lucas, 1998). At the same time, epidemiological evidence (Strachan, 1989) initiated the concept that the immune system also is "programmed for life" during the neonatal period. It is argued that an obsession with cleanliness and hygiene in infancy may result in a predominance of Th2 over Th1 cells and an increased risk of atopic diseases such as asthma, whereas a "mucky" environment may produce a more favorable long term balance between the Th2 and Th1 cells and decrease the risk of these diseases (Hamilton, 1998).

Living in a hygienic environment, the breast-fed, in contrast to the non-breast-fed, infant acquires types of long-term immunological attributes (enhanced vaccine responses, down-regulation of atopy and transplant rejection) similar to those associated with children living in "mucky" environments. Furthermore, the chances of survival over the first few months of life in "mucky" environments are greatly reduced in the absence of exclusive breast-feeding (Hanson *et al.*, 1996). Obviously, the influence of breast-feeding in optimising the long-term development of the immune system requires further consideration. Since, physiological and dietary factors can modulate the specific and non-specific immune factors in milk, controlling the infant's immune development through breast milk offers a physiological alternative to the proposed "mucky" vaccination of all infants born into hygienic environments. However, the subtle interactions between maternal dietary factors, such as LCPUFA, vitamins and minerals, and the mother's ability to produce milk that promotes the optimal development of her neonate's immature immune system requires further research. These subtle interactions between the mother and her offspring need to be identified so that the development of the immune system is programmed to maximize survival in the adult nutritional and pathological environment. Mother's milk could then be considered to be "the gold standard functional food".

A consideration of the maternal modulation of specific and non-specific immune components of colostrum and mature milk reinforces the reasoning behind Carl Linnaeus' choice of the term Mammalia in 1758. He chose Mammalia from a number of other suitable terms because he wanted to emphasize that this class of animals evolved mammary glands so that they could feed their *own* young (Schiebinger, 1993). Linnaeus supported breast-feeding and opposed wet nursing, then commonly practiced by Swedish parents of higher socioeconomic status.

ACKNOWLEDGMENTS. The authors would like to acknowledge the assistance of Miss Christine Chan and the support of the NH & MRC, Lotteries Commission of Western Australia and the Women and Infants Research Foundation. Post-graduate scholarships were provided by Medela Inc. USA (MDC) and GRDC/Meadow Lea Foods Ltd. (LRM). All studies conducted by the authors were approved by the Committee for Human Rights of The University of Western Australia.

References

Adlerberth, I., Carlsson, B., de Man, P., Jalil, F., Khan, S.R., Larsson, P., Mellander, L., Svanborg, C., Wolf, A.E., and Hanson, L.A. 1991. Intestinal colonization with Enterobacteriaceae in Pakistani and Swedish hospital-delivered infants. *Acta Paediatr. Scand.* **80(6–7)**:602.

American Academy of Pediatrics. 1997. Breastfeeding and the use of human milk. *Pediatrics.* **100(6)**:1035.

Arthur, P.G., Kent, J.C., and Hartmann, P.E. 1994. Metabolites of lactose synthesis in milk from diabetic and nondiabetic women during lactogenesis II. *J. Pediatr. Gastroenterol. Nutr.* **19(1)**:100.

Atkinson, S.A. 1995. Determinants of milk volume and composition: E. Effects of gestational stage at delivery on human milk components, in: *Handbook of Milk Composition* (R.G. Jensen, ed.), p. 222, Academic Press, San Diego.

Atwood, C.S. and Hartmann, P.E. 1992. Collection of fore- and hind-milk from the sow. *J. Dairy Res.* **59**:287.

Barker, D.J.P. 1994. *Mothers, Babies and Diseases in Later Life.* BMJ Publishing Group, London.
Bilger, B. 1996. Beyond the pail. *The Sciences.* **Nov./Dec.**:16.
Boediman, D., Ismail, D., Iman, S., Ismangoen, and Ismadi, S.D. 1979. Composition of breast milk beyond one year. *J. Trop. Pediatr. Environ. Child Health.* **25(4)**:107.
Brams, M. and Maloney, J. 1983. "Nursing-bottle caries" in breast-fed children. *J. Pediatr.* **103(3)**:415.
Brand Miller, J.C., Miller, J.J., McVeagh, P., and Bull, S. 1994. The oligosaccharide composition of human milk: Temporal and individual variation. *J. Pediatr. Gastroenterol.* **19**:371.
Brandon, M.R., Husband, A.J., and Lascelles, A.K. 1975. The effect of glucocorticoid on immunoglobulin secretion into colostrum in cows. *Aust. J. Exp. Biol. Med. Sci.* **53(1)**:43.
Brodribb, W. 1997. *Breastfeeding Management in Australia.* Merrily Merrily Enterprises Pty. Ltd., Melbourne.
Burr, G.O. and Burr, A.M. 1929. A new deficiency disease produced by the rigid exclusion of fat from the diet. *J. Biol. Chem.* **82**:345.
Campbell, B. and Petersen, W.E. 1955. Use of protective principles in milk and colostrum in prevention of disease in man and animals. *Lancet* **75**:494.
Casey, C.E., Smith, A., and Zhang, P. 1995. Minerals, ions and trace elements in milk: C. Microminerals in human and animal milks, in: *Handbook of Milk Composition* (R.G. Jensen, ed.), p. 622, Academic Press, San Diego.
Chandra, R.K. 1993. Nutrition and the immune system. *Proc. Nutr. Soc.* **52**:77.
Chang, S.-J. 1990. Antimicrobial proteins of maternal and cord sera and human milk in relation to maternal nutritional status. *Am. J. Clin. Nutr.* **51(2)**:183.
Chang, S.J. and Kirksey, A. 1990. Pyroxidine supplementation of lactating mothers: Relation to maternal nutritional status and vitamin B-6 concentrations in milk. *Am. J. Clin. Nutr.* **51(5)**:826.
Coghlan, A. 1999. Milk hormone faces growing opposition. *New Scient.* **2170**:13.
Collier, R.J. 1997. Strategies for POSILAC® use in dairy herds, in: *Milk Composition, Production and Biotechnology* (R.A.S. Welch, D.J.W. Burns, S.R. Davis, A.I. Popay, and C.G. Prosser, eds.), p. 295, CAB Intetnational, New York.
Collins, F.D., Sinclair, A.J., Royle, J.P., Coats, D.A., Maynard, A.T., and Leonard, R.F. 1971. Plasma lipids in human linoleic acid deficiency. *Nutr. Metab.* **13(3)**:150.
Cowie, A.T., Forsyth, I.A., and Hart, I.C. 1980. Hormonal control of lactation. *Monogr. Endocrinol.* **15(I–XIV)**:1.
Cox, D.B., Owens, R.A., and Hartmann, P.E. 1996. Blood and milk prolactin and the rate of milk synthesis in women. *Exp. Physiol.* **81**:1007.
Daly, S.E.J., DiRosso, A., Owens, R.A., and Hartmann, P.E. 1993a. Degree of breast emptying explains changes in fat content, but not fatty acid composition, of human milk. *Exp. Physiol.* **78**:741.
Daly, S.E.J., Owens, R.A., and Hartmann, P.E. 1993b. The short-term synthesis and infant-regulated removal of milk in lactating women. *Exp. Physiol.* **78**:209.
Davies, D.T., Holt, C., and Christie, W.W. 1983. The composition of milk, in: *Biochemistry of Milk* (T.B. Mepham, ed.), p. 71, Elsevier, Amsterdam.
De Passillé, A.M.B., Rushen, J., Foxcroft, G.R., Aherne, F.X., and Schaefer, A. 1993. Performance of young pigs: Relationships and periparturient progesterone, prolactin, insulin of sows. *J. Anim. Sci.* **71**:179.

Dettwyler, K.A. 1995. Time to wean: The hominid blueprint for the natural age of weaning in modern human populations, in: *Breastfeeding: Biocultural perspectives* (P. Stuart-Macadam and K.A. Dettwyler, eds.), p. 39, Aldine De Gruyter, New York.

Dewey, K.G., Heinig, M.J., Nommsen, L.A., and Lönnerdal, B. 1991. Maternal versus infant factors related to breast milk intake and residual milk volume: the DARLING study. *Pediatrics* **87(6)**:829.

Duchmann, R., Kaiser, I., Hermann, E., Mayet, W., Ewe, K., and Meyer zum Buschenfelde, K.H. 1995. Tolerance exists towards resident intestinal flora but is broken in active inflammatory bowel disease (IBD). *Clin. Exp. Immunol.* **102(3)**:445.

Ely, F. and Peterson, W.E. 1941. Factors involved in the ejection of milk. *J. Dairy Sci.* **24**:211.

Emmett, P.M. and Rogers, I.S. 1997. Properties of human milk and their relationship with maternal nutrition. *Early Human Dev.* **49**:7S.

Fetherston, C.M., Lee, C.S., and Hartmann, P.E. 2001. Mammary gland defence: Implications for the prevention and treatment of mastitis during lactation in women, in: *Advances in Nutritional Research, Vol 10: The Immunological Properties of Milk* (B. Woodward and H.H. Draper, eds.), pp. 167–198, Plenum Press, New York.

Florin, T.H.J. 1997. Food, microbial ecology and mucosal immunity. *Proc. Nutr. Soc. Aust.* **21**:14.

Folley, S.J. 1956. *The Physiology and Biochemisrty of Lactation*, Oliver and Boyd, Edinburgh.

Gillen, F.D., Reiner, D.S., and Gault, M.J. 1985. Cholate-dependent killing of Giardia Lamblia by human milk. *Infect. Immun.* **47(3)**:619.

Goldman, A.S., Chheda, S., and Garofalo R. 1998. Evolution of immunologic functions of the mammary gland and the postnatal development of immunity. *Pediatr. Res.* **43(2)**:155.

Goldman, A.S. and Goldblum, R.M. 1995. Defense agents in milk: A. defense agents in human milk, in: *Handbook of Milk Composition* (R.G. Jensen, ed.), p. 727, Academic Press, San Diego.

Green, B. 1984. Composition of milk and energetics of growth in marsupials. *Symp. Zool. Soc. London.* **51**:369.

Gregory, M.E., Rowland, S.J., Thompson, S.Y., and Kon, V.M. 1965. Changes during lactation in the composition of the milk of the African black rhinoceros (*Diceros bicornis*). *Proc. Zool. Soc. London.* **145**:327.

Guy, M.A., McFadden, T.B., Cockrell, D.C., and Besser, T.E. 1994. Effects of unilateral prepartum milking on concentrations of immunoglobulin G_1 and prolactin in colostrum. *J. Dairy Sci.* **77**:3584.

Hachey, D.L., Thomas, M.R., Emken, E.A., Garza, C., Brown-Booth, L., Adolf, R.O., and Klein, P.D. 1987. Human lactation: Maternal transfer of dietary triglycerides labelled with stable isotopes. *J. Lipid Res.* **28**:1185.

Hahn-Zoric, M., Carlsson, B., Jeansson, S., Ekre, H.P., Osterhaus, A.D., Roberton, D., and Hanson, L.Å. 1993. Anti-idiotypic antibodies to poliovirus antibodies in commercial immunoglobulin preparations, human serum, and milk. *Pediatr. Res.* **33(5)**:475.

Hamilton, G. 1998. Let them eat dirt. *New Scient.* **2143**:26.

Hamosh, M. 1998. Protective function of proteins and lipids in human milk. *Biol. Neonate.* **74**:163.

Hanson, L.Å., Wiedermann, U., Ashraf, R., Zaman, S., Adlerberth, I., Dahlgren, U., Wold, A., and Jalil, F. 1996. Effects of breastfeeding on the baby and on its immune system. *Food Nutr. Bull.* **17(4)**:384.

Harris, W.S., Connor, W.E., and Lindsey, S. 1984. Will dietary ?-3 fatty acids change the composition of milk? *Am. J. Clin. Nutr.* **40**:780.

Hartmann, P.E. 1973. Changes in the composition and yield of the mammary secretion of cows during the initiation of lactation. *J. Endocrinol.* **59**:231.

Hartmann, P.E., Atwood, C.S., Cox, D.B., and Daly, S.E.J. 1995a. Endocrine and autocrine strategies for the control of lactation in women and sows, in: *Intercellular signalling in the mammary gland* (C.J. Wilde, M. Peaker, and C.H. Knight, eds.), p. 203, Plenum Press, New York.

Hartmann, P.E. and Kulski, J.K. 1978. Changes in the composition of the mammary secretion of women after abrupt termination of breast feeding. *J. Physiol.* **275**:1.

Hartmann, P.E., Rattigan, S., Saint, L., and Supriyana, O. 1985. Variation in the yield and composition of human milk. *Oxf. Rev. Reprod. Biol.* **7**:118.

Hartmann, P.E., Sherriff, J.L., and Kent, J.C. 1995b. Maternal nutrition and milk synthesis. *Proc. Nutr. Soc.* **54**:379.

Hartmann, P.E., Sherriff, J.L., and Mitoulas, L.R. 1998. Homeostatic mechanisms that regulate lactation during energetic stress. *J. Nutr.* **128**:394S.

Hartmann, P.E., Smith, N.A., Thompson, M.J., Wakeford, C.M., and Arthur, P.G. 1997. The lactation cycle in the sow: Physiological and management contradictions. *Livest. Prod. Sci.* **50**:75.

Harzer, G. and Haschke, F. 1989. Micronutrients in human milk, in: *Micronutrients in Milk and Milk-based Food Products* (E. Renner, ed.), p. 125, Elsevier Science Publishers, New York.

Heesom, K.J., Souza, P.F.A., Ilic, V., and Williamson, D.H. 1992. Chain-length dependency of interactions of medium-chain fatty acids with glucose metabolism in acini isolated from lactating rat mammary glands. *Biochem. J.* **281**:273.

Innis, S.M. and Kuhnlein, H.V. 1988. Long chain *n*-3 fatty acids in breast milk of Inuit women consuming traditional foods. *Early Hum. Dev.* **18**:185.

Insull, Jr., W., Hirsch, J., James, T., and Ahrens, Jr., E.H. 1959. The fatty acids of human milk. II. Alterations produced by manipulation of caloric balance and exchange of dietary fats. *J. Clin. Invest.* **38(1)**:443.

Isaacs, C.E., Thormar, H., and Pessolano, T. 1986. Membrane disruptive effect of human milk: Inactivation of enveloped viruses. *J. Infect. Dis.* **154**:966.

Iverson, S.J., Bowen, W.D., Boness, D.J., and Oftedal, O.T. 1993. The effect of maternal size and milk energy output on pup growth in grey seals (*Halichoerus grypus*). *Physiol. Zool.* **66**:61.

Iyer, S. and Lönnerdal, B. 1993. Lactoferrin, lactoferrin receptors and iron metabolism. *Eur. J. Clin. Nutr.* **47(4)**:232.

Jensen, R.G. 1995. Determinants of milk volume and composition: F. Miscellaneous factors affecting composition and volume of human and bovine milks, in: *Handbook of Milk Composition* (R.G. Jensen, ed.), p. 237, Academic Press, San Diego.

Jensen, R.G., Bitman, J., Carlson, S.E., Couch, S.C., Hamosh, M., and Newburg, D.S. 1995. Milk lipids: A. Human milk lipids, in: *Handbook of Milk Composition* (R.G. Jensen, ed.), p. 495, Academic Press, San Diego.

Jensen, R.G. and Newburg, D.S. 1995. Milk lipids: B. Bovine milk lipids, in: *Handbook of Milk Composition* (R.G. Jensen, ed.), p. 543, Academic Press, San Diego.

Jobe, A.H., Newnham, J., Willet, K., Sly, P., and Ikegamim. 1998. Fetal versus maternal and gestational age effects of repetitive antenatal glucocorticoids. *Pediatrics* **102(5)**:1116.

Karimbakas, J., Langkamp-Henken, B., and Percival, S.S. 1998. Arrested maturation of granulocytes in copper-deficient mice. *J. Nutr.* **128**:1855.

Kent, J.C., Mitoulas, L., Cox, D.B., Owens, R.A., and Hartmann, P.E. 1999. Breast volume and milk production during extended lactation in women. *Exp. Physiol.* **84(2)**:435.

Kneebone, G.M., Kneebone, R., and Gibson, R.A. 1985. Fatty acid composition of breast milk from three racial groups from Penang, Malaysia. *Am. J. Clin. Nutr.* **41**:765.

Koletzko, B., Aggett, P.J., Bindels, J.G., Bung, P., Ferré, P., Gil, A., Lentze, M.J., Roberfroid, M., and Strobel, S. 1998. Growth, development and differentiation: A functional food science approach. *Br. J. Nutr.* **80**:5S.

Konner, M. and Worthman, C. 1980. Nursing frequency, gonadal function, and birth spacing among ! Kung hunter-gatherers. *Science*. **207**:788.

Kulski, J.K. and Hartmann, P.E. 1981. Changes in human milk composition during the initiation of lactation. *Aust. J. Expl. Biol. Med. Sci.* **59**:101.

Kulski, J.K., Hartmann, P.E., Martin, J.D., and Smith, M. 1978. Effects of bromocriptine mesylate on the composition of the mammary secretion in non-breast feeding women. *Obstet. Gynaecol.* **52**:38.

Kulski, J.K., Smith, M., and Hartmann, P.E. 1981. Normal and caesarian section delivery and the initiation of lactation in women. *Aust. J. Exp. Biol. Med. Sci.* **59**:405.

Lawrence, R.A. 1994. *Breastfeeding: A Guide to the Medical Profession.* Mosby, St. Louis.

Linzell, J.L. and Peaker, M. 1971. The effects of oxytocin and milk removal on milk secretion in the goat. *J. Physiol.* **216**:717.

Lovelady, C.A., Lönnerdal, B., and Dewey, K.G. 1990. Lactation performance of exercising women. *Am. J. Clin. Nutr.* **52(1)**:103.

Lucas, A. 1998. Programming by early nutrition: An experimental approach. *J. Nutr.* **128**:401S.

Lyons, W.R. 1958. Hormonal synergism in mammary growth. *Proc. Royal Soc. (Lond)* **149**:303.

Makrides, M., Neumann, M.A., and Gibson, R.A. 1996. Effect of maternal docosahexaenoic acid (DHA) supplementation on breast milk composition. *Eur. J. Clin. Nutr.* **50**:352.

McFadden, T.B., Besser, T.B., and Barrington, G.M. 1997. Regulation of immunoglobulin transfer into mammary secretions of ruminants, in: *Milk Composition, Production and Biotechnology* (R.A.S. Welch, D.J.W. Burns, S.R. Davis, A.I. Popay, and C.G. Prosser, eds.), p. 133, CAB Intetnational, New York.

McVeagh, P. and Brand Miller, J. 1997. Human milk oligosaccharides: only the breast. *J. Paediatr. Child Health.* **33**:281.

Miles, E.A. and Calder, P.C. 1998. Modulation of immune function by dietary fatty acids. *Proc. Nutr. Soc.* **57**:277.

Minchin, M.K. 1985. *Breastfeeding Matters: What We Need To Know About Infant Feeding.* Alma Publications, Sydney.

Miyauchi, H., Kaino, A., Shinoda, I., Fukuwatari, Y., and Hayasawa, H. 1997. Immunomodulatory effect of bovine lactoferrin pepsin hydrolysate on murine splenocytes and Peyer's patch cells. *J. Dairy Sci.* **80(10)**:2330.

Molenaar, A.J., Davis, S.R., and Wilkins, R.J. 1992. Expression of ?-lactalbumin, and lactoferrin genes is heterogeneous in sheep and cattle mammary tissue. *J. Histochem. Cytochem.* **40**:611.

Molenaar, A.J., Kuys, Y.M., Davis, S.R., Wilkins, R.J., Mead, P.E., and Tweedie, J.W. 1996. Elevation of lactoferrin gene expression in developing, ductal, resting, and regressing parenchymal epithelium of the ruminant mammary gland. *J. Dairy Sci.* **79**:1198.

Moneret-Vautrin, A.D. 1994. Food allergy and intolerance in infancy, in: *Infant Nutrition* (A.F. Walker and B.A. Rolls, eds.), Chapman and Hall, London.

Moore, D.M., Vogl, A.W., Baimbridge, K., and Emerman, J.T. 1987. Effect of calcium on oxytocin-induced contraction of mammary gland myoepithelium as visualized by NBD-phallicidin. *J. Cell Sci.* **88**:563.

Moore, S.E. 1998. Nutrition, immunity and the fetal and infant origins of disease hypothesis in developing countries. *Proc. Nutr. Soc.* **57**:241.

Muller, L.D. and Ellinger, D.K 1981. Colostral immunoglobulin concentrations among breeds of dairy cattle. *J. Dairy Sci.* **64(8)**:1727.

National Health and Medical Research Council 1996. *Infant Feeding Guidelines for Health Workers*. Australian Government Publishing Service, Canberra.

Neifert, M.R., McDonough, S.L., and Neville, M.C. 1981. Failure of lactogenesis associated with placental retention. *Am. J. Obstet. Gynecol.* **140(4)**:477.

Neubauer, S.H., Ferris, A.M., Chase C.G., Fanelli, J., Thompson, C.A., Lammi-Keefe, C.J., Clark, R.M., Jensen, R.G., Bendel, R.B., and Green, K.W. 1993. Delayed lactogenesis in women with insulin-dependant diabetes mellitus. *Am. J. Clin. Nutr.* **58(1)**:54.

Newburg, D.S. 1997. Do the binding properties of oligosaccharides in milk protect human infants from gastrointestinal bacteria? *J. Nutr.* **127(5)**:980S.

Newberg, D.S. and Neubauer, S.H. 1995. Carbohydrates in milks: analysis, quantities, and significance, in: *Handbook of Milk Composition* (R.G. Jensen, ed.), p. 273, Academic Press, San Diego.

Newman, J. 1995. How breast milk protects newborns. *Sci. Am.* **273(6)**:58.

Nicholas, K., Simpson, K., Wilson, M., Trott, J., and Shaw, D. 1997. The Tammar Wallaby: A model to study putative autocrine-induced changes in milk composition. *J. Mam. Gland Biol. Neoplasia.* **1(3)**:285.

Oftedal, O.T. 1993. The adaptation of milk secretion to the constraints of fasting in bears, seals and baleen whales. *J. Dairy Sci.* **76**:3234.

Oftedal, O.T. and Iverson, S.J. 1995. Comparative analysis of nonhuman milks: A. Phylogenetic variation in the gross composition of milks, in: *Handbook of Milk Composition* (R.G. Jensen, ed.), p. 749, Academic Press, San Diego.

Outteridge, P.M., Williams, R.C., and Lascelles, A.K. 1968. Local immunity on the mammary gland following the infusion of a staphylococcal cell-toxoid vaccine. *Res. Vet. Sci.* **9(5)**:416.

Patton, S. and Jensen, R.G. 1976. *Biomedical Aspects of Lactation.* Pergamon Press, Oxford.

Peaker, M. 1977. Mechanisms of milk secretion: milk composition in relation to potential difference across the mammary epithelium. *J. Physiol.* **270**:489.

Peaker, M. and Wilde, C.J. 1987. Milk secretion: Autocrine control. *News Physiol. Sci.* **2**:124.

Peterson, J.A., Patton, S., and Hamosh, M. 1998. Glycoproteins of the human milk fat globule in the protection of the breast-fed infant against infections. *Biol. Neonate.* **74**:134.

Petrie, L., Acres, L.S., and McCarteny, D.S. 1984. The yield of colostrum and colostral gammaglobulins in beef cows and the absorption of colostral gammaglobulins by beef calves. *Can. Vet. J.* **25**:273.

Prentice, A. 1985. The influence of maternal parity on breast-milk composition, in: *Composition and Physiological Properties of Human Milk* (J. Schaub, ed.), p. 309, Elsevier, Amsterdam.

Prentice, A.M., Paul, A., Prentice, A., Black, A., Cole, T., and Whitehead, R.G. 1986. Cross-cultural differences in lactational performance, in: *Human Lactation 2: Maternal and Environmental Factors* (M. Hamosh and A.S. Goldman, eds.), p. 13, Plenum, Press, New York.

Prohaska, J.R. and Failla, M.L. 1993. Copper and Immunity, in: *Human Lactation—A Comprehensive Treatise, Volume 8: Nutrition and Immunology* (D.M. Klurfield, ed.), p. 309, Plenum Press, New York.

Prosser, C.G. and Hartmann, P.E. 1983. Comparison of mammary gland function during the ovulatory menstrual cycle and acute breast inflammation in women. *Aust. J. Exp. Biol. Med. Sci.* **61**:277.

Read, W.W.C., Lutz, P.G., and Tashjian, A. 1965. Human milk lipids: II. The influence of dietary carbohydrates and fat on the fatty acids of mature milk. A study in four ethnic groups. *Am. J. Clin. Nutr.* **17**:180.

Renner, E., Schaafsma, G., and Scott, K.J. 1989. Micronutrients in milk, in: *Micronutrients in Milk and Milk-based Food Products* (E. Renner, ed.), p. 1, Elsevier Science Publishers, New York.

Riordan, J. and Auerbach, K.G. 1993. *Breastfeeding and Human Lactation*. Jones and Bartlett, Boston.

Saarinen, U.M. and Kajosaari, M. 1995. Breastfeeding as prophylaxis against atopic disease: Prospective follow-up study until 17 years old. *Lancet* **346**:1065.

Saha, K., Garg, M., Rao, K.N., Thirupuram, S., and Gipta, M.M. 1987. Lymphocyte subsets in human colostrum with special reference to the of undernourished mother. *J. Trop. Pediatr.* **33**:329.

Sanders, T.A.B., Ellis, F.R., and Dickerson, J.W.T. 1978. Studies of vegans: The fatty acid composition of plasma choline phosphoglycerides, erythrocytes, adipose tissue and breast milk, and some indicators of susceptibility to ischemic heart disease in vegans and omnivore controls. *Am. J. Clin. Nutr.* **31**:805.

Sanders, T.A.B. and Reddy, S. 1992. The influence of a vegetarian diet on the fatty acid composition of human milk and the essential fatty acid status of the infant. *J. Pediatr.* **120**:71S.

Schiebinger, L. 1993. Why mammals are called mammals: Gender politics in eighteenth-century natural history. *Am. Hist. Rev.* **98**:382.

Stelwagen, K. and Lacy-Hulbert, S.J. 1996. Effect of milking frequency on milk somatic cell count characteristics and mammary secretory cell damage in cows. *Am. J. Vet. Res.* **57(3)**:902.

Strachan, D.P. 1989. Hay fever, hygiene and household size. *Br. Med. J.* **299**:1259.

Streuli, C.H. and Edwards, G.M. 1998. Control of normal mammary epithelial phenotype by integrins. *J. Mammary Gland Biol. Neoplasia* **3(2)**:151.

Thomson, D. 1983. *Children of the Wilderness*. Gordon and Gotch, Melbourne.

Thompson, B.J. and Smith, S. 1985. Biosynthesis of fatty acids by lactating human breast epithelial cells: An evaluation of the contribution to the overall composition of human milk fat. *Pediatr. Res.* **19(1)**:139.

Thompson, M.J., Arthur, P.G., and Hartmann, P.E. 1997. Bioluminescent methods for determining metabolites in micro-samples of pig plasma. *Eur. J. Clin. Chem. Clin. Biochem.* **35(11)**:861.

Thormar, H., Isaacs, C.E., Brown, H.R., Barshatzky, M.R., and Pessolano, T. 1987. Inactivation of enveloped viruses and killing of cells by fatty acids and monoglycerides. *Antimicrob. Agents Chemother.* **31(1)**:27.

Viverge, D., Grimmonprez, L., Cassanas, G., Bardet, L., and Solere, M. 1990. Discriminant carbohydrate components of human milk according to donor secretor types. *J. Pediatr. Gastroenterol. Nutr.* **11(3)**:365.

Welch, R.A.S., Burns, D.J.W., Davis, S.R., Popay, I.A., and Prosser, C.G. 1997. *Milk Composition, Production and Biotechnology*. CAB International, New York.

Whittlestone, W.G. 1953. Variation in the fat content of milk throughout the milking process. *J. Dairy Res.* **20**:146.

Wilde, C.J., Addey, C.V.P., Boddy, L.M., and Peaker, M. 1995. Autocrine regulation of milk secretion by a protein in milk. *Biochem. J.* **305**:51.

Xanthou, M. 1998. Immune protection of human milk. *Biol. Neonate.* **74**:121.

Chapter 19

Milk Banking: The Influence of Storage Procedures and Subsequent Processing on Immunologic Components of Human Milk

Ruth A. Lawrence

1. Introduction

The tremendous value of human milk for the newborn infant and especially for the prematurely born infant is well established (Lawrence, 1999). For centuries, the problem of providing human milk for infants whose mothers cannot provide it or for infants whose mothers can provide it but are unable to feed their infant at the breast has challenged technology. The historic method of solving the problem was to have another woman who was lactating nurse the infant. This was originally called "wet nursing" and is now called "cross nursing" (Lawrence, 1999a). This approach was abandoned decades ago as fewer and fewer women were lactating and formulas became available. It is now considered obsolete and sometimes dangerous because of the risk of transmitting an undiagnosed infectious disease to the suckling surrogate infant in spite of the fact that milk directly from the breast is the ideal method to maintain 100% of the constituents.

Early experimentation to make human milk readily available and to improve the shelf life of human milk began with methods of evaporation and drying it to replace wet nursing (Emerson, 1922). Talbot and others abandoned efforts to maintain a registry of acceptable wet nurses (Talbot, 1928). A system was developed to collect milk from lactating women in nursing bottles which were picked

Ruth A. Lawrence • University of Rochester School of Medicine and Dentistry, Rochester, New York 14642 up daily by public health nurses. It was sterilized in a hot water bath for at least 20 minutes and refrigerated prior to use. Donors were medically screened and paid a small fee for their milk (Emerson, 1933). In 1927, the process was changed. Women came to the milk bank and pumped their milk under supervision. The volume of milk collected increased by 94% in the first year of this new procedure (Jones, 1928), and the system of processing and storing human milk gradually replaced wet nursing. As refrigeration became possible, the storage of human milk provided by donor mothers resulted in the development of milk banking which has existed around the world since the 1920's (Chapin, 1923). Freezing then became practical and extended the shelf life of milk by months (Emerson, 1933).

Milk banks now are dwindling because the requirements for preparation and storage are very demanding and expensive. Milk banks have worked to overcome criticism of possible contamination and alteration of milk constituents during processing, and universal standards have evolved (Human Milk Banking Association of North America, 1996). This chapter reviews the impact of various steps in the process of collection, storage and preparation of human milk for consumption by infants and especially by infants who are ill or premature. The focus is on the immunological components of human milk.

2. Scope

The regulations controlling the preparation of pumped human milk are different for milk provided for a mother's own infant than for milk intended for an unrelated infant. In the hospital setting, a mother may provide her own milk, and its preparation and storage is not regulated by the hospital. The requirements are designed, however, for donor milk which means milk collected from someone who is not the infant's biologic mother (Human Milk Banking Association of North America, 1996). The factors which are important to the stability of the milk sample include the method of collection, the collection container and the storage container if the milk is transferred. Storage temperature and the process for sterilization or pasteurization have the greatest impact.

The immunologic constituents of human milk that are considered here include living cells and antibodies together with other immunologic proteins as well as the anti-inflammatory factors (Goldman, 1993). These constituents, summarized in Table 1, have been carefully described in previous chapters. Not all of the factors relevant to milk immunology have been studied in relation to collection and storage processes but this discussion includes all of the available work. Additional processes that affect the ultimate immunologic benefits to the recipient infant include the processes of thawing and warming and the process of administration by cup, bottle, or feeding tube (Lawrence, 1999b).

Table 1. Antimicrobial agents and immunomodulating factors in human milk

Constituent	Primary function
Proteins	
Lactoferrin	Fe^{++} chelation
Lysozyme	Degrade peptidoglycans
Fibronectin	Opsonins
Secretory IgA	Antigen binding
Mucin	Fragments are opsonins
C_3	Anti-rota virus
Oligosaccharides	Receptor analogues
Lipids	Disrupt enveloped viruses
Cytokines	
Interleukin-1-beta	Activates T cells
Interleukin 6	Enhances IgA production
TNF-alpha	Enhances SC production
TGF-beta	Enhances iso type switching to $IgA^{+}B$ cells

Adapted from Goldman (1993).
Lawrence (1999b). Reproduced with permission.

3. Procedures and Containers for Collection and Storage

The infant is the most effective pump. All other collection procedures are subject to variability. Drip milk collected while the infant suckles at the other breast or collected in a nesty cup, a device worn over the breast inside the brassiere between feedings, is very low in fat and significantly contaminated (Lawrence, 1999b). Thus, drip milk is never recommended for feeding an infant and, indeed, is not suitable for use in milk assays. Manual expression, a process by which the mother expresses milk with her hands, is very successful for some women. This milk is surprisingly uncontaminated if the mother is careful to wash her hands and her breast. Some women find it very difficult and are able to express only a few milliliters. Very few women are able to express milk with the fat content of milk obtained by an actively suckling infant or a good electric pump by manual expression because they cannot produce sufficient stimulus to the alveolus by hand.

A number of devices have been developed over the years to extract milk from the breast. There is a small hand pump, referred to as the bicycle horn pump (original design circa 1830) which can be found in the Corning Glass Medical Museum in Corning, New York (Lawrence, 1999a). This device looks like a bicycle horn with a rigid phalange that fits over the nipple and areola. There is room for only a few milliliters of milk in the well of this pump and the suction bulb that develops the negative pressure is quickly contaminated. This device is not recommended because of the contamination and the low fat content of the milk obtained. There are other hand pump models which avoid contamination by locating the collecting device significantly below the breast with no chance of the milk refluxing back over the breast or into the suction bulb. These hand devices work well in a fully lactating woman especially if she uses it while the baby is nursing on the opposite breast. They are useful for pumping milk for one's own infant, for example, in the situation of a working woman who pumps once or twice a day while at work. They are usually not satisfactory when pumping to donate milk or when the infant cannot breast-feed (Franz, 1994).

The ideal artificial collection agent is an electric pump that cycles the negative pressure with a rhythmic action simulating suckling, which provides good flow and good fat content. The breast phalange and tubing must, at a minimum, be sterilized in a dishwasher to prevent bacterial contamination, if they are not sterilized by boiling. Electric pumps vary in design and each one has special features, but for the purpose of this discussion they will be considered together.

The storage container that the milk is placed in can influence the composition of milk to some degree (Goldblum *et al.*, 1981). The total cell content is diminished as the cells, especially mononuclear phagocytes, adhere to the walls of a Pyrex glass container. It has been observed, however, that if the milk stands in the Pyrex container under refrigeration for eight hours or longer, the cells will again go into suspension and the count returns to normal. In contrast, milk cells have been reported not to cling to the walls of polyethylene or polypropylene containers (Goldblum *et al.*, 1981). The advantages of a glass container are that glass is rigid and does not ordinarily leak. It also tolerates heating and freezing, and can be used directly to feed the infant without further transfer of the milk. A polypropylene container can also be used for pasteurization and freezing and, if of rigid construction, can be used to feed the infant. Water-soluble constituents including IgA remain stable in both glass and polypropylene. Rigid containers are easier to handle than flexible soft plastic bags. Glass, of course, has the potential for breakage. Table 2 illustrates the effect of the container on milk constituents at 4 hours and 24 hours of storage at 4°C. It should be noted that the cell count at 4 hours was similar in all containers but after 24 hours of storage is increased over the count at 4 hours in glass. This is presumably because the cells are no longer adherent to the glass. Functional activity of the cells as indicated by mitogen responses of lymphocytes and phagocytic activity of macrophages and neutrophils has been reported to be reduced after 24 hours of storage but not at 4 hours according to Goldblum *et al.* (1981). The levels of lysozyme, lactoferrin and sIgA diminished between 4 and 24 hours in all containers. The reductions were greater for lysozyme and lactoferrin in Pyrex and polypropylene and greater for sIgA in polyethylene.

Table 2. Effect of container type on milk constituents after 4 hour and 24 hour storage

	Pyrex	Polypropylene	Polyethylene	
			Bags	Rigid
Colostrum	NC	NC	NC	NC
Volume	NC	NC	NC	NC
Mature milk cells				
— numbers	↑	↑	NC	NC
— functions	↓	↓	↓	↓
Proteins				
— lactoferrin	sli↓	sli↓	NC	NC
— lysozyme	↓	NC	↓	↓
— sIgA	NC	NC	NC	NC
— total IgA	NC	NC	NC	NC
Antibodies to E. coli	NC	NC	↓	↓
Water sol vitamin C	NC	↓		
Fat sol vitamins	NC	NC	NC	NC

NC—no change, sli—detectable but with little functional effect.
Adapted from Goldblum *et al.* (1981).
Lawrence (1999b). Reproduced with permission.

Colostrum has been found to be remarkably stable in all types of containers (Goldblum *et al.*, 1982). In a study on the effects of the storage container and time on the immunologic factors in colostrum, aliquots were collected by electric pump at 48 to 72 hours postpartum and stored in polyethylene bags, polypropylene or Pyrex for 4 or 24 hours at 4°C. Lactoferrin and lysozyme were quantified by electro-immunodiffusion and secretory IgA was determined by immunoassay. The viable lymphocytes, macrophages and neutrophils were counted. In addition, the proliferative response to phytohemagglutinin and the phagocytic activity of macrophages were measured. There was no change in the concentrations of lysozyme, lactoferrin or secretory IgA by storage in any of the three types of containers. Although polyethylene containers markedly decreased sIgA antibodies to E. coli somatic antigens, no effect was apparent on the total secretory IgA concentration. It is noteworthy that the initial studies occurred at four hours after collection and not at time zero. The values were compatible with levels measured in fresh human milk.

Why these immunologic factors are more stable in colostrum than in mature milk is not known. It is known, however, that their concentration, as well as that of total protein is greater in colostrum. Goldblum *et al.* (1982) have suggested that colostrum preserves certain immunologic components in the fluid phase by preventing their degradation or their adsorption onto the surface of containers.

Whether the addition of small amounts of colostrum enhances the preservation of these immunologic factors in mature milk was being investigated by Goldblum *et al.* (1982) but is not reported yet.

4. Cooling Effects

There is considerable variability in the effect of temperature on the different constituents of human milk, some being quite stable and others being quite labile. Temperatures to which milk intended for banking may be exposed range from −80°C to 100°C.

The impact of storage at 15°C, 25°C and 38°C has been reported by Hamosh *et al.* (1997). They found that the pH decreased two units within 24 hours at all temperatures and in all samples. Samples were collected at one month and at five to six months postpartum (Hamosh *et al.*, 1996). Milk was collected at home by Hamosh *et al.* (1996) between 8 AM and 10 AM using an electric pump, emptying both breasts. The collection was mixed and 50 ml was placed in a sterile polypropylene tube. One milliliter was placed immediately on ice as a home control. The rest of the sample was maintained at ambient temperature (18°C–25°C) during transport to the laboratory. The trip took 78 ± 7.8 minutes. Mothers in the second group pumped at the hospital. A second 1 ml sample was taken as a laboratory control. Milk was stored at 15°C, 25°C, and 38°C in stoppered glass tubes. A 0.6 ml sample was removed in duplicate at 1, 2, 4, 8, and 24 hours for analysis and at 4, 8, and 24 hours for quantitation of bacterial growth. The ambient temperature of the mothers' homes, hospital, and their cars was similar (20°C to 25°C). At the laboratory, the control specimens were immediately stored at −70°C (Hamosh *et al.*, 1996).

Hamosh *et al.* (1996) found that proteolysis was minimal at 15°C and at 25°C in the first 24 hours. By 24 hours proteolysis was measurable at 38°C. Lipolysis began within the first hour and increased throughout the first 24 hours to 8%. The level of free fatty acid increased to 440% over levels measured immediately after expression within one hour of exposure to 38°C and rose to 710% at 24 hours. Lavine and Clark (1987) had previously reported similar findings and reported that free fatty acids increased over time when human milk was stored at 25°C. These investigators measured lipid and free fatty acids after storage at 25°C, 4°C and −11°C for 6, 12, 24, and 48 hours. Additional samples were stored at −11°C and −70°C for 1, 2, 4, 6, or 8 weeks. The effects of temperature, length of storage and total milk lipids on release of individual fatty acids were measured. There was no measurable lipolysis at −70°C. Free fatty acids increased with storage time and temperature. Total lipid was not a factor during the first 48 hours of storage, but when milk was stored for more than 48 hours at −11°C, the release of free fatty acids increased with lipid content. Arachidonic, linoleic and other long-chain polyunsaturated fatty acids were released in greater proportion than other fatty acids, which tended to decrease in concentration. The significance of non-esterified fatty acid accumulation during storage requires further study. The digestive enzymes lipase and amylase were reported by Hamosh *et al.* (1997) and Ellis and Hamosh (1992) to be stable in human milk stored at 15–38°C for 24 hours.

In studies by Pardou *et al.* (1994) bacterial growth was minimal at cool temperatures (15°C) throughout a 24 hour period. At 25°C bacterial growth was minimal for the first 4 to 8 hours and then slowly increased. At 38°C there was considerable bacterial growth during the first 4 hours. Hamosh *et al.* (1996) considered storage at 25°C to be safe for four hours and storage at 15°C to be safe for 24 hours. Storage at 38°C was not considered safe. The effect of refrigeration at 0°C to 4°C on a wide range of constituents of human milk has been reported by a number of investigators. The results are summarized by constituent in Table 3. Bacterial growth decreased over a 24 hour period when stored at 4°C in both lightly and heavily contaminated samples. The bacterial count by colony forming

Table 3. Impact of storage temperature on immunological properties of milk

	0–4°C	−20°C
Total sIgA	NC	NC
Lactoferrin	NC	NC
Lysozyme	NC	NC
Fibronectin	NC	
Mucid	NC	NC
C_3 complement		
Bifidum factor (oligo saccharide)	NC	NC
Glycoproteins		
Cell count		
— number	NC	↓
— function	NC	↓
Bacterial growth	NC	NC
Inhibition of E. coli	adherence to Hep 2 cells	NC
Cytokinesis—no studies		
Nucleotides—appear to be stable as have been used to fortify formulas without destruction		

NC—no change.

Lawrence (1999b). Reproduced with permission.

units (CFU/ml) decreased in all contaminated units as illustrated by a change from 2.85×10^3 CFU/ml at extraction time to 1.45×10^3 CFU/ml after 8 days of refrigeration in the most heavily contaminated unit. This is because the anti-infective properties that destroy bacteria continue to function under refrigeration but are inactive during freezing. Pardou *et al.* (1994) consider storage at 0°C to 4°C for up to eight days satisfactory. This assumes that the specimen was collected carefully and is minimally contaminated. No specimen of human milk freshly collected is sterile, and ordinarily specimens collected for immediate feeding are not cultured each time.

In work by Pittard and Bill (1981), macrophages and neutrophils were decreased in number but lymphocyte numbers were unchanged. Cooling to a temperature between 0°C and 4°C for 48 hours greatly reduces the activity of cells in milk. According to Pardou *et al.* (1994), who examined the milk after 8 days of storage, other constituents of human milk such as lactose, lipids and IgA, remain stable at this temperature. Fat content as determined by the creamatocrit decreases slightly within hours when stored at room temperature but is stable for 14 days when the temperature is reduced to 4°C. Serum stimulated and serum independent lipolytic activity was significantly decreased when milk was stored at 25°C or at 4°C for 8 days. Free fatty acid accumulation appears to be a balance between the rate of triglyceride hydrolysis and the rate of fatty acid consumption both at 4°C and at higher temperatures (Jensen, 1995a,b; Jensen and Jensen, 1992).

From a practical standpoint, the impact of storing milk at room temperature is important for the mother who pumps at work or when separated from her infant and does not have access to refrigeration before returning home. Room temperature (25°C) is considered safe for four hours because there is minimal growth of bacteria and minimal loss of nutrients and protective properties such as immune proteins and enzymes (Hamosh *et al.*, 1996). Thus, it can be concluded that a mother who saves her milk daily and who does so with reasonable cleanliness can store her milk at room temperature temporarily without difficulty. When refrigerated at 0°C to 4°C, milk is safe from the standpoint of bacteria and lipolysis for up to 72 hours and in some cases for longer periods (Hamosh *et al.*, 1996). When milk is collected and stored for a sick or premature infant, the issues are not only safety but the preservation of important constituents (Jensen and Jensen, 1992). It has been concluded that milk refrigerated for 72 hours will retain most of its constituents and contain minimal bacteria provided the specimen was collected with good technique (Sosa and Barress, 1987). Storage at room temperature and with refrigeration has not been reported to measurably affect the immunoprotective qualities of human milk.

If milk must be stored for a longer period, fresh freezing is utilized, i.e., the mother freezes the sample immediately after collection (Silprasert *et al.*, 1986). Milk that is banked for use by another infant (i.e., donor milk) must be fresh frozen immediately at home. Following transport to the milk bank, it is thawed, pooled, pasteurized and refrozen until used (Human Milk Banking Association of North America, 1996).

5. Freezing

Freezing has been used as a practical method for storing human milk in milk banks and in hospitals as well as at home. However, freezing conditions vary. Home freezers generally have a limit of −20°C, whereas laboratory freezers ordinarily operate at −70°C. A problem with home freezers is that they usually have an automatic defrost cycle which may cause the formation of ice crystals in the milk (Lawrence, 1999b). This may damage some milk components although this possibility has not been specifically studied. From the standpoint of an infant receiving a feeding or two of previously frozen and thawed mother's milk each day, the loss of some constituents poses little problem since the infant will also receive fresh milk at the breast for the remaining feedings. However, the effect of freezing on milk provided to a hospitalized child at every feeding raises some significant questions. A summary of the effects of freezing is provided in Table 4. The data have been collected from various reports in the literature and conditions have varied (Pittard and Bill, 1981; Friend *et al.*, 1983; Clark *et al.*, 1984; Wardell *et al.*, 1984; Silprasert *et al.*, 1986). In summary, freezing appears to cause no changes not already precipitated by contamination or by the nature of the container.

It is difficult to separate effects of freezing from those of thawing, but it is thought that thawing is associated with lipolysis, demulsification, and protein denaturation. Freezing has been noted to break the emulsion between milk fat globules and the aqueous fraction (Keenan and Patton, 1995). In addition, lipid may adhere to the walls of the container and not be recovered by gentle shaking (Goldblum *et al.*, 1981). It appears that slow freezing affords essentially the same protection to milk constituents as quick freezing does, and it is considerably simpler (Friend *et al.*, 1983).

When milk is frozen at −20°C for 12 months or indefinitely at −70°C, the appearance, odor, and smell of the milk appears unchanged, and it remains palatable for the newborn (Friend *et al.*, 1983; Keenan and Patton, 1995; Neville, 1995). The only documented changes are a reduced cell count and a reduction in cell viability to almost zero as assessed by trypan blue exclusion (Pittard and Bill, 1981). Freezing also disrupts the structure of the milk lipid globule. When frozen milk is thawed, the fat globules are clumped. Freezing appears to disrupt the globule membrane allowing the core triacylglycerols to emerge and coalesce (Keenan and Patton, 1995). The activity of the enzymes, lipase and bile salt stimulated lipase, is not affected by freeze-thaw cycles (Jensen, 1995), and

Table 4. Impact of freezing temperatures on immunological constituents of human mature milk

	−20°C	−70°
Volume	NC	NC
Bacterial growth	NC	NC
Cells		
number	↓	↓
function	↓	↓
Lipid	NC-↓ in contaminated milk	NC
Creamatocrit	NC-freezing and thawing X 2-↓	
Total sIgA	NC-↓ in contaminated milk	
Proteins	± denaturalization on thawing	
Enzymes		
Lipase	NC	NC
Lactoperoxidase	NC	NC
Serum-stimulated lipolytic activity	NC	NC
Serum-independent lipolytic activity	NC	NC
Lipolysis	±	±

NC—no change.
Lawrence (1999b). Reproduced with permission.

lipolysis can be detected in milk stored frozen for more than a month (Berkow *et al.*, 1984). Not until freezing has occurred for over a month does lipolysis occur at −20°C (Berkow *et al.*, 1984). It is recommended that banked milk be stored at −70°C (Berkow *et al.*, 1984).

Some mothers report that fresh frozen milk samples collected with the intention of providing for their infant while they return to work have spoiled or become rancid (Lawrence, 1999a). This may occur because the large amounts of lipase in the milk of some women digest milk fat to short and medium chain fatty acids (C4 to C12). This is analogous to the theory that high lipase levels are a cause of breast milk jaundice (Poland *et al.*, 1980). These storage problems have been alleviated by scalding the milk immediately after collection and before freezing. This process involves heating the milk to a temperature just below boiling. Degradation of triglycerides has been attributed by McKay and Beachan (1995) to certain strains of pseudomonad bacteria. In some cases rancid or off-flavored milk may result from the action of these bacteria, which are able to grow at 7°C (*Pseudomonas fluorcens* and *fragi*). A culture of the rancid milk will clarify this point. The problem can be solved by treating the mother for *Pseudomonas* and/or by flash heating the milk to destroy the bacteria before freezing.

6. Heat Treatment

Because of the risk of bacterial growth in stored milk, the impact of sterilization or pasteurization becomes a significant issue. With fresh collected and stored milk, there is always a risk of transmission of pathogenic bacteria or, in the case of donor milk, bacteria which serve as pathogens to the recipient infant. An infant who receives his or her own mother's milk usually receives the appropriate antibodies at the same time. Originally, human banks relied on screening the donors for illness and confirmed the purity of milk by random culturing of samples as a check on technique. That approach is no longer tenable because of the risk not only of HIV but of other infectious viruses, as well as other microorganisms. It has been mandated by the Human Milk Banking Association of North America (HMBANA) (Human Milk Banking Association of North America, 1996), in conjunction with the Food and Drug Administration (FDA), that all samples of donor milk (i.e., milk collected from a mother who is not the biologic mother) must be pasteurized. Many studies have been conducted for commercial purposes on the treatment of bovine and other mammalian milks. Loss of cells and alteration of immunoprotective qualities has not been of concern for bovine milk, as this is marketed only for its nutritive value. A few studies have been conducted on the best way to sterilize human milk for consumption by other infants while preserving its constituents (Ford *et al.*, 1977; Evans *et al.*, 1978; Williamson *et al.*, 1978; Goldblum *et al.*, 1984; Wardell *et al.*, 1984; Van Zoeren-Grobben *et al.*, 1987; Carbonare *et al.*, 1996). Further work on the preservation of human milk in the freeze-dried form would be valuable. Heating at any sterilizing temperature destroys the cells. There is also some reduction in secretory IgA (sIgA) and enzyme activity. The process recommended by HMBANA for pasteurization is as follows: after proper preparation* in a tightly closed glass container, the container is submerged in a well-agitated or shaking water bath preheated to a minimum of 56°C and maintained at this temperature for 30 minutes (Human Milk Banking Association of North America, 1996).

Premature infants fed raw, pasteurized or boiled human milk for three consecutive weeks were evaluated by Williamson *et al.* (1978). There was a reduction in fat absorption and the amount of nitrogen retained when the milk was heated by either process. The mean weight gain was one-third greater in those infants fed raw milk. There was no change in the retention of calcium, phosphorous or sodium. The effects of heat processing on human milk are summarized

---

* Proper preparation begins with milk collection following thorough hand washing by the mother and the use of clean pumping equipment. Milk is collected in a sterile glass or plastic container, tightly covered, frozen at −20°C and then transported to the Milk Bank under refrigeration. At the bank, milk is thawed, pooled and placed in glass storage bottles in appropriate single feeding aliquots (4 or 8 ounces).

Table 5. Impact of heat processing on the constituents of human milk

	Pasteurization 56°C, 30 min	Rapid hi-temp 72°C, 15 sec	Boiled
Volume bacterial growth	0	0	0
Cells			
number	↓	↓	0
function	0	0	0
Lipid	NC	NC	↓
Creamatocrit	NC		↓
Total sIgA	±	NC	↓
Proteins	NC	NC	↓
Enzymes			
Lipase (BSSL)	↓↓	↓↓	↓↓↓↓
Lactoperoxidase	↓		↓↓
Serum-stimulated lipolytic activity	↓		↓↓
Serum-independent lipolytic activity	↓		↓↓
Lipolysis (8% in 24 hr at 15°C)	↓		↓↓

NC-no change.
Lawrence (1999b). Reproduced with permission.

in Table 5 in which the effect of formal pasteurization at 56°C for 30 minutes is compared with bringing the sample to high heat for 15 seconds (i.e., rapid, high temperature treatment) (Goldblum *et al.*, 1984). Rapid treatment involved heating to 72°C or to 87°C for 1, 3, or 15 seconds on a plate heat exchanger with rapid cooling to 2°C (in less than 3 seconds). At least 5 seconds of heat treatment were necessary to sterilize the milk. This technique did not reduce vitamin content, lactoferrin, sIgA, or lysozyme activity. Boiling (i.e., raising the temperature of milk to a full rolling boil at 100°C for 1 minute) decreases enzyme activity, serum stimulated lipolytic activity and general lipolytic capacity. There is also a decrease in secretory IgA activity. All of these factors are thought to be important to the immunoprotection provided by milk (Goldblum *et al.*, 1982; Goldman, 1993). It is for this reason that the HMBANA has recommended a careful pasteurization process (i.e., 56°C for 30 minutes).

Microwave radiation is not recommended for the warming of infant feedings because it is a possible source of constituent loss (Quan *et al.*, 1992; Carbonare *et al.*, 1996). Lysozyme activity, total IgA and specific secretory IgA to *Escherichia coli* serotypes 01, 04 and 06 were determined before and after exposure of fresh frozen milk samples to heat by microwave for 30 seconds at low or at high power settings. All components were decreased, especially at high power. Total IgA was unchanged at low power (Quan *et al.*, 1992). IgA was decreased by 98% and lysozyme activity by 96%. As a result, the growth of *E.*

coli was greatly increased, especially in milk that had been subjected to higher microwave temperatures. There was also a reduction in *E. coli* specific IgA. Kerner *et al.* (1987) and Silprasert *et al.* (1986) concluded that microwave irradiation of human milk reduces the immunologic properties of human milk. Nevertheless, the ability to inhibit *Escherichia coli* adherence to HEp-2 cells was unaffected by microwave radiation (Rayol *et al.*, 1993; Carbonare *et al.*, 1996).

7. Other Stresses Upon Milk

Other stresses on stored and artificially fed human milk include exposure to light and the settling of fat in a feeding tube or syringe en route to the premature infant (Stocks *et al.*, 1985). Exposure to daylight has been found to reduce the amount of vitamins A, C (44%) and B_6 (19%) (Bates *et al.*, 1985; McKay and Beachan, 1995; Lawrence, 1999b). Vitamins D and E are not affected. In this work the experimental design included temperatures of 25°C, 4°C, −11°C, −20°C, and −70°C, and length of time of exposure included 8, 24, and 72 hours as well as 1, 4, 8, and 16 weeks (Moffatt *et al.*, 1987). Exposure to therapeutic phototherapy when the feeding tube or pumping device was within 18 inches of the light source in the treatment of neonatal jaundice caused a 53% loss of vitamin C (Bates *et al.*, 1985). While vitamins are not part of the defense mechanism against infection directly, they are important to overall nutrition.

An additional hazard of tube feeding is the loss of fat which clings to the walls of the tube as the slow moving milk settles in the feeding system (Moffatt *et al.*, 1987). There is also a loss of protein (Neville, 1995). This loss is greater when milk is supplied by gravity feeding than by bolus feedings using a Holter pump (Stocks *et al.*, 1985), and includes a proportional loss of immunoglobulins. At the same time settling does not change the effectiveness of the suspended immunoglobulins.

Ultrasonic homogenization has been evaluated by Desai *et al.* (1987), Martinez *et al.* (1987) and Rayol *et al.* (1993), who found that fat loss was reduced by decreasing creaming and reducing the size of the fat globule. The fat globule thus did not cling to the feeding device and was available to the recipient infant. Because milk contains not only calories and macronutrients but a great number of other bioactive components, Hamosh (1988) has expressed concern that this highly structured fluid may lose some unique immunologic properties when the fat and aqueous fraction are homogenized. For example, homogenization of bovine milk activates lipoprotein lipase with consequent rapid production of free fatty acids and marked alteration in the nature of the lipid ingested by the infant. Such an effect in human milk could change the pH and interfere with the enzyme activities (Hamosh, 1988).

8. Summary

The immunoprotective constituents of human milk are stable when stored at room temperature for 8 hours, when stored at 0°–4°C for three days, or when frozen at –20°C for 12 months. They are also stable during pasteurization at 56°C for 30 minutes. Sonification may reduce levels of sIgA and lysozyme and the ability of milk to inhibit growth of *E coli*. The number of cells in human milk is reduced by storage, freezing, pasteurizing, microwaving and sonification, and the functional capacity of surviving cells is also reduced.

References

Bates, C.J., Liu, D.-S., Fuller, N.J., and Lucas, A. 1985. Susceptibility of riboflavin and vitamin A in breast milk to photodegradation and its implications for the use of banked breast milk in infant feeding. *Acta Paediatr. Scand.* **74**:40.

Berkow, S.E., Freed, L.M., Hamosh, M., Bitman, J., Wood, D.L., Happ, B., and Hamosh, P. 1984. Lipases and lipids in human milk: Effect of freeze-thawing and storage. *Pediatr. Res.* **18**:1257.

Carbonare, S.B., Palmeira, P., Silva, M.L., and Carneiro-Sampaio, M.M. 1996. Effects of microwave radiation, pasteurization and lyophilization on the ability of human milk to inhibit *Escherichia coli* adherence to HEp-2 cells. *J. Diarrhoeal Diseases Res.* **14**:90.

Chapin, H.D. 1923. The operation of a breast milk dairy. *J. Am. Med. Assoc.* **81**:200.

Clark, R.M., Hundrieser, K.H., Ross, S., and Brown, P.R. 1984. Effect of temperature and length of storage on serum-stimulated and serum-independent lipolytic activities in human milk. *J. Ped. Gastroenterol. Nutr.* **3**:567.

Desai, I.D., Martinez, F.E., Davidson, A.F.G., Anderson, J.D., Nakai, S., and Radcliffe, A. 1987. Ultrasonic homogenization and changes in the beneficial properties of human milk. *Fed. Proc.* **46**:3.

Ellis, L.A. and Hamosh, M. 1992. Stability of digestive enzymes in expressed human milk. In: *Mechanisms Regulating Lactation and Infant Nutrient Utilization*, p. 389, Wiley-Liss, Inc.

Emerson, P.W. 1922. The collection and the preservation of human milk: The preliminary report. *J. Am. Med. Assoc.* **78**:641.

Emerson, P.W. 1933. The collection and preservation of human milk VI. A preliminary note on the freezing process. *J. Pediatr.* **2**:472.

Evans, T.J., Ryley, H.C., Neale, L.M., Dodge, J.A., and Lewarne, V.M. 1978. Effect of storage and heat on antimicrobial proteins in human milk. *Arch. Dis. Child.* **53**:239.

Ford, J.E., Law, B.A., Marshall, V.M., and Reiter, B. 1977. Influence of the heat treatment of human milk on some of its protective constituents. *J. Pediatr.* **90**:29.

Franz, K. 1994. *Breastfeeding Product Guide*, Geddes Productions, Sunland, CA.

Friend, B.A., Khem, M.S., Long, C.A., and Vaughn, L.A. 1983. The effect of processing and storage on key enzymes, B vitamins, and lipids of mature human milk. I. Evaluation of fresh samples and effects of freezing and frozen storage. *Pediatr. Res.* **17**:61.

Goldblum, R.M., Dill, C.W., Albrecht, T.B., Alford, E.S., Garza, C., and Goldman, A.S. 1984. Rapid high-temperature treatment of human milk. *J. Pediatr.* **104**:380.

Goldblum, R.M., Garza, C., Johnson, C.A., Harrist, R., Nichols, B.L., and Goldman, A.S. 1981. Human milk banking. I. Effects of container upon immunologic factors in mature milk. *Nutr. Res.* **1**:449.

Goldblum, R.M., Goldman, A.S., Garza, C., Johnson, C.A., and Nichols, B.L. 1982. Human milk banking. II. Relative stability of immunologic factors in stored colostrum. *Acta Paediat. Scand.* **71**:143.

Goldman, A.S. 1993. The immune system of human milk: Antimicrobial, antiinflammatory and immunomodulating properties. *Pediatr. Infect. Dis. J.* **12**:664.

Goldman, A.S. and Goldblum, R.M. 1995. Defense agents in milk, in: *Handbook of Milk Composition*, (R.G. Jensen, ed.), p. 727, Academic Press, San Diego.

Hamosh, M. 1988. Letter to the Editor. *J. Ped. Gastroenterol. Nutr.* **7:2**:307.

Hamosh, M., Ellis, L.A., Pollock, D.R., Henderson, T.R., and Hamosh, P. 1996. Breastfeeding and the working mother: Effect of time and temperature of short-term storage on proteolysis, lipolysis, and bacterial growth in milk. *Pediatrics* **97**:492.

Hamosh, M., Henderson, T.R., Ellis, L.A., Mao, J.-I., and Hamosh, P. 1997. Digestive enzymes in human milk: Stability at suboptimal storage temperatures. *J. Ped. Gastroenterol. Nutr.* **24**:38.

Human Milk Banking Association of North America. 1996. Guidelines for establishment and operation of a donor human milk bank, Sandwich, MA.

Jensen, R.G. 1995a. Determinants of milk volume and composition, in: *Handbook of Milk Composition*, (R.G. Jensen, ed.), p. 254, Academic Press, Inc.

Jensen, R.G. 1995b. Determinants of milk volume and composition: Miscellaneous factors affecting composition and volume of human and bovine milks, in: *Handbook of Milk Composition*, (R.G. Jensen, ed.), p. 237, Academic Press, Inc.

Jensen, R.G. and Jensen, G.L. 1992. Specialty lipids for infant nutrition. I. Milks and formulas. *J. Pediatr. Gastroenterol. Nutr.* **15**:232.

Jones, K. 1928. The mothers milk bureau of Detroit, *Pub Health Nurse*, p. 142.

Keenan, T.W. and Patton, S. 1995. The structure of milk: Implications for sampling and storage, in: *Handbook of Milk Composition*, (R.G. Jensen, ed.), p. 5, Academic Press, Inc., San Diego.

Kerner, J.A., Quan, R., Yang, C., Lewiston, N., Sunshine, P., and Stevenson, D.K. 1987. Effects of microwave thawing (MT) on anti-infective factors in human milk (HM). *Ped. Res.* **21**:430A.

Lavine, M. and Clark, R.M. 1987. Changing patterns of free fatty acids in breast milk during storage. *J. Ped. Gastroenterol. Nutr.* **65**:769.

Lawrence, R.A. 1999a. *Breastfeeding: A Guide for the Medical Profession*, 5th Edition, C.V. Mosby.

Lawrence, R.A. 1999b. The storage of human milk and the influence of procedures on immunological components of human milk. *Acta Paedicatrica Scandanavia* **430**:1.

Martinez, F.E., Desai, I.D., Davidson, A.G.F., Nakai, S., and Radcliffe, A. 1987. Ultrasonic homogenization of expressed human milk to prevent fat loss during tube feeding. *J. Ped. Gastroenterol. Nutr.* **6**:593.

McKay, D.B. and Beachan, I.R. 1995. The effect of temperature on degradation of triglycerides by a pseudomonad isolated from milk: Free fatty acid accumulations as a balance between rates of triglyceride hydrolysis and fatty acid consumption. *J. Appl. Bacteriol.* **79**:651.

Moffatt, P.A., Lammi-Keefe, C.J., Ferris, A.M., and Jensen, R.G. 1987. Alpha and gamma tocopherols in pooled mature human milk after storage. *J. Ped. Gastroenterol. Nutr.* **6**:225.

Neville, M.C. 1995. The structure of milk: Implications for sampling and storage. Sampling and storage of human milk, in: *Handbook of Milk Composition*, (R.G. Jensen, ed.), p. 63, Academic Press, Inc.

Pardou, A., Serruys, E., Mascart-Lemone, F., Dramaix, M., and Vis, H.L. 1994. Human milk banking: Influence of storage processes and of bacterial contamination on some milk constituents. *Biol. Neonate* **65**:302.

Pittard, W.B. and Bill, K. 1981. Human milk banking—effect of refrigeration on cellular components. *Clin. Pediatr.* **20**:31.

Poland, R.L., Schultz, G.E., and Garg, G. 1980. High milk lipase activity associated with breast milk jaundice. *Pediatr. Res.* **14**:1328.

Quan, R., Yang, C., Rubinstein, S., Lewiston, N.J., Sunshine, P., Stevenson, D.K., and Kerner, J.A. 1992. Effects of microwave radiation on antiinfective factors in human milk. *Pediatrics* **89**:667.

Rayol, M.R.S., Martinez, F.E., Jorge, S.M., Goncalves, A.L., and Desai, I.D. 1993. Feeding premature infants banked human milk homogenized by ultrasonic treatment. *J. Pediatr.* **123**:985.

Silprasert, A., Dejsarai, W., Keawvichit, R., and Amatayakul, K. 1986. Effect of storage of the crematocrit and total energy content of human milk. *Human Nutrition: Clinical Nutrition* **40C**:31.

Sosa, R. and Barress, L. 1987. Bacterial growth in refrigerated human milk, *AJDC* **41**:111.

Stocks, R.J., Davies, D.P., Allen, F., and Sewell, D. 1985. Loss of breast milk nutrients during tube feeding. *Arch. Dis. Child.* **60**:164.

Talbot, F.A. 1928. An organization for supplying human milk. *New Engl. J. Med.* **199**:610.

Van Zoeren-Grobben, D., Schrijver, J., Van Den Berg, H., and Berger, H.M. 1987. Human milk vitamin content after pasteurisation, storage, or tube feeding. *Arch. Dis. Child.* **62**:161.

Wardell, J.M., Wright, A.J., Bardsley, W.G., and D'Souza, S.W. 1984. Bile salt-stimulated lipase and esterase activity in human milk after collection, storage, and heating: Nutritional implications. *Pediatr. Res.* **18**:4:382.

Williamson, E., Finucane, H., Ellis, H., and Gamsu, H.R. 1978. Effect of heat treatment of human milk on absorption of nitrogen, fat, sodium, calcium, and phosphorus by preterm infants. *Arch. Dis. Child.* **53**:555.

Index

MIX
Papier aus verantwortungsvollen Quellen
Paper from responsible sources
FSC® C105338

If you have any concerns about our products,
you can contact us on
ProductSafety@springernature.com

In case Publisher is established outside the EU,
the EU authorized representative is:
Springer Nature Customer Service Center GmbH
Europaplatz 3, 69115 Heidelberg, Germany

Printed by Libri Plureos GmbH
in Hamburg, Germany